FOUNDATIONS™ T

Martin Engineering amplifies the contents of this book by offering a series of training programs.

4. Proper distance from hazard.

Foundations™ Training can be presented at the customer's site or a neutral location and focuses on helping plant personnel run cleaner, safer, and more productive conveyors.

"The content, the demonstrations and exercises shown, the written material, and the post workshop tools were excellent. I am anxious to use them to make my plant's conveyors perform better."

"I have been working around belt conveyors for 34 years...I can't believe I learned more in the last 7 hours than I did in the last 34 years."

For more information or to reserve Foundations™ training, contact us today.

foundations@martin-eng.com | 800-544-2947 | 309-852-2384

FOUNDATIONS™ TRAINING

Basic Training

Topics Covered:

- Belt Conveyors
- Conveyor Components
- Conveyor Safety
- Principles of Fugitive Material

Audience:

Employees who are unfamiliar with conveyor belt systems

- 2-hour session
- Can be presented by Martin® Trainer or site manager/supervisor (materials provided)

Operations & Maintenance

Topics Covered:

- Introduction to Material Handling
- Conveyor Safety
- Belt Alignment
- Dust Management
- Belting & Splices
- Transfer Point Improvement
- Belt Cleaning

Audience:

Employees and managers of employees in charge of operating and maintaining belt conveyor systems

- One day seminar (6-8 hours)
- Second-day custom program available – includes site walk down with or without class
- Online option

Advanced Training

Topics Covered:

- Cost Saving Strategies
- Total Material Control
- Conveyor Design & Material Science
- Performance Measurements
- Dust Control
- Controlling Fugitive Material
- Payback from System Improvements

*Customized for your plant's conveyor problems, including a walk down of your conveyors

Audience:

Engineers & Plant Managers

- 1 or 2 day program

May qualify for Continuing Education Units (CEUs) or Professional Development Hours (PDHs) or Parts 46/48 Annual Refresher Training

" Probably the best on-site, 1-day class I've had."

" Very relevant information to take back to my plant to assist in addressing problem areas."

" Well presented, simple explanations for complex problems."

Martin® trainers can provide conveyor-specific training, allowing attendees to execute the newly-learned concepts on their own conveyor belts.

FOUNDATIONS™

The Practical Resource for Cleaner, Safer, More Productive
Dust & Material Control

Fourth Edition

FOUNDATIONS™

The Practical Resource
for Cleaner, Safer, More Productive
Dust & Material Control

Fourth Edition

by

R. Todd Swinderman, P.E.
Andrew D. Marti
Larry J. Goldbeck
Daniel Marshall
&
Mark G. Strebel

Martin Engineering Company
Neponset, Illinois
U.S.A.

Application of the information and principles in this book should be carefully evaluated to determine their suitability for a specific project.

For assistance in the application of the information and principles presented here on specific conveyors, consult Martin Engineering or other knowledgeable engineers.

Disclaimer

1. Martin Engineering publishes this book as a service to the bulk-material handling industry. The book is provided for general information purposes only and is not intended to provide comprehensive knowledge pertaining to the control of fugitive materials in bulk-material handling operations. The opinions expressed herein are those of the authors and represent a consensus of the authors regarding the topics discussed.

2. Pictures, graphics, tables, and charts contained in this book are used to convey specific points and therefore may not be technically correct or complete in every detail. Fictitious names and data provided in this book are intended to convey concepts and any similarity thereof to actual entity names or data is purely coincidental and unintentional.

3. This book is provided without any representations or warranties as to the accuracy or completeness of the content of the book. Without limiting the generality of the foregoing, the "Safety Concerns" sections of this book are intended to highlight specific safety issues and should not be considered as inclusive of all safety concerns related to bulk-material handling operations.

4. To the maximum extent permitted by applicable law, IN NO EVENT SHALL MARTIN ENGINEERING OR THE AUTHORS BE LIABLE FOR PERSONAL INJURY OR FOR ANY INDIRECT, SPECIAL, OR CONSEQUENTIAL DAMAGES WHATSOEVER ARISING OUT OF OR IN ANY WAY CONNECTED TO THIS BOOK, INCLUDING WITHOUT LIMITATION ANY DAMAGES ARISING OUT OF THE APPLICATION OF THE INFORMATION, PRINCIPLES, OR OTHER CONTENTS IN THIS BOOK. In all events, to the maximum extent allowed by law, Martin Engineering's and the authors' aggregate liability for claims relating to the book shall be limited to the cost of replacing this book.

5. Information presented in this volume is subject to modification without notice. Martin Engineering reserves the right to make corrections, deletions, or additions to the book without prior notice or obligation to replace previously published versions. If an error is found or you wish to provide input to future editions, please contact Marketing Manager, Martin Engineering at info@martin-eng.com, by phone: 309-594-2384, or by fax: 309-594-2432.

FOUNDATIONS™

ISBN 978-0-9717121-1-9

Library of Congress Control Number: 2007942747

Copyright © 2009 Martin Engineering

Second Printing: February 2011

Third Printing: February 2012

Fourth Printing: September 2013

Fifth Printing: February 2015

Part Number L3271-4

Cover Photo © Lester Lefkowitz/Corbis Corporation

Printed in the United States of America.
Printer: Worzalla Publishing Company, Stevens Point, WI

Metric / Imperial Measurements

Metric measurements, with their common Imperial conversions, are used throughout the book except where the original source information is specified in Imperial units. In that case, the actual Imperial units are used with approximate metric conversions.

The comma has been used throughout this book as the decimal marker in metric measurements, which is the current practice in standards published by the International Organization for Standardization (ISO).

Martin Engineering Company
One Martin Place
Neponset, Illinois 61345-9766 USA
800-544-2947 or 309-594-2384
Fax: 309-594-2432
info@martin-eng.com
www.martin-eng.com

Section One

FOUNDATIONS OF SAFE BULK-MATERIALS HANDLING

Section Two

LOADING THE BELT

Section Three

RETURN RUN OF THE BELT

Section Four

DUST MANAGEMENT

Section Five

LEADING-EDGE CONCEPTS

Section Six

CONVEYOR MAINTENANCE

Section Seven

THE BIG PICTURE OF BULK-MATERIALS HANDLING

> *"The control of dust and spillage is not only a science but also an art; Martin has mastered both."*

Working in our basement workshop in 1944, my father, Edwin F. Peterson, invented an answer to problems in bulk-materials handling: the ball-type industrial vibrator. Trademarked Vibrolator®, his invention provided the foundation for the success of Martin Engineering. Since then, we have expanded around the world: We own and operate business units in Michigan, Brazil, China, Germany, India, Indonesia, Mexico, and South Africa, with licensees in Australia, Canada, and Chile. In Europe, we also have branch offices in France, Turkey, and the United Kingdom. We have over 800 employees around the world; our dedicated people made it possible to achieve sales exceeding $135M USD in 2011.

Following my father's example, Martin remains steadfastly innovative in solving problems in bulk-materials handling. Intrinsic to Martin's values is our quest to make the industrial-material handling environment cleaner, safer, and more productive. For nearly 65 years, we have worked to improve our global environment by controlling dust and spillage in bulk-materials handling, "going green" long before it became popular. The control of dust and spillage is not only a science but also an art; Martin has mastered both.

Martin, the pioneer and world leader in the development of engineered belt-cleaning and belt-sealing systems, continues to be an innovator with capabilities to meet industry needs around the world. Our Center for Bulk Materials Handling Innovation— opened in June 2008—demonstrates our commitment to improving the industry. Located at our world headquarters in Neponset, Illinois, USA, the Center for Innovation (CFI) is part pure-science research laboratory and part industrial-product development center. It is focused on helping our customers understand and solve their bulk-materials handling problems. Controlling dust and spillage is a goal that comes true everyday for our customers with our *Absolutely No Excuses Guarantee*.

It is with great pleasure that I offer the fourth edition of *Foundations™: The Practical Resource for Cleaner, Safer, More Productive Dust & Material Control* to those involved with bulk-materials handling.

Edwin H. Peterson
Chairman, Board of Directors
Martin Engineering

From its beginning in the 1940s, Martin Engineering has been committed to safety, excellence, and innovation in bulk-materials handling. From our earliest vibrators to today's leading-edge conveyor concepts, our strategic focus on continuous improvement represents the growth and evolution of practices for the control of material movement and the improvement of belt conveyor operations. Many of our improvements are based on our fundamental belief that in order to have clean, safe, and productive bulk-materials handling, companies must ensure total control of material on their belt conveyors.

With the opening of the Center for Bulk Materials Handling Innovation at our world headquarters in Neponset, Illinois, USA, Martin has made a significant commitment and a substantial statement for continued improvement to the bulk-materials handling industry. It is our intent to share the knowledge gained from this basic science and applied practical research facility with universities, associations, and customers. With these resources and commitment, Martin will continue to be the leader in providing information and developing technologies to control dust, spillage, and fugitive materials.

Based upon our experience and research, we intend to provide an extensive educational program that will include interactive online training, workshops, seminars, certification programs, university-accredited courses, and technical presentations at association conferences. We will work closely with participating universities to bring the bulk-materials handling industry the most current research, the most reliable technical data, and the best practices. Martin will continue to present information around the world through our books and workshops, all focused on helping the industries handling bulk materials to "Think Clean" while improving efficiency, productivity, and safety.

As we all know, we work in a dangerous global industry, and we cannot yell loud enough to voice our concern. There are too many injuries and fatalities in our business that could be prevented. We must bring safe conveyor operations to the forefront. Early in the revision of *Foundations 3* for this, the fourth edition, the authors made a conscious decision to move our emphasis on safety to the beginning section of the book. In addition to the material in the Safety chapter, the reader will find additional information and callouts on safety dispersed throughout the book. Most chapters include a separate section entitled "Safety Concerns" for the reader to consider. Quite simply, we must do a better job protecting the lives of those who work in our industry.

By reading this edition of *Foundations*™, *The Practical Resource for Cleaner, Safer, More Productive Dust & Material Control*, you will learn more about Martin's expertise and philosophy. We hope you find it useful in making your conveyor operations cleaner, safer, and more productive.

Scott E. Hutter
President and CEO
Martin Engineering

The question most often asked by engineers I meet in my world travels is, "What are the changes you have seen in the bulk-materials handling industry over these past 50 years?"

What has changed?

A. Computers: First and foremost is the computer: It tracks, records, and protects everything we do. Fifty years ago, design priorities for conveyors were contained in books; today, they are in computer databases. As a result, mechanical engineers are graduating who can make drawings and calculations only with the assistance of the computer.

> *"What are the changes you have seen in the bulk-materials handling industry over these past 50 years?"*

Advanced computer programming permits engineers to see "virtual imagery" and designers to see visual representations of the flow characteristics of a material. These programs allow multiple design possibilities so engineers can obtain the most desirable results at each transfer point in the conveyor.

However, it is important to remember that the computer-generated model is valid only when an engineer or technician who has hands-on experience with the bulk material in question has reviewed it. In addition, the experienced engineer or technician must approve the conveyor components, keeping in mind that they will eventually need to be serviced, so they need to be designed to be accessible and serviced safely.

B. Hands-on Experience: Today, most plants operate 7 days a week, 24 hours a day. Management may expect a plant to double production with a crew that is reduced by half. These plants will need to fix conveyor problems as they occur, but operators cannot stop production because of the need to meet quotas.

Back when new engineers were placed with experienced colleagues in apprentice programs, they were exposed to real-world situations. They acquired experience in rapidly analyzing and fixing breakdowns to allow the plant to resume production.

As a result of modern engineering techniques, bulk-materials handling systems are being designed by detailers with little or no hands-on experience. While plants and mines are trying to keep the mechanical equipment working and producing, operators are forced to accept a design that has few provisions for servicing the system. As I look at modern plants with the same engineering designs that were used 30 years ago, I wonder who will be able to maintain, on a daily basis, the components that are prone to fail or wear out, especially when the design has no provision to allow those necessary repairs to be made easily and quickly.

Improvements in design can best be accomplished by those with hands-on conveyor experience. Experienced operators have the knowledge to guide a design team to anticipate hardware failures and provide the means to fix problems with the least possible downtime. They must be allowed and encouraged to contribute to the design.

C. Environmental and Safety Regulations: Other differences are increased environmental and safety rules and regulations that must now be incorporated into design priorities. These regulations have become as important as the conveyor's production requirements. It has been obvious to me, since the 1950s, that "Durt" and dust were a safety / health / profit-loss issue that needed to be addressed during the initial design, instead of waiting until regulators threatened fines or shut down operations to eliminate the "Durt" hazard.

What is in the Future?

A. Clean Conveyors: I see lots of blue sky; I see clean material handling from conveyor belts that are washed and dried. These super clean systems will be designed by today's fraternity of dedicated engineers who will work with safety, environmental, and process technicians to develop designs that are kind not only to the people who work on them, but also to the environment. The goal is to design systems that meet all governmental regulations for a safe, clean operation.

B. Conveyor Serviceability: Conveyor design will be based on quick, easy, safe, and intuitive service. Emerging safety regulations in the future may permit certified conveyor technicians (CCTs) to service conveyor belts during operation. Similar to current safety regulations that allow certified electricians to inspect an electrical control box without terminating power, CCTs may be able to service operating conveyors, resulting in fewer accidents, greater production, and higher profit. Furthermore, with improved serviceability better controlling "Durt" problems (the real cause of the majority of conveyor-related accidents), CCTs will significantly increase conveyor production availability.

C. Cleanliness, Safety, and the Environment: There will be an increasing demand by individuals, communities, and governmental agencies for conveyor operations to be clean, safe, and environmentally sound. I see those operators who choose to ignore these demands being either shut down or subject to heavy governmental fines.

Past 50–Future 50

There have been many changes over the past 50 years that have made bulk-materials handling cleaner, safer, and more productive: use of computers to make ideal designs for each transfer point, design teams utilizing the knowledge-base of operators to optimize the design, and regulations to make bulk-materials handling cleaner and safer for both the workers and the environment.

There is still a lot of room for positive changes during the next 50 years to make the industry even cleaner, more serviceable, more productive, and safer. Improvements will continue to be made that will revolutionize bulk-materials handling. Hopefully, the next edition of *Foundations*™ will describe even more of these advancements.

"Think Clean!"

Dick Stahura
Product Application Consultant

DEDICATION

It is an honor for us to dedicate the fourth edition of *Foundations™: The Practical Resource for Cleaner, Safer, More Productive Dust & Material Control* to the following:

Employees in Industries Handling Bulk Materials: Much thought was given to the safety chapter and the safety section of each chapter. Therefore, we dedicate this book to the employees around the world who were, unfortunately, injured while working around conveyor belts. We especially dedicate this book to the employees who lost their lives and to their families. Although tragic, these misfortunes broaden awareness about the dangers associated with conveyor belts and the importance of safety.

R. Todd Swinderman, P.E.: Todd is a driving force behind many of the current innovations in our industry. He has been instrumental in developing consistent standards for the industry through the Conveyor Equipment Manufacturers Association (CEMA). Serving as one of the association's officers and as former President and CEO of Martin Engineering, Todd's guidance, influence, and professional engineering experience have touched every facet of conveyor operations throughout the world.

The Peterson Family and the Employees of Martin Engineering (Past, Present, and Future): For more than 65 years, members of the Peterson Family have devoted their lives to the improvement of those industries handling bulk materials by making the work environment cleaner, safer, and more productive. Martin employees have always had the best interest of the customers, the industry, and the company in mind. They worked countless hours to achieve the vision of the company's founding father. Through their dedication and tenacity, the employees helped continue the Martin tradition of leading the world in conveyor research and innovation.

Andrew D. Marti

Larry J. Goldbeck

Daniel Marshall

Mark G. Strebel

SECTION 1
FOUNDATIONS OF SAFE BULK-MATERIALS HANDLING

1

Figure 1.1

Control of material movement is critical to clean, safe, and productive material handling.

Chapter **1**

TOTAL MATERIAL CONTROL

In this Chapter...

In this chapter, we describe some of the problems that occur as a result of fugitive materials: reduced operating efficiency, plant safety, product quality, and employee morale along with increased maintenance cost and scrutiny from outside agencies. We also identify the costs from this loss. As a way to address these problems, we discuss the need for total material control, which forms the basis for this book (**Figure 1.1**).

A bulk-materials handling operation is designed to accept the input of a certain amount of raw material and to reliably deliver that same amount of material at a predetermined rate to one or more points at the other end of the process.

Unfortunately, this seldom happens. Material losses, spillage, emissions, flow restrictions, and blockages can all occur in the handling process, resulting in the loss of production and creating other, associated, problems. These problems will cost billions of dollars annually across industries handling bulk materials worldwide.

This book seeks to identify many of the causes of material-handling problems and suggest practical strategies, actions, and equipment that can be applied to help increase efficiency in materials handling. This is a concept called Total Material Control®.

Total Material Control and TMC are registered trademarks of Engineering Services & Supplies PTY Limited (ESS), a Martin Engineering licensee, located in Currumbin, Australia *(Reference 1.1)*.

CONVEYORS AND FUGITIVE MATERIALS

Escape of materials from conveyors is an everyday occurrence in many plants. It occurs in the forms of spillage and leakage from transfer points or carryback that adheres to the belt past the discharge point and drops off along the conveyor return. It also occurs in the form of airborne dust that is carried off the cargo by air currents

and the forces of loading and then settles on structure, equipment and the ground. Sometimes the nature of the problems of a given conveyor can be determined from the location of the pile of lost material (**Figure 1.2**). Carryback falls under the conveyor, spillage falls to the sides, and dust falls on everything, including systems and structures above the conveyor. However, many conveyors show all of these symptoms, making it more difficult to place the blame on one type of problem (**Figure 1.3**).

Another problem besetting materials-handling operations is flow restrictions. A materials-handling plant is designed to operate at a certain rate of throughput. While much attention has been paid to the cost of spillage, the cost of restricted throughput and delayed production cannot be ignored.

Chute or bin blockages can bring a production process to a standstill, causing delays that cost thousands of dollars per hour in downtime and in lost opportunities. Chute blockages often cause material boilover, with materials overflowing the

Figure 1.2

The source of the fugitive material can sometimes be determined from the location of the pile of lost material.

Figure 1.3

Many bulk-materials handling belt conveyors show all symptoms of spillage, carryback, and airborne dust, making it more difficult to identify any one source or apply any one remedy.

chute. Chute or bin hang-ups often cause sudden material surges, in which amounts of material suddenly drop through the vessel and onto the receiving belt. Both boilovers and surges are major contributors to spillage. Material lying under the head end of a conveyor is often mistakenly identified as carryback, when it can actually result from surges and boilovers. Carryback will generally be fine material, so the presence of lumps greater than 10 millimeters (0.39 in.) will often pinpoint the cause of the fugitive material as a surge or boilover.

PROBLEMS FROM FUGITIVE MATERIALS

Results of Fugitive Materials

Fugitive materials have been around plants since conveyors were first put into operation; therefore, their presence is often accepted as a part of the industry. In fact, maintenance and production employees who are regularly assigned to cleaning duties may see this work as a form of "job security."

As a result, the problem of materials escaping from bulk-materials handling systems is often regarded with resignation. While it is recognized as a mess and a hazard, it is believed that no effective, practical, real-life systems have been developed to control it. Therefore, spillage and dust from leaky transfer points and other sources within plants are accepted as routine, unalterable courses of events. Fugitive materials become a sign that the plant is operating: "We're making money, so there's fugitive material."

At one time, pollution—whether from smokestacks or from conveyor transfer points—was seen as a sign of industrial strength. Now these problems are recognized as an indication of possible mismanagement and waste. This pollution and waste offer an opportunity for improvements in both efficiency and bottom line results.

Left unchecked, fugitive materials represent an ever increasing drain on a conveyor's, hence a plant's, efficiency, productivity, and profitability. Materials lost from the conveyor system cost the plant in a number of ways. The following are just a few:

A. Reduced operating efficiency

B. Increased conveyor maintenance costs

C. Reduced plant safety

D. Lowered employee morale

E. Diminished product quality

F. Heightened scrutiny from outside agencies and other groups

These costs will be more thoroughly covered in the sections that follow.

Reduced Operating Efficiency

It can be said the most expensive material in any operation is the material spilled from the belt. At a clean plant, "all" the material is loaded onto a conveyor belt at one end and then it is "all" unloaded at the other end. The material is handled only once: when it is placed on the belt. This, of course, equates to high efficiency: The plant has handled the material as little as possible. Material that has spilled or otherwise become fugitive, on the other hand, is material that has been received, processed (to some extent), and then lost. It has been paid for, but there will be no financial return.

In fact, fugitive material may prove to be a continuing drain: It degrades equipment, such as conveyor idlers, over time, and it might require additional labor to reprocess it before it can be returned to the system— if it can be returned to the process. However, once fugitive, it may be contaminated and unsuitable for return to the system. If fugitive material cannot be reclaimed, efficiency decreases more dramatically. In many places, even basic materials such as limestone or sand that fall from the belt are classified as hazardous waste and must be disposed of at a significant cost.

1

Fugitive materials also prove to be a drain in efficiency by requiring additional labor to clean up. Production materials can be handled by large machinery in significant quantities in large batches, in massive bucketfuls, and by the railcar load, often automatically or under remote controls. Fugitive materials, in contrast, are usually picked up by a skid steer, an end loader, or a vacuum truck—or the old-fashioned way, by a laborer, one shovel at a time.

Increased Conveyor Maintenance Costs

The escape of materials from conveyors leads to any number of problems on the conveyor system itself. These problems increase maintenance expenses.

The first and most visible added expense is the cost of cleanup. This includes the cost for personnel shoveling or vacuuming up material and returning it to the belt (**Figure 1.4**). In some plants, cleanup means a man with a shovel; in others, the cost is escalated, because it includes equipment hours on wheeled loaders, "sucker" trucks, or other heavy equipment used to move large material piles. A factor that is harder to track, but that should be included, is the value of other work not being performed because personnel have their attention diverted to cleanup activities. This delay in maintenance activities may result in catastrophic failures and even additional expense.

As materials escape, they accumulate on various conveyor components and other nearby equipment. Idlers fail when clogged or buried under materials (**Figure 1.5**). No matter how well an idler is constructed, fines eventually migrate through the seal to the bearing. Once the bearings seize, the constant movement of the belt across the idler can wear through the idler shell with surprising rapidity, leaving a razor-sharp edge on the seized roll, posing a threat to the life of the belt (**Figure 1.6**). "Frozen" idlers and pulleys increase the friction against the belt, consuming additional power from the conveyor drive motor.

Seized idlers create other even greater risks, including the possibility of fires in the system. A coal export facility in Australia suffered damage from a fire on a main in-loading conveyor. The fire was caused by a seized roller and fueled by accumulated spillage. The fire destroyed much of the head end of the conveyor, causing the failure of the 1600-millimeter (60-in.) belt and burning out the electrical cables and controls. Repairs were completed in four days to restore operation, but the total cost of the fire was estimated at $12 million USD.

Another risk is that material buildup on the face of pulleys and idlers can cause the belt to run off center (**Figure 1.7**). An accumulation of materials on rolling components can lead to significant belt-tracking problems, resulting in damage to the belt and other equipment, as well as the risk of injury to personnel.

Figure 1.4

For some plants, the cost of cleanup includes the cost of operating vacuum ("sucker") trucks and other heavy equipment.

Figure 1.5

Fugitive material can bury the load zone, resulting in idler failures, belt fires, and belt mistracking.

Figure 1.6

Idlers fail when clogged or buried under material. The motion of the belt across "frozen" idlers will wear rollers to knife-like edges.

1

A mistracking belt can move over into the conveyor structure and begin abrading the belt and the structure. If this condition is not noticed right away, great lengths of valuable belting can be destroyed, and the structural steel itself can be destroyed. Belt wander creates interruptions in production, as the belt must be stopped, repaired, and retrained prior to resuming operations.

A particularly ugly circumstance is that fugitive materials can create a problem and then hide the evidence. For example, accumulations of damp materials around steel conveyor structures can accelerate corrosion, while at the same time making it difficult for plant personnel to observe the problem (**Figure 1.8**). In a worst-case scenario, this can lead to catastrophic damage.

What is particularly troubling about these problems is that they become self-perpetuating: Spillage leads to buildups on idlers, which leads to belt wander, which leads, in turn, to more spillage. Fugitive materials truly create a vicious circle of activities—all of which increase maintenance costs.

Reduced Plant Safety

Industrial accidents are costly, in terms of both the health of personnel and the volume and efficiency of production. In 2005, the National Safety Council in the United States listed $1,190,000 USD as the cost of a work-related death; the cost of a disabling injury assessed at $38,000 USD includes wage and productivity losses, medical expenses, and administrative expenses. These figures do not include any estimate of property damage, and should not be used to estimate the total economic loss to a community.

Statistics from the Mine Safety and Health Administration (MSHA) in the United States indicate that roughly one-half of accidents that occur around belt conveyors in mines are attributable to cleanup and repairs required by spillage and buildup. If fugitive materials could be eliminated, the frequency at which personnel are exposed to these hazards would be significantly reduced. Excessive spillage can also create other, less obvious, safety hazards.

In Australia, a Department of Primary Industries safety seminar advised that in the six-year period from 1999 to 2005, a total of 85 fires were reported on conveyor belts in underground coal mines in the state of New South Wales. Of these, 22 were identified as attributable to coal spillage, and 38 to conveyor tracking. Included among the twelve recommendations of the report were: "Improve belt tracking" and "Stop running the conveyors in spillage."

In 2006 in the United States, a conveyor belt fire in an underground coal mine caused two deaths. The cause of this fire was attributed to frictional heat from a mistracking belt that ignited accumulations of coal dust, fines, and spillage, along with grease and oil.

Many countries now enforce regulatory safety procedures on companies. Included is the requirement to conduct hazard analyses on all tasks. Codes of practice in design and in plant operation require that once a hazard has been identified, it must be acted upon. The hierarchy of controls

Figure 1.7

Material buildup on the face of pulleys and idlers can cause the belt to mistrack, resulting in damage to the belt and other equipment.

Figure 1.8

Accumulations of damp material around steel conveyor structures can accelerate corrosion, while at the same time making it difficult for plant personnel to observe the problem.

for hazards will usually advise that the most appropriate action will be to "design out the hazard." The control will depend on the severity of the hazard and the layout of the existing equipment.

Lowered Employee Morale

While the specific details of an individual's job have much to do with the amount of gratification received at work, the physical environment is also a significant influence on a worker's feelings toward his or her workplace.

A clean plant provides a safer place to work and fosters a sense of pride in one's workplace. As a result, employees have better morale. Workers with higher morale are more likely to be at work on time and to perform better in their assignments. People tend to feel proud if their place of work is a showplace, and they will work to keep it in that condition. It is hard to feel proud about working at a plant that is perceived as dirty and inefficient by neighbors, friends, and, especially, the workers themselves.

It is recognized that jobs involving repetitive and unrewarding tasks, such as the cleanup of conveyor spillage, have the highest levels of employee absenteeism and workplace injuries. It is a mind-numbing exercise to shovel away a pile of spillage today, knowing that the pile will be back again tomorrow.

Diminished Product Quality

Fugitive materials can contaminate the plant, the process, and the finished product. Materials can be deposited on sensitive equipment and adversely influence sensor readings or corrupt tightly controlled formulas.

Fugitive materials impart a negative image for a plant's product quality and set a bad example for overall employee efforts. The most universal and basic tenet of many of the corporate "Total Quality" or other quality improvement programs popular in recent years is that each por-

tion of every job must be performed to meet the quality standard. Each employee's effort must contribute to, and reflect, the entire quality effort. If employees see that a portion of the operation, such as a belt conveyor, is operating inefficiently—making a mess and contaminating the remainder of the plant with fugitive material—they will become used to accepting less than perfect performance. A negative attitude and lax or sloppy performance may result. Fugitive materials provide a visible example of sloppy practices that corporate quality programs work to eliminate.

Heightened Scrutiny from Outside Agencies and Other Groups

Fugitive materials act as a lightning rod: They present an easy target. A billowing cloud of dust draws the eye and the attention of concerned outsiders, including regulatory agencies and community groups. Accumulations of materials under conveyors or on nearby roads, buildings, and equipment sends a message to governmental agencies and insurance companies alike: The message is that this plant is slack in its operations and merits additional inspections or attention.

If a plant is cited as dirty or unsafe, some regulatory agencies can mandate the operation be shut down until the problems are solved. Community groups can generate unpleasant exposure in the media and create confrontations at various permit hearings and other public gatherings.

A clean operation receives less unwanted attention from regulatory agencies; it is also less of a target for environmental action groups. Cost savings can result from fewer agency fines, lower insurance, reduced attorney's fees, and less need for community relations programs.

The Added Problem of Airborne Dust

Serious concerns arise when dust becomes airborne and escapes from conveyor systems. Dust is a greater problem than spillage: Whereas spillage is contained on the plant's ground, airborne dust particles are easily carried off-premises (**Figure 1.9**).

In its series, *Best Practice Environmental Management in Mining*, Environment Australia (the Australian government's equivalent to the US Department of the Environment) issued a report on dust control in 1998 *(Reference 1.2)*. The report analyzed the sources of airborne dust in various mineral processing plants. The report indicated that the primary sources of dust were as follows:

Crushing.................................. 1-15 percent

Screening................................5-10 percent

Stockpiling............................. 10-30 percent

Reclaiming .... 1-10 percent

Belt Conveyor Systems.......... 30-60 percent

In the Clean Air Act, the United States Environmental Protection Agency (EPA) is required by law to reduce the level of ambient particulates. Most bulk-materials handling facilities are required to maintain respirable dust levels in enclosed areas below two milligrams per cubic meter (2.0 mg/m^3) for an eight-hour period. Underground mining operations may soon be required to meet levels of 1.0 mg/m^3. Failure to comply with air-quality standards can result in stiff penalties from federal, state, and local regulatory agencies.

The Occupational Safety & Health Administration (OSHA) in the United States has determined that airborne dust in and around equipment can result in hazardous working conditions. When OSHA or MSHA inspectors receive a complaint or an air sample that shows a health violation, litigation may follow.

Respirable dust, particles smaller than 10 microns in diameter, are not filtered out by the natural defenses of the human respiratory system and so penetrate deeply into the lungs—where they can get trapped and lead to serious health problems. These health issues could be seen in the workforce and might even occur in neighborhood residents.

A frightening possibility that can arise from airborne dust is the risk of dust explosions. Dust can concentrate to explosive levels within a confined space. One incident of this nature—while tremendous in repair, replacement, regulatory fines, and lost productivity costs—can result in the greatest cost of all: the cost of someone's life.

ISO 14000 and the Environment

The continuing globalization of commerce promises more unified standards. Just as ISO 9000 developed by the International Organization for Standardization (ISO) has become a worldwide standard for codifying quality procedures, the development of ISO 14000 will set an international agenda for an operation's impact on the environment. ISO 14000 prescribes voluntary guidelines and specifications for environmental management. The program requires:

A. Identification of a company's activities that have a significant impact on the environment

B. Training of all personnel whose work may significantly impact the environment

C. The development of an audit system to ensure the program is properly implemented and maintained

Regulatory Limits

While no regulatory agency has established specific limits on the amount of fugitive materials allowed—the height of a pile beside the conveyor or the amount of carryback under an idler—there have been limits specified for quantities of airborne dust. OSHA has determined Permissible

Figure 1.9

Airborne dust is a serious concern as it escapes from conveyors and transfers.

Exposure Limits (PELs) and Threshold Limit Values (TLVs) for about 600 regulated substances.

These regulations specify the amount of dust allowed, as expressed in parts per million parts of air (ppm) for gases and in milligrams per cubic meter (mg/m³) for particulates such as dust, smoke, and mist. It is the company's responsibility to comply with these standards or face penalties such as regulatory citations, legal action, increased insurance rates, and even jail time.

These OSHA procedures note that inspectors should be aware of accumulations of dust on ceilings, walls, floors, and other surfaces. The presence of fugitive materials serves as an alarm to inspectors and drives the need for air sampling to address the possibility of elevated quantities of airborne dust.

While ISO and other agencies/groups continue to push for regulatory limits, these limits will continue to differ from country to country. It seems safe to say that environmental regulations, including dust control, will continue to grow more restrictive around the world. These guidelines will almost certainly be extended to include fugitive materials released from conveyors.

ECONOMICS OF MATERIAL CONTROL

How a Little Material Turns into Big Problems

Fugitive materials escaping from conveyors present a serious threat to the financial well-being of an operation. The obvious question is: "How can it cost so much?" A transfer point spills only a very small fraction of the material that moves through it. In the case of a transfer point on a conveyor that runs continuously, a little bit of material can quickly add up to a sizable amount. Relatively small amounts of fugitive materials can accumulate to large quantities over time (**Table 1.1**).

In real life, fugitive materials escape from transfer points in quantities much greater than four grams per minute. Studies performed in Sweden and the United Kingdom examined the real losses of fugitive materials and the costs of those losses.

Research on the Cost of Fugitive Materials

In a report titled *The Cost to UK Industry of Dust, Mess and Spillage in Bulk Materials Handling Plants*, eight plants in the United Kingdom handling materials such as alumina, coke, limestone, cement, and

Accumulation of Fugitive Material Over Time					
Fugitive Material Released	**Accumulation**				
	Hour	**Day**	**Week**	**Month**	**Year**
	(60 minutes)	**(24 hours)**	**(7 days)**	**(30 days)**	**(360 days)**
"packet of sugar" (4 g) per hour	4 g (0.1 oz)	96 g (3.4 oz)	672 g (1.5 lb$_m$)	2,9 kg (6.3 lb$_m$)	34,6 kg (75.6 lb$_m$)
"packet of sugar" (4 g) per minute	240 g (8.5 oz)	6,2 kg (13.8 lb$_m$)	43,7 kg (96.3 lb$_m$)	187,2 kg (412.7 lb$_m$)	2,2 t (2.5 st)
"shovel full" 9 kg (20 lb$_m$) per hour	9 kg (20 lb$_m$)	216 kg (480 lb$_m$)	1,5 t (1.7 st)	6,5 t (7.2 st)	77,8 t (86.4 st)
"bucket full" 20 kg (44 lb$_m$) per hour	20 kg (44 lb$_m$)	480 kg (1056 lb$_m$)	3,4 t (3.7 st)	134,4 t (15.8 st)	172,8 t (190 st)
"shovel full" 9 kg (20 lb$_m$) per minute	540 kg (1200 lb$_m$)	13 t (14.4 st)	90,7 t (100.8 st)	388,8 t (432 st)	4665,6 t (5184 st)

Table 1.1

china clay were examined. The costs have been adjusted to reflect an annual increase for inflation. This study, compiled for the Institution of Mechanical Engineers, established that industrial fugitive materials add costs amounting to a one percent loss of materials and 40 pence ($0.70 USD) per ton of throughput. In short, for every ton carried on the conveyor, there is a loss of 10 kilograms (or 20 lbs lost for every short ton of material), as well as substantial additional overhead costs.

This overall cost was determined by adding four components together. Those components included:

A. The value of lost material (calculated at one percent of material)

B. The cost of labor devoted to cleaning up spillage, which averaged 12.8 pence ($0.22 USD) per ton of output

C. The cost of parts and labor for additional maintenance arising from spillage, which averaged 8.6 pence ($0.15 USD) per ton of output

D. Special costs peculiar to particular industries, such as the costs of reprocessing spillage and the cost of required medical checkups for personnel due to dusty environments, representing 19.7 pence ($0.33 USD) per ton of output

Note: This loss includes fugitive materials arising from problems such as spillage and conveyor belt carryback along with fugitive materials windblown from stockpiles.

A similar study of 40 plants, performed by the Royal Institute of Technology in Sweden, estimated that material losses would represent two-tenths of 1 percent of the material handled, and the overall added costs would reach nearly 13 Swedish Krona ($2.02 USD) per ton.

It is interesting that in both of these surveys, it was actual material loss, not the parts and labor for cleanup and maintenance, which added the largest cost per transported ton. However, the indirect costs of using labor for time-consuming cleanup

duties rather than for production are not included in the survey. Those figures would be difficult to calculate.

It is easier to calculate the actual costs for the disruption of a conveying system that, for example, lowers the amount of material processed in one day. If a belt runs 24 hours a day, each hour's production loss due to a belt outage can be calculated as the amount and the market value of material not delivered from the system's total capacity. This affects the plant's revenues and profits.

The Economics of Material Control

The cost of systems to control fugitive materials is usually considered three times during a conveyor's life. The first is during system design; the second, at system startup; and the third, during ongoing operations, when it is discovered the initial systems did not prevent fugitive materials.

It is often very difficult, with new installations, to predict the precise requirements for material control. In most cases, only a guess can be made, based on experience with similar materials on similar conveyors, indexed with "seat of the pants" engineering judgments. An axiom worth remembering is this: "A decision that costs $1 to make at the planning stage typically costs $10 to change at the design stage or $100 to correct on the site." The lesson: It is better to plan for worst-case conditions than to try to shoehorn in additional equipment after the initial system has been found to be underdesigned.

The details of conveyor transfer points, such as the final design and placement of chute deflectors, are sometimes left to the start-up engineer. It may be advantageous to allow the suppliers of specialized systems to be responsible for the final (on-site) engineering, installation, and start up of their own equipment. This may add additional start-up costs, but it is usually the most effective way to get correct installation and single-source responsibility for equipment performance.

1

Plants are often constructed on a rate of cost per ton of fabricated steel. Even if the best materials-handling controls are not put into place at the time of design, it costs little extra to ensure structures and chutes are installed that will allow for the installation of superior systems at some future date. The consequences of penny-wise, pound-foolish choices made in the initial design are the problems created: fugitive materials and chute plugging, compounded by the additional expense of doing it over again.

RECORD-KEEPING FOR TOTAL MATERIAL CONTROL

A great deal of attention is paid to the engineering of key components of belt conveyors. Too often, other factors affecting the reliability and efficiency of these expensive systems are ignored. The cost of fugitive materials is one such factor.

Record-keeping on the subject of fugitive materials is not part of standard reporting done by operations or maintenance personnel. The amount of spillage, the frequency of occurrence, maintenance materials consumed, and labor costs are rarely totaled to arrive at a true cost of fugitive materials. Factors—such as cleanup labor hours and frequency; the wear on conveyor skirting and conveyor belting; the cost of idler replacement including purchase price, labor, and downtime; even the extra power consumed to overcome stubborn bearings seizing from accumulations of materials—should all be calculated to determine a true cost of fugitive materials. Components whose service-life may be shortened by fugitive materials, such as idlers, pulley lagging, and the belt itself, should be examined to determine service-life and replacement cycle.

Computerized maintenance programs could easily include a field for cause of failure of any replaced parts. Pull-down prompts in these programs should include causes such as spillage, dust ingress, water ingress, and wear from material abrasion

(for rollers). This would allow computer-generated reporting of cost versus cause of component failures. This program should include data on belt-cleaning and belt-sealing devices, so accurate costs can be determined for the system installed.

Some contract maintenance services maintain conveyor databases on customers' conveyors, recording system specifications, details of equipment status, and service procedures performed. This information is helpful in scheduling preventative maintenance activities and in determining when outside resources should be utilized. This information can be used to better manage an operation's equipment and budget.

The measurement of fugitive materials at transfer points is difficult. In an enclosed area, it is possible to use opacity measuring devices to judge the relative density of dust in the air. For transfer points in the open, dust measurement is more challenging, although not impossible.

A basic technique is to clean a defined area and weigh or estimate the weight or volume of material cleaned and the time consumed in cleaning. Follow up is then conducted with repeat cleanings after regular intervals of time. Whether this interval should be weekly, daily, or hourly will depend on plant conditions.

What will be more difficult to determine is the point of origin of the lost materials. Fugitive materials can originate from conveyor carryback, spillage due to belt wander, skirt-seal leakage, spillage from loading surges or off-center loading, leakage through holes in chutework caused by corrosion or missing bolts, or even from floors above.

The individual making a fugitive material study has to bear in mind the number of variables that may influence the results. This requires the survey to be conducted over a reasonable time frame and include most of the common operating conditions, including: environmental conditions, operating schedule, material moisture content,

1

and other factors that create or complicate problems with fugitive materials.

Record-keeping of the amount of spillage—and of the costs of labor, parts, and downtime associated with it—should be a key part of the management information system for the operation of belt conveyors. Only when armed with such records covering a period of operation will an engineering study of fugitive materials and recommendations for total material control seem reasonable.

For many conveyor systems, the costs associated with lost materials will easily justify corrective measures. In most cases where adequate records have been kept, it has been shown that a modest improvement in material control will rapidly repay the costs of installing improved systems. Savings in labor expense alone often offsets the cost in less than one year of any retrofit equipment installed.

ADVANCED TOPICS

The Management of Risk and the Risk for Management

Many countries are starting to hold management personally accountable for failing to mitigate conditions such as spillage and dust resulting from poorly designed, operated, or maintained conveyors. In Australia, for example, the maximum penalty for failing to take corrective action to a known problem that causes death or grievous bodily harm is a $60,000 (AUD) fine and two years in prison for the manager, as well as a $300,000 (AUD) fine levied against the company. There is no doubt that a substantial number of accidents around conveyors are directly related to cleaning spillage and carryback, and it is also known that there are methods and products to control these problems. Consequently, any manager who chooses to ignore these problems and, as a result, risks the health of workers runs the risk of these penalties.

Using a standard industrial "Hazard Analysis" format—to determine the prob-

ability and consequence level of "hazards" experienced in cleaning spillage and carryback from under and around conveyors— provides a determination of the risk for employees and managers (**Table 1.2**).

Most conservative operators and maintenance people would evaluate the "Probability" of a safety incident taking place when cleaning spillage and carryback from under and around belt conveyors as "B: Has happened or near miss has happened" or "C: Could occur or I have heard of it happening," with the "People Consequence" rated as "2: Serious Injury." By moving these "Probability" and "People Consequence" values to the "Level of Risk Reckoner," a rating of Level 5 or Level 8 places the cleanup activities in the category of "Extreme Risk" of a serious injury (**Table 1.2**).

These ratings demonstrate a situation in which the risk management for the operations manager means proper diligence must be exercised by putting systems in place to eliminate or minimize these hazards.

Consequently, managers must do all within their power to eliminate the occasions (such as conveyor cleanup) that put employees in harm's way, both for the well-being of their employees and for the reduction of their own personal risks.

THE OPPORTUNITY FOR TOTAL MATERIAL CONTROL

In Closing...

When the costs created by fugitive materials are understood, it becomes obvious that controlling materials at conveyors and transfer points can provide major benefits for belt conveyors and to the operations that rely on these conveyors. This control has proven difficult to achieve—and more difficult to retain.

A planned and maintained approach is needed to aid in total material control. This is an opportunity to reduce costs and to increase efficiency and profitability for many operations.

Total material control means that materials are kept on the belt and within the system. Materials are moved—where they are needed, in the condition they are needed, at the flow rate they are needed—without loss or excess energy consumption, and without premature equipment failures or excessive maintenance costs. Total material control improves plant efficiency and reduces the cost of ownership.

This book presents many concepts that can be used in a program to achieve total material control for belt conveyors.

Looking Ahead...

This chapter about Total Material Control, the first chapter in the section Foundations of Safe Bulk-Materials Handling, introduced the need for and benefits of reducing spillage and dust. The following chapter, Safety, continues this section and explains the importance of safe practices around bulk-materials handling equipment as well as ways in which total material control will increase safety in the plant.

REFERENCES

1.1 Engineering Services & Supplies PTY Limited. Australian Registration #908273, Total Material Control and Registration #716561, TMC.

1.2 Environment Australia. (1998). *Best Practice Environmental Management in Mining: Dust Control*, (ISBN 0 642 54570 7).

Table 1.2

Risk Matrix System				
Step 1: Determine Probability		**Step 2: Determine Consequence (Higher of the Two)**		
Probability		**People Consequence**	**Plant, Property, Productivity, & Environmental Consequence**	
A	Daily: Common or frequent occurance	1	Fatality, permanent disability	Extreme danger, extreme business reorganization. Major environmental damage.
B	Weekly: Has happened or near miss has happened	2	Serious injury or illness (lost time)	High-level damage, significant business reorganization. Serious environmental damage.
C	Monthly: Could occur or I have heard of its happening	3	Disability or short-term injury (lost time)	Medium-level damage, serious production disruption. Reversible environmental damage.
D	Annually: Not likely to occur	4	Medical treatment injury	Low-level damage, slight production disruption. Minor environmental damage.
E	Once in 5 Years: Practically impossible	5	First aid or no injury	Negligible damage, minimal production disruption. No environmental damage.

Step 3: Level of Risk "Reckoner"—Calculate Risk

	A	B	C	D	E
1	1 EXTREME	2 EXTREME	4 EXTREME	7 EXTREME	11 SIGNIFICANT
2	3 EXTREME	5 EXTREME	8 EXTREME	12 SIGNIFICANT	16 MODERATE
3	6 EXTREME	9 SIGNIFICANT	13 SIGNIFICANT	17 MODERATE	20 MODERATE
4	10 SIGNIFICANT	14 SIGNIFICANT	18 MODERATE	21 LOWER	23 LOWER
5	15 SIGNIFICANT	19 MODERATE	22 LOWER	24 LOWER	25 LOWER

Typical example of a risk matrix system as used in Australia, origin unknown.

2

Figure 2.1

Belt conveyors feature fast-moving bulk materials and "pinch" points, characteristics that create risks for personnel who are working on or near them.

Chapter 2

SAFETY

2

In this Chapter...

In this chapter, we focus on the importance of safety practices and training for those working on or around conveyors. Potential causes of accidents are examined along with the costs of accidents, both direct and indirect. General safety procedures are described as well as specific safety practices. The importance of proper training, for both new employees and veterans, and appropriate content for such training are also discussed. The chapter concludes with a review of personal responsibility required for preventing accidents on and around conveyors.

Conveyors apply large amounts of mechanical energy to what is basically a giant rubber band, stretched tight and threaded through a maze of components (**Figure 2.1**). This band is burdened with a heavy load and then pulled at high speed. The forces applied are significant and potentially dangerous. By their very nature, belt conveyors feature fast-moving bulk materials and "pinch" points. These characteristics can create risks to personnel who are working on or near belt conveyors. Anyone who may be required to come into close proximity of a belt conveyor must always be aware of the power of that system and maintain a healthy respect for its potential to injure or kill an untrained or unwary individual.

A conveyor system may have a drive of 450 kilowatts of power (600 hp)—as well as all the inertia and potential energy of its load of tons of material. It is easy to see how a moving conveyor belt will easily win a tugging match with a worker, resulting in the chance for serious injury or death.

All forms of bulk transport—from mine haul trucks to trains to ships—carry their own hazards and safety concerns. While they also present some risks, properly designed, operated, and maintained conveyors can provide a safe and effective method of material movement.

Accidents can happen, but they can also be prevented. Conveyor safety begins with designs that avoid foreseeable hazards. Management must specify equipment that is safe and easy to maintain, and workers must follow the rules. The establishment and maintenance of safe practices in the design, construction, operation, and maintenance of conveyors and conveyor transfer points will aid greatly in the prevention of accidents. Proper training is a key to promoting safety for workers whose jobs bring them into the vicinity of belt conveyor systems.

THE RECORD AND THE PROBLEMS

How Conveyor Safety Relates to Dust and Spillage

Fugitive material from belt conveyors increases safety risks in many ways (**Figure 2.2**). Fugitive material creates the need for personnel to clean and to perform maintenance on and around conveyors. Placing personnel in close proximity to the moving belt creates the opportunity for an inadvertent contact to turn into a serious injury or death (**Figure 2.3**).

Figure 2.2

Fugitive material increases safety risks for personnel who work on and around conveyors.

Figure 2.3

Personnel working near a moving belt can create the opportunity for an inadvertent contact to turn into a serious injury or death.

2

Conveyors and Safety: A Look at the Record

Because of the size of their material cargos, the speed of their operation, and the amount of energy they consume and contain, conveyors pose a unique set of hazards. As a result, conveyors have shown to be a leading cause of industrial accidents, including serious injuries and fatalities.

A report from the Mine Safety and Health Administration (MSHA) in the United States examined conveyor accidents in metal/nonmetal mines recorded over the four-year period from 1996 to 2000. The MSHA report *(Reference 2.1)* listed the following worker activities related to those accidents:

A. Working under or next to poorly guarded equipment

B. Using hand or tool to remove material from moving rolls

C. Trying to free stalled rolls while conveyor is moving

D. Attempting to remove or install guards on an operating conveyor

E. Attempting to remove material at head or tail pulley while belt is in operation

F. Wearing loose clothing around moving belt conveyors

G. Not blocking stalled conveyor belt prior to unplugging (both flat and inclined belts) as energy is stored in a stalled conveyor belt

H. Reaching behind guard to pull V-Belt to start conveyor belt

An analysis of the document *(Reference 2.1)*, covering a total of 459 accidents of which 22 were disabling and 13 were fatal, shows 192 (42 percent) of the reported injuries (including 10 fatalities) occurred while the injured worker was performing maintenance, lubricating, or checking the conveyor. Another 179 (39 percent) of the reported injuries (including 3 fatalities) occurred while the subject was cleaning and shoveling around belt conveyors (**Table 2.1**).

The MSHA report found no differentiation in the likelihood of accidents based on age, experience, or job title of the accident victim.

A preliminary study of 233 fatal mining accidents in the United States during the years 2001 to 2008 revealed there were 48 fatalities in 47 incidents involving conveyors *(Reference 2.2)*. The data were compiled from reports by the Mine Safety and Health Administration of the US Department of Labor.

The activities most frequently listed as leading to the conveyor-related fatalities were listed as Maintenance, e.g., replacing idlers or belting or clearing blockages with 35 percent (or 17 deaths), with Cleanup, e.g., shoveling or hosing spillage or clearing buildup at an idler second with 27 percent (or 13 deaths). Many of these fatalities resulted from the victim becoming caught in the moving belt by getting too close to an unguarded pinch point or working on a moving conveyor.

Table 2.1

1996-2000 MSHA Conveyor Accident Data			
Cause of Injury	Fatal	Non-Fatal	Total
Caught in Moving Belt	10	280	290
Maintenance, Lubrication, or Inspection	10	182	192
Cleaning and Shoveling	3	176	179
Total*	13	446	459

Note: The total figure in each column is not the sum of that column, as there may be multiple causes cited for any specific accident (Reference 2.1).

CONVEYORS CAN BE DEADLY

As he was walking the length of the 45-meter- (150-ft-) long conveyor belt that ran from the crusher to the storage bin, he noticed that the take-up pulley was caked with gunk. He'd worked in the mill for 10 years now, and for all that time the scraper on the head pulley that was supposed to keep the belt clean seemed to constantly need adjusting. Oh well. He'd get to that in a minute. First to clean the take-up pulley.

The crusher operator was away from his post, and both the crusher and the conveyor were shut down. He walked back to the control box, turned on the conveyor, and then got a front-end loader and drove it back to the take-up pulley. He raised the bucket of the loader high enough so that he could stand in it and reach the pulley. Grabbing a tool—an ordinary garden hoe with a handle cut down to 40 centimeters (16 in.)—he climbed in the bucket and began to scrape the gunk off the pulley. He'd done it this way several times before and it was so much faster when the pulley was turning. And the belt was only moving 2,3 meters per second (450 ft/min)—that's only about 8 kilometers per hour (5 mph).

Before he even knew what was happening, something seemed to grab first the tool and then his arm and yanked him up out of the bucket. The pain of his arm being crushed lasted only until his shoulder and neck smashed into the conveyor structure. He died instantly. He was 37.

This story is not fiction. It is also not the only one that could have been selected. Numerous such case histories are available. Conveyors kill all too frequently—or amputate, cut, or crush.

Because conveyors are so common, it is easy to take them for granted and become complacent. And at about 8 kilometers per hour (5 mph), they don't seem to be moving that fast. It's easy to believe that the victim in this story could have simply dropped the tool when it was caught. Not so.

Eight kilometers per hour (5 mph) works out to about 2,2 meters per second (7.3 ft/s). At that speed, if your reaction time is three-fourths of a second, your hand would travel 1,6 meters (5.5 ft) before you let go of the tool. Even if your reaction time is a fast three-tenths of a second, your hand would travel over 660 millimeters (26 in.). That's far enough to become entangled if you're using a short tool. And if it's a loose piece of clothing that gets caught, then it doesn't matter what your reaction time is. It's already too late.

Reprinted with permission from the "Safety Reminder" newsletter, published by Ontario Natural Resources Safety Association (Reference 2.4). (Measurement conversions added by Martin Engineering.)

In South Africa, the Conveyor Belt Systems Safety in Mines Research Advisory Committee's report examined more than 3000 accidents (including 161 belt conveyor fatalities) between 1988 and 1999 *(Reference 2.3)*. In findings that echoed the above-cited MSHA report, the document noted that "people working on moving conveyors, inadequate guarding, and ineffective locking out stand out as major causes of conveyor accidents." According to the report, injuries most frequently result from people working at the tail pulley, head pulley, idlers, and loading chute.

The Cost of Conveyor Accidents

Some of the direct costs arising from accidents—such as medical treatment, lost wages, and decreased productivity—can be identified. Less evident expenses associated with accidents are known as "indirect" or "hidden" costs and can be several times greater than the value of direct costs. These hidden costs include:

A. Expense and time of finding a temporary replacement for the injured worker

B. Time used by other employees to assist the injured worker

C. Time used by supervision to investigate the mishap, prepare accident reports, and make adjustments to work schedules

D. Property damage to tools, materials, and equipment

E. Delays in accomplishment of tasks

In 2005, the National Safety Council in the United States listed $1,190,000 USD as the cost of a work-related death; the cost of a disabling injury was assessed at $38,000 USD. Their accounting includes wage and productivity losses, medical expenses, and administrative expenses. The figures do not include any estimate of property damage.

It is easy to see that even a slight reduction in the number of conveyor-related accidents can save significant amounts of money for an operation.

CONVEYOR SAFETY PRACTICES

General Conveyor Safety Practices

There are certain safety practices that should be observed regardless of the design of the conveyor or the circumstances of its operation. They include:

A. Lockout / tagout / blockout / testout procedures

Lockout / tagout / blockout / testout procedures must be established for all energy sources of the conveyor belt, as well as conveyor accessories and associated process equipment. Bulk materials on the belt may present danger from falling lumps or potential energy that can cause the conveyor to move even when the system has been de-energized. Lockout and tagout alone may not be enough to ensure a worker's safety; therefore, it is important that after lockout and tagout, the worker blocks out the conveyor (blockout) and tests to make sure it cannot move (testout). These procedures should be followed before beginning any work—whether it is construction, installation, maintenance, or inspection—in the area.

B. Inspection/maintenance schedule

A formal inspection and maintenance schedule must be developed and followed for the material-handling system. This program should include emergency switches, lights, horns, wiring, and warning labels, as well as the conveyor's moving parts and accessory components.

C. Observance of operating speed and capacity

The design operating speed and capacity for the conveyors, chutes, and other material-handling equipment must not be exceeded.

D. Safety "walk around"

All tools and work materials must be removed from the belt and chute before restarting a conveyor. A safety "walk around" is recommended prior to resuming conveyor operation.

E. Emergency controls

All emergency controls must be close to the system, easy to access, and free of debris and obstructions.

F. Personal protective equipment (PPE) and attire

Appropriate personal protective equipment and attire, in accordance with local requirements (often including a hard hat, safety glasses, and steel-toe shoes) must be worn when in the area of the conveyor. Loose or bulky clothing is not allowed; nor is loose long hair or jewelry.

G. Safe practices while the system is in operation

2

It is important to never poke, prod, or reach into a conveyor or other material-handling system or attempt to clean or adjust rollers or other components while the system is in operation.

H. No personnel on conveyor

Personnel should never be allowed to sit on, cross over, or ride on an operating material-handling conveyor. (In some parts of the world, man-riding belt conveyors are the accepted method for workers to reach their assigned areas; in other regions, it is strictly forbidden. For simplicity's sake, this volume will hold man-riding conveyors separate from its discussion of belt conveyors for bulk materials.)

Safety Standards for Conveyor Design and Operation

The above practices are not intended to replace the more detailed safety guidelines published by the American Society of Mechanical Engineers (ASME) in ASME Standard B20.1-2006 *Safety Standard for Conveyors and Related Equipment* and B15.1 *Safety Standard for Mechanical Power Transmission Apparatus* in the United States—by appropriate regulatory and safety agencies around the world, or the rules of a specific plant. Consult those references, as well as the safety instructions provided by the manufacturers of specific systems.

In Australia, Australia Standard (AS) specification AS1755-2000 *Conveyor Safety Equipment* applies to the design, construction, installation, and guarding of conveyors and related systems for the conveyance of materials.

These references and/or their national or international equivalent should be consulted as a guide for the design and construction of any belt conveyor system.

Conveyor Electrical Systems and Safety

The electrical systems of conveyors often involve high voltages and complex control and communication systems. The electrical trade is almost always considered a separate group within a plant's maintenance department. Only workers who are specially trained and certified should work on conveyor electrical supply and control systems.

Pre-Job Safety Assessment

Prior to commencing any work on belt conveyors, it is recommended that a pre-job safety assessment be performed. This assessment should include all equipment that may be interlocked with the actual piece of equipment to be serviced. This pre-job assessment should make sure that the entire area is safe for workers and the proper equipment is available to perform the work safely. In addition, the pre-job safety inspection should include close inspection of the surrounding work area for fire hazards, tripping hazards, or falling objects.

One item often overlooked in the pre-job safety inspection is the coordination of all employee activities on the belt conveyor. For example, workers in the head chute changing belt-cleaner blades could be injured if workers changing the belt were to move the belt.

Lockout/Tagout Procedures

A crucial part of a conveyor safety program is the lockout/tagout procedure. In the United States, lockout/tagout is an Occupational Safety & Health Administration (OSHA) requirement; MSHA has adapted a similar version of this rule. To achieve complete safety in the face of the potential

Figure 2.4

Lockout/tagout rules require that power to the conveyor system (and any accessory equipment) be shut down, locked, and tagged by the person who will be performing work on the system.

2

energy stored in belt tension or elevated bulk materials, the additional components of blockout and testout are recommended.

The lockout/tagout rules require that power to the conveyor system (and any accessory equipment) be shut down, locked, and tagged by the person who will be performing work on the system (**Figure 2.4**). Only the person who locked out the system can unlock it. This prevents someone from starting the conveyor belt unknowingly while someone else is working on it.

Typical lockout/tagout procedures follow:

A. Own lock

Each worker is required to place a personal lock on the de-energizing switch or switches. This may require one lock bar or multiple lock bars (**Figure 2.5**).

B. Own key

Only the employee who places each lock has the key to that lock, and only that employee can remove the lock.

C. Multiple locks

If a number of employees are working in a given area, each should place a lock on the power source. Some equipment will have numerous locations that may require lockout.

D. Own tag

Each employee who placed a lock should also place a tag that includes the employee's name and contact information.

Blockout Procedures

Even when a belt conveyor has been properly locked out and tagged out, there may still be significant amounts of tension or potential energy present in the system. One easy-to-understand scenario is if an inclined belt had an emergency shutdown with material loaded on the belt, the weight of the material will cause the belt to roll backward. Both the movement of the belt and the potential cascade of material off the downhill end of the conveyor cause risk of injury to the unlucky or unwary employee.

Lifting the gravity take-up's counterweight might not release these tensions: This method should not be relied upon. Properly installed brakes and backstops may help prevent this roll back. However, a plant should not rely on the backstops or brakes to prevent a belt from moving on its own. There have been instances in which the belt has moved due to the internal tensions by the belt stretch.

Blocked chutes, material trapped at load zones, material under the belt, or bad bearings may stall the belt sufficiently to create considerable belt tensions. The belt may move in either direction, based on the conditions present at the time of the work; these conditions can and do change as the work progresses.

If employees are required to be on the belt or near pinch points on the conveyor, the belt should be physically restrained from moving under its own power. This

Figure 2.5

Placing a lock on the conveyor's power system will prevent someone from starting the conveyor belt unknowingly while someone else is working on it.

Figure 2.6

The belt-clamping device should be secured to a structural member of the conveyor capable of restraining the expected forces, not to an idler.

is called "blocking" the conveyor belt, or blockout. Belt clamps, chains, and come-alongs (ratchet lever hoists) can be used to physically restrain the belt by securing the blocking device to a structural member of the conveyor capable of restraining the expected forces (**Figure 2.6**).

It is recommended that engineered equipment be purchased that will securely clamp to the belt to prevent movement.

Testout Procedures

A testout procedure provides a final check to make sure the system is secure and de-energized before work proceeds. It is a good practice to try to start the belt conveyor or interlocked equipment after the lockout lock has been placed. This should include both local start/stop stations and the system's remote controls. This ensures that the correct breakers were de-energized.

Equipment Guards

It is important that pinch points—on rotating equipment like head pulleys and on equipment that allows sudden movements, like gravity take-ups—be equipped with guards to prevent accidental or unwise encroachment by employees (**Figure 2.7**).

It is becoming more common for conveyors to be totally enclosed in guard barriers along walkways to protect personnel who are required to use the walkways (**Figure 2.8**). These guards are fabricated of metal mesh or screen that allows observation of moving parts and pinch points without allowing personnel to get so close as to pose a risk of injury (**Figure 2.9**).

While each nation has its individual requirements that must be met when it comes to the proper placement of guards on equipment, local and industry requirements should also be thoroughly investigated and implemented.

At the same time, it must be remembered that service access must be provided to the various pieces of equipment. The physical

Figure 2.7

Pinch points on rotating equipment, like head pulleys, and on equipment that allows sudden movements, like gravity take-ups, should be equipped with guards to prevent accidental or unwise encroachment by workers.

Figure 2.8

It is becoming more common for conveyors to be totally enclosed in guard barriers along walkways, to protect personnel who are required to use the walkways.

Figure 2.9

Guards are fabricated of metal mesh or screen that allows personnel to observe the conveyor's moving parts and pinch points without getting so close as to risk an injury.

Figure 2.10

To protect personnel working near the belt, the conveyor should be equipped with "pull-rope" emergency stop switches.

barriers installed to shield this equipment must be carefully designed, or they will interfere with maintenance efficiency. Removing the guards should require a special tool to prevent untrained individuals from entry into the system. The guards must be sufficient to prevent an employee from reaching in, or around, the guards to gain access to the pinch point.

After service procedures are completed, it is important that guards be returned to position prior to restarting the conveyor.

Emergency Stop Switches

To protect personnel working near the belt, the conveyor should be equipped with "pull-rope" emergency stop switches. These safety switches should be conveniently mounted along conveyor catwalks and right-of-ways (**Figure 2.10**). The system should run the full length of both sides of the conveyor if there is access, or a walkway, on both sides of the belt. The switches should be wired into the belt's power system so that in an emergency, a pull of the rope interrupts the power to the conveyor, stopping the belt.

In 1995, MSHA in the United States alerted mine operators about potential failures of emergency-stop pull-cord systems along conveyors. After tests of over 1100 systems, MSHA noted a failure rate of two percent. MSHA attributed these problems to several factors:

A. Spillage around the switch that prevents deactivation of the conveyor

B. Broken pull cords or excessive slack in cords

C. Frozen pivot bearings where the switch shaft enters the enclosure

D. Failure of electrical switches inside the enclosure

E. Incorrect wiring of switch or control circuits

The solution to this problem is proper service attention and testing drills similar to school fire drills, when the operation of the conveyor safety equipment can be checked. These tests should be performed monthly.

Safety Signage

Safety stickers and warning labels should be affixed at pinch points, service access doors, and other hazardous areas on conveyor equipment (**Figure 2.11**). It is the responsibility of the manufacturer to supply—and whenever possible, apply—safety warnings to equipment. These signs must be kept clean and legible, and should be reapplied appropriately to suit changes in equipment or procedures. It is the responsibility of operations management to replace worn, damaged, or unreadable safety warnings. It is the responsibility of the employee to comply with safety signage.

Signs warning of equipment that can be remotely started should be placed prominently along the conveyor. Many times the conveyor belts start automatically or when triggered by an operator in a remote control room, nowhere near the conveyor.

Chutes that have flow aid devices such as air cannons, that can seriously injure employees when discharged, must be clearly marked. This situation also requires the chute to be marked with restrictions for vessel entry. Flow aids must be de-energized and properly locked out / tagged out / blocked out / tested out and chutes cleaned prior to chute entry.

Safety stickers and signs are available from reputable manufacturers of conveyors and related equipment, as well as safety supply houses. In the United States, the Conveyor Equipment Manufacturers Association (CEMA) has a variety of safety

Figure 2.11

Safety stickers and warning labels should be affixed at pinch points, service access doors, and other hazardous areas on conveyor equipment.

and precautionary labels available for bulk-material conveyors and several common accessories. These labels can be viewed and ordered from the organization's website at http://www.cemanet.org

Signage incorporating pictograms may be required to protect all workers in plants where multiple languages are spoken (**Figure 2.12**). If required, translation of the messages of safety signage should be performed locally to ensure accurate meaning. (The International Organization for Standardization has a goal to eliminate words that need to be translated and use signs that incorporate only pictures.)

Safety Considerations at New Conveyor Startup

Start-up time for new conveyors can be one of the most dangerous times, because the system may not behave as expected, and safety equipment such as startup horns and signals may not work properly. It is recommended that multiple spotters be stationed along the conveyor route prior to starting the belt. They should have radio communication with the control room or person starting the belt conveyor.

Adjustments that may be required to get the belt running smoothly should be made only while the conveyor is properly locked out / tagged out / blocked out / tested out. Making even a simple adjustment while the belt is running may result in an accident.

Hazardous Thinking

Many operations forbid working on the conveyor system when it is running. Nevertheless, the same operations require working around the conveyor while it is in operation. Workers come to the conveyors for inspection and maintenance procedures and to clean up fugitive material.

As Richard J. Wilson (*Reference 2.5*), of the Bureau of Mines Twin Cities Research Center, noted:

Most procedures require locking out the main power switch at the head pul-

ley or control room. As this could be quite some distance from the work site, compliance could require a considerable amount of time and effort. It is not difficult to imagine maintenance personnel rationalizing that it is all right to quickly perform some routine repair work without locking out the belt when implementing the lockout procedure would take much longer than the job itself.

This "shortcut" creates the opportunity for accidents and injuries.

SAFETY TRAINING

The South African report (*Reference 2.3*) noted: "Most accidents can be attributed to a lack of an understanding of the inherent risks of conveyor systems and the safe use of such systems."

The best approach to accident prevention is a well-designed safety program combined with effective, and repeated, training.

Figure 2.12

Signage incorporating pictograms may be required to protect all workers in plants where multiple languages are spoken.

Conveyor Training for New Employees

Conveyor belt safety must start with the newest employee. The tendency is to send "the new kid"—the least-experienced, most-recently hired—out to do "the job that nobody wants:" clean up around moving belt conveyors. Prior to assigning a new employee the task of working around a conveyor belt, the employee should attend a minimum of four hours of classroom instruction specific to belt conveyors.

The Conveyor Training Course

Each plant should have in place a program of training for individuals whose positions will require them to work on or near conveyors (**Figure 2.13**). This program will discuss the risks of, and safe procedures for, working in the vicinity of belt conveyors. The training should include a thorough understanding of the variety of belt conditions that can influence belt operations, fugitive material, and personnel safety. By understanding different conditions of the conveyor, accidents can be reduced.

This comprehensive training should include as a minimum the following:

A. General safety practices around belt conveyors

B. Personal protective equipment (PPE)

C. Proper personal grooming and apparel

D. Proper shoveling techniques

E. Safe inspection and maintenance practices

F. Conditions of the conveyor that cause problems (leading to maintenance and safety issues)

G. Belt selection to match structure and conditions

H. Identification of the sources of fugitive material

I. Elimination of fugitive material (dust and spillage)

J. Belt-tracking procedures

Refresher and Reminder Training

In addition to new employees, veteran employees also need training. Senior employees have probably had little or no training about conveyor belts or conveyor belt safety. Seminars such as Martin Engineering's Foundations™ Workshop have proven effective in providing focused conveyor training to personnel ranging from operations and maintenance workers to conveyor engineers and plant managers. The corresponding Foundations™ Certification Program provides a self-guided conveyor learning experience.

Periodic reminders between the regular training and retraining sessions are also beneficial. Agencies such as the US MSHA are a source for case histories on conveyor belt accidents that can provide effective refresher training materials.

THE IMPORTANCE OF PERSONAL RESPONSIBILITY

In many ways, plant safety is like plant cleanliness: They are both matters of attitude. Plant management can set a tone for the overall operation; however, it is the response of each individual worker that will have the greatest impact on a plant's safety record.

Safety is not just the responsibility of a plant's safety department or a governmental agency. Rather, it is the responsibility of each worker to ensure safety for oneself and for one's co-workers.

Personal responsibility includes:

A. Use of personal safety equipment, including dust masks and respirators,

Figure 2.13

Each plant should have in place a program of training for individuals whose positions will require them to work on or near conveyors.

hearing protection, hard hats, and steel-toe shoes

B. Attention to safety practices

C. Good standards of housekeeping to provide a clean and safe work area

D. Thorough review of equipment manuals to learn safe operations and maintenance procedures

E. Willingness to stop unsafe actions by other workers and to coach others in proper safety procedures

ADVANCED TOPICS

"New Generation" Guarding Systems and Conveyor Stopping Time

There is a question as to whether or not the "new generation" guarding technologies can react quickly enough to prevent injury when compared to conventional barrier guarding. The determining factors are the detection range of the equipment, the guarding device's speed of response, the distance of the protection device from the nearest moving part, the conveyor belt's stopping time, and the person's speed of movement.

Almost every conveyor design program calculates the stopping time for the conveyor based on control of the belt tension for optimizing the drive components. Instantaneous stopping, starting, or reversing of the belt can cause serious dynamic problems with the conveyor and accessories. Even on short conveyors, where dynamic concerns are not considered, it is common for a belt to take 5 seconds to come to a stop. On long systems, where dynamic issues are a concern, it is common for the conveyor to take 30 seconds to come to a stop. Simple reaction time for a person is usually defined as the time required for an observer to detect the presence of a stimulus. Given that the reaction time for the average person is approximately 0.2 seconds and that a conveyor belt will take time to stop, there is more than enough time to entrap a person in the conveyor.

Therefore, it is highly likely that if a worker has violated the safety rules and/or systems and becomes entrapped, the conveyor cannot be stopped in time to prevent serious injury or death.

DOUBLE BENEFIT OF CONVEYOR SAFETY

In Closing...

Is it possible to eliminate accidents involving conveyors? Probably not. However, we can work toward zero accidents with a two-pronged approach:

A. Training new-hires and veteran employees about how to work safely on and around belt conveyors

B. Eliminating many of the problems that require the employee to work in close proximity to belt conveyors

Bear in mind that the cost of one accident can easily exceed the cost of bringing a training program into the organization, or even the annual salary of a full-time training person.

Providing training on conveyor operations and maintenance to improve conveyor safety has the potential to also improve operating results. In fact, training in safe conveyor operations provides the best of both worlds: It presents an opportunity to provide worker safety and simultaneously improve a facility's operating efficiency.

Looking Ahead...

This chapter about Safety, the second chapter in the section about Foundations of Safe Bulk-Materials Handling, follows Total Material Control and explains how accidents can be caused by a lack of total material control. The next three chapters focus on Conveyors 101, beginning with Conveyor Components, followed by The Belt and, finally, Splicing the Belt.

2

REFERENCES

2.1 Padgett, Harvey L. (2001). *Powered Haulage Conveyor Belt Injuries in Surface Areas of Metal/Nonmetal Mines, 1996–2000*. Denver, Colorado: MSHA Office of Injury and Employment Information.

2.2 Maki, D. Michele, PhD. 2009. Conveyor-Related Mining Fatalities 2001-2008: Preliminary Data. Unpublished Report for Martin Engineering.

2.3 Dreyer, E., and Nel, P.J. (July 2001). *Best Practice: Conveyor Belt Systems*. Project Number GEN-701. Braamfontein, South Africa: Safety in Mines Research Advisory Committee (sic) (SIMRAC), Mine Health and Safety Council.

2.4 Ontario Natural Resources Safety Association. *Safety Reminder*, newsletter. P.O. Box 2040, 690 McKeown Avenue, North Bay, Ontario, Canada, B1B 9PI Telephone: (705) 474-SAFE.

2.5 Wilson, Richard J. (August, 1982). *Conveyor Safety Research*. Bureau of Mines Twin Cities Research Center.

2.6 Giraud, Laurent; Schreiber, Luc; Massé, Serge; Turcot, André; and Dubé, Julie. (2007). *A User's Guide to Conveyor Belt Safety: Protection from Danger Zones*. Guide RG-490, 75 pages. Montréal, Quebec, Canada: IRSST (Institut de recherche Robert-Sauvé en santé et en sécurité du travail), CSST. Available in English and French as a free downloadable PDF: http://www.irsst.qc.ca/files/documents/PubIRSST/RG-490.pdf.

2

3

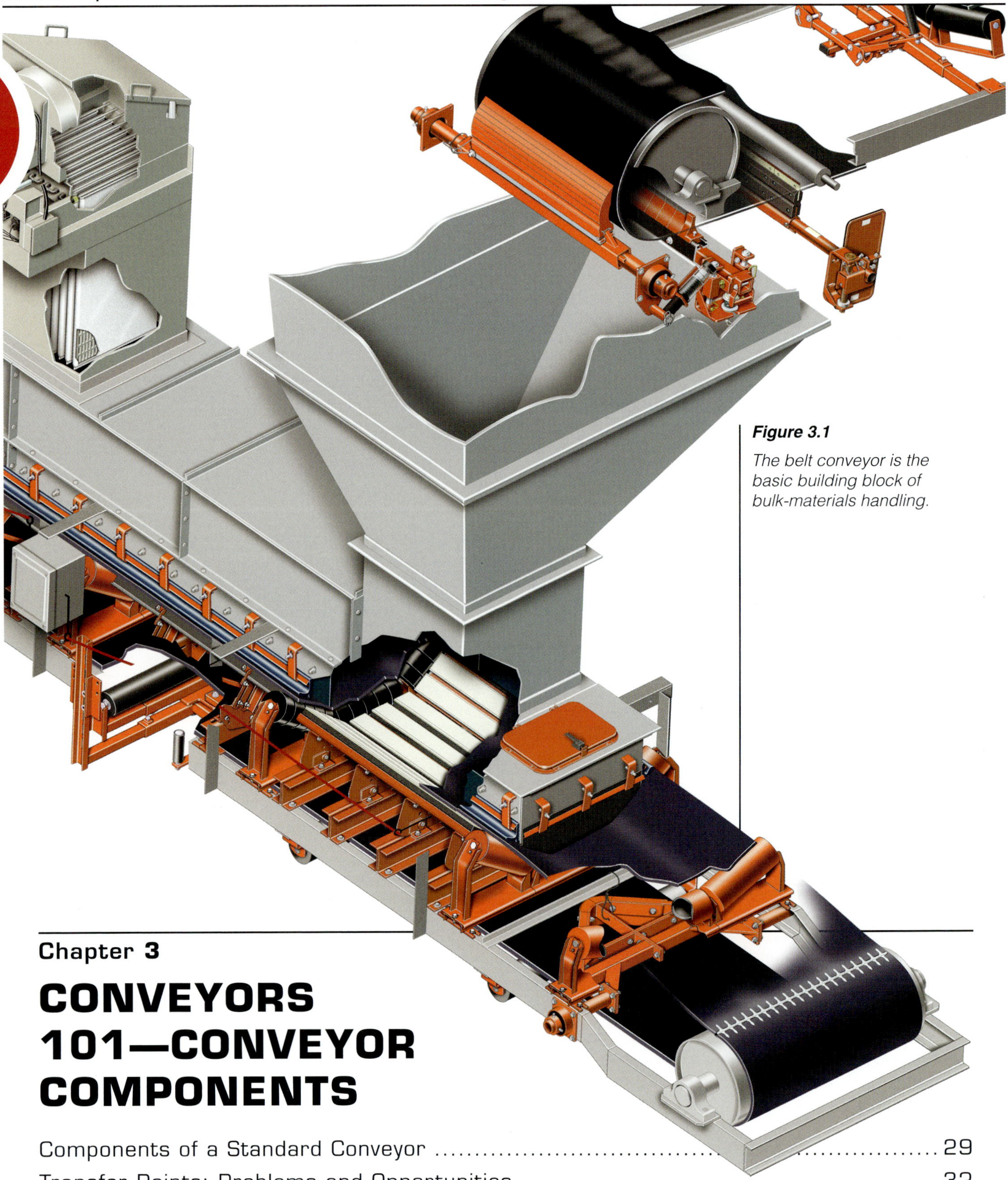

Figure 3.1

The belt conveyor is the basic building block of bulk-materials handling.

Chapter 3

CONVEYORS 101—CONVEYOR COMPONENTS

In this Chapter

In this chapter, we describe the most common components of the belt conveyors used in handling bulk materials and provide an equation for calculating the power required to operate the conveyor belt. Three methods of designing transfer points are explained, along with the benefits of using a systems-engineering approach to aid in material control.

Belt conveyors have been used for decades to transport large quantities of material over long distances. Conveyors have proven time and time again that they are a reliable and cost-effective method for material movement. Belt conveyors can transport materials up steep inclines, around corners, over hills and valleys, across bodies of water, above ground, or below ground. Belt conveyors integrate well into other processes, such as crushing, screening, rail-car- and ship-loading and unloading, and stockpile and reclaim operations.

Belt conveyors have shown the ability to transport materials that vary from large, heavy, sharp-edged lumps to fine particles; from wet, sticky, slurry to dry, dusty powder; from run-of-mine ore to foundry sand; from tree-length logs to wood chips, and even potato chips.

Of all materials-handling systems, belt conveyors typically operate with the lowest transport cost per ton, the lowest maintenance cost per ton, the lowest energy cost per ton, and the lowest labor cost per ton. These advantages may not be realized if careful consideration of the bulk material to be handled and of the overall process is not taken when specifying a conveyor system.

Many designs of conveyor systems are available. Many are developed to solve unique and difficult problems specific to a particular industry or bulk material. With any choice, certain design fundamentals can determine the success or failure in controlling the spillage or dusting of fugitive material.

This chapter will describe the most common components used in bulk-materials handling.

COMPONENTS OF A STANDARD CONVEYOR

The Basics

For many plants, the belt conveyor is the basic building block of bulk-materials handling (**Figure 3.1**). In essence, a belt conveyor is a large reinforced rubber band stretched around two or more pulleys, traveling at a defined rate of speed, carrying a specified quantity of material(s). Complications arise as the line of travel becomes inclined or curved, when the conveyor must be added into a sophisticated process or plant, or when it needs to meet material feed-rate requirements or other operational constraints.

A belt conveyor is a relatively simple piece of equipment (**Figure 3.2**). Its basic design is so robust that it will convey material under the most adverse conditions—overloaded, flooded with water, buried in fugitive material, or abused in any number of other ways. The difference, however, between a belt conveyor that is correctly engineered, operated, and maintained and a dysfunctional system usually becomes quickly apparent in the system's operating and maintenance costs.

Common belt conveyors for bulk materials range in width from 300 millimeters (12 in.) to 3000 millimeters (120 in.), with belts 5000 millimeters (200 in.) wide seen in applications such as iron ore pellet plants. The conveyors can be any length. Cargo capacity is limited by the width and speed

Figure 3.2

A belt conveyor is a relatively simple piece of equipment.

3

of the conveyor belt, with conveyors often moving several thousand tons of material per hour, day in and day out.

Every bulk-materials handling belt conveyor is composed of six major elements:

A. The belt

Forms the moving surface upon which material rides

B. The pulleys

Support and move the belt and control its tension

C. The drive

Imparts power to one or more pulleys to move the belt

D. The structure

Supports and aligns the rolling components

E. The belt-support systems

Support the belt's carrying and return strands

F. The transfer systems

Load or discharge the conveyor's cargo

Another part of every conveyor is the ancillary equipment installed to improve the system's operation. This would include such components as take-ups, belt cleaners, tramp-iron detectors, skirtboards and seals, belt-support systems, plows, safety switches, dust-suppression and dust-collection systems, and weather-protection systems.

Components of a Standard Belt Conveyor

Although each belt conveyor is somewhat different, it shares many common components (**Figure 3.3**). A conveyor consists of a continuous rubber belt stretched between terminal pulleys. One end is the tail. This is usually where the loading of the cargo takes place, but loading may take place anywhere along the length of the conveyor, and conveyors with multiple load zones are relatively common. The other end of the conveyor is called the head. The cargo is usually discharged at the head, but with the use of plows or trippers, the load may discharge anywhere along the conveyor's length.

The belt is supported along the top (carrying) side with flat or troughing rollers called idlers. Troughing rollers form the belt into a U-shape that increases the cargo capacity of the conveyor. On the bottom (return) side of the conveyor, where the belt returns to the loading point, the belt is supported with return idlers. The rolling components are mounted in frames and supported by a steel structure called the stringers. In some applications, such as underground or overland conveyors, the rolling components of the conveyor are mounted on suspended wire ropes.

Figure 3.3

Although each belt conveyor is somewhat different, all conveyors share many common components.

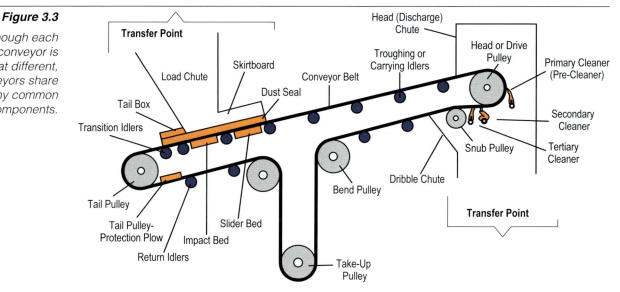

Usually electrically powered, conveyors' drive motors are most often located to turn the conveyor's head pulley. The motor(s) can be located at any point along the conveyor. Multiple motors are often used on long or heavily-loaded conveyors.

A tensioning device, called a take-up, is used to make sure that the belt remains tight against the drive pulley to maintain the required tension in the belt to move the belt and cargo. Most common is an automatic tensioning device referred to as a gravity take-up, which uses a counterweight to create tension in the belt. The gravity take-up is often installed near the drive pulley on the return side of the belt. Bend pulleys are used to direct the belt into the take-up pulley, which is attached to the counterweight of the gravity take-up.

Another type of pulley, called a snub pulley, is often placed immediately after the head pulley on the return side of the belt to increase the contact of the belt with this pulley, allowing a smaller drive pulley to transmit the required tension to the belt.

The cargo is usually loaded near the tail end in an area referred to as the loading zone. The components of the loading zone will likely consist of a loading chute, tail pulley, idlers, belt-support systems, skirtboards, wear liners, dust seals, entry seals, and exit seals.

A conveyor's head, or discharge, end will usually consist of the head pulley, a discharge chute along with a belt-cleaning system, a dribble chute, and other equipment to monitor and maintain flow.

A transfer point is where the bulk material moves from one piece of equipment to another. A transfer point can be either a loading or discharge zone, or in the case where one conveyor is feeding another, one transfer point can contain both the loading and discharge zones. However, a transfer point could also be where a belt feeds another bulk-materials handling or processing system or where the belt receives cargo from another bulk-materials handling or processing system. These systems might be storage vessels of any type; cargo carriers such as trucks, railcars, barges, or ships; or other pieces of process equipment.

Depending on the conveyed material, a variety of other ancillary equipment may also be installed along the run of the conveyor or in the transfer point at either end.

Conveyor Drive Power

As discussed above, conveyors are driven by the attachment of a motor to the drive pulley. Determining the power requirement of the conveyor (that is, specifying the size of the motor required) is a question of the tension to drive the belt and the speed of the belt.

The sixth edition of Conveyor Equipment Manufacturers Association's (CEMA) *BELT CONVEYORS for BULK MATERIALS* provides an equation, originally developed by Deutsches Institut für Normung (DIN), for determining the power requirement of a "Basic Conveyor."

The factors with the largest influence on conveyor power requirements are the weight of the cargo and the amount of elevation the cargo must be lifted. Friction of the various conveyor components is normally a small part of the power requirement. When the belt is horizontal, this friction becomes the most important consideration. The tension is used to find the power required to operate the conveyor belt (**Equation 3.1**). The ΔT value is the sum of all positive and negative tensions in the system that must be transferred to the belt by the drive pulley to overcome the resistance of the components and transport the load. Each of the calculations for the component tension can be found in this book or in the sixth edition of CEMA's *BELT CONVEYORS for BULK MATERIALS*. The tension added to lift the actual material and conveyor belt is calculated in CEMA's *BELT CONVEYORS for BULK MATERIALS*.

Once the tension for each component is found, they can be added to arrive at

Equation 3.1

Calculation for Power Required

$$P = \Delta T \cdot V \cdot k$$

Given: 5400 newtons (1200 lb$_f$) of tension is added to a belt traveling at 3 meters per second (600 ft/min). **Find:** The power required to operate the conveyor belt.

	Variables	Metric Units	Imperial Units
P	Power Required	kilowatts	horsepower
ΔT	Net Tension Added to the Belt	5400 N	1200 lb$_f$
V	Belt Speed	3 m/s	600 ft/min
k	Conversion Factor	1/1000	1/33000

Metric: $P = \dfrac{5400 \cdot 3}{1000} = 16$

Imperial: $P = \dfrac{1200 \cdot 600}{33000} = 22$

P	Power Required	16 kW	22 hp

a total power requirement. A reasonable safety factor should be added to the power requirements to ensure the conveyor belt is not operated near or at an overload condition. This value should be used as a MINIMUM when selecting drive components.

The Value of the Conventional Conveyor

There are a variety of advanced conveyor systems that provide alternatives for material handling. *(See Chapter 33: Considerations for Specialty Conveyors.)* For general purposes, the conventional troughed-idler belt conveyor is the performance standard and the value leader against which these other systems must be evaluated. Troughed-idler belt conveyors have a long history of satisfactory performance in challenging conditions.

The overall success of a belt conveyor system greatly depends on the success of its transfer points. If material is loaded poorly, the conveyor will suffer damage to the belt, to its rolling components, and/

or to its structure, decreasing its operating efficiency. If material is allowed to escape, this fugitive material will cause numerous maintenance problems, again leading to reduced production efficiency and increased operating and maintenance costs.

TRANSFER POINTS: PROBLEMS AND OPPORTUNITIES

The Challenge of the Transfer Point

A typical transfer point is composed of metal chutes that guide the flow of material (**Figure 3.4**). It may also include systems to regulate the flow, to properly place the flow within the receiving structure (whether belt, vessel, or other equipment), and to prevent the release of fugitive material.

Transfer points are typically installed on conveyors for any or all of the following reasons:

A. To move the material to or from storage or process equipment

B. To change the horizontal direction of the material movement

C. To divert the flow to intermediate storage

D. To allow effective drive power over a distance that is too long for a single conveyor

The method and equipment for loading the belt contribute much toward prolong-

Figure 3.4

A typical transfer point is composed of chutes that guide the flow of material into the receiving belt, vessel, or other process system.

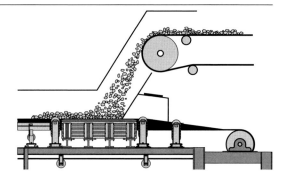

ing the life of the belt, reducing spillage, and keeping the belt running in the proper path. The design of chutes and other loading equipment is influenced by conditions such as the capacity, the characteristics of material handled, the speed and inclination of the belt, and the number of loading zones on the conveyor.

To minimize material degradation and component wear, an ideal transfer point would place the specified quantity of material on the receiving belt by loading:

A. In the center of the belt

B. At a uniform rate

C. In the direction of belt travel

D. At the same speed as the belt is moving

E. After the belt is fully troughed

F. With minimum impact force

At the same time, it would provide adequate space and/or systems for:

A. Edge sealing and back sealing

B. Carryback removal

C. Fugitive material management

D. Inspection and service

But in Real Life...

Achieving all these design goals in a single transfer-point design is difficult. The accommodations required by actual circumstances are likely to lead to compromises. The loading point of any conveyor is nearly always the single most critical factor in the life of the belt. It is here where the conveyor belt receives most of its abrasion and practically all of its impact. It is at the conveyor transfers where the forces that lead to spillage or dust creation act on the material and the belt. An optimal transfer point is essential to the conveyor's throughput and control of fugitive material (**Figure 3.5**).

The problem is that transfer points are the center for the interaction of many and often conflicting requirements, some arising from the materials passing through and others from the belts that run into and out of the load zones. Material characteristics, air movements, and impact levels add forces that must be addressed by any system designed to prevent the escape of fugitive material. In addition, many requirements imposed by the plant's overall process will subject transfer points to additional forces and limitations.

The Engineering of Transfer Points

There are three basic approaches taken to the design of transfer points. The first and most common is the conventional method of drafting a solution using "rules of thumb" to fit the master layout of conveyors. This would be the drafted solution (**Figure 3.6**). The second method is to specify the critical components of the transfer point and design the overall conveyor layout to minimize transfer-point problems. This is the specified solution (**Figure 3.7**).

Figure 3.5

A proper transfer point is essential to the conveyor's control of material.

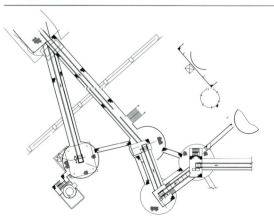

Figure 3.6

The first and most common method of designing a transfer point is the conventional method of drafting a solution using "rules of thumb" to fit the master layout of conveyors.

3

The third method is an engineered solution. This method is used to analyze the characteristics of the bulk material and produce custom-engineered chutes, which minimize the disruption of the bulk material trajectory and place the material on the next belt in the proper direction and at the speed of the receiving belt. This third class of transfer points is typified by specifications that require the bulk material to be tested for its flow properties. The transfer of material from one belt to another is engineered using fluid mechanics to minimize the dust, spillage, and wear. This engineering can be done for new construction, or it can be done as re-engineering for existing transfer points (**Figure 3.8**).

Specifications for an engineered transfer point should include:

A. Material characteristics and flow rates

B. Minimum performance requirements in terms of hours of cleanup labor and/or amount of spillage per hours of operation

C. Maximum budget requirements for annual maintenance and periodic rebuild at supplier-specified intervals

D. Ergonomic requirements of access for cleanup and maintenance

E. Engineering drawings and specifications for wear parts as well as complete maintenance manuals

SYSTEMS ENGINEERING

One Step Forward, One Step Back...

Unfortunately, improving the operation of complex systems like conveyor transfer points is not a question of solving one narrowly-defined problem. Rather, the attempt to solve one problem in these sophisticated operating systems typically uncovers or creates another problem. This second problem can prove as difficult to solve as the original, if not more.

It is never easy to achieve total material control, because fugitive materials problems often combine multiple causes and multiple effects. For example, a new edge-sealing system may provide an immediate improvement in the prevention of material spillage from a transfer point. However, if there is no wear liner present inside the chute, the force and/or the weight of material on the skirting will create side pressure that abuses the new seals, leading to abrasion and premature failure. Eventually, the amount of spillage returns to its previous unacceptable level. Spillage will continue to extract its high price from the efficiency of the conveyor and the overall operation.

The Systems Approach

The key to any engineering improvement is a detailed solution that encompasses all components of the problem. The costs of undertaking such a systems approach will prove higher than those of upgrading any single component. However, the return on investment will justify the expense.

Talking "systems engineering" is easy; it is the application of this approach that proves difficult. Development of a comprehensive approach requires knowledge of the material; understanding of the process;

Figure 3.7

The second method is to identify the critical components of the transfer point, which would be the specified solution.

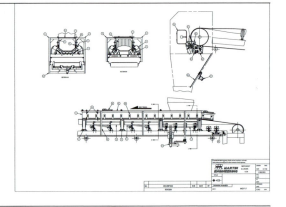

Figure 3.8

An engineered transfer-point solution is a method used to analyze the characteristics of the bulk materials to produce custom-engineered chutes to meet the requirements of a specific operation.

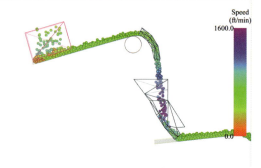

3

SAFETY CONCERNS

The design of any bulk-materials handling system must begin with a concern for safety. Every consideration must be taken for the safety of the personnel tasked with the everyday operation and maintenance of the system. Emergency pull cords, zero speed switches, rolling component guarding, and walkways with handrails and guards are just a few basic safety features that must be included in conveyor systems. Proper lockout / tagout / blockout / testout procedures are a requirement for any work to be performed on or around the conveyor system.

Proper design of loading and discharge points will assist with the safe operation and maintenance of conveyor systems. Total containment of transferred material keeps fugitive material from blocking walkways, stairs, and ladders and creating a hazard. Eliminating airborne dust will reduce health risks to employees and maintenance/repair costs to conveying equipment.

commitment of the resources to properly engineer, install, and operate a system, and consistent maintenance to keep that system operating at peak efficiency and to achieve total material control.

BELT CONVEYORS: SIMPLE AND COMPLEX

In Closing...

Belt conveyors handling bulk materials are simple machines, governed by the universal laws of physics. However, they are also complex, as they are vulnerable to a variety of uncertainties arising from the large quantities of otherwise unconfined material moving through them and the energy they impart to that material. Poorly contained, this material in motion can spread across a facility as spillage, carryback, and dust, reducing efficiency, shortening equipment life, and raising costs. Understanding the basics of material properties and equipment performance, and applying the remedies discussed in this book, can provide significant improvements in materials-handling efficiency and profitability.

Looking Ahead...

This chapter about Conveyor Components, the third chapter in the section Foundations of Safe Bulk-Materials Handling, is the first of three chapters pertaining to Conveyors 101. Chapters 4 and 5 continue this section about the basics of belt conveyors and the control of material for reduced dust and spillage, describing The Belt and Splicing the Belt.

REFERENCES

3.1 Any manufacturer and most distributors of belting can provide a variety of materials on the construction and use of their specific products, as well as on conveyor belts in general.

3.2 Conveyor Equipment Manufacturers Association (CEMA). (2005). *BELT CONVEYORS for BULK MATERIALS, Sixth Edition*. Naples, Florida.

3.3 The website http://www.conveyor-beltguide.com is a valuable and non-commercial resource covering many aspects of belting.

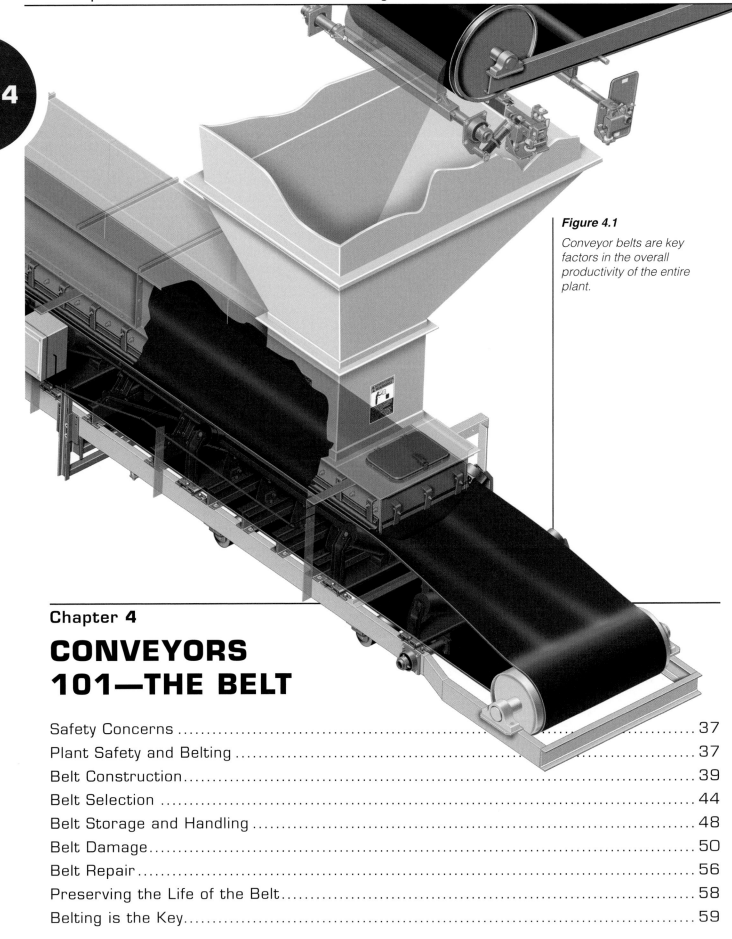

4

Chapter **4**

CONVEYORS 101—THE BELT

In this Chapter...

This chapter continues the discussion of the foundations of safe bulk-materials handling and the basics of conveyors by focusing on the construction and proper use of belts. Considerations for belt selection are included, along with the importance of proper storage and handling. In addition, various types of belt damage are discussed, as well as methods to repair and preserve the life of the belt.

A belt conveyor system is composed of many components; however, none is more important than the belt (**Figure 4.1**). The belting represents a substantial portion of the cost of the conveyor, and its successful operation may be the key factor in the overall productivity of the entire plant in which the conveyor is located. Therefore, the belting must be selected with care, and all possible measures should be employed to safeguard its usefulness.

This chapter focuses on the heavy-duty belting typically used in bulk-materials handling. The most common types of belting for bulk handling are made with covers of rubber or polyvinyl chloride (PVC) and an internal tensioning carcass of synthetic fabric or steel cables.

PLANT SAFETY AND BELTING

Conveyor Belt Fire Resistance

A conveyor belt fire is a significant risk. The belting itself can burn; however, it is the length and movement of the belt that poses risks that a belt can spread a fire over a great distance within a facility in a very short time.

Fires on conveyor belting are most commonly ignited by the heat generated from friction induced by a pulley turning against a stalled (or slipping) belt or by the belt moving over a seized idler. Other conveyor fires have occurred when hot or burning material is inadvertently loaded onto the belt. Best practices for minimizing the fire risk of any conveyor belt include:

A. Conducting regular belt examinations

B. Removing all accumulations of combustible materials along the conveyor belt

C. Correcting potential sources of fire such as seized rollers, overheated bearings, or belt misalignment

With this risk of fire, compounded by the toxic gases, thick smoke, and noxious fumes that can result from a conveyor fire, belting is regulated in those applications where

---

⚠ SAFETY CONCERNS

As with any moving machinery, the conveyor belt must be treated with respect and with the knowledge that it does cause injuries. There are a number of risks for personal injury from the conveyor belt that can lead to death or serious injury. Most of these concerns arise from the movement of the belt through the conveyor systems, such as contact burns and the risk of entrapment from contact with the moving belt. Particular care should be taken when observing a moving belt to look for damaged areas or check its tracking.

Rolls of belting are large and unwieldy and should be handled carefully. When transported, they should be secured so they cannot break loose, and they should be maneuvered with the appropriate equipment and safeguards.

Repairing a belt exposes workers to heavy lifting, sharp tools, and industrial chemicals. Appropriate lockout / tagout / blockout / testout procedures should be followed before beginning any work on or around the conveyor.

Appropriate PPE must be used, and manufacturers' procedures for handling chemicals must be followed.

4

these conditions are most dangerous—underground mining and, particularly, underground coal mining. Many countries replaced the earlier requirement for flame-retardant belting, belting that passes a smaller Bunsen Burner Test, by mandating the use of self-extinguishing belting in underground applications.

It should be noted that ALL conveyor belts will burn given sufficient heat and airflow. However, standardized laboratory tests have been accepted by governmental regulatory bodies to measure and categorize the burning characteristics of different conveyor belts. In very general terms, "self-extinguishing" is considered to be a belt that will not propagate a fire in a laboratory setting once the ignition source has been removed.

Self-extinguishing belting is higher in cost than flame-retardant belting. This cost premium is generally 10 to 50 percent, but it will vary depending on the carcass construction and cover gauges.

With the exception of the United States, fire-safety standards for conveyor belting are similar in the largest coal mining countries, including Australia, Canada, China, Germany, India, Indonesia, Poland, Russia, and South Africa. In Germany, for instance, strict requirements were implemented more than 30 years ago.

There are numerous international regulatory and advisory organizations and governmental agencies that provide guidance and direction. Those agencies include, but are not limited to: British Standards Institution (BSI), Conveyor Manufacturers Equipment Association (CEMA), Deutsches Institu für Normung (DIN), European Standards (EN), and the International Organization for Standardization (ISO).

The tests in these and most other countries include:

A. Drum Friction Test (DIN 22100 et. al)

The drum friction test measures whether the surface temperature remains under a required maximum after a specific time and under a specific tension. The test procedure simulates a belt slipping over a jammed pulley or a pulley rotating under a stationary conveyor belt. To pass this test, the surface temperature of the belting must remain below 325 degrees Celsius (617° F) with no flame or glow visible.

B. Surface Resistance Test (ISO 284/EN 20284/DIN 20284)

An electrostatic charge may build up on the conveyor belt surface and ignite a mixture of flammable gases and air. By keeping the surface resistance of the belting low (making the belt cover more conductive), the conveyor belt allows the charge to flow freely, eliminating the risk of sparks.

C. High-Energy Propane Burner Test (EN 12881)

To determine whether a conveyor belt will propagate fire, a belt sample 2,0 to 2,5 meters long by 1200 millimeters wide (80 to 100 in. by 48 in.) is ignited by a propane burner. After the ignition-source has been removed, the flames must self-extinguish within a certain amount of time, leaving a defined area of belt undamaged.

D. Laboratory Scale Gallery Test (DIN 22100 and 22118)

A specimen of belting 1200 millimeters long by 120 millimeters wide (48 in. by 4.8 in.) is placed over a propane burner. After the ignition source is removed, the flames must self-extinguish, and a defined undamaged length must remain.

In the United States, conveyor belt flammability standards can be considered less stringent than those used in other countries, because a total-system approach is taken to fire suppression. The US regulations include not only conveyor belting but air monitoring and motor-slip detection devices.

The present flame resistance requirement in the United States for applications for everything except underground coal mines, as published in the Code of Federal Regulations (CFR), is quite simple:

Bunsen Burner Test. (CFR Part 30 Section 18.65)

A small (approximately 150 by 12 millimeters) (6 by ½ in.) piece of a belt is held over a Bunsen burner flame for one minute, after which time the flame is removed and an airflow applied for three minutes. After a set length of time, the duration of the flames is recorded. The average of four samples must not burn for more than one minute or exhibit afterglow for more than three minutes.

This test, which was implemented per the 1969 Federal Coal Mine Health and Safety Act, is similar to a standard for underground conveyor belts that was in force in Europe until the mid 1970's. However, with the advent of the more stringent regulations, flame-retardant belts have been allowed to be used in Europe only in applications above ground; self-extinguishing belts are required underground.

The United States now has a more stringent standard for flame-resistant belts in underground coal mines. In December 1992, the US Department of Labor, Mine Safety and Health Administration (MSHA), proposed a new rule for conveyor belt test requirements (Federal Register, Vol. 57, No. 248) that would bring safety standards up to the international level. Nearly ten years later, in July 2002, this proposed rule was withdrawn. The reasons cited for withdrawal were that the number of conveyor belt fires had significantly declined and that improvements in belt monitoring were being implemented.

Recommendations from the Mine Improvement and New Emergency Response (MINER) Act of 2006 resulted in a new rule for underground coal mines (CFR Part 30 Section 14.20) effective December 2008 that includes the Belt Evaluation Laboratory Test (BELT), a laboratory-scale flame resistance test based on the work done for the 1992 proposed rule. In order for a belt to pass the BELT method, it must have improved fire-resistant capability, which greatly limits flame propagation.

The test requires that three belt samples, approximately 152 by 23 centimeters (60 by 9 in.), be placed in a test chamber 168 centimeters (66 in.) long by 456 centimeters (18 in.) square. After applying the burner flame to the front edge of the sample for 5 minutes and the flame is extinguished, each tested sample must exhibit an undamaged portion across its entire width.

At the time of this writing, the final rule published by MSHA requires conveyor belts placed in service in underground coal mines to be more flame resistant than those previously required beginning December 31, 2009. The rule also requires existing belting to be replaced within ten years. MSHA or a reputable belting supplier can be contacted for additional, updated information.

Other Belting Safety Concerns

Other standards are sometimes in place. Some countries have even more stringent requirements regarding, for example, the belt's toxicity, hygiene, or cover roughness. The exact specifications can be found in the standards in a given geographic region or industry. Procedures and standards are offered under DIN, EN, ISO, BSI, CEMA, and other standards. Of course, it is imperative that the belting be compatible with the materials to be transported on it.

BELT CONSTRUCTION

The Belt Carcass

Conveyor belting is composed of two parts: the inside carcass and the outside covers. The carcass is the most important structural section of the belt, as it contains the tensile member to handle the load of cargo carried on the conveyor. The primary purpose of the carcass is to

4

transmit the tension necessary to lift and move the loaded belt and to absorb the impact energy unleashed by the material as it is loaded onto the belt. No matter what belt-support system is employed, if the belt carcass cannot handle the initial impact energy, the belt will fail prematurely. The carcass must be adequate to allow proper splicing techniques and strong enough to handle the forces that occur in starting, moving, and stopping the loaded belt. The carcass also provides the stability necessary for proper support between idlers and for maintaining alignment.

Most carcasses are made of one or more plies of woven fabric, although heavy-duty belting may incorporate parallel steel cables to replace some or all of the fabric. Carcass fabric is usually made of yarns woven in a specific pattern. The yarns that run lengthwise, parallel to the conveyor, are referred to as warp yarns and are the tension-bearing members. The transverse or cross fibers are called weft yarns and are primarily designed for impact resistance, mechanical fastener holding, load support, and general fabric stability.

Years ago, conveyor belts typically used yarns made of cotton as the textile reinforcement. For improved cover adhesion and abuse resistance, a breaker fabric was often placed between the cover and the carcass. Throughout the 1960s and 1970s, carcass reinforcements underwent a change. Today, most belt carcasses are made with manmade fabrics such as nylon, polyester, or a combination of the two. These fabrics are superior to the older natural fabrics in nearly all respects, including strength, adhesion, abuse resistance, fastener holding, and flex life. Presently, fabrics incorporating aramid fibers are used for some applications in conveyor belting. The aramid fabrics offer high strength, low elongation, and heat resistance. Breaker fabrics are rarely used with these man-made fabrics, because little or no improvement is achieved.

Carcass Types

There are four types of belt carcasses:

A. Multiple-ply belting

Multiple-ply belting is usually made up of two or more plies, or layers, of woven cotton, rayon, or a combination of these fabrics, bonded together by an elastomer compound. Belt-strength and load-support characteristics vary according to the number of plies and the fabric used. The multiple-ply conveyor belt was the most widely used belt through the mid-1960s, but today it has been supplanted by reduced-ply belting.

B. Reduced-ply belting

Reduced-ply belts consist of carcasses with either fewer plies than comparable multiple-ply belts or special weaves. In most cases, the reduced-ply belt depends on the use of higher-strength synthetic textile fibers concentrated in a carcass of fewer plies to provide higher unit strength than in a comparable multiple-ply belt. The technical data available from belt manufacturers generally indicate that reduced-ply belting can be used for the full range of applications specified for multiple-ply belting.

C. Steel-cable belting

Steel-cable conveyor belts are made with a single layer of parallel steel cables completely embedded in the rubber as the tension element. The carcass of steel-cable belting is available in two types of construction. The all-gum construction uses only the steel cables and rubber; the fabric-reinforced construction has one or more plies of fabric above and/or below the cables but separated from the cables by the cable rubber. Both types have appropriate top and bottom covers. Steel-cable belting is produced using a broad range of cable diameters and spacing, depending primarily on the desired belt strength. Steel-cable belting is often used in applications requiring operating tensions beyond the range of fabric belts. Another application is on conveyors where, due to limitations in the distance the take-up system can travel, the belting cannot be allowed to stretch significantly.

D. Solid-woven belting

This type of belting consists of a single ply of solid-woven fabric, usually impregnated and covered with PVC with relatively thin top and bottom covers. The surface of PVC belts is often rough on purpose to aid in conveying on inclines, but the rough surface makes belt cleaning more difficult. The abrasion resistance of PVC is lower than rubber, so some solid-woven belts are made with a combination of a PVC core and rubber covers.

Top and Bottom Covers

Covers protect the carcass of the belt from load abrasion and any other conditions that could contribute to belt deterioration. The top and bottom covers of the conveyor belt provide very little, if any, structural strength to the belt. The purpose of the top cover is to protect the carcass from impact damage and wear. The bottom cover provides a friction surface for driving and tracking the belt. Usually, the top cover is thicker than the bottom cover and more durable for abrasion, impact damage, and wear, due to its increased potential for damage. Abrasion and cutting may be so severe that a top cover as thick as 18 millimeters (0.75 in.) or more is required. In any case, the goal of cover selection is to provide sufficient thickness to protect the carcass to the practical limit of carcass-life.

The covers can be made of a number of elastomers, including natural and synthetic rubbers, PVC, and materials specially formulated to meet special application requirements such as resistance to oil, fire, or abrasion.

Users might be tempted to turn a belt over when the carrying side has become worn. In general, it is better to avoid inverting the belt after deep wear on the top side. Turning the belt over presents an irregular surface to the pulley, resulting in poor lateral distribution of tension, and may lead to belt wander. Another problem is that there may be cargo fines embedded into what was formerly the belt's carrying surface; when the belt is turned over, this material is now placed in abrasive contact with pulley lagging, idlers, and other belt-support systems. In addition, after years of being troughed in one direction, the belt tends to take a "set" (a predisposition to a direction) and will resist the necessary reversing of trough needed to invert the belt. Sometimes this can take weeks to overcome and can lead to belt-tracking problems.

A specific mention must be made of the practice of some belting manufacturers of stamping their logo into the carrying surface of the belt (**Figure 4.2**). Even when near the belt edge, this recessed area becomes a trap for conveyed material, and the roughness of the area can abuse the belt-cleaning and sealing systems under which the embossed area will pass. It is recommended that users specify that these supplier logos be positioned on the return, non-carrying side of the belting.

Aspect Ratio

Whereas some belts have the same cover thickness on both sides, most belts are fabricated with the pulley-side cover somewhat thinner (lighter in gauge) than the carrying side of the belt, because of the difference in wear resistance needed. The difference in thickness between the top and bottom covers is referred to as a belt's aspect ratio. However, the difference in thickness between the two covers cannot be too great, or the belt may cup.

Figure 4.2

It is a bad practice for the belt manufacturer to emboss a logo into the carrying side of the belt, because carryback can become entrapped in the logo.

4

The problem with belts with poorly-designed aspect ratios is that the larger mass of rubber will shrink more than the smaller. Consequently, if a belt has an inordinately large top-to-bottom cover ratio, and the top cover shrinks due to age, exposure to ultraviolet light, or other factors, the belt will cup up, reducing the area of the bottom cover in contact with the idlers. This will make it more difficult to keep the belt running in alignment. This problem is most likely to occur when, in the interest of getting a thick top cover to extend service-life, a plant orders belting with a top cover that is too thick for the bottom cover. To provide consistent shrinkage and more consistent tracking, an aspect ratio of 1.5-to-1 is recommended for belts up to 900 millimeters (36 in.), with a 2-to-1 aspect ratio recommended for belts from 1000 to 1600 millimeters (42 to 60 in.). For belts above 1600 millimeters (60 in.) a 3-to-1 aspect ratio is recommended. Belting with a 3-to-1 aspect ratio is suitable for many purposes and is the ratio mostly commonly stocked at belting distributors.

Cleats, Ribs, Chevrons, and Lugs

Raised elements are sometimes used on a belt surface to assist in the carrying of material (**Figure 4.3**). These cleats, ribs, chevrons, and lugs are generally used to allow a conveyor to carry material at a higher angle of incline than would generally be possible with a flat belt. This is particularly useful with lumps or stones that could easily roll down an unobstructed incline.

Figure 4.3

Cleats, ribs, chevrons, and lugs are raised elements in a belt's surface that allow it to carry material at a higher angle of incline.

Cleats, or ribs, can be seen as walls or shelves installed perpendicular to the lines formed by the belt edges. Chevrons are in a V-shaped arrangement. Lugs are individual "islands" or pillars in the belt's surface. All are available in a variety of patterns and styles, with heights determined by the application. They can be molded integrally into the surface during the belt's original manufacture, or they can be bolted or vulcanized to the surface of the belting.

Bear in mind that the taller the cleats, ribs, chevrons, or lugs are, the more vulnerable they are to damage and the harder the belt is to clean and seal.

One way to increase traction between the belt and the conveyed material is to use a top cover that features inverted chevrons. Instead of extending above the belt cover, inverted chevrons are recessed into the top cover, like the tread on a tire. The grooves are cut into the belt cover with a router; the grooves can be at a chevron angle or straight across the belt 90 degrees to the edge. This design allows greater success in cleaning and sealing the belt with traditional systems, although it is possible to fill the recessed cleats with material.

Grades of Belting

Various national and international bodies have established rating systems for the belting used in general-purpose bulk-materials handling. Designed to provide a reference for end users as to what grades to use in different applications, the ratings specify different laboratory-test criteria without providing any guarantee of performance in a specific application.

In the United States, the Rubber Manufacturers Association (RMA) has established two standard grades of belting covers. RMA Grade I belting meets higher rubber tensile and elongation requirements, typically indicating improved cut and gouge resistance over the performance of Grade II covers. It should be noted that grade rating does not necessarily denote overall abrasion resistance.

The International Organization for Standardization (ISO) has similarly established a grading system under ISO 10247. This standard includes Category H (Severe Cut and Gouge Service), Category D (Severe Abrasion Service), and Category L (Moderate Service). Category H is roughly comparable to RMA Grade I; Categories D and L approximate RMA Grade II belting.

In addition, there are belting types manufactured to meet specific requirements of stressful applications, such as service with hot materials, in underground mines, or with exposure to oil or chemicals. As with most things, it is best to acquire an understanding of the operating conditions and then consult with reputable suppliers before selecting categories of belting.

Abrasion Resistance in Belting

There are two types of abrasion that occur on conveyor belts. One is caused by the material rubbing against the belt cover. As a woodworker sanding an object, this wear is relatively even under the influence of the material pressing against the surface. The actual rate of abrasion will depend on the nature of the material, as modified by the density of material loading and the speed of the belt. This is called impingement damage.

A more aggressive form of abrasion is the damage to the surface by sharp-edged materials that cut or gouge the belt. This is generally called impact damage.

There are two types of tests used to measure belt-cover wear. One is the ISO 4649 Types A and B Abrasion Test Methodology (formerly DIN 53516). This test uses a sample of the rubber cover and holds it against a rotating abrasive drum for a fixed interval. The cover sample is weighed before and after to calculate volume loss. The lower the number (the less material lost), the more resistance to abrasion.

A second method of testing is the Pico Abrasion Test, also referred to by the American Society for Testing and Materials

(ASTM) as ASTM Test Method D2228. In this test, tungsten carbide knifes are used to abrade a small sample of the belt cover. As above, the sample is weighed before and after the procedure, and the weight loss is calculated. Results are given as an index, so the higher the number, the better the abrasion resistance.

Most references caution that neither test should be seen as a precise prediction of actual performance in field applications.

New Developments in Belting

A recent innovation is the development of energy-efficient belt covers. Called Low Rolling Resistant (LRR) Covers, these bottom covers reduce the tension required to operate the belt, because there is less roller indentation resistance as the belt moves over the idlers. According to manufacturers, this belting can produce operating energy consumption savings of 10 percent or more. The savings occur where the rubber belting meets the conveyor system's idlers. The energy-efficient cover has less rolling resistance, because the bottom cover returns to a flat configuration more quickly than conventional belts, which deform as they go over the conveyor's rolling components.

The manufacturers note that the benefits of this energy-efficient cover can best be realized on long horizontal conveyors utilizing wide, fully-loaded belts carrying high-density material, where the friction of the system is dominated by idler-related resistances. The LRR compound is at a premium cost over other cover compounds. However, on those installations where the benefits can be fully realized, the compound compensates for its additional expense through a reduction in power costs and, on new systems, by allowing the conveyor to be equipped with smaller motors, pulleys, gear boxes, shafts, bearings, idlers, and steel structures.

Users cannot assume that an LRR cover will reduce operating expenses, nor can they just specify LRR, as each belt is

4

compounded for a specific application. The relationship between the power consumption with an LRR belt and temperature conditions is not linear, and there is typically a small window of application. A specific LRR bottom cover designed to save energy at 20 degrees may cost more to operate at 0 or 30 degrees, so each belt must be designed for the climate conditions of each application.

Another new development in belt construction is the use of non-stick belt covers, to prevent belt-borne carryback. This belting is created by applying a non-stick coating to the belting to prevent the accumulation of carryback material on the belt. This coating should reduce the need for belt cleaning, thereby extending belt-life by reducing cover wear. This coating is also resistant to oil and grease and unaffected by weathering and aging. It should be noted that conventional belt cleaners (scrapers) should be removed if using non-stick belting, as the "cleaning edge" of even "soft" urethane pre-cleaners might remove the coating.

Conveyors are designed as a system, and any changes from the original belt specification can adversely affect the operation of a conveyor. Belting manufacturers should be consulted to determine which belting type is most appropriate for any application.

Cut Edge or Molded Edge

There are two methods to create the edges on a belt: molded edges or cut edges.

A molded-edge belt is manufactured to the exact width specified for the belt, so the edges of the belt are enclosed in rubber. As a result, the carcass fabrics are not exposed to the elements. Because a molded-edge belt is made for a specific order, it will probably require a longer lead-time and is generally more costly than a cut-edge belt.

A cut-edge belt is manufactured and then cut or slit down to the specified width

required to fulfill the order. Using this method, the manufacturer may hope to fill two or three customer orders out of one piece of belting produced. As a result, this makes cut-edge belting more cost-effective (hence economical) to manufacture, so this type of belt has become more common. The slitting to the specified width may occur at the time of manufacture, or it may be done when a belt is cut from a larger roll in a secondary operation, either at the manufacturer or at a belting distributor.

A cut-edge belt can be cut down from any larger width of belting. This makes it more readily available. However, there are some drawbacks. At the cut edge(s) of the belt, the carcass of the belt is exposed; therefore, the carcass is more vulnerable to problems arising from abusive environmental conditions in storage, handling, and use. In addition, the slitting process is vulnerable to problems. Dull slitting knives can lead to problems such as belt camber—a curve in the edge of the belt. In addition, there are the unknowns that come with buying used or re-slit belting, including its age, environmental exposure, and application history.

Steel-cable belting is manufactured to a pre-determined width, so it has molded edges. Fabric-ply belting is available with either a molded edge or a cut edge.

BELT SELECTION

Specifying a Belt

The selection and engineering of the proper belting is best left up to an expert, who might be found working for a belting manufacturer or distributor or as an independent consultant. A properly specified and manufactured belt will give optimum performance and life at the lowest cost. Improper selection or substitution can have a catastrophic consequence.

There are a number of operating parameters and material conditions that should be detailed when specifying a conveyor belt. Material conditions to be detailed include:

A. Thickness

Limit variations in thickness to a sliding scale of +/− 20 percent for thin covers such as 2,4 millimeters (0.094 in.), and +/− 5 percent for cover gauges greater than 19 millimeters (0.75 in.).

B. Camber or bow

Limit camber or bow to one-quarter of one percent (0.0025). This allows a camber or bow dimension of +/- 25 millimeters in 10 meters (0.75 in. in 25 feet). Camber is a convex edge of the belt; bow is a concave edge of the belt. RMA defines bow (and camber) as the ratio of the distance, midway between two points along the belt edge that are 15 to 30 meters (50 to 100 feet) apart, between the actual belt edge and a tape or string stretched straight between the two points. To express this in percent, calculate the ratio in hundredths and multiply by 100. For example, if 30 meters of belting was out of true by 450 millimeters (by 0,45 meters), this would equal a camber of 1.5 percent. In Imperial measurements, a distance of 18 inches (1.5 ft) over 100-foot length of belting would be a 1.5 percent camber.

C. Belt surface

Specify the belt surface to be smooth, flat, and uniform +/- 5 hardness points. Hardness is measured in the United States in Shore A Durometer. Readings range from 30 to 95 points—the higher the durometer number, the harder the compound. The International Rubber Hardness Degrees (IRHD) scale has a range of 0 to 100, corresponding to elastic modulus of 0 (0) and infinite (100), respectively.

D. Manufacturer's mark

Require the manufacturer's mark to be eliminated or molded into the bottom rather than the top cover, where it will not interfere with belt cleaning and sealing systems.

Operating parameters to be detailed when specifying a conveyor belt include:

A. Hours of operation loaded and unloaded

B. Details of the transfer point, including trough angle and transition distance, as well as information on material trajectory, drop height, and speed

C. Description of material to be handled as completely as possible, including lump sizes and material temperature range

D. Description of belt-cleaning system to be used

E. Description of chemical treatments (e.g., de-icing agents or dust suppressants) to be applied

F. Description of atmospheric contaminants (from nearby processes or other sources)

G. Specification of local weather extremes that the belt must withstand

Know Your Structure, Know Your Belt

Placing any belt on a conveyor structure without understanding the characteristics of the belt will impair the performance of the system and reduce the performance of the belt. There can be problems in the form of mistracking, shortened belt-life, damaged splices, unscheduled downtime, and added maintenance expenses.

A detailed analysis of the conveyor structure and rolling components is required to ensure that the belt used on the system is the right choice. It is recommended that all parameters be fully understood prior to selecting and installing a belt on an existing structure. It is always wise to consider the advice of belting suppliers.

Compatibility with Structure and Rolling Components

Buying belting is like buying clothing. To fit best, it must be tailored to the existing

4

construction. Conveyor belts are designed for different capacities, lengths, widths, trough angles, and tensions. A belt must be compatible with the conveyor structure, and there is more to compatibility than belt width.

Unfortunately, this is not commonly understood at a plant's operating level. Too often, there is a "belting is belting" philosophy in place. This originates from an incomplete understanding of the complexity of the belting equation. This philosophy becomes practice at times when there is a need to economize or to provide a faster return to service. The typical response in these cases is to use belting from stock, either a leftover piece or a spare belt found in maintenance stores, or to use belting readily available from an outside source, like a belting distributor or a used-equipment dealer.

It is a false economy to use a "bargain" belt that is not fully compatible with the conveyor system. Incompatibility of belt to structure is a common problem leading to poor belt performance and a poor return on the belting investment. This incompatibility could well be the most common cause of the tracking problems seen on conveyors where a replacement belt has been installed or pieces have been added to the existing belting. Understanding the basics of compatibility is essential to ensuring good performance of the belt and conveyor.

Specifying a conveyor belt is an important undertaking. It is in an operation's best interest to allow an expert to take ownership of this part of the conveyor process. This expert will be familiar with the capabilities of the belts provided by manufacturers and know the proper questions to ask.

Belt-Tension Rating

Each belt is rated as to its strength—the amount of pulling force that it will withstand. The strength of a belt (or more accurately, the tension it is able to withstand) is rated in the United States in Pounds per Inch of Width, commonly abbreviated as

PIW. In other parts of the world, the belt is rated in ultimate breaking-strength in the metric units of newtons per millimeter (N/mm) or kilonewtons per meter (kN/m).

The strength rating is a function of the reinforcement included in the carcass of the belt and the number of, and type of, material in the fabric plies, or, if it is a steel-cable belt, the size of the cables. As noted above, a belt's top and bottom covers provide very little of the belt's strength or tension rating.

The belt's strength, either carcass tension rating or ultimate breaking-strength, represents the amount of force that can be applied to the belting. Putting greater demands in the form of material load, take-up weight, and incline gravity against this belt would cause severe problems, including the possibility of breaking the belt. The higher the rated tension of the belt, the more critical the compatibility of the belt with the structure and rolling components becomes.

Each conveyor structure will require a belt with a specific tension rating. Factors affecting this decision are:

A. Length of structure

B. Incline angle of the conveyor

C. Desired capacity

D. Width of belt

E. Drag and inertia of rolling components

Minimum Bend Radius

Belting is designed with a minimum pulley size specified by the manufacturer. Bending a belt over a radius that is too small can damage the belt. This may result in separation of plies, ply failure, or cracking of the belt's top cover. Inadequate pulley size can also lead to the pullout of mechanical splices. The minimum pulley diameter is determined by the number and material of plies, whether steel or fabric-reinforced, the rated tension of the belt, and the thickness of the top and bottom cover.

When a conveyor system is originally designed, the desire to use a thicker belt on a conveyor system (to extend belt-life in the face of high impact levels in the loading zone, for example) may require the installation of larger-diameter pulleys.

A common mistake occurs when an operation notices some type of surface damage to the carrying side of the belt. The immediate reaction is to install an even thicker belt on the conveyor in the expectation of getting a longer service-life. If the thicker belt has a minimum pulley size that is larger than the pulleys on the structure, the belt may actually yield a shorter life, worsening the problem the thicker cover was selected to solve.

Trough Angle

Belts are troughed to allow the conveyor to carry more material. As the trough angle is increased, more material can be carried. All flat rubber or PVC belts can be formed into a trough by idlers. The type of belt carcass, the thickness of the belt, the width of the belt, and the tension rating of the belt determine the maximum trough angle. On belting manufacturers' technical data sheets, troughability is typically shown as the minimum belt width allowed for the various trough angles.

Exceeding the maximum trough angle of a particular belt can cause the belt to permanently deform into a cupped position. Cupping can make a belt difficult to seal, difficult to clean, and almost impossible to track. As the cupping increases, the surface contact between the conveyor's rolling components and the belt is reduced, diminishing the ability of the rolling component to steer the belt properly.

If the belt's troughability is exceeded, the belt may not form the trough correctly, creating sealing and tracking problems. If a belt is too stiff and will not properly trough, it will not steer (track) properly through the system. This will quickly evolve into spillage off the sides of the conveyor and damage to the edges of the belt (**Figure 4.4**).

Another problem that may occur if the belt's troughing capability is exceeded is damage to the top and bottom covers and to the carcass in the idler-junction area.

In addition, if the belt's troughability is not compatible with the troughing idlers, it might take more power to operate the conveyor than originally designed.

Transition Distance

The belt travels across the tail pulley in a flat position. As the belt leaves the tail pulley and moves into the loading zone, the belt edges are elevated, forming the trough

Figure 4.4

Exceeding its troughing capability can result in damage to the belt.

Figure 4.5

Transition idlers are used to raise the belt edges and form the trough that carries the cargo.

Figure 4.6

Junction-joint failure is caused by an improper transition distance (center of terminal pulley to first full troughing roll).

where the material is carried (**Figure 4.5**). This trough is formed with transition idlers—idlers set at angles between flat and the conveyor's final trough angle.

There is a similar, but reverse, transition area at the conveyor's head pulley, where the conveyor is taken from a troughed to a flat profile just before it reaches the discharge point.

As the belt is formed into a trough, the outer edges of the belt are stretched more than the center of the belt. If the transition is made over too short a distance, damage may occur in the idler-junction areas of the belt—the points over the intersection between the flat central roller and the angle wing roller (**Figure 4.6**).

It is common to see a conveyor that has a transition area that is shorter than what is required. There are a number of reasons for this: faulty engineering or failure to

Figure 4.7

Belting should be rolled carrying side out on a core with a square opening.

Figure 4.8

Storing conveyor belting on the ground is a bad practice that can lead to damage.

understand the importance of transition, lack of space, or desire to reduce costs. So it is even more critical not to increase the problem by applying a replacement belt that requires a longer transition distance.

It might be possible to lengthen a conveyor's transition area. There are two ways to do this. One is moving the tail pulley back to extend the distance before the load zone. The second method is adapting a two-stage transition area, where the belt is partially troughed before it enters the loading zone and then completes its transition to the final trough angle after the cargo has been loaded. *(See Chapter 6: Before the Loading Zone.)*

More commonly, however, circumstances, such as lack of available space and limits of available budget, preclude lengthening a conveyor's transition area. The most common solution is to make sure the belting is suitable for the existing transition distance. It may not be the most economical solution when all costs—such as loading problems, increased edge tension, and belt damage—are included. A poorly designed transition area will increase costs and decrease the life of the belt.

BELT STORAGE AND HANDLING

The conveyor belt has long been the most economical and most efficient form of bulk-materials handling for many industries. However, if this important part of the plant is to perform as expected, it must be carefully stored and handled from the time of its manufacture until the time of its installation on the conveyor system. Improper storage techniques can lead to a damaged belt that will perform poorly when installed on the conveyor structure. As the length of storage time increases, and as the size of the roll of belting increases, so does the importance of following the correct procedures. The costs for handling, shipping, and storing the conveyor belt are minor compared to the purchase price of the belt; therefore, the correct procedures should be followed to protect the investment.

4

The following are the key storage and handling guidelines:

A. Rolled on a core

As the belt leaves the manufacturer or the supplier, the belt should be rolled carrying side out on a core with a square opening (**Figure 4.7**). The core gives the belt protection from being rolled into a diameter that is too small and protection when the belt is lifted through the center. It also provides a means for unrolling the belt onto the conveyor. The core size is determined by the manufacturer, based on the type, width, and length of the roll of belt. The core size can be smaller than the belt's minimum pulley diameter, as the rolled belt is not in a tensioned state. The lifting bar should be square to closely match the square opening in the core.

B. Properly supported

The conveyor belt should never be stored on the ground (**Figure 4.8**). Ground storage concentrates the weight of the roll onto the bottom surface. The belt carcass is compressed in this small area and not compressed equally from side to side. The carcass may be stretched more on one side or the other. This is a likely cause of belt camber, a banana-like curvature of the belt, running the length of the belt.

Under no circumstances should a roll of belting be stored on its side (**Figure 4.9**). The weight of the roll may cause that side of the belt to expand, creating camber problems. Moisture may migrate into the carcass through the cut edge of the belt, creating carcass problems or belt camber.

The belt should be supported in an upright position on a stand, off the ground (**Figure 4.10**). This places the stress of one-half of the roll's weight on the core, relieving the load on the bottom. This support stand can be utilized during shipping to better distribute the weight of the belt. The support stand can then

be utilized in the plant for storage, or the belt can be transferred to an in-plant storage system that properly supports the roll. It is important that the roll be properly supported from the time of manufacture to the time of installation.

Figure 4.9

Storing a roll of belting on its side may lead to problems with camber.

Figure 4.10

Supporting the roll of belting on a cradle or stand will prevent uneven stress, whereas storing it on the ground is bad practice.

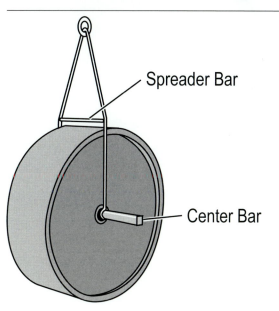

Spreader Bar

Center Bar

Figure 4.11

Using a spreader bar when lifting a roll of belting prevents damage to the belt edges.

4

C. Rotated on its stand

If the support stand is designed correctly, the roll of belt can be randomly rotated every 90 days. This will more evenly distribute the load throughout the carcass. The reel of belting should have been marked at the factory with an arrow to indicate the direction of rotation. Rotating the belt in the opposite direction will cause the roll to loosen and telescope.

D. Properly protected

During shipping and storage, the roll of belt should be covered with a tarp or wrapped in an opaque water-resistant material. Covering the roll of belting protects it from rain, sunlight, or ozone. The covering should remain in place during the entire storage process.

The roll of belt should be stored inside a building to protect it from the environment. The storage area should not contain large transformers or high-voltage lines that may create ozone and affect the belt. The building does not have to be heated, but it should be relatively weather tight.

E. Lifted correctly

When lifting a roll of belt, a square lifting bar of the correct size should be placed through the core. Slings or chains of the correct size for the weight of the roll should be used. A spreader bar should be utilized to prevent the chains or slings from damaging the edges of the conveyor belt (**Figure 4.11**).

Additional guidelines are given in ISO 5285; belting manufacturers can provide guidance for their specific products.

BELT DAMAGE

Extending Belt Life

As noted above, the cost of the belt will easily exceed the cost of other conveyor components and may reach the point where it approaches the cost of the steel conveyor structure. A key to providing a reasonable return on the investment in belting is avoiding damage and prolonging its service-life. Obviously, all systems installed around the conveyor—whether to feed it, receive material from it, or assist in its material transport function—must be designed to present the minimum risk to the belt.

Damage to belting can be a major drain on the profitability of operations using conveyors. This expense, which occurs regularly in plants around the world at costs of thousands of dollars, can often be prevented. Unfortunately, relatively little effort is put into the analysis of the life of the belt and the reasons for a belt not reaching its optimum life, because of the difficulty of identifying and measuring all the variables that affect belt-life.

The types of belt damage can be divided into two groups: normal wear and avoidable damage. Wear due to the normal operation of the conveyor can be managed and minimized to prolong the belt-life, but a certain amount of wear is considered acceptable. Perhaps avoidable damage cannot be totally prevented, but it can be minimized through proper equipment design and maintenance management.

Figure 4.12

Impact damage is caused by large, sharp material striking the belt, resulting in a random nicking, scratching, or gouging of the top cover.

Figure 4.13

Entrapment damage is usually seen as two grooves, one on each side of the belt, near the edge where the belt runs under the steel conveyor skirtboard.

4

The first step in preventing belt damage is to identify the cause or source(s) of the damage. A step-by-step analysis can almost always lead to the "culprit."

Types of Belt Damage

The following is a brief review of the major types of belt damage:

A. Impact damage

Impact damage is caused by large, sharp conveyed material striking the top cover of the belt. The result of this impact is a random nicking, scratching, or gouging of the top cover (**Figure 4.12**). A large frozen lump of coal may cause this type of damage. If the impact is severe enough, the belt can actually be torn completely through. This type of damage is usually seen under crushers or in mines on conveyors handling run-of-mine (ROM) material.

Long material drops without some method to help the belt absorb the energy can also lead to impact damage. *(See Chapter 8: Conventional Transfer Chutes and Chapter 10: Belt Support.)*

B. Entrapment damage

Entrapment damage is usually seen as two grooves, one on each side of the belt, near the edge where the belt runs under the steel conveyor skirtboard (**Figure 4.13**). Many times this damage will be blamed on pressure from a skirtboard-sealing system. However, extensive study has shown this type of belt damage is more likely due to the entrapment of conveyed material between the sealing system and the conveyor belt.

This material entrapment occurs when the belt is allowed to sag below the normal belt line and away from the sealing system. Material becomes wedged into this "pinch point," forming a spearhead to gouge or abrade the surface of the belt as it moves past (**Figure 4.14**). This leads to any of several negative events:

a. Scalloping

The trapped material will form a high-pressure area, causing excessive wear on the sealing system (seen as scalloping in the seal at each idler).

b. Grooves

Grooves will be worn along the entire length of the belt under the skirtboard (**Figure 4.15**).

c. Material spillage

Material will be forced off the sides of the belt, leading to piles of material spillage under the load zone.

Figure 4.14

Material becomes wedged into this "pinch point," forming a spearhead to gouge or abrade the surface of the belt as it moves past.

Figure 4.15

Trapped material can wear grooves along the entire length of the belt under the skirtboard.

Figure 4.16

Damage to the belt edge is a sign of the belt mistracking into the conveyor structure.

4

Material entrapment can also be caused when the skirting is placed inside the chutewall in the path of the material flow. Not only does this arrangement cause material entrapment and belt damage, it also reduces the cross-sectional area of the chutewall, in turn reducing conveyor capacity. This same damage can be seen when installing leftover or used belting as a dust seal, as the carcass is more abrasive than the belt cover and will wear the cover away. Incorrectly installed wear liners can also cause entrapment points, creating this type of wear.

One way to prevent sag is to use bar support systems to support the belt and stabilize the path in the entire skirted area. *(See Chapter 10: Belt Support.)*

C. Belt-edge damage

Edge damage is usually seen as frayed edges on one or both sides of the belt (**Figure 4.16**). If edge damage is not identified and corrected, it can be severe enough to actually reduce the width of the belt to a point where it will no

longer carry the rated capacity of the conveyor.

Belt mistracking is probably the leading cause of belt-edge damage. There are numerous reasons why a belt might mistrack. These causes range from out-of-alignment conveyor structures, off-center belt loading, accumulations of material on rolling components, or even the effect of the sun on one side of the belt.

There are many techniques and technologies that can be used to train the belt. These would include laser surveying of the structure, adjustment of idlers to counter the belt's tendency to mistrack, and installation of self-adjusting belt-training idlers that use the force of the belt movement to steer the belt's path.

The key to good belt tracking is to find the cause for the mistracking and then remedy that cause, rather than spending time and money turning one idler one direction and another idler a different direction in pursuit of better tracking. *(See Chapter 16: Belt Alignment.)*

D. Belt delamination

Another form of damage seen at the belt edge is delamination, in which the plies of the carcass separate, or the covers pull away from the carcass (**Figure 4.17**). This can be caused by belting being wrapped around pulleys that are too small. The entry of moisture, chemicals, or other foreign materials into the edge of the belt can contribute to this problem.

E. Worn top cover

Damage to the top cover is seen when the top cover of the belt is worn in the load carrying area of the belt or even across the entire top (**Figure 4.18**). Several factors can contribute to worn top covers.

One cause can be abrasion from material loading. There is an abrasive or grinding action on the belt cover created

Figure 4.17

Another form of damage seen at the belt edge is delamination, in which the plies of the carcass separate or the covers pull away from the carcass.

Figure 4.18

Abrasion from material loading will be seen as wear in the top cover in the load carrying area of the belt.

from the material falling onto the moving belt.

Another cause can be carryback. This is material that clings to the conveyor belt past the discharge and then drops off along the conveyor return. If not controlled, this fugitive material can build up on the ground, in confined spaces, and on rolling components. These accumulations can quickly build to a point where the belt runs through a pile of fugitive material that wears away the top cover. This damage will happen more quickly when the materials have sharp-edged particles and higher abrasion levels.

Faulty belt-cleaner selection and improper cleaner mounting can also lead to top-cover damage. Belt cleaners must be mounted properly to avoid chattering. Belt-cleaner chatter can quickly remove the top cover of the belt if not corrected immediately.

Research has shown that even properly installed belt-cleaning systems can cause some wear on the cover of the belt. This would qualify as a portion of the "normal wear" of the belt. With properly tensioned cleaning devices, this wear is modest and has been shown to be less than the abrasion from one idler seizing due to material buildup.

Slow moving, feeder-type belts that convey materials from vessels under high "head loads" can also suffer top-cover damage. Reducing this downward pressure from the material load onto the belt will reduce the potential for damage.

F. Rips and grooves from foreign objects

Damage in the form of rips and grooves is caused by stray pieces of metal, ranging from packing-crate strapping to the teeth from loader buckets (**Figure 4.19**).

These metal pieces can become wedged into the conveyor structure, forming a knife to gouge or slit the belt. This damage can be the most difficult type to con-

trol, because it occurs very quickly and often with catastrophic effects. There are a number of ways to minimize, but not totally eliminate, the amount of "tramp iron" in the material flow. These methods include grizzly screens, metal detectors, and video monitors. Regardless of the effectiveness of the precautions, the belt is still vulnerable.

G. Belt-cleaner damage

Conveyor belts often have a paradoxical relationship with belt cleaners. Cleaning systems are required to remove carryback, which reduces fugitive material along the conveyor and so preserves the

Figure 4.19

Tramp iron, ranging from packing-crate strapping to the teeth from loader buckets, can become wedged in the conveyor structure and damage the belt.

Figure 4.20

Impacts from belt-cleaner blade "chatter" can put nicks in the cover of the belt.

Figure 4.21

Bending the belt around too small a pulley can cause cracks in the top cover perpendicular to belt travel.

4

life of the belt; however, belt cleaners can also have negative effects. Like any foreign object, cleaners can damage a belt, particularly when the cleaning system is poorly applied or poorly maintained. Damage can come from too much pressure or out-of-alignment installation. Chattering cleaner blades can remove pieces from the belt surface (**Figure 4.20**).

Any damage on the surface of the belt or on the cleaning edge of the belt cleaner can create additional vibration, expanding the movement and perpetuating the cycle.

H. Top-cover cracking

Short random cracks in the top cover running perpendicular to the direction of belt travel may be caused by a mismatch between the belt and the pulley diameters (**Figure 4.21**).

Each belt, depending on the manufacturer, number of ply, reinforcing

materials, and thickness, will require a different minimum-bend radius. This type of damage occurs if the belt is not matched to all pulley diameters of the structure. Bending a belt in too small of a radius will cause stress on the top cover. This undue stress on the top cover will cause the rubber to crack, exposing the reinforcing materials of the belt, which can lead to damage to the belt's internal carcass.

Any change from the original belt specifications should also be done in conjunction with a rerun of the belt-specification program which includes a study of the conveyor's pulley diameters and tension required to drive the belt.

Installing a thicker belt on an existing system to improve life by preventing other types of belt damage, such as impact, may dramatically shorten the life of the new belt if the diameters of the conveyor pulleys are smaller than recommended by the manufacturer. It is important to always check with the belt manufacturer to ensure this design parameter is met.

I. Heat damage

Conveying hot materials may also cause the top cover to crack or the plies to separate. The cracks from heat damage may run either parallel or perpendicular (or both) to the direction of belt travel (**Figure 4.22**). If the conveyed material is hotter than the belt can handle, holes may be burned through the belt. Using high-temperature belting may reduce this heat cracking and increase belt-life. The only true solution is to cool the material prior to conveying it or to use some other method of moving the material, at least until it is cooled sufficiently.

J. Junction-joint failure

As the belt moves along the conveying system, going from flat to troughed at the terminal pulleys, the outer one-third of the belt is required to travel farther than the center one-third of the belt.

Figure 4.22

Carrying hot material may lead to cracks in the belt's top cover.

Figure 4.23

Junction-joint failure can be seen as a "W" or "M" shape in the belt as it passes over a return roller.

Thus, the outer one-third of the belt must stretch more than the inner one-third. If this stretching takes place in too short of a distance, the belt can become damaged at the point where the outer wing rolls meet the center flat roll. This damage is termed junction-joint failure.

Junction-joint failure appears as small stretch marks running the entire length of the belt in the areas which pass over the points where the wing rollers and the flat rollers meet. (These stretch marks run parallel to the belt, approximately one-third of the belt width in from each edge). In early stages, it can be seen as a "W" or "M" shape in the belt as it passes over a return roller (**Figure 4.23**). This type of belt damage may be so severe as to actually tear the belt into three separate pieces.

Junction-joint failure is caused by too large a gap between idler rollers and enough tension or load to force the belt to deform into the gap. A transition distance that is too short and/or an idler-junction gap of more than 10 millimeters (0.4 in.) or twice the belt thickness may cause junction-joint failure. Belt thickness, reinforcing materials, materials of construction, and trough angles all determine the transition distance of a specific belt. When designing a new system, contemplating a change in belt specifications, or increasing trough angles, it is important to check with the belting manufacturer to ensure that proper transition distances are maintained at both the head and tail pulleys.

K. Belt cupping

A cupped belt happens when the belt has a permanent curvature across its face, perpendicular to the line of travel (**Figure 4.24**).

Belt cupping can be caused by heat, by transition distances not matched to the belt, or by too severe a trough angle for the type of belt being used. Another cause of cupping is over-tensioning the belt. The presence of chemicals such as deicers or dust surfactants can also cause a belt to cup up or down, depending on whether the chemical shrinks or swells the elastomer in the belt's top cover. Aspect ratios that are too great (where the top cover is too thick for the bottom cover) can also cause a belt to cup.

Cupped belts are extremely difficult to track, as the frictional area, the surface where the belt contacts the rollers, is drastically reduced.

L. Belt camber

Camber is a longitudinal curve in the belt when the belt is viewed from the top. The Rubber Manufacturers Association defines camber as the convex edge of the belt; the concave side of the belt is called the bow (**Figure 4.25**). If a belt is composed of more than one section,

Figure 4.24

A cupped belt will not lie down on the idlers.

Figure 4.25

Viewed from the top, camber is a longitudinal curve in the belting.

4

it may have more than one camber or even conflicting cambers.

This type of damage can be created during manufacture or from improper storing, splicing, or tensioning of the belt (**Figure 4.26**). Proper storage and handling are essential from the time of manufacture to the time of installation.

These curves in the belt produce tracking problems that are often confused with a crooked splice. Camber and bow will produce a slow side to side movement; a crooked splice produces a more rapid "jump" in the belt's tracking. However, a crooked splice has a short

area of influence, whereas the curve of a camber or bow is from one end of a belt section to the other.

BELT REPAIR

Repair of Conveyor Belting

For most operations, conveyor belt-life is measured in years. To achieve the lowest operating cost, inspection of the belt should be a scheduled maintenance procedure. Any belt damage noted during these inspections should be repaired promptly to prevent small problems from becoming big trouble. Damage to a belt can permit the entrance of moisture or foreign materials into the belting, and thus promote premature failure of the belt. To preserve the belt, it is important to make prompt and effective repairs of any damage.

Vulcanized repairs can be made during scheduled maintenance outages when sufficient conveyor downtime is available to allow the long time required to make a vulcanized joint. In nearly all cases, a vulcanized repair requires removal of a complete section of belt and then either re-splicing the remainder or adding an additional piece of belting, often called a "saddle."

Fortunately, many forms of damage lend themselves to relatively simple methods of repair. Repairable forms of damage include:

A. Grooves in which the top cover is worn away by abrasion from material or a foreign object

B. Longitudinal rips in which the belt is slit by a fixed object, such as a metal bar becoming wedged into the conveyor structure

C. Profile rips in which a small tear in the belt edge extends inward

D. Edge gouges in which blunt objects tear chunks of rubber out of the belt edge, generally caused by the mistracking of the belt into the conveyor's structure

Figure 4.26

Belt camber can be created during manufacture or by improper storing, splicing, or tensioning of the belt.

Figure 4.27

Specialized adhesives can be used to repair a damaged belt.

Repairs can be made with self-curing, adhesive-like repair materials to keep moisture or foreign material out of the carcass. Mechanical fasteners are another method for repairing damaged belting to restore service without significant downtime and extend the service-life of expensive belts.

Belt Repair Using Adhesives

Adhesives provide a cost-effective means to repair conveyor belting with a high quality bond. Use of adhesive compounds will save downtime and money in maintenance budgets without requiring heavy vulcanizing equipment or creating obstructions with repair hardware in the belt (**Figure 4.27**). Adhesive repair compounds offer simple solutions for belt maintenance that are durable, reliable, and easy to use. There are a number of products available to do this. They include solvent-based contact cements, heat-activated thermoplastics, and two-component urethane elastomers.

All of these systems require some degree of surface preparation, ranging from a simple solvent wipe to extensive grinding or sandblasting. Some may need an application of a primer to improve adhesion.

Most commonly used for standard cold-vulcanized splices, solvent cements are also used for bonding repair strips and patches over damaged areas.

Thermoplastic compounds are "hot melts" that are heated to a liquid state and then harden as they cool, forming a bond. As they cool quickly from their application temperature of 120 to 150 degrees Celsius (250° to 300° F), the repair must be performed quickly, before the adhesive returns to the hardened (non-adhesive) condition. Problems encountered with thermoplastic adhesives include the possibility of shrinkage as the adhesive cools and the risk that high-temperature operations or cargo may cause a softening of the adhesive, leading in turn to failure of the repair.

Urethane products are typically two-component systems that the user can mix and then spread like cake frosting directly onto the area to be repaired. They typically achieve operating strength in a short period, one to two hours, but will continue to cure for eight to twelve hours until full cure strength is reached.

All adhesive systems offer fairly simple applications, assuming the instructions are followed. Of course, it is critical that the adhesive manufacturer's instructions be followed carefully as to surface preparation, component mixing, pot life, application technique, and cure time. The length of time for an operating cure and full cure may provide the basis for selecting any particular product.

It is important to get the profile of the repaired area down to match the profile of the original belt in order to preserve the repair and avoid more damage to the belt.

It is also important to identify and resolve the cause of the problem, removing the obstruction or correcting the mistracking that led to the belt damage in the first place. Otherwise, the resumption of operations after repair merely initiates a waiting period until the damage recurs and the repair must be made again.

Mechanical Fasteners for Belt Repair

Because of their comparative ease of installation, mechanical splices are often used in emergency repair situations when a new piece of belting must be added to an old belt or when a belt must be patched or a rip closed (**Figure 4.28**). In these cases, the mechanical fasteners are used as a "band-aid" to cover damage and close a

Figure 4.28

Damaged belts can be repaired using mechanical fasteners.

4

hole, allowing the conveyor to begin running again.

Mechanical splices can be used effectively for belt repair, providing care is used to properly install and recess the fasteners. Of course, the problem with all temporary repairs is that too quickly the "temporary" part becomes forgotten. The system is running; the plant personnel have moved on, at least mentally, to solving other problems. It must be remembered that these repairs

Figure 4.29

Claw-type fasteners can be hammered in place for a fast repair of a ripped belt.

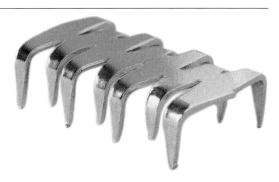

Figure 4.30

Damaged belting can be repaired with rip-repair fasteners installed in a pattern that alternates two- and three-bolt fasteners.

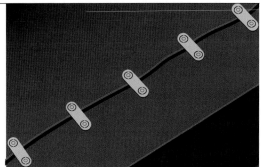

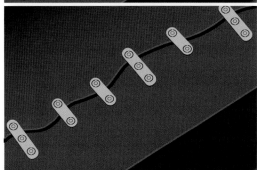

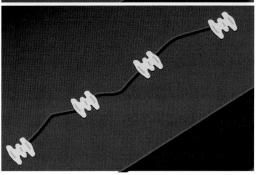

are only temporary stopgaps and are not designed for permanence. It is always important to solve the root cause of the problem in order to prevent recurrence.

Recovery from belt damage does not have to involve lengthy downtime. Mechanical rip repair fasteners offer an inexpensive and fast repair. They can be installed with simple tools and without discarding any belting. As soon as the fasteners are in place, the belt can be returned to service without waiting for any "cure" time. They can be installed from the top side of the belt, without removing the belt from the conveyor.

One-piece, hammer-on "claw" style fasteners can provide temporary rip repair where the speed of repair and return to operation is critical (**Figure 4.29**). These rip-repair fasteners can also be used to fortify gouges and soft, damaged spots in the belting to prevent these spots from becoming rips. For repair of jagged ("zig-zag") rips, splice suppliers recommend alternating two- and three-bolt fasteners (**Figure 4.30**) along the repair. The larger (two-bolt) side of the three-bolt fasteners should be placed on the "weaker" flap side of the rip to provide greater strength. For straight rips, standard two-bolt mechanical fasteners are acceptable.

PRESERVING THE LIFE OF THE BELT

Rip-Detection Systems

Increasing numbers of operations are acting to preserve the life of their belting by installing rip-detection systems. In the event of a rip in the belt, these systems sound an alarm and/or automatically shut down a conveyor.

These systems are designed for the situation in which a belt-length rip—arising from a piece of tramp iron or a wedged lump of material that slits the belting into two independent, or nearly independent, pieces—would require the entire replacement of a costly belt. Without a rip-detection system, a belt rip may continue for hundreds or thousands of feet.

Rip-detection systems are most commonly seen on very expensive, production-critical conveyors. In these circumstances, the operation would be shut down for the time required to acquire and install a new belt or make a repair to a belt-length rip.

The rip-detection systems are particularly valuable on long conveyors, where the 60 meter (200 ft) length typically damaged before the conveyor can be shut down is an insignificant loss compared to the value of the length of belting saved by the automatic shutdown.

Although there are differences in operating principles for the various rip-detection systems, each system basically embeds a sensor or signaling agent at places in the conveyor belt. As the belt moves, these indicators pass over the detection points—typically installed at the places where it is likely for a conveyor rip to occur: the loading zone and discharge. When the rip in the belt causes the signal to be interrupted, the alarm is sounded and the belt stopped.

These systems will minimize belt damage and allow the plant to reduce the amount of belting it needs to keep in inventory.

Monitoring of Conveyor Belting

As operating plants push to extend production periods, windows of opportunity for maintenance continue to shrink. Conveyor supply and support companies can now help to accommodate this situation by providing tools that give a better understanding of the condition of conveyor belting.

In addition to the rip-detection systems mentioned above, there are services which will provide a comprehensive monitoring of the condition of a belt. Factors analyzed include the condition of the carcass; the condition and wear of the top cover, including its thickness and its estimated remaining life; and the condition of the splice(s).

Early detection, mapping, and monitoring of damaged areas and of splice strength allow planners to schedule maintenance windows in advance and extend the service-life of the conveyor systems under their control.

BELTING IS THE KEY

In Closing...

The conveyor system is a key to the efficiency of an entire operation; the belting is the key to a conveyor's productivity. Consequently, preservation of the capabilities and life of the belt is essential. Considering the size of the initial investment in conveyor belting, the importance of preserving a belt through regular inspection and repair activities cannot be overstated. The relatively minor costs for careful inspection and belt repair and the somewhat more significant expense of a conveyor outage to allow that repair to be made will be paid back many times over by an extended belt-life.

Looking Ahead...

This chapter, The Belt, was the second chapter regarding basics of conveyors in the section Foundations of Safe Bulk-Materials Handling. The next chapter, Splicing the Belt, concludes this section, describing various types of belt splices and their impact on fugitive materials.

REFERENCES

4.1   Any manufacturer and most distributors of belting can provide a variety of materials on the construction and use of their specific products, as well as on conveyor belts in general.

4.2   Conveyor Equipment Manufacturers Association (CEMA). (2005). *BELT CONVEYORS for BULK MATERIALS, Sixth Edition*. Naples, Florida.

4.3   The website http://www.conveyor-beltguide.com is a valuable and non-commercial resource covering many aspects of belting.

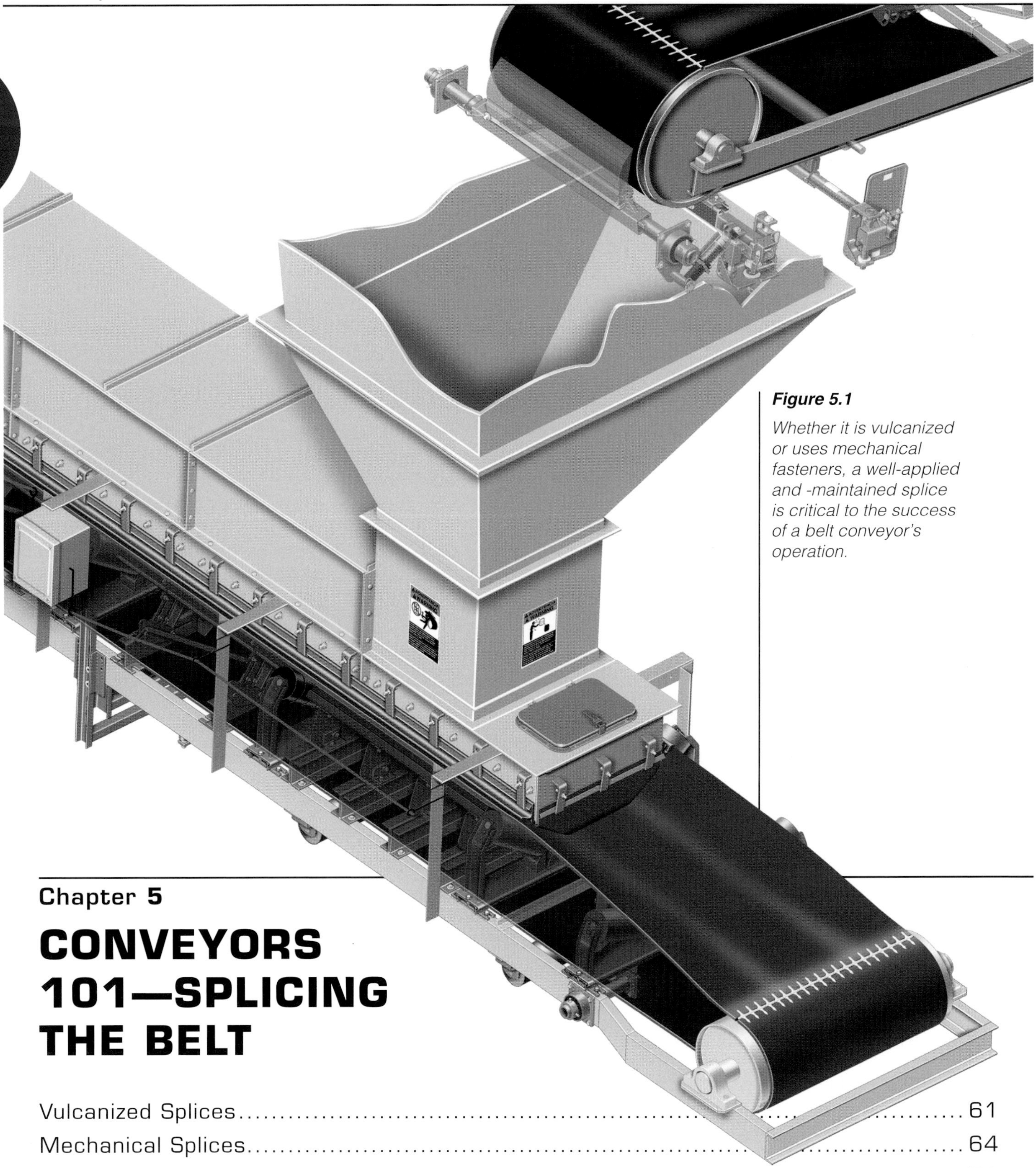

5

Figure 5.1

Whether it is vulcanized or uses mechanical fasteners, a well-applied and -maintained splice is critical to the success of a belt conveyor's operation.

Chapter 5

CONVEYORS 101—SPLICING THE BELT

5

In this Chapter...

In this chapter, we will discuss the methods used to join the belt, including mechanical splices and hot and cold vulcanization (**Figure 5.1**). The chapter will review the advantages and disadvantages of the various systems while emphasizing the importance of proper installation, inspection, and maintenance.

Conveyor belting is shipped from the factory on a roll, and before its use, the two ends of the belt must be joined together to provide a continuous loop. The two methods for splicing the ends of the belt together are vulcanization and mechanical fasteners.

Both techniques are employed around the world. In North America, mechanically splicing the belt is more prevalent; outside North America, vulcanization is more common. For reasons that will be discussed in this section, vulcanization is preferred for the control of fugitive materials; however, in many cases, the need to return a conveyor to service promptly will mandate the use of mechanical splices.

VULCANIZED SPLICES

Vulcanization is the process of curing raw rubber by combining the rubber with additives in the presence of heat and pressure ("hot" vulcanization). Bonding of the belt ends with adhesives is known in the belting trade as "cold" vulcanization.

Vulcanization is generally the preferred method of belt splicing, due to the superior strength, longer service-life, and cleaner operation it offers. Vulcanized splices are really the only option for the long-term performance of high-tension steel-cord belting. Operations that require frequent additions or removal of sections of belting, such as extendable underground belts or conveyors with limited take-up capabilities that require shortening of the belt to maintain tension, are not suitable for vulcanized splices.

Due to their superior strength, vulcanized splices allow the application of maximum belt tension, resulting in better pulley-to-belt traction. A vulcanized splice has no internal weaving, braiding, sewing, welding, or other mechanical link. The splice is solely dependent on rubber adhesion to the carcass or steel cords, as the tensile members of the splice, textile plies or steel cords, do not touch each other. Adhesion is obtained through use of an intermediary rubber or rubber-like material called tie gum, installation gum, or cement.

Steps in Vulcanizing a Belt

Step-by-step procedures for vulcanized splices vary between manufacturers (**Figure 5.2**). In general, there are three steps:

A. Preparation of the belt ends

In the first step for a fabric carcass belt, the ends are cut at the correct angle and then stripped or pulled apart to expose the various plies to be joined. Care must be taken not to damage the plies or cords. The process for a steel-cable belt involves cutting back the rubber cover (**Figure 5.3**).

B. Application of cement, gum, or other intermediary material

The second step provides the buildup of the layers, much like the making of a sandwich, which will form the completed splice. For steel-cable belts, the cords are overlapped, and then appropriate bonding agents are applied to the exposed cables. Fill and cover rubbers are

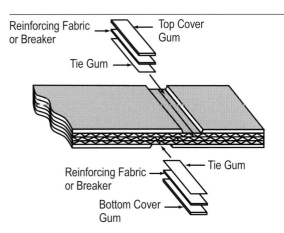

Figure 5.2

A vulcanized splice is made from a sandwich of belting and additional materials.

5

then laid in place, and the belt is cured in the same manner for both steel-cable and fabric belts.

C. Curing of the splice

The assembled materials are pressed together and cured, through the application of heat, pressure, and/or time, to form the finished splice.

Typically, the materials used for a vulcanized splice—cement, tie gum, strings of rubber called noodles, or cover stock, all depending on belt style and construction—are available in kit form. Kits from the belt manufacturer are sometimes preferred, although there are generic kits available for the most common belt grades. The materials in the kit are perishable; they have a specified shelf life in storage and a limited "pot life" when they are mixed into the ready-to-apply state.

There are two types of vulcanization: hot and cold. In hot vulcanization, the layers of a belt are stripped in a stair-step or finger fashion and overlapped with glue and rubber. A heated press or "cooker" then ap-

plies heat and pressure to "vulcanize" the belt into an endless loop. In cold vulcanization (technically called chemical bonding), the belt's layers are joined with an adhesive or bonding agent that cures at room temperature. Vulcanization, particularly hot vulcanization, is usually performed by outside contractors who have the specialized equipment and expertise to perform the required procedure.

Hot Vulcanization

In hot vulcanization, a special press (**Figure 5.4**) applies both heat and pressure to the splice to cure the intermediary and cover materials into a high-strength joint. The press applies pressure consistently across all surfaces. Pressure can range from 34 to 1200 kilopascal (5 lb_f/in.^{2} to 175 lb_f/in.^{2}), depending on the belt. Cooking temperatures range from 120 to 200 degrees Celsius (250° to 400° F), depending on belt type and rubber compound. The time required to cure will depend on belt thickness and compound: Belt manufacturers normally include time and temperature tables in splicing manuals. Although the equipment is automated, the process may require constant human attention to achieve the best results. Portable vulcanizing presses for curing the splice are available in sizes to match various belt widths. Small fabric belt splices can often be cured in a single setup. Larger fabric belt splices can be cured in two, three, or more settings of the vulcanizing press without problems. With steel-cord belts and finger splices, it is important that the press be large enough to cure the splice in a single setup to avoid undesirable rubber flow and cord displacement.

When the vulcanization procedure, or "cook," is completed, the resulting splice should be inspected for any visible defects that might indicate a weakness. It is common practice to grind or buff away any surplus rubber from the splice to improve the performance of the joint as it passes through belt cleaners and other conveyor components.

Figure 5.3

The splicing process for a steel cable belt involves cutting back the rubber cover.

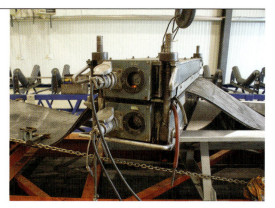

Figure 5.4

Heat and pressure supplied by a special press are used to form a "hot vulcanized" splice.

Cold Vulcanization (Chemical Bonding)

In cold vulcanization, the belt is joined using adhesives or bonding agents that will fuse the ends of the belt together to form a continuous loop.

In cold splicing, the joint is not cured in a press. The belt ends are carefully laid together in proper alignment with the adhesive, and full contact is achieved with hand rollers, pressure rollers, or hammering in a prescribed pattern. The bond can often be improved by simply putting weights on the belt during the cure interval. Most cold vulcanization cements require at least four hours for a usable cure and 24 hours for a full cure. Best results are achieved by following the manufacturer's recommendations. The belting manufacturer is the best source of information on proper vulcanization techniques and materials.

Splice Pattern

Vulcanized splices require the cutting away of layers of rubber covers and fabric carcass to let the belt ends be overlapped and joined. In general, the geometry of a splice can be the same whether the joint will be vulcanized hot or cold.

Bias splices are most common, as the angle increases the length of the bonding surface and reduces stress on the splice as it wraps around the conveyor's pulleys. The bias angle also serves to reduce the chance of tearing open the leading edge of the splice. This bias angle is generally 22 degrees; most vulcanizing presses are manufactured with this bias angle built in (**Figure 5.5**).

A splice design that is seeing increasing acceptance, particularly with high-tension fabric belts, is the finger splice (**Figure 5.6**). This design involves cutting the two ends of the belt into a number of narrow triangular "fingers." The fingers—each typically 30 or 50 millimeters (1-3/16 or 2 in.) wide at the base and between 850 to 1200 millimeters (33 to 48 in.) in length,

depending on belting specifications—are interlaced, and then the hot vulcanization is performed. Finger splices must be performed in a single "cook."

For all splice designs, it is important that the overlapped areas and any materials added to the joint be properly installed to minimize damage to the finished splice from belt-cleaning systems or other components.

Advantages of Vulcanization

Even though a vulcanized splice is more expensive and time-consuming to perform, it is usually an excellent investment. It provides a strong joint able to withstand high levels of belt tension. Splices performed by reputable firms will feature high-quality materials and workmanship and are typically guaranteed. Since a vulcanized splice

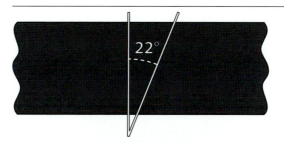

Figure 5.5

Bias angles are generally 22 degrees.

Figure 5.6

Used on high-tension belting, a finger splice can provide the best mix of splice strength and life.

5

chemically bonds the belt into an endless piece with no possibility of material sifting through the splice, from the standpoint of control of fugitive material, vulcanization is the splice of choice. A properly-performed vulcanized splice will not interfere with rubber skirting, idler rolls, belt-support structures, or belt cleaners.

Cold vulcanization offers some advantages over hot vulcanization. There is no heating source nor press required, the equipment is easier to transport, and no special electricity is required. Therefore, cold vulcanized splices can be performed even at remote sites where access is difficult and power is unavailable. Only small hand tools are required, so the cost to purchase and maintain the splicing equipment is low.

Hot and cold vulcanized splices take roughly the same amount of time to prepare the belt and complete the joining process; however, the cold splice may require more downtime than hot vulcanization due to the long cure time of the adhesive bond.

The finger splice can provide the best mix of splice strength and dynamic life in applications on high-tension belting. This system keeps all the factory belt plies in place, without any steps cut into the belt. A finger splice can be cut square or on a bias across the belt.

Disadvantages of Vulcanized Splices

The disadvantages of vulcanization that must be considered are the higher initial cost and the length of time required to perform the splice when compared to a mechanical splice. The peeling back of layers of belting to prepare for both hot vulcanization and cold chemical bonding can be difficult. It can take over 24 hours to return a conveyor to service by the time the splice is prepared, heated, and cooled sufficiently to allow the finished joint to be handled or even longer for the cold chemical bond to cure.

This added time to complete a vulcanized splice will be particularly troubling

(and expensive) in cases where an emergency repair is required to allow the resumption of operations. In this case, the delay required for hiring and bringing on-site an outside crew and equipment increases the cost of the downtime by extending the outage and adding "emergency response" surcharges.

Due to time and cost pressures, vulcanized belt splices cannot be justified in applications where frequent extensions or retractions of the conveyor length are required. The same is true where the take-up allowance does not allow enough belting for a vulcanized splice, and a short section of belting, often called a saddle, must be added, requiring two splices.

Vulcanizing can be more difficult and less reliable on older, worn belts. In applications on conveyors that are utilized in the process of transporting hot materials, it is important that all material be discharged from the belt prior to stopping the belt. Hot material left on a stopped belt can "bake" a splice and reduce its life.

Installation of a vulcanized splice can consume a considerable length of belting, as much as 2,4 to 3 meters (8 to 10 ft) in some cases, particularly when a bias splice is used on a wide belt. This installation may require a longer belt to be purchased or a new section of belt, or saddle, to be added.

When designing new conveyor systems that will incorporate vulcanized belts, it is wise to include a take-up pulley mechanism, designed to take up slack in the belt. The take-up pulley should have sufficient movement to account for belt stretch, thus avoiding the need to shorten the belt with a time-consuming new splice.

MECHANICAL SPLICES

Mechanical Fasteners

Today there are many types of mechanical fasteners available for belt splicing. They all work on the principle of joining the two ends of the belt together with a hinge-and-pin or plate design. Mechanical fasteners are now fabricated from a variety of materials to resist corrosion and wear and to match the conditions of application.

For many years, mechanical splices were considered the low-quality alternative to vulcanization as a method of joining the belt. Recent developments have moved mechanical fasteners into a better position versus vulcanization. These innovations include the use of thinner belts (made possible by the use of synthetic materials in belting), improvements in design and materials used in fasteners to increase strength and reduce wear, and the development of tools to recess the profile of the splice.

Types of Mechanical Splices

Mechanical fasteners for bulk material handling belts are available as hinged fasteners or plate fasteners, with options within each group.

Hinged Fasteners

In hinged-fasteners splices, a strip composed of top and bottom plates joined on one side by metal loops is placed on each of the two belt ends (**Figure 5.7**). These strips are attached to the belt by staples, bolts, or rivets. The belt is then joined by passing a linking pin through the alternating hinge loops.

Hinged fasteners are usually supplied in continuous strips to fit standard belt widths. These strip assemblies ensure proper spacing and alignment. The strips are fabricated so pieces can be snapped off to fit non-standard belt widths.

The chief advantage of hinged fasteners is that the belt can be separated by removing the linking pin. This way the belt can be shortened, extended, removed from the structure, or opened to allow maintenance on conveyor components.

Hinged fasteners provide several other benefits. Installation on the two belt ends can be done separately and even performed off-site. While it is not a recommended

practice to join belts of different thicknesses—because of problems it can create, including sealing, tracking, and cleaning—hinged loop fasteners would allow different thicknesses of belting to be combined using fasteners matched to their respective belt halves.

Solid-Plate Segments

A second type of mechanical splice is performed with plate fasteners (**Figure 5.8**). This class of fastener makes a strong, durable joint with no hinge gap for fines. Plate fasteners are effective in the most rugged conveying applications in mines, quarries, and steel mills. In applications where the belt is thicker than 22 millimeters (7/8-in.), plate fasteners are the only choice for mechanical fastening. Solid-plate segment fasteners are intended for permanent joints only and are not recommended for belts in applications that require opening of the joint to change belt length or location.

Figure 5.7

Supplied in continuous strips to match the belt width, hinged fasteners are joined together with a linking pin.

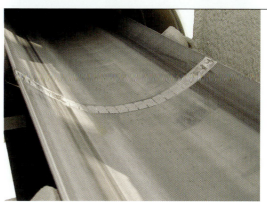

Figure 5.8

Applied by rivet, staple, or bolt, solid-plate fasteners provide a sturdy joint for heavy-duty applications.

5

Solid-plate segment fasteners are typically provided as individual pieces packed loose in a box or bucket. The plate segments are installed from one belt edge to the other using staples, rivets, or bolts.

Bolted solid-plate fasteners have some unique advantages. They can be applied diagonally across the belt to allow use on pulleys that are smaller than the size recommended for the fastener. They can also be installed in a V-shaped pattern (**Figure 5.9**), which may be the only choice for using fasteners to join the thick, high-tension belts designed for vulcanization.

One problem with bolt-fastened solid-plate segment fasteners is that they typically use only two bolts on each plate, with one on each side of the splice. Tightening down on the ends of the splice means the leading and trailing edges are more compressed than the middle of the plate. This allows the middle to crown, creating a wear point

in the fastener and in belt cleaners or other systems that contact the belt as it moves on its path through the conveyor.

Riveted solid-plate fasteners are designed for the most demanding, highest-tension applications. The multi-point attachment on each side of the hinge provides the highest holding power of any mechanical fastener. They can be installed without power tools, using a hammer to set the rivets and break off any heads above the belt. This is an advantage in remote or underground locations.

A problem arises if the conveyor uses pulleys that are smaller than 300 millimeters (12 in.) in diameter. In this case, solid-plate fasteners may be too large to bend around the pulley, causing components of the splice to pull out or break.

Flexible-Matrix Plate Splices

One additional splicing technique is flexible-matrix plate splices. This system uses self-tapping screws driven through an H-shaped (or perhaps I-beam-shaped) hinge matrix. To form this joint, the two belt ends are skived on a bias down to the fabric carcass and then inserted into the open ends of the H-shaped reinforced-rubber hinge matrix. The matrix (which runs the full width of the belt) is then fastened to the belting using up to 240 screws per meter of belt width.

This system offers relatively quick and easy installation, using only skiving tools and a power screwdriver (**Figure 5.10**). The splice can be installed in any weather and in locations where a vulcanizing press or other splicing tools would be difficult to use. It requires no cure time and can be used for joining or repairing belts. If it is used for a temporary repair, the matrix can be removed and reused.

A benefit of the flexible-matrix splice is that it is leak-proof, as there are no openings between belt ends or holes in the belting that allow material to sift through.

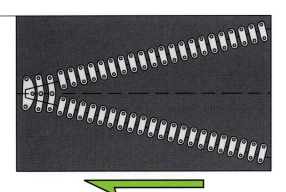

Figure 5.9

A "V"-shaped splice may be the best choice for joining high-tension belts with mechanical fasteners.

Figure 5.10

A flexible-matrix splice offers relatively simple installation using only simple tools. The splice uses self-tapping screws through an H-shaped matrix to join the two belt edges.

This system is presently used for joining fabric carcass belts; the supplier is attempting to develop and secure approval for use on steel-cable belts.

Selecting the Proper Fastener

Most fasteners are available in a range of sizes. In all cases, the manufacturer's recommendations should be checked to ensure the fastener size is matched to the pulley sizes and belt thickness.

If the belt is to be skived in order to countersink the fastener down to the surface of the belt, this skived thickness should be considered when thinking about fastener size. The fastener should be selected based on the diameter of the smallest pulley in the system.

Fasteners are available in a variety of different metals to meet the requirements of special applications. These properties include non-sparking, non-magnetic, abrasion-resistant, and/or corrosion-resistant materials. Hinge pins are available in a similar selection. The manufacturer should be contacted for the proper recommendation for any specific application.

Training for selection and installation of splices should be carried out by qualified personnel. When installed in accordance with manufacturers' instructions, mechanical splices can provide an economical method of joining the belt. When incorrectly specified or applied, mechanical splices can create expensive and recurrent problems.

Proper Installation of Fasteners

Mechanical splices can be installed relatively easily by plant personnel; however, as a consequence, they can be easily misapplied, particularly by untrained personnel or in an emergency "get running in a hurry" situation. It is critical that plant personnel be trained in the proper installation of mechanical fasteners.

It is a common but incorrect practice to stock only one size of mechanical splices in the maintenance supply room. Over the years, the specifications for the belts used within a plant may have changed, but the mechanical fasteners kept on hand in the storeroom have stayed the same, which can lead to a variety of problems, including splice failure and damage to conveyor components. Installing a mechanical fastener properly requires using the correct fastener, proper tools, and attention to detail.

Squaring the Belt Ends

Where belt ends are joined with mechanical fasteners, the first requirement of a good joint is usually that the belt ends be cut square. Failure to do so will cause the belt's splice area to run to one side of the structure at all points along the conveyor. This is usually seen as a quick side-to-side motion as the spliced area passes over any point on the structure. Using the belt edge as a squaring guide is not recommended, as the belt edge may not be straight. Used belting may have an indistinct edge due to wear, so one of the following procedures is recommended:

A. Centerline method

To find the belt's average centerline, measure from one belt edge to the other at five points along the belt, each roughly 300 millimeters (12 in.) farther from the end of the belt. Mark a series of points at the center of the belt, and connect these points using a chalk line or ruler to determine the average centerline (**Figure 5.11**).

Draw the cut line by using a square. Draw a line across the belt perpendicular to the average centerline. This line can be used for the cut line (**Figure 5.12**).

B. Double-arc method

For greater accuracy, or on belts with worn edges, a "double intersecting arc" method can be employed. After establishing an average centerline as above, pick a point on the centerline two to three times the belt width from the belt

5

Figure 5.11

To determine the belt's average centerline, connect the center of five measuring points across the belt.

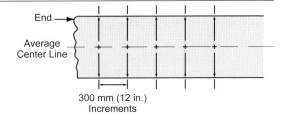

Figure 5.12

Use a square to draw a cut line across the belt perpendicular to the average centerline.

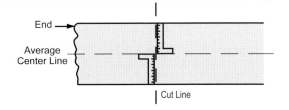

Figure 5.13

Step One: First draw an arc from a pivot point two or three times the belt width from the belt end.

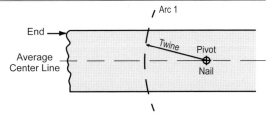

Figure 5.14

Step Two: Then draw a second arc so that it intersects the first arc near the belt edge on both sides.

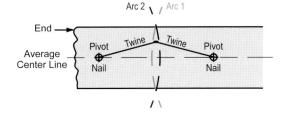

Figure 5.15

Step Three: Finally, mark the line to cut on by connecting the points where the two arcs intersect near both edges of the belt.

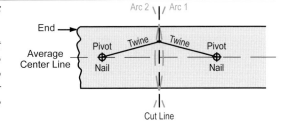

Figure 5.16

To check if the cut line is straight across the belt, measure diagonally from points the same distance on both sides of the belts. The diagonal lines should be the same length.

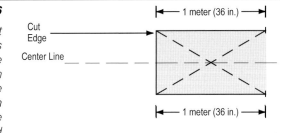

end. Using a string with a nail on the centerline as a pivot point, draw an arc across the belt so the arc crosses the edge of the belt on both sides (**Figure 5.13**). Now, create a second pivot point on the centerline much closer to the belt end. Strike a second arc across the belt, this one facing the opposite direction, so the second arc crosses the first arc on both sides of the average centerline near the belt edges (**Figure 5.14**). Draw a line from the intersection of the arcs on one side of the belt to the intersection of the arcs on the other side (**Figure 5.15**). This new line is perpendicular to the centerline of the belt and becomes the cut or splice line.

Checking the Accuracy of the Squared Ends

Regardless of which method is used, checking the accuracy is necessary. To check the accuracy of the squared end, measure a given distance (say, 1 meter or 36 in.) away from the line on both sides of the belt. Then measure diagonally from these new points to the end of the cut on the opposite side of the belt, so a diagonal measure is taken. The two diagonal lines should intersect on the belt's centerline, and the diagonal lines should be the same length (**Figure 5.16**).

The Importance of Skiving

For a mechanical splice to function in a transfer point and allow effective sealing and cleaning, both the top and bottom splice sections must be recessed sufficiently into the belt to keep the belt thickness constant and the splice surface smooth, to avoid damage to components and the splice.

Cutting down the covers of the belt, typically called skiving, mounts the fasteners closer to the fabric of the belt carcass for a firmer grip (**Figure 5.17**). Skiving requires the top and bottom covers be cut down to the belt carcass. As the carcass provides the strength of the belt, and the top and bottom covers provide very little strength, this will not reduce the integrity of the belt or splice. Great care is required when skiving the belt, as any damage to the carcass of the belt can weaken the splice and therefore reduce the strength of the belt. When the splice is properly recessed, the metal components of the mechanical hinge will move without incident past potential obstructions such as impact bars, rubber-edge skirting, and belt-cleaner blades. Skiving is recommended to ensure the integrity of the belt, splice, and other conveyor components. Skiving the belt reduces noise in the conveying operation, as clips are now recessed and do not "click clack" against the idlers as the belt moves through the system.

Skiving equipment can be purchased from most splice suppliers.

Dressing a Mechanical Splice

If for some reason, such as limited belt thickness, belt damage, or limited time to complete a repair, it is impossible to properly recess a mechanical splice by skiving, the splice can be dressed. This can be done by either lowering its projecting surfaces by grinding or submerging the raised surfaces by encapsulating.

With the first approach, grinding away the high spots will ensure the leading edge or bolts and rivets do not protrude above the splice. Care must be taken when grinding the splice to avoid digging into the belt or removing too much of the splice.

The second approach is the encapsulation of the splice in a material to protect both it and the cleaner from impact damage (**Figure 5.18**). This is usually accomplished with an adhesive or elastomer applied like putty onto the belt and splice. Although the cleaning system will still have to ride up and over the mechanical clips, the splice surface will be smoother, without obstacles like fastener heads in the cleaner's path. The downside of this procedure is that because the mechanical splice is covered, the joint is harder to inspect and repair.

Notching the Trailing Side

To protect the corners of the belt at the splice, it is often useful to notch, or chamfer, the corners of the belt at the joint. On one-direction belts, it is necessary to notch only the trailing belt end. The notch is cut in the belt from the first fastener on each end of the splice out to the belt edge at a 60-degree angle. The notch will help prevent the corners of the belt from catching on the conveyor structure and damaging the splice or tearing the belt (**Figure 5.19**).

Figure 5.17

If properly recessed, the top of the mechanical fastener will be even with or lower than the top of the belt.

Figure 5.18

Dressing the splice will protect both the mechanical fasteners and belt cleaners. This can be done by either lowering its projecting surfaces by grinding or submerging the raised surfaces by encapsulating. Top photo: before encapsulation. Bottom photo: after encapsulation.

5

Getting Along with Belt Cleaners

Mechanical belt fasteners sometimes conflict with aggressive belt-cleaning systems, especially where hardened metal blades are used. Many operators prefer to use non-metallic (e.g., urethane) belt-cleaner blades on belts with mechanical splices for fear of wearing or catching the splice and ripping it out. Most of these types of problems with belt-cleaning systems can be traced to improper selection or installation of mechanical splices.

New developments help make mechanical fasteners more scraper-friendly. One is the development of tools for skiving that easily remove a uniform strip of belt cover material, leaving a smooth, flat-bottomed trough with a rounded lip to receive the splice. These devices are much faster and safer than earlier methods that used knives or grinders.

A second development is new designs for "fastener friendly" cleaners, offering special blade shapes, materials, and mounting methods that minimize impact problems with fastener plates. The recent introduction of "scalloped" mechanical clips, designed to allow belt-cleaner blades to "ramp" over the plates without damage to the cleaner or the splice, offers the possibility of improvement in durability of both blade and fastener.

There are no empirical studies on the wear of splices due to interaction with bulk material and with the cleaning and sealing systems. If good installation and maintenance practices are observed, the cleaning and sealing systems and the splices should be chosen on the basis of the performance required, rather than any worries over their life expectancy.

Advantages of Mechanical Splices

The principal advantage of mechanical splicing is that it allows the belt to be separated easily. This separation of the splice allows extension or shortening of the belt in applications like mining; and it enables service to other conveyor components, such as pulley lagging, idlers, or impact cradles to be more easily completed.

An additional advantage of mechanical fasteners is that they minimize repair downtime. These splices can often be installed in an hour or two, whereas a vulcanized joint can easily take a full day or more to complete. Fasteners are easily installed by available plant maintenance personnel, using only hand tools or simple portable machines; in contrast, vulcanizing usually requires calling independent contractors with specialized equipment. The fastened joint will cost a few hundred dollars and consume only a few millimeters of belting, whereas the vulcanized joint can cost several thousand dollars and consume several meters of belting.

Mechanical fasteners provide a splice that is simple to perform and easy to inspect. If regularly inspected, a mechanical splice will normally provide notice of an impending failure. Mechanical splices are low cost and can be stored for long periods. They allow quick installation and enable easy lengthening or shortening of the belt.

It is important to make sure that fastener selection follows the recommendations of manufacturers of both the belt and the fasteners.

Disadvantages of Mechanical Splices

If the materials to be conveyed are hot, the transmission of heat through a metal fastener may be a factor that leads to the selection of a vulcanized splice. When the material temperature exceeds 121 degrees

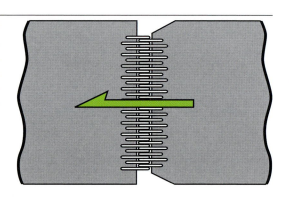

Figure 5.19

The belt's trailing edge should be cut away at the splice to prevent the corners catching on obstructions.

Celsius (250° F), the amount of heat passing through the metal fastener into the belt carcass can weaken the fibers, ultimately allowing the fastener to pull out. In these applications, a vulcanized splice would be preferable.

Failure to inspect fasteners and the resulting failures of those fasteners can result in severe belt damage. If the fasteners begin pulling out on a portion of the splice width, longitudinal ripping of the belt may occur. When belt and fasteners have been properly selected, pullout is usually due to insufficiently tight bolts or worn hooks or plates. Plate-type mechanical fasteners typically allow the replacement of individual plates, which, if performed when damage is first observed, may eliminate the need to cut out and replace the entire joint.

Using the wrong size or type of mechanical fastener can greatly reduce the operating tension capacity of a belt. The extra thickness of a mechanical splice not properly recessed or of the wrong specification will make sealing the transfer point almost impossible. Splices that are oversized and too thick to pass through the transfer-point area can catch on the wear liner or skirtboard, abusing the splice and shortening its life. These splice issues sometimes require the wear liner and skirtboard to be higher above the belt, allowing more material to reach the edge-sealing system. This, in turn, results in accelerated wear and spillage. Often, the fasteners used in the splice will not be properly trimmed, and these extended rivets or bolts can catch on other components like skirtboard-sealing systems or belt cleaners.

⚠ SAFETY CONCERNS

Any untrained individual attempting to use belt splicing equipment for vulcanization or mechanical fastening runs the risk of injury as well as the likelihood of creating a poor splice. A significant risk of an improperly performed belt splice is the likelihood that the splice will fail under the tension of application conditions. Splice failure could result in personnel injury and equipment damage.

Pre-work inspections should be completed, and manufacturer's instructions should be followed, when using any splicing tools, machinery, or chemicals.

All chemicals, including solvents, primers, and cements, should be stored and handled properly, in accordance with manufacturer instructions, including special attention to shelf-life limitations.

Proper protective clothing, including suitable gloves and eye protection, should be worn, and the work area should be properly ventilated.

Sharp hand tools and power grinders are used for cutting the rubber and preparing the joint: They pose cut and abrasion hazards to the workers.

Conveyor-belt splicing is often performed under somewhat hazardous conditions, such as inside underground mines, on inclined or elevated structures, or in areas with limited access. As always, proper lockout / tagout / blockout / testout procedures are required. Proper blockout by clamping the belt to the structure is necessary to prevent any belt movement. Clamping fixtures that are engineered for the specific size and weight of the belt being spliced should be purchased from reputable suppliers.

5

Most, if not all, mechanical splices will allow some small quantity of the conveyed material to filter through the joint itself. This material will fall along the run of the conveyor, resulting in cleanup problems and the potential for damage to idlers, pulleys, and other conveyor components. Plate-type fasteners, in a well-made joint, are quite free of material leakage. The hinge-type fasteners are all subject to problems with fine materials sifting through the joint; this problem is eliminated with vulcanized splices.

While it does provide greater strength, the V-shaped splice does have its costs. It can require up to 3 meters (10 ft) of belting for completion. That may be a significant amount of expensive belting to be discarded.

Mechanical splices are used on fabric belts for making the belt endless or for repairing rips and holes; however, on steel-cable belts, they can be used only for temporary repairs.

SAFE SPLICE DESIGN

Both mechanical and vulcanized splices must be designed with factors of safety when compared to the expected belt tension. These design factors for mechanical fasteners are built into the manufacturers' selection tables. Vulcanized splices on high-tension steel cable belts are often individually designed by the belt manufacturer or consultant. Failure to match the splice with the belt and account for the correct service and safety factors can result in catastrophic splice failures leading to injury, death, loss of production, and equipment damage.

MAINTENANCE AND INSTALLATION STATIONS

Some operations develop what is called a belt-splicing station along the conveyor. Here, tools and equipment are stored for splice maintenance, and the space and work surfaces are available for splice instal-lation. This may also be the point at which a new belt is pulled onto the conveyor.

A splice station should be located where there is plenty of room, ideally including workspace on both sides of the conveyor structure. The station should provide protection for the belt from climate conditions and fugitive material. The space should be placed at a point where there is a distance of at least five belt widths of straight conveyor stringer on either side of the point where the splice will be made. Power should be readily available, including outlets for hand tools.

INSPECTION AND MONITORING

Splice Inspection and Service

Where bolt-style fasteners are used, it is important that the plates be kept properly tightened. The most practical way to achieve this is to tighten the bolts so that the rubber behind the plate slightly swells. Care must be taken not to over-tighten fasteners or "bury" the plates into the belt cover, as this could cause damage to the plies of the belting. Manufacturers generally suggest retightening the fasteners after the first few hours of operation, again after the first few days of operation, and then at intervals of two or three months of operation.

Splices should normally be inspected on a weekly basis, with replacement of any fasteners that look worn, watching for crosswise breaks on the back of fasteners, and checking for fastener pullout.

Splice-Monitoring Systems

Newer technologies are now available which allow the remote evaluation of splices by measuring any elongation of the splice. These systems are based on the principle that the lengthening of a splice is an indication of an impending failure. The system is installed on vulcanized belts by placing small magnetic targets into the belting at a set distance on either side of splices; if the belt has mechanical fasteners in place, the system can use those as the targets. The system will monitor the distance between the paired targets each time a splice passes the scanner. This distance is measured, and if a splice falls outside the set limits, the monitoring system will shut down the belt or alert plant personnel to check the problem. In addition, the system can help identify if a clip has suffered severe damage and must be replaced.

THE IMPORTANCE OF THE SPLICE

In Closing...

Whether it is vulcanized or uses mechanical fasteners, a properly designed, well-applied and maintained splice is critical to the success of a belt conveyor's operations. Improper application of a splice will shorten the life of the belt and interfere with the conveyor's operating schedule and efficiency. Care in applying the proper splice in the correct fashion will provide benefits for the entire plant. In the words of an old axiom: "If you don't have the time to do it right, how are you going to find the time to do it over?"

Looking Ahead...

This chapter, Splicing the Belt, explaining how delayed or improper splicing can allow fugitive materials to escape from the belt, concludes the section Foundations of Safe Bulk-Materials Handling. The following chapter begins the section related to Loading the Belt and addresses the area Before the Loading Zone, looking at tail pulleys and transition areas.

REFERENCES

5.1   Any manufacturer and most distributors of belting can provide a variety of materials on the construction and use of their specific products, as well as on conveyor belts in general.

5.2   Conveyor Equipment Manufacturers Association (CEMA). (2005). *BELT CONVEYORS for BULK MATERIALS, Sixth Edition*. Naples, Florida.

5.3   The website http://www.conveyor-beltguide.com is a valuable and non-commercial resource covering many aspects of belting.

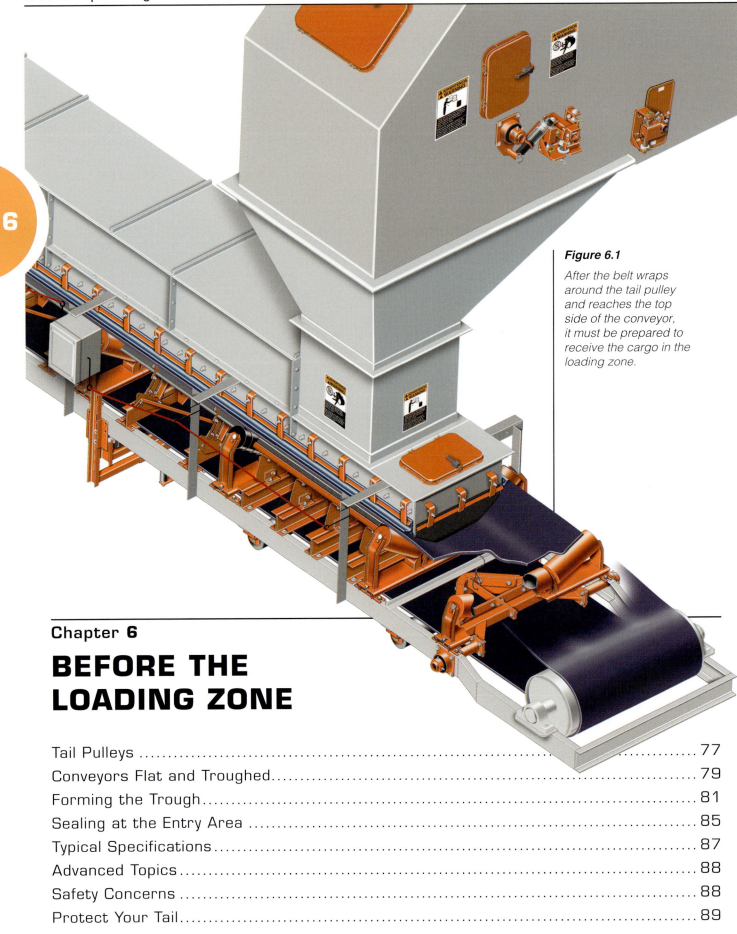

6

Figure 6.1

After the belt wraps around the tail pulley and reaches the top side of the conveyor, it must be prepared to receive the cargo in the loading zone.

Chapter **6**

BEFORE THE LOADING ZONE

In this Chapter...

Before a conveyor belt can be loaded, it must be transformed into the shape that will carry the bulk materials. This chapter examines transition areas, the areas between the terminal pulleys and the first fully-troughed idlers, looking at how a belt is formed into a trough and the importance of utilizing the proper transition distance to change the belt shape. The chapter also examines tail pulleys, looking at proper pulley configuration and location. Also considered are techniques to prevent the escape of fugitive material at the tail of the conveyor.

At the end of the conveyor's return run, the belt wraps around the tail pulley and moves up onto the top, or carrying side. It is here the belt must be prepared to receive cargo before entering the loading zone (**Figure 6.1**). These preparations include stabilizing the belt path, centering the belt on the structure, shaping the belt into the desired profile for load carrying, and sealing the back and edges of the load zone to prevent spillage.

Care must be taken in accomplishing these tasks in order to minimize fugitive material—material that has escaped from the conveyor and the process—both in the transition area and the conveying process, as well as preserving the equipment and preparing the conveyor for maximum efficiency.

In many plants, this area of the conveyor system is notorious for spillage problems and for employee injuries. This chapter will discuss ways to properly transform the belt from flat to troughed for loading and back to flat at the discharge without increasing fugitive material and employee risk.

TAIL PULLEYS

Centering the Belt

Having the belt in the center of the conveyor structure as it goes into the loading zone is critical. If the belt is not properly centered when it receives the load, the force

of loading will increase belt misalignment and compound other problems encountered on the conveyor's carrying side.

The area between the pulley and the load zone is too short and the belt has too much tension to allow for mistracking correction here, so belt-centering devices are installed on the conveyor's return to make sure the belt is centered as it enters (and exits) the tail pulley. If the belt, centered at its entry to the tail pulley, mistracks between the pulley and the load zone, most likely the problem is that the pulley is out of alignment. If the tail pulley is straight and a tracking device is installed, the belt should be centered in the load zone. *(See Chapter 16: Belt Alignment.)*

Tail Pulleys: Wings and Wraps

Wing-type tail pulleys are often installed as a method to reduce the risk of belt damage from the entrapment of lumps of material between the belt and the pulley. Wing pulleys feature vanes that resemble the paddle wheel on a steamboat (**Figure 6.2**). This design allows material that would otherwise become trapped between a solid pulley and the belt to pass through the pulley face. Between the pulley's crossbars are inclined, valley-shaped recesses that prevent fine or granular material from being caught between the tail pulley and the return belt. These valleys provide a self-cleaning function; that is, there is little surface area on which material can accumulate, and the rotation of the pulley throws the material off the pulley's face (**Figure 6.3**). If the

Figure 6.2

Wing pulleys incorporate vanes like the paddle wheel of a steamboat.

conveyor is likely to spill some of its cargo onto the return belt, the wing pulley can act as an effective device for removing this spillage without belt damage, although proper sealing of the belt at the load zone along with installation of a pulley-protection plow is the preferred solution.

Figure 6.3

The valleys in the wing pulley serve to cast out material accumulations.

Figure 6.4

Lumps of material can be caught in the valleys of the wing pulley.

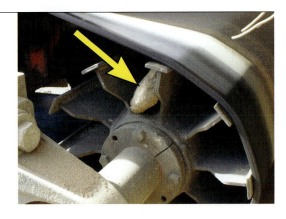

Figure 6.5

A wrapped pulley will eliminate the vibration of the wing pulley while still preventing material accumulation.

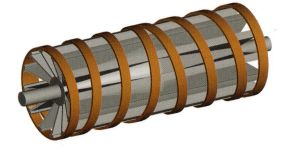

Figure 6.6

A steel band wrapped in a spiral around the pulley can eliminate some of the problems of the wing pulley.

Wing-type pulleys are also seen on gravity take-ups, where they offer the same benefits and limitations.

Despite their design intention, wing-type tail pulleys are still subject to buildup and entrapment and often do not provide the desired protection. They are most successful on slow-moving belts where cleaning and sealing are not critical requirements. Larger lumps of material can become wedged in the "wings" of the pulley, potentially causing the damage the pulley was designed to avoid (**Figure 6.4**).

Wing-type tail pulleys with less than the manufacturer's recommended minimum bend radius can cause damage to the conveyor belt's carcass.

The most significant drawback of wing pulleys is the oscillating action they introduce to the belt path. The wings on the pulley introduce a pulsating motion that destabilizes the belt path and adversely affects the belt-sealing system. It is counterproductive to design a transfer point that emphasizes belt stability to minimize fugitive material and then install a winged tail pulley that introduces instability into the system. The Conveyor Equipment Manufacturers Assocation (CEMA) recommends that wing pulleys not be used on belts traveling over 2,25 meters per second (450 ft/min).

A better choice than the conventional winged tail pulley is a spiral-wrapped tail pulley (**Figure 6.5**). These pulleys have an additional steel strip wrapped in a spiral around the pulley circumference. The steel band is wrapped over the top of the wings in two spirals, converging in the center from each end of the pulley. Wrapping the band(s) of steel around the wing pulley allows the pulley to provide the self-cleaning function but eliminates the "bounce" imparted to the belt.

Spiral-wrapped wing pulleys are sometimes installed as original equipment on new conveyor installations. Existing wing pulleys can be upgraded with narrow, 50 to 75 millimeters (2 to 3 in.), steel strips welded around the outside edge of the wings (**Figure 6.6**).

The best solution to preventing material buildup on tail pulleys is to use a solid flat steel pulley, protected by a cleaning device located directly in front of the pulley (**Figure 6.7**). This diagonal or V-type plow should be located just prior to the tail pulley on the non-carrying side of the belt to remove any fugitive material that may be carried on the inside of the belt. *(See Chapter 15: Pulley-Protection Plows.)*

Crowned Pulleys

A straight-faced pulley is the same diameter across the face of the pulley. A crown-faced pulley changes in diameter from the outer edges to the center of the pulley, with the center being slightly larger than the edges (**Figure 6.8**).

Crowned pulleys are sometimes used at the conveyor tail as it is widely believed that the crowned face will improve the tracking of the belt as it goes around the pulley and into the loading zone. However, this is not always true, and there are instances where the crown face of the pulley can actually damage the belt.

Crown-faced pulleys should never be used in a high-tension area of the belt. This is usually the driven pulley. The driven pulley may be at the head end, the tail end, or, with a center drive, anywhere along the return side of the conveyor. In these high-tension areas, the additional diameter in the center of the pulley adds additional stress to the center of the belt and may cause carcass and lagging damage. The exception to this is when the rated tension of the belt is 35 kilonewtons per meter (200 PIW) or less; then a crowned pulley may be used anywhere in the system.

In lower tension areas of the belt, crowned-faced pulleys may have a slight influence on belt tracking. However, if there are serious problems with the belt such as belt cupping, belt camber, or idler-junction

failure, no amount of pulley crown will track the belt. It is always best to identify the cause of the mistracking and cure the problem. *(See Chapter 16: Belt Alignment.)*

Crown pulleys also present problems with belt cleaners mounted on the face of the discharge pulley.

CONVEYORS FLAT AND TROUGHED

Flat Belts

Many bulk materials can be carried on flat belts. Flat belts are particularly common for materials with a steep angle of repose, the angle that a freely formed pile of material will make. Materials with angles of repose above 30 degrees are materials suitable for flat belts and range from irregular, granular, or lumpy materials like coal, stone, and ore, to sluggish materials that are typically irregular, stringy, fibrous, and interlocking, such as wood chips and bark (**Figure 6.9**). When maintaining the same edge distance with materials with a low angle of repose, the volume of material conveyed is reduced; therefore, materials with a low angle of repose generally require the belt to be troughed.

Figure 6.7

A plow is positioned in front of the pulley to protect the pulley from lumps of material.

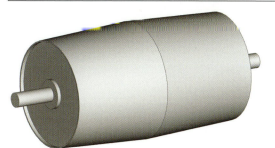

Figure 6.8

A crowned pulley is slightly larger in the center than at the edges.

Flat belts are especially effective when the load, or a portion of the load, is to be discharged from the belt at intermediate points by plows or deflector plates.

Belt feeders use flat belts almost exclusively. This is because feeders are generally very short, and they must fit into operations where there is little room to form the belt into a trough. Feeders typically operate with very high loads and use heavy-duty idlers. Many feeder belts can reverse direction. To move a large material load, feeder belts often run at high tension, making it difficult to trough a belt. In

addition, the high head load of belt feeders makes sealing difficult. This difficulty can be overcome by leaving generous edge distances and operating at slower speeds; then spillage from these belt feeders can be controlled. In many cases, these belts are equipped with a skirtboard and a sealing system along their full length. Other feeder belts incorporate dual chutewall design, where a space is left between the interior chutewall installed with a wear liner and an outside chutewall that includes the belt's edge seal (**Figure 6.10**).

Flat belts do not require the transition areas or suffer the transition problems encountered by troughed conveyors. However, most of the other conveyor components and problems discussed in this book will apply to flat belt conveyors.

Troughed Conveyors

For most materials and most conveyors, the forming of the belt into a trough provides the benefit of a generous increase in the belt's carrying capacity (**Figure 6.11**).

Typical Trough Angles

The standard trough angles in Europe are 20, 30, and 40 degrees; in North America, trough angles of 20, 35, and 45 degrees are common (**Figure 6.12**). However, with an ever-increasing global economy, one might find conveyors of any trough angle in any location around the world. At one time, the 20-degree trough was standard, but the deeper troughs have become more common as improvements in belt design and construction allow greater bending of belt edges without premature failure. In some special applications, such as high-tonnage mining, catenary idlers with a 60-degree trough are used to reduce spillage and impact damage.

Longer, higher speed conveyors may require the use of a thick belt, often with steel cords in the carcass. As a result, these belts may have less trough capability. Because of the lower bend requirements and the resulting reduction of stress in the belt,

Figure 6.9

Materials with a high angle of repose can be carried on flat belts. For materials with a low angle of repose, carrying the same volume of material would result in spillage off the belt; therefore, materials with a low angle of repose generally require the belt to be troughed.

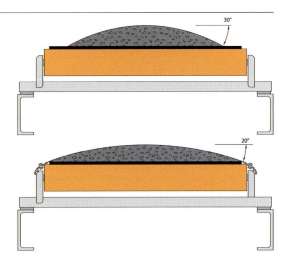

Figure 6.10

Some belt feeders incorporate a dual wall system with space between the wear liner and the outside chutewall.

Figure 6.11

Forming the belt into a trough generally increases the capacity of the conveyor.

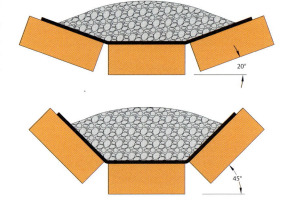

a 20-degree trough permits the use of the thickest belts, thereby allowing the heaviest materials and largest lumps to be carried.

Troughing angles steeper than 20 degrees are usually specified when the material has a low angle of repose. Higher troughing angles are suitable for a very broad range of applications. Higher trough angles work best when allowances are made for constraints such as limitations in transition distance and the requirement for exposed edge distance for skirt sealing.

While they offer the benefit of greater capacity, troughed belts also present some limitations. An area that can present problems, if not properly considered during the conveyor design and belting specification process, is the longer transition distance required to prevent stress at the belt edges. Other disadvantages of the steeper trough angles include a greater vulnerability to the effect of wind and higher potential for damage to the belt.

Troughing the belt typically makes a positive contribution to belt tracking. Another benefit of a troughed belt includes improved ability to contain material as a result of reduced edge spillage and loss due to wind.

Generally speaking, the selection of the angle of trough to be used on any conveyor is in most cases determined by the requirement to use the least expensive, and hence the narrowest, belt possible to transport the required tonnage of material.

FORMING THE TROUGH

Transition

On a typical conveyor, the belt is troughed for the carrying portion of its journey and returned to a flat configuration for the return run. Consequently, at a terminal (head or tail) pulley, the belt must be converted from flat to troughed shape or from troughed back to flat. This changing of the belt's profile is commonly called the transition (**Figure 6.13**). Transitions exist

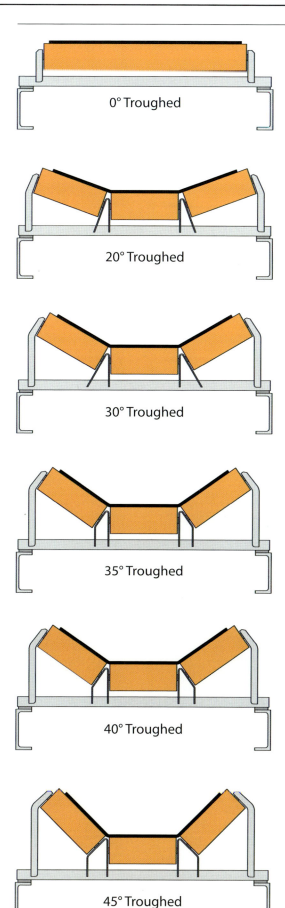

Figure 6.12

The standard angles for conveyor trough vary around the world.

0° Troughed

20° Troughed

30° Troughed

35° Troughed

40° Troughed

45° Troughed

at the tail (loading) and head (discharge) pulley locations of a troughed conveyor and can occur in other areas of the conveyor, such as at a tripper head.

The distance from the centerline of the terminal pulley to the first fully-troughed

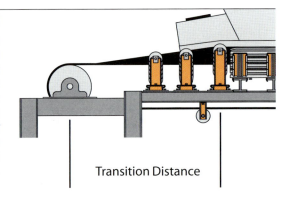

Figure 6.13

The distance from the terminal (tail or head) pulley to the first fully-troughed idler is called the transition distance; the belt is changed from flat to troughed or from troughed to flat in this area.

Transition Distance

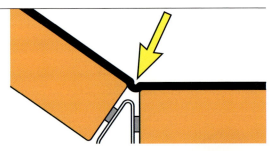

Figure 6.14

Too short a transition distance can result in creases at the junction where the center roller joins the wing rollers.

Figure 6.15

The damage caused by junction-joint failure appears in two parallel lines above the joint between the center roller and wing rollers.

Figure 6.16

One sign of junction-joint failure can be the belt assuming a "W" or "M" configuration.

idler is called the transition distance. This area poses more potential risk to the belt than any other area of the conveyor. In changing from a flat belt to a fully-troughed profile, tension at the sides of the belt is greater than at the center. The outer one-third of the belt must stretch and travel farther than the center one-third. This can cause the splice, either mechanical or vulcanized, to fail at the belt edges. In addition, the plies of the belt can separate due to this stress.

The transition distance, the spacing allowed for this change in the belt's contour, must be sufficient at each terminal pulley. Otherwise, the belt will experience extreme stress in the idler junctions (the points on a troughed idler set where the horizontal roller meets the inclined rollers). Because the outer third of the belt is stretched farther, the result may be a crease in the idler junction that may eventually tear along the entire length of belt (**Figure 6.14**). Also, if the elasticity of the carcass is slightly exceeded, the belt may not tear but rather stretch beyond its limits, leading to belt-tracking problems. If the transition distance is too short, an excessive difference between edge and center tensions can overcome the belt's lateral stiffness. This can force the belt down into the trough, so it buckles through the center or catches in the idler junctions where the rollers of the idler join (**Figure 6.15**). The first sign of idler-junction failure will be noticed as a "W" or "M" fold or shape in the belt's return side (**Figure 6.16**). The increased edge tension seen in the belt from having too short a transition area will place an increased load on the outer bearings of the troughing idlers and could lead to premature idler failure.

Belt tension can be kept within safe limits by maintaining the proper transition distance between the pulley and the first fully-troughed idler, thus minimizing the stress induced in the belt.

To properly support the belt at these transitions, idlers with intermediate angles should be used between the terminal pulley and the first fully-troughed idler. These transition idlers will allow the belt to gently change its profile to the proper trough angle. Strain on the belt at the idler junction is then minimized since it has been spread over several idlers and a greater distance.

Transition Distance

The distance required for a belt's transition varies with the amount of troughing required, the belt thickness, the construction of the belt, the type of carcass (steel cable or fabric), and the rated tension of the belt. A transition distance must be selected to provide at least the minimum distance for the belt selected.

The heavier the belt carcass, the more it will resist being placed in a troughed configuration and the longer the required transition distance. This is easy to understand if one remembers that a string stretched down the center of the conveyor will be shorter than the string placed on the outside edge of the idlers. The outer edges of the belt must travel farther than the middle of the belt. The higher the trough angle, the more the edges are stretched and the greater the distance required to reach that angle.

The transition distance required is a function of the construction of a belt. When engineering a new conveyor, the belting should be selected to match the material load and conveyance length characteristics of the conveyor. The transition distance of the system would then be designed to match the requirements of the selected belting. However, a more likely scenario is that, due to space constraints and cost

considerations, the belting will be selected to match the transition distance engineered into the steel conveyor structure. Either way, however, the belting manufacturer should be consulted when determining the recommended transition distance.

In the case of replacement belting for existing conveyors, the belt should be selected to match the transition distance provided in the conveyor structure. In no case should a belt be placed on a conveyor where the transition distance is too short for the belt.

It is highly recommended that the supplier of the belt be contacted to ensure that the transition distance of the existing structure is compatible with the belt. Charts identifying the recommended transition distance as a function of the rated belt tension for both fabric and steel cord belts at the various trough angles are published in manufacturers' literature and by CEMA in *BELT CONVEYORS for BULK MATERIALS, Sixth Edition*.

Transition Idlers

Depending on the distance, one or more transitional idlers should be used to support the belt between the terminal pulley and the first fully-troughed idler.

It is good practice to install several transition idlers supporting the belt to gradually change from a flat profile to a fully-troughed contour (**Figure 6.17**). Transition idlers can be manufactured at specific intermediate angles (between flat and fully troughed), or they can be adjustable to fit various positions (**Figure 6.18**). For example, it would be good practice to place a 20-degree troughing idler as a transition

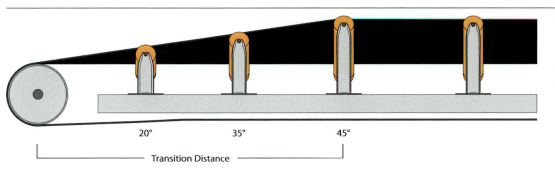

Figure 6.17

Several transition idlers should be installed between the pulley and the first fully-troughed idler.

idler forward of a 35-degree troughing idler, and both a 20-degree and a 35-degree idler in front of a 45-degree idler. CEMA recommends that all transition idlers use metal rollers.

It is also important to the stability of the belt and the sealability of the transfer point that the transition idler closest to the terminal pulley be installed so the top of the pulley and the top of the idler's center roll are in the same horizontal plane. This is referred to as a full-trough transition.

Half-Trough Pulley Depth

To shorten the required transition distance, the conveyor designer may be tempted to use a "half-trough" transition arrangement which calls for the raising of the tail pulley. By elevating the pulley so its top (where the belt comes off the pulley) is in line with the midpoint of the wing rollers (rather than in line with the top of the center roller), the required transition distance can be reduced roughly in half (**Figure 6.19**). This technique is usually employed to shorten the transition distance to avoid an obstruction or to save a small amount of conveyor length.

In the past, this half-trough arrangement has been accepted by CEMA and belting manufacturers as a way to avoid excessive strain at the idler joint as a belt transforms, particularly when fitting a conveyor into a limited space. However, problems can arise with the half-trough design, including the belt lifting off the idlers when it is traveling unloaded (**Figure 6.20**). While a half-troughed belt is being loaded, peaks and surges in the rate of material flow will dramatically change the belt line and prevent the transfer point from being effectively sealed. These shifts in the belt line create a "pumping action" that acts as a fan to push out airborne dust. In addition,

Figure 6.18

Some idler frames are manufactured with multiple mounting holes for the wing rollers to allow their use as transition idlers.

Figure 6.19

In the half-trough design, (bottom), raising the tail pulley so the belt is in line with the midpoint of the wing rollers allows the transition distance to be shortened. However, this can create other problems.

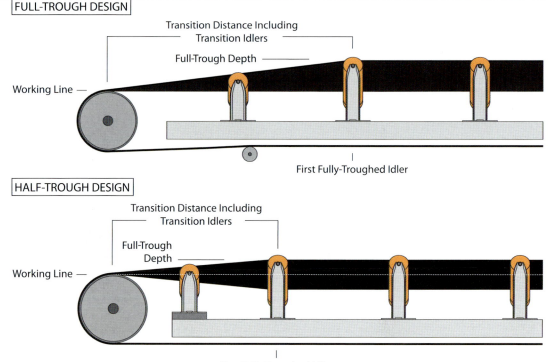

FULL-TROUGH DESIGN

Transition Distance Including Transition Idlers

Full-Trough Depth

Working Line

First Fully-Troughed Idler

HALF-TROUGH DESIGN

Transition Distance Including Transition Idlers

Full-Trough Depth

Working Line

First Fully-Troughed Idler

this design can cause the belt to buckle in the transition area. Loading the belt when it is deformed in this manner makes effective sealing impossible and increases belt wear due to increased levels of impact and abrasion on the "high spots" in the belt.

Solving the problems created by a half-trough pulley depth is more complicated than merely lowering the tail pulley to line up with the center roll of the idlers. The minimum transition distance must be maintained: As the pulley moves down, it must also be moved further away from the first fully-troughed idler. If this is not possible, other modifications should be made, such as the reduction of the trough angle in the loading area to shorten the required transition distance. The belt can then be changed to a higher trough angle outside the load zone. Another approach would be to adopt a very gradual transition area. Both of these techniques are discussed below in Advanced Topics.

A better practice, and the current CEMA recommendation, is to use the fully-troughed arrangement where the pulley is in line with the top of the center roll in the idler set. This requires a longer transition distance, but greatly improves the stability of the belt as it enters the loading zone and, as a result, improves the sealability of the transfer point.

Loading in the Transition Area

Loading the belt while it is undergoing transition is bad practice and should be avoided. The area where the load is introduced to the belt should begin no sooner than the point where the belt is fully troughed and properly supported by a slider bed or by the midpoint of the first set of fully-troughed idlers. A better solution is to introduce the load 300 millimeters (12 in.) or so beyond this fully-troughed point in order to accommodate any bounce-back of material caused by turbulence.

If loading is performed while the belt is still transitioning into the troughed configuration, the load is dropped onto a slightly

larger area with non-parallel sides. This larger area increases pressure on the side skirts and increases wear on the belt and liners as the belt forms its full trough. In addition, since the belt in the transition area is changing in contour, it will not have the stable belt profile required for effective sealing.

Material bouncing off other material and the walls of the chute can deflect behind the intended load point. Therefore, the trajectory of the material needs to be designed to contact the belt far enough in front of the tail pulley to prevent material from flooding the transition area. The provision of adequate belt support in the loading zone ensures that the belt maintains the flat surface that is critical for effective sealing.

SEALING AT THE ENTRY AREA

Sealing Systems

Sealing of the belt entry in the load zone is often a problem (**Figure 6.21**). The turbulence of material as it is loaded can cause some particles to bounce or roll backward toward the tail of the conveyor.

Figure 6.20

Using the half-trough design can cause the belt to rise off the idlers (particularly when unloaded).

Figure 6.21

Providing an effective seal at the back of the loading zone can be a problem.

6

Material will bounce back out of the load zone and roll down the conveyor, accumulating on the pulley, the bearing supports, or on the floor near the tail pulley.

Figure 6.22

Simple back seals of rubber or plastic have generally proven ineffective.

Figure 6.23

If the tail seal is forced too tightly against the belt, a "belt cleaning effect" will scrape material off the belt to pile up at the back of the load zone.

Figure 6.24

If the conveyor is inclined, material will roll down the belt and pile up on the floor.

Figure 6.25

Sealing at the corners of the loading zone is difficult, allowing material to spill and dust to leak out.

In an attempt to solve this problem, a sealing system of some sort is applied at the back of the loading chute. Typically, this seal is a curtain or wall, fabricated from a sheet of plastic or rubber (**Figure 6.22**). This seal can create as many problems as it solves.

If the seal at the belt entry into the chutework is applied too loosely, material will continue to escape out the back of the loading zone, down the transition area, and onto the floor. If a sealing system is placed tightly enough against the belt to prevent leakage out the back of the loading zone, the seal may instead act as a belt cleaner. In this instance, the seal will scrape any material that has remained adhered to the belt during the journey back from the discharge pulley. The material removed by this "belt cleaner effect" will then accumulate at the point where the belt enters the loading zone (**Figure 6.23**) or, if the conveyor is inclined, roll down the belt to pile up at the tail of the conveyor (**Figure 6.24**).

Sealing the corners of the chute behind where the belt is loaded is difficult due to high material pressures and significant air movement that can carry dust out of the transfer point (**Figure 6.25**). This difficulty in sealing the entry area is compounded by any dynamic vibrations in the belt line created by fluctuations in belt tension resulting from "peaks and valleys" in material loading or from the use of a wing-type tail pulley. Wing pulleys should be avoided for this reason.

Multiple-Barrier Sealing Box

An effective approach is to seal the area behind the load zone with a multiple-barrier sealing box (**Figure 6.26**). Attached to the back wall of the loading chute, this box connects the chute to the area where the belt is flat as it crosses the tail pulley. In the ideal situation, the tail box would be extended to the flat surface of the tail pulley making sealing more effective and easier to maintain (**Figure 6.27**). A tail-sealing box is often installed on the transition area when retrofitting existing conveyors. For new conveyors, this is not recommended, because it is difficult to seal the transition bend.

A sealing strip is installed on the outside of the back wall, the wall closest to the tail pulley, of the sealing box (**Figure 6.28**). Deflected by and in the direction of belt motion, this strip forms a one-way seal that prevents material escape out the back of the loading zone and off the tail of the conveyor. As this strip lies on the belt with only gentle pressure, it avoids the belt cleaner effect. Material adhering to the belt can pass under the seal without being "cleaned" off the belt.

On its sides, the box should be fitted with a low-maintenance, multiple-layer skirtboard seal to prevent material spillage over the edges of the belt. The tail-sealing box should incorporate the beginning of the conveyor, so the sealing strip runs continuously from the tail-sealing box to the exit end of the skirtboard (**Figure 6.29**). This continuous seal eliminates the problems with sealing the high-pressure corners of the impact zone.

The top of the tail-sealing box should include an access door to allow the return of any fugitive material to the conveyor (**Figure 6.30**).

TYPICAL SPECIFICATIONS

A. Transition

The conveyor design should incorporate sufficient transition distance and transition idlers to allow the belt to be fully troughed before any material is loaded onto the belt.

B. "Full trough" height

The tail pulley should be placed at a "full trough" height, so that the belt coming off the pulley is in line with the center roller of the troughed idler set.

C. Tailgate-sealing box

A tailgate-sealing box with an effective seal at the tail pulley end of the enclo-

Figure 6.26

Tail-sealing boxes are installed to prevent material from falling out the back of the conveyor loading zone.

Figure 6.27

A tail-sealing box is often installed on the transition area when retrofitting existing conveyors. For new conveyors, this is not recommended because it is difficult to seal the transition bend.

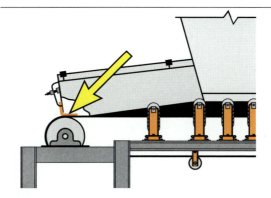

Figure 6.28

A sealing strip should be installed on the outside of the back wall of the sealing box, closest to the tail pulley.

Figure 6.29

The belt edge-sealing systems should extend in a continuous strip from tail-sealing box to the end of the skirted area.

Figure 6.30

An access door in the tail-sealing box will allow any spilled material to be returned to the conveyor.

sure should be installed on the fully troughed belt to prevent the escape of fugitive material from the back of the load zone.

D. Skirtboard-sealing strip

Effective sealing on the belt edges at the sides of the tail-sealing box will be provided by extending the skirtboard-sealing strip from the box through the transfer load zone as a single continuous sealing strip, without a joint or seam that might leak material.

ADVANCED TOPICS

Two-Stage Transition Areas

For many years, the recommendation has been that the belt should be fully troughed before the load is introduced. A variation on this thinking is the idea that it is more critical that the belt be stable (e.g., not undergoing transition) when it is loaded than that it be fully troughed when it is loaded. Given a conveyor where there is a very short distance between the tail pulley and the loading zone, it may be better to partially trough the belt in the area between the tail pulley and the load zone, and then complete the transition after the belt has been loaded (**Figure 6.31**). In order for this to provide any benefit, the belt line must be stabilized with improved support structures, such as impact cradles and side-support cradles, and the edges must be effectively sealed after the belt's initial troughing. Raising the belt edges to the final trough angle can be performed after the load is on the belt.

For conveyors with inadequate space in the traditional transition area between the tail pulley and the load chute, this method provides the benefit of a higher troughing angle without creating the instability of loading while the belt is undergoing transition.

Gradual Transitions

Another method to deal with the problem of too short a transition distance is the use of a very gradual transition. Rather than risk damage by troughing the belt too quickly, the belt is troughed over an extended distance, the length of the load zone. This makes the change so gradual as to be almost unnoticeable.

In one case, the belt profile was changed from flat to a 35-degree trough over a 12-meter (40-ft) transition area. This conveyor was perhaps a special circumstance with a long transfer point incorporat-

⚠ SAFETY CONCERNS

Because of the numerous pinch points in the area of the conveyor tail, employees can easily become caught in the moving belt at this location. In addition, with this area contributing to spillage problems, shoveling accidents are likely to occur. Proper lockout / tagout / blockout / testout procedures should be used prior to performing any maintenance or cleanup work in this area. Employees should never work on or shovel onto a moving conveyor.

Proper guards and safety labels on all rotating equipment and pinch points are required; guards should not be removed or disabled while the conveyor is in operation.

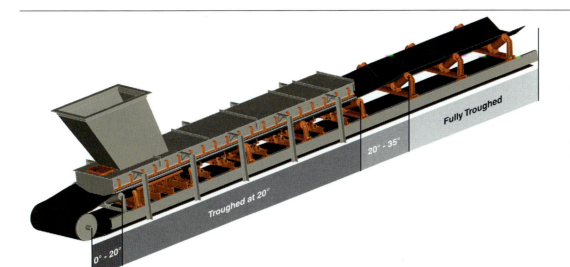

Figure 6.31

In a two-stage transition area, the belt is partially troughed before it is loaded. After the cargo is placed, the belt goes through another transition that increases the trough angle.

6

ing multiple load zones, a thick belt, and minimal distance between the tail pulley and the first loading zone. The key for this technique is to maintain belt support to provide a straight line for the trough. Rather than use specially designed components, this gradual troughing change was accomplished by installing conventional components in a "racked" or slightly out-of-alignment fashion. Belt-support cradles incorporating sufficient adjustment to accommodate a deliberate "out-of-alignment" installation were used in combination with troughing idlers with adjustable angles.

PROTECT YOUR TAIL

In Closing...

While the tail section of a conveyor belt is relatively simple, and the components are generally taken for granted, this section of the conveyor could be one of the most important. If care is not taken in alignment, tail-pulley type, transition distance, and sealing, the negative effects can downgrade the performance of the entire conveyor system.

Looking Ahead...

This chapter, Before the Loading Zone, began the discussion of Loading the Belt by examining tail pulleys and transition areas along with techniques to prevent the escape of fugitive material at the tail of the conveyor. The following chapters will address other aspects of belt loading, beginning with Air Control.

REFERENCES

6.1 Conveyor Equipment Manufacturers Association (CEMA). (2005). *BELT CONVEYORS for BULK MATERIALS, Sixth Edition*. Naples, Florida.

6.2 The website http://www.conveyor-beltguide.com is a valuable and non-commercial resource covering many aspects of belting.

6.3 Any manufacturer and most distributors of conveyor products can provide a variety of materials on the construction and use of their specific products.

7

Figure 7.1

A key in controlling the amount of dust that escapes from the transfer point is to minimize and control the flow of air as it passes through the transfer point.

Chapter 7

AIR CONTROL

In this Chapter...

In this chapter, we look at the importance of controlling air movement to control the escape of airborne dust. Equations to calculate the amount of displaced air, induced air, and generated air are presented, along with their relationship to the total airflow. Methods to measure air velocity and volume are also discussed. Four design parameters, along with several additional techniques, for controlling the air to minimize airborne dust are explained. Finally, maintenance and safety concerns are included.

Conveyor loading zones and discharge points are prime sources for the creation and release of airborne dust. The amount of dust created in a transfer point depends on a number of factors, including the nature of the material carried, the height of drop onto the belt, and the speeds and angles of unloading and loading belts. A key in controlling the amount of dust that escapes from the transfer point is to minimize and control the flow of air as it passes through the transfer point (**Figure 7.1**).

DUST AND AIR MOVEMENT

To Control Dust, Control Air

As materials move on a conveyor and through the transfer point, they carry a stream of air in and with them. With sufficient velocity, this air stream can pick fine particles out of the material body and carry them along with the materials, or it can spread them outside the enclosures of the conveyor.

The conditions that determine whether or not fine materials become airborne are air velocity, particle size, and cohesion of the bulk materials. These characteristics contribute to the amount of dust generated by the following intuitive, relative relationship: The amount of dust generated is proportional to air velocity, as divided by the factors of particle size and material cohesiveness (**Figure 7.2**). Where one or more of these parameters is a given, the ability to control dust depends on altering one or both of the other characteristics. If air velocity is increased, but particle size and cohesiveness remain constant, then airborne dust will increase. If air velocity remains constant, and particle size or cohesiveness is increased, the amount of airborne dust will be reduced. If velocity remains constant, and particle size or cohesiveness is decreased, then the amount of airborne dust will increase.

When the size of particles being conveyed cannot be changed, the velocity of the air or the cohesive force of the particles must be altered in order to minimize the emission of dust. *(See Chapter 19: Dust Suppression for information about changing the cohesive force of particles.)* Control of the air movement into and out of a conveyor transfer point will not reduce the dust created inside that transfer point, but it will have a significant effect on the amount of dust that is carried out of the transfer point. Limiting the positive pressure released by a transfer point will have significant benefits in the control of fugitive materials.

Air Movement through Transfer Points

The volume of air that moves through a transfer is directly related to the size of the transfer-point enclosure, the openings in the enclosure, and the presence of other process equipment. The cost of the components of a conveyor's dust-management system is directly related to the volume of air that is pulled through the system. Therefore, an understanding and control of air movement is fundamental to efficient and economical dust control through the design of the transfer point.

$$\frac{\text{Dust}}{\text{Generated}} \; \alpha \; \frac{\text{Air Velocity}}{\text{Particle Size} \bullet \text{Cohesiveness}}$$

Figure 7.2

Relationship in Creating Airborne Dust

Ideally, a slight negative pressure is to be desired inside the enclosure. This condition would pull air into the enclosure so that fines and airborne dust are retained in the structure, rather than carried out. Typically, this is difficult, if not impossible, to achieve consistently without an active dust-collection system. The airflow, created by the equipment above the transfer point and the movement of materials through the transfer point, creates a positive pressure through the system, creating an outward flow of air from the transfer point. This is most true in a conveyor loading zone, because the impact of materials on the receiving belt drives out the air with a significant "splash." The greater the impact, the stronger the air current away from the impact area will be. If this positive pressure is not addressed with control of material flow, adequate pressure relief, or dust-collection systems, the particles of dust will be carried out of the transfer point on the outward flow of air.

DISPLACED, INDUCED, AND GENERATED AIR

Calculating the Airflow

Airflow can be measured or calculated. The following is offered as a theoretical, yet workable method. The conditions of any specific combination of conveyor design and material flow can affect the results significantly.

There are three sources for the air movement that might be present around or through a given transfer point: displaced air, induced air, and generated air.

The total airflow in a given transfer point can be calculated by adding the displaced air, induced air, and generated air (**Equation 7.1**).

Displaced Air

The first category is displaced air. A simple explanation of displaced air starts with a coffee cup. When coffee is poured into this cup, the air inside is displaced by the coffee. This same effect occurs when materials enter a loading chute: The air that filled the chute is pushed out, displaced by the materials. The amount of air displaced from the chute is equal to the volume of materials placed into the chute. The movement of materials through a transfer point will always produce displaced air that can be calculated by the load (amount of materials conveyed) and bulk density (**Equation 7.2**).

Induced Air

Induced air is present in conveyor loading zones whenever bulk materials are moving, because bulk materials have a certain amount of entrapped air and carry a small amount of air with them as they travel on the belt. As the materials leave the head pulley in a normal trajectory, the material stream expands, pulling air into

Equation 7.1
Total Airflow Calculation

$$Q_{tot} = Q_{dis} + Q_{ind} + Q_{gen}$$

Given: A transfer point is attached to a crusher that generates 0,77 cubic meters per second (1625 ft³/min) of air. The displaced air is 0,06 cubic meters per second (133 ft³/min) and the air induced is 0,055 cubic meters per second (117 ft³/min). **Find:** The total air movement.

	Variables	Metric Units	Imperial Units
Q_{tot}	Total Airflow	cubic meters per second	cubic feet per minute
Q_{dis}	Displaced Air	0,06 m³/s	133 ft³/min
Q_{ind}	Induced Air	0,055 m³/s	117 ft³/min
Q_{gen}	Generated Air (If Present)	0,77 m³/s	1625 ft³/min
Metric: $Q_{tot} = 0{,}06 + 0{,}055 + 0{,}77 = 0{,}885$			
Imperial: $Q_{tot} = 133 + 117 + 1625 = 1875$			
Q_{tot}	Total Airflow	0,885 m³/s	1875 ft³/min

the new voids. Each particle of material gives energy to an amount of air, pulling the air along with the material stream. When the product lands and compresses back into a pile, this induced air is released, causing substantial positive pressure flowing away from the center of the load zone. If this positive pressure is not addressed with proper transfer-point design or relief systems, the dust fines will be carried out of the system on the current of air.

An example of induced air would be when one turns on the water in a shower, the stream of water from the shower head spreads out. This moving water pulls a volume of air along with it. This air current can be noticed in the movement of the shower curtain towards the flow of water.

The factors influencing the amount of induced air at a conveyor transfer point include the quantity of materials, the size of the particles of conveyed materials, the speed of the belt, the height of the material drop, and the size of the opening(s) in the head chute which allow air to be taken into the enclosure. Induced air can be calculated from these factors (**Equation 7.3**).

The most controllable factor in controlling induced air is the size of the opening in the head chute (A_u) through which the air induction occurs. The smaller the opening(s) for air to enter the system, the smaller will be the value of A_u and the smaller the volume of air that will escape or need to be exhausted. (Note: A_u is the size of the belt entry to the head chute enclosure, rather than the size of doors downstream at the loading zone or on the skirted section of the receiving conveyor.)

An easy and cost-effective method to reduce the amount of induced air is to reduce the size of all openings in the head chute. This would include sealing the open areas where the belt enters and leaves the head chute, as well as placing seals on pulley shafts and putting doors over belt cleaner and other inspection openings.

Generated Air

Other sources of moving air are those devices that feed the conveyor load zone. This includes equipment such as crushers, wood chippers, hammer mills, or any device with a turning motion that creates a fan-like effect, pushing air into the transfer point. While not present in all transfer points, this generated airflow can be the most severe of all air movements.

Other devices that must be considered, if present, would be air cannons, vibrators,

---

$$Q_{dis} = \frac{k \cdot L}{\rho}$$

Equation 7.2

Displaced Air Calculation

Given: A transfer chute is carrying 180 tons per hour (200 st/h) of material that is 800 kilograms per cubic meter (50 lb$_m$/ft³). **Find:** The displaced air.

Variables		Metric Units	Imperial Units
Q_{dis}	Displaced Air	cubic meters per second	cubic feet per minute
L	Load (amount of material conveyed)	180 t/h	200 st/h
ρ	Bulk Density	800 kg/m³	50 lb$_m$/ft³
k	Conversion Factor	0,277	33.3

Metric: $Q_{dis} = \dfrac{0,277 \cdot 180}{800} = 0,062$

Imperial: $Q_{dis} = \dfrac{33.3 \cdot 200}{50} = 133$

Q_{dis}	Displaced Air	0,062 m³/s	133 ft³/min

and compressed-air hoses, used to promote material flow. This type of air movement can be measured using air velocity and volume gauges, such as Pitot tubes and manometers. It might be simpler for the end-user to contact the equipment's manufacturer to obtain a calculation of the air output by various pieces of equipment. Depending upon the crusher, manufacturers estimate the air generated for various types (**Table 7.1**).

Since generated air can be a significant amount, the amount of generated air should be obtained from the manufacturer of the equipment, or the flow can be calculated by multiplying the exhaust area times the air velocity, measured while the equipment is in operation.

AIR VELOCITY AND VOLUME

Air Velocity

Air flows from a high- to a low-pressure zone because of the pressure difference. While there are a number of variables that can cause dust to remain in the material steam—including particle sizes, material cohesiveness, and moisture content—in general, dust particles have a pickup velocity in the range of 1,0 to 1,25 meters per second (200 to 250 ft/min). That means that air moving over a bed of material at this speed can pick up dust off the surface and carry it away.

A good design parameter for the sizing of the exit chutes of load zones is to maintain the exit area air velocity below

---

Equation 7.3

Induced Air Calculation

$$Q_{ind} = k \cdot A_u \cdot \sqrt[3]{\frac{RS^2}{D}}$$

Given: A transfer chute is carrying 180 tons per hour (200 st/h) and has an open-end area of 0,046 square meters (0.5 ft²). The material with average diameters of 0,075 meters (0.25 ft) falls 1,25 meters (4 ft). **Find:** The induced air.

Variables		Metric Units	Imperial Units
Q_{ind}	Volume of Induced Air	cubic meters per second	cubic feet per minute
A_u	Head Chute Open Area	0,046 m²	0.5 ft²
R	Rate of Material Flow	180 t/h	200 st/h
S	Height of Material Free Fall	1,25 m	4 ft
D	Average Material Diameter	0,075 m	0.25 ft
k	Conversion Factor	0,078	10

Metric: $Q_{ind} = 0,078 \cdot 0,046 \cdot \sqrt[3]{\dfrac{180 \cdot 1,25^2}{0,075}} = 0,055$

Imperial: $Q_{ind} = 10 \cdot 0.5 \cdot \sqrt[3]{\dfrac{200 \cdot 4^2}{0.25}} = 117$

Q_{ind}	Volume of Induced Air	0,055 m³/s	117 ft³/min

Table 7.1

Approximate Levels of Generated Air from Various Types of Crushers		
Type of Crusher	**At Crusher Feed per 300 millimeters (12 in.) Opening Width**	**At Crusher Discharge per 300 millimeters (12 in.) Conveyor Width**
Jaw Crusher	850 m³/h (500 ft³/min)	850 m³/h (500 ft³/min)
Gyratory Crusher	850 m³/h (500 ft³/min)	1700 m³/h (1000 ft³/min)
Hammermills and Impactors	850 m³/h (500 ft³/min)	2550 m³/h (1500 ft³/min)

1,0 meters per second (200 ft/min). Higher velocities may allow the air current to pick up particles of material and hold them in suspension in the air, making containment, collection, or suppression more difficult.

Checking Air Velocity and Volume

The quantity of air flowing through the transfer point each minute can be calculated from measurements (**Equation 7.4**). To calculate the volume of moving air, multiply the measured air speed leaving each open area of the transfer point—including the belt exit, tail box, sides of the belt, dust pickups, and other openings—by the area of each opening. These air flows are then added to produce a total air flow. These measurements must be taken while the transfer point is in operation. The air velocity measurements can be performed with a relatively inexpensive handheld anemometer; the area can be measured with a tape measure (**Figure 7.3**).

As additional airflow through the transfer-point enclosure can be produced by crushers, vibrating screens, feeders, and other process and handling equipment, it will be necessary to measure the air velocity while these devices are in operation as well.

This air volume calculation should be compared to the computations of the air volume (**Equation 7.1**). If a major discrepancy exists, the airflow calculated from measured air velocity (**Equation 7.4**) should ALWAYS be used.

CONTROLLING THE AIR

Controlling Air Movement

A complete system to control dust at conveyor transfer points is based upon four design parameters:

A. Limit the amount of air entering the enclosure

Preventing air from entering the enclosure at the head pulley of the discharging conveyor is possible without sophisticated or expensive changes. Conventional rubber curtains can be installed at the belt's entrance and exit, and other openings, such as around the pulley shafts, can be sealed. Perhaps the easiest thing that can be done to limit the intake of air at the discharge end of conveyors is to make sure inspection doors are all closed.

Figure 7.3

Air measurements can be performed with a relatively inexpensive handheld anemometer and a ruler.

$Q_{tot} = A \cdot V$			

Equation 7.4

Air Quantity Calculation

Given: The air velocity coming out of a transfer point is measured at 4,3 meters per second (850 ft/min). The transfer-point enclosure has a total cross-sectional area of 0,19 square meters (2 ft²).
Find: The total airflow.

	Variables	Metric Units	Imperial Units
Q_{tot}	Total Air Movement	cubic meters per second	cubic feet per minute
A	Cross-Sectional Area of Transfer-Point Enclosure	0,19 m²	2 ft²
V	Velocity of Air	4,3 m/s	850 ft/min
Metric: $Q_{tot} = 0{,}19 \cdot 4{,}3 = 0{,}81$			
Imperial: $Q_{tot} = 2 \cdot 850 = 1700$			
Q_{tot}	Total Air Movement	0,81 m³/s	1700 ft³/min

B. Limit the spreading of the material stream

As it moves through the transfer point, each particle or lump of material acts on the air in the enclosure, carrying some of the air along with it. Keeping the materials in a consolidated stream as they leave the head pulley and move through the transfer point can be done with deflectors or engineered hoods and spoons. A deflector may create material flow problems, whereas engineered hoods and spoons are less likely to create flow problems. The more materials and the faster the movement, the greater is the need for an engineered chute. *(See Chapter 22: Engineered Flow Chutes.)*

C. Limit the material drop height

In a conventional conveyor discharge, the materials free fall. This disperses the materials, making the stream larger and able to take more air with it, because air fills the voids created within the spreading materials. When the materials land on the next belt, the entrained air is pushed away from the pile, creating a positive pressure. The further the materials fall, the greater the force of landing; hence, the greater will be the outward pressure of air. Limiting the drop height addresses this problem. Limiting the drop height usually involves moving the conveyors closer together. This is an incredibly complicated process to implement once a conveyor is installed; however, it is relatively easy to minimize the drop height in the system design.

D. Limit the air speed inside the enclosure to below the pickup velocity of the dust particles

Conventional conveyor enclosures behave like large ducts moving air. As such, the cross-sectional area of the duct, formed by the conveyor chute and skirtboard, can be increased or decreased to change the velocity of the air flowing through the enclosure. *(See Chapter 11: Skirtboards, especially Advanced Topics for sample problems to determine velocity.)*

Hood and Spoon Systems

Preventing the materials from spreading out when they leave the discharge pulley will significantly reduce the amount of air that is pulled in as induced air. Chutes employing a "hood and spoon" design, to confine the stream of moving materials, reduce the airflow (**Figure 7.4**). The hood minimizes the expansion of the material body, deflecting the stream downward. The spoon provides a curved loading chute that provides a smooth line of descent, so the materials slide down to the receptacle, whether that is a vessel or the loading zone of another conveyor. The spoon "feeds" the materials evenly and consistently, controlling the speed, direction, and impact level of the materials in the load zone. Paradoxically, the design of the hood and spoon depends upon gravity and friction to maintain the speed of the material flow through the chute. In some installations, there may not be sufficient drop height to use this technique to control dust.

By reducing the velocity and force of material impact in the load zone, to approximate the belt speed and direction, this system mitigates splash when materials hit

Figure 7.4

By confining the material stream, a "hood and spoon" chute minimizes the air entrained with the material, so it reduces airborne dust.

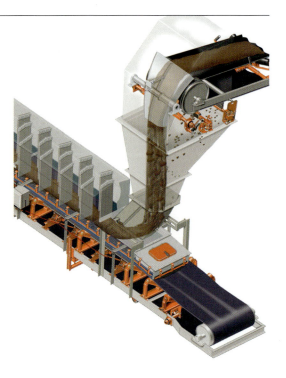

7

the receiving conveyor. Therefore, there is less dust and high velocity air escaping. As the materials are deposited gently on the belt, there is minimal tumbling or turbulence of the materials on the belt. There is less impact in the loading zone, which will reduce impingement damage to the belt. Because there is minimal tumbling or turbulence of the materials and lower side forces, the skirted length can be shorter and sealed more effectively.

Gravity and the flow of materials will tend to keep the hood and spoon from building up and plugging the chute. Sometimes, there is not enough space to include both the hood and spoon in the design. In some cases with free-flowing materials, only a spoon is used to change the direction of the stream, to minimize belt abrasion and skirt side pressure. Spoons are prone to backing up, or flushing, if the characteristics of the bulk materials are variable. Some compensation can be designed into the spoon for variability of materials.

The main perceived drawback to using the "hood and spoon" concept is the price of these specially-designed components. Even so, where they can be applied and maintained, a full cost analysis will show significant cost-saving benefits in reduced dust, spillage, and belt wear.

This "hood and spoon" system works best when the material stream is kept as close as possible to continuous flow. The design minimizes the amount of expansion of the material profile, to reduce induced air and provide consistent flow. As the materials fall, the gravity-induced increase in speed allows the gradual reduction in the cross-sectional area of the chute without increasing the risk of plugs inside the chute. Variations in the rate of materials loading onto the belt may conflict with the design of the ideal hood and spoon, so some compromises in the chute design may be required.

Hood and spoon designs are a typical feature of engineered flow chutes, devel-

oped using material properties and continuum mechanics, and verified through Discrete Element Modeling (DEM) method. The success of this system may well eliminate the need for "baghouse" dust-collection systems in some operations. *(See Chapter 22: Engineered Flow Chutes.)*

Settling Zones

Settling zone is the name used for the covered skirtboard length including, if required, an additional enclosed volume of the load zone after the product has been placed onto the belt. The settling zone is usually an enlarged portion of the covered skirtboard area at the transfer point (**Figure 7.5**). This extra volume slows the air and allows most of the dust to settle and cleaner air to escape.

The size of a settling zone should be determined by six factors: width and speed of the belt, chute width, amount of airflow, depth of the material bed, and diameter of the largest lump of material that may pass through the settling zone. As any one or more of these factors increase, the size of the settling zone must also increase. Calculations to determine the size of a settling zone are for air space only—the area above the cargo. When calculating the cross-sectional area of the chute exit, subtract the area occupied by the body of material to find the area of the settling zone. *(See Chapter 11: Skirtboards for sample problems in calculating the proper size of a skirted area, including settling zone.)*

Figure 7.5

The settling zone is usually an enlarged portion of the covered skirtboard area at the transfer point that slows the air and allows the conveyed product to settle and cleaner air to escape.

In addition to increasing the size of the settling zone, another way to slow down the air in the settling zone is the installation of rubber curtains as baffles. *(See Chapter 18: Passive Dust Control for more information about dust curtains.)*

SYSTEM MAINTENANCE

For effective control of air within (and escaping from) a transfer point, it is important that holes be closed, whether the opening arises from rust or wear or from an opened door. Maintenance of the components, such as wear liners and deflectors, inside the transfer point is critical to minimizing disruption of the materials and airflow.

Because of the need to control air movement—and the resulting dust—to maintain a clean, safe, and productive workplace, many companies outsource maintenance of passive and active dust-control systems to specialty contractors.

TYPICAL SPECIFICATIONS

Typical specifications have been developed for the design of a transfer point's skirted area (including tail seal, skirtboard, and settling zone) as appropriate for control of spillage and air movement. *(See Chapter 6: Before the Loading Zone and Chapter 11: Skirtboards.)*

Figure 7.6

Installation of a piece of old belting as a curtain between the carrying run and the return side can reduce air induction at the belt entry.

ADVANCED TOPICS

Changing Head Chute Open Areas and Drop Heights to Minimize Induced Air

Volume of induced air (Q ind) is a function of the open-end area (A_u), rate of flow (R), drop height (S), and average material diameter (D) (**Equation 7.3**). The open-end area and the drop height are the only things that can realistically be varied. Those two variables have different mathematical impacts on the induced air. A reduction in open-end area of 5 percent will yield a 4.27 percent reduction in induced air. However, a 5 percent reduction in drop height yields only a 3.42 percent reduction in induced air. The cost to reduce open-end area is usually far less than the cost to reduce drop height. This lower cost and higher effect makes reducing open-end area to limit the amount of air flowing into a conveyor belt a priority.

It should be noted that if the open-end area is reduced by 5 percent and the drop height is reduced by 5 percent, the induced air is reduced by 6.84 percent.

Restricting Air Movement at the Entry to the Head Chute

In addition to the techniques already mentioned, another method employed to minimize induced air is to cover the inbound portion of the conveyor for several feet before it enters the head chute. This increases the resistance to air entering the openings and, thus, reduces airflow.

One technique to reduce air induction at the belt entry is the installation of a piece of old belting as a curtain between the carrying run and the return side (**Figure 7.6**). Placed across from one chutewall to the other, this curtain acts as a wall, enclosing the head pulley and reducing air movement.

SAFETY CONCERNS

It is important to follow the established safety rules for personal protective equipment (PPE), confined-space entry, and exposure to dust created by bulk-materials handling in the workplace.

In applications where the danger of explosion or fire exists, the established procedures for minimizing the risk should be followed.

AIR CONTROL ≈ DUST CONTROL

In Closing...

Dust is carried out of a transfer point by the current of air created by the passage of bulk materials through that transfer point (**Figure 7.7**). Although there will be dust created without air currents, the escape of dust will be minimized without a current of air. The more control a transfer point (or an entire operation) establishes over air movement, the more control it will have over the escape of airborne dust.

Looking Ahead...

This chapter, Air Control, is the second chapter in the section of Loading the Belt, following the topics of tail pulleys and transition areas in Before the Loading Zone. The following two chapters continue the discussion in this section of reducing spillage and dust by focusing on material control: Chapter 8 looks at Conventional Transfer Chutes, and Chapter 9 examines Flow Aids.

Figure 7.7

Dust is carried out of a transfer point by the current of air created by the passage of bulk materials through that transfer point.

REFERENCES

7.1 Conveyor Equipment Manufacturers Association (CEMA). (2005). *BELT CONVEYORS for BULK MATERIALS, Sixth Edition*. Naples, Florida.

7.2 Any manufacturer and most distributors of conveyor products can provide a variety of materials on the construction and use of their specific products.

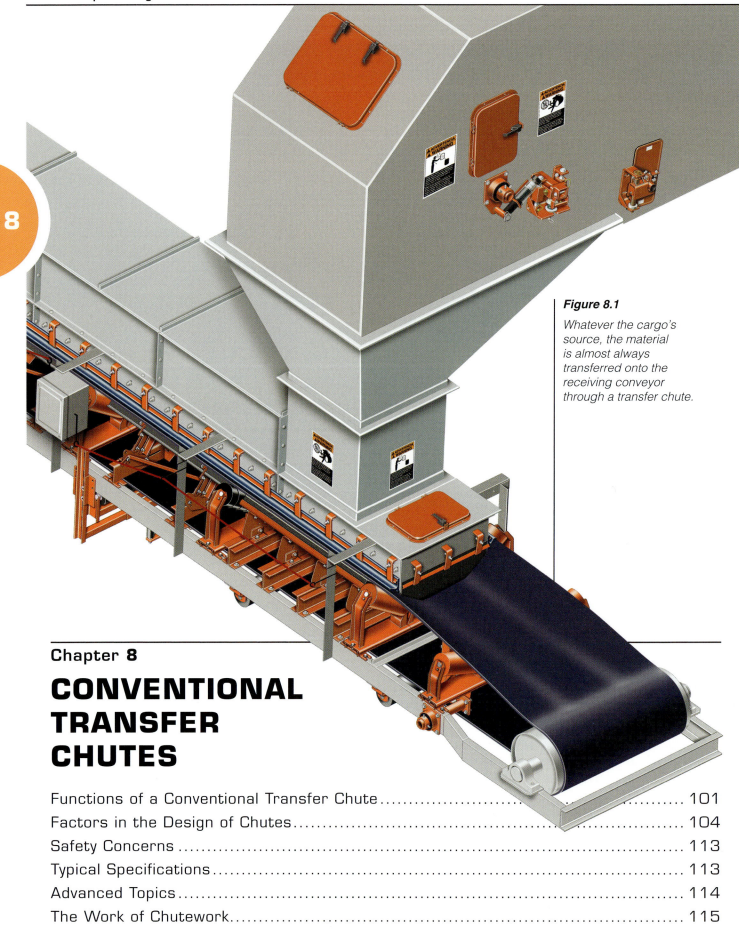

8

Figure 8.1

Whatever the cargo's source, the material is almost always transferred onto the receiving conveyor through a transfer chute.

Chapter 8

CONVENTIONAL TRANSFER CHUTES

In this Chapter...

In this chapter, we focus on conventional transfer chutes: their function, design, and specifications. We discuss a variety of methods that can be used to safely manage material flow, decrease wear, and control airflow to minimize dust and spillage and preserve the life of the chute. An equation for calculating valley angles is also included.

A conveyor receives its cargo from other conveyors, storage containers, feeders, mobile equipment, rail cars, or other materials-handling systems. Although the sources may vary, the materials are almost always transferred to the receiving conveyor through a device called a transfer chute (**Figure 8.1**). This chapter covers conventional transfer chute design.

Because each material and each application has its own characteristics, an effective transfer chute must be more than just a hollow vessel through which material is channeled. A well-designed chute will control the flow path of the material, prevent blockages, and minimize spillage and dust, thereby reducing plant maintenance costs. The designer of an effective chute must take into consideration not only the bulk-material characteristics, which may vary over time, but also the material's interaction with various parts of the overall system.

FUNCTIONS OF A CONVENTIONAL TRANSFER CHUTE

A conventional transfer chute accomplishes its purpose when it achieves the following objectives (**Figure 8.2**):

A. Provide the transfer of the bulk material at the specified design rate without plugging

B. Protect personnel from injury

C. Minimize escape of fugitive materials

D. Return belt scrapings to the main material flow

E. Be service-friendly

Because conveyors usually do not stand alone but are part of complex systems, compromise is often necessary during design. Consequently, these objectives are not absolute requirements but rather the goals for the design of an effective transfer chute.

There are many "rules of thumb" for designing conventional transfer chutes based on experience and engineering principles. Sometimes these rules overlap or conflict. Chute design is a combination of science and art, so it is always wise to consult a conveyor engineer experienced in design systems for specific bulk-materials handling applications. *(See Chapter 22: Engineered Flow Chutes for a discussion about advanced chute design.)*

Transferring the Material

The primary function of a transfer chute is to reliably transfer the bulk material at the specified rate of flow. If the material will not flow reliably through the chute, then meeting any or all other objectives is irrelevant.

Bulk materials should flow through a transfer chute evenly and consistently. A transfer chute that places surges of material onto the conveyor belt poses a number of problems for the conveyor system. Periodic heavy deposits of material on the belt may cause the center of gravity to shift and the belt to track off-center. Surge loading also has the potential to over-stress the components of the conveyor system, particularly the drive motor or the belt-support system, and may lead to plugging problems if the cross-sectional area of the chute is too small.

Figure 8.2

A well-designed conventional transfer chute provides the transfer of the bulk material at the specified design rate without plugging, while minimizing risk to personnel and the escape of fugitive material.

New methods, such as computer-based Discrete Element Modeling (DEM) method, are now available to verify that material will flow reliably. The vast majority of conventional chutes are still designed based on long-used "rules of thumb."

Protecting Personnel

While open transfers are common in some industries such as aggregate and underground mining, the trend in conventional chute design is to enclose the transfer point as much as possible from the discharge pulley to some distance along the receiving conveyor. Simply enclosing the transfer point is an effective way to contain the bulk material, reduce the escape of fugitive materials, limit noise, and prevent the exposure of personnel to the conveyor's numerous pinch points.

Minimizing the Escape of Fugitive Materials

The size of the enclosure is often based on the space available, which can lead to a less than desirable design. The transfer chute should be large enough to allow any service that might be required. It should also be large enough to reduce dust emissions by allowing sufficient volume to reduce the positive pressure and the velocity of the air flowing in and through the transfer.

There are a number of interrelated design elements that affect the creation of fugitive materials in the form of dust and spillage. A key factor in reducing material escape is the placement of the cargo in the center of the belt.

Off-center loading—placing the cargo predominantly on one side of the belt—is a problem at many transfer points that contributes to generation of fugitive materials (**Figure 8.3**). The problem is most common on non-linear transfer points, where the material's direction of travel is changed. Off-center loading can also be found on in-line transfer points, where material has accumulated within the transfer chute or when changes in material characteristics (such as moisture content, particle size, or speed) have altered the material's trajectory, resulting in material being piled deeper on one side of the receiving belt. This displacement causes tracking problems and may result in spillage over the edge of the belt outside the transfer point (**Figure 8.4**).

Although the ideal is to design a transfer chute to prevent the problems associated with off-center loading, there are solutions that can be implemented within the loading zone to compensate for it. Training idlers and other belt-aligning systems are limited in their ability to counter the effects of off-center loading. Installation of corrective measures, such as deflectors or flow aids within the loading zone, in combination with belt-aligning systems, provides an effective approach. *(See Chapter 16: Belt Alignment for more information.)*

A number of fixtures—such as deflectors, liners, baffles, shapers, screens, grizzly bars, or rock boxes—can be placed within the transfer chute to help direct the flow of material and provide a balanced loading pattern; they are discussed later in this

Figure 8.3

Off-center loading—placing the cargo predominantly on one side of the belt—is a problem at many transfer points that contributes to generation of fugitive materials.

Figure 8.4

Off-center loading results in material being piled deeper on one side of the receiving belt, leading in turn to tracking problems and material spillage.

chapter. The geometry of loading gates or chutework should be calculated during the design of the chute, based on expected material flow patterns, to promote centering of the load.

Returning Belt Scrapings to the Material Flow

Belt cleaners are installed at the discharge pulley to remove residual material that has adhered to the belt beyond the discharge point.

The material removed by cleaners should be returned to the main material flow so that it does not build up on the walls of the head chute or other components. Consequently, a large dribble chute that encloses the belt-cleaning system with steep walls is usually required to accommodate the removed material and direct it back into the main material stream. Carryback has high adhesion, so whenever possible, the dribble chute should have steep, almost vertical walls.

Accomplishing this design objective may require the use of oversize chutes, low-friction chute liners, and/or auxiliary devices such as vibrating dribble chutes, air cannons, and scavenger conveyors. *(See Chapter 14: Belt Cleaning.)*

When designing a transfer, it should be kept in mind that the shallowest angle is the valley angle between two chutewalls (**Figure 8.5**). The steeper the valley angles need to be to minimize the adherence of carryback, the steeper the wall angles must be. To achieve a given valley angle, wall angles with even steeper pitch(es) are needed. Whenever possible, the corners should be rounded to reduce opportunities for the buildup of fines.

Being Service Friendly

Designing the transfer chute so that components can be easily accessed for service is critical to efficient maintenance. Often this is as simple as designing the structure to accommodate the preferred location of components or providing a means for

lifting heavy sections of chute wall or other components to be serviced. Many suppliers provide service-friendly arrangements of their components only to have these features canceled out by the design of the structure or by the placement of utility piping and conduits or other components (**Figure 8.6**).

Simply providing sufficient space for access and setting the work platforms at heights convenient for service will go a long way toward making a transfer chute service-friendly. The Conveyor Equipment Manufacturers Association's (CEMA) *BELT CONVEYORS for BULK MATERIALS, Sixth Edition*, provides recommended clearances around chutes. *(See also Chapter 26: Conveyor Accessibility.)*

It is often necessary to put scaffolds or work platforms inside the transfer chute for

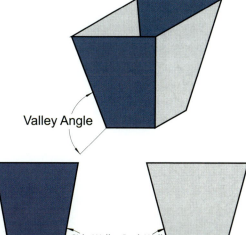
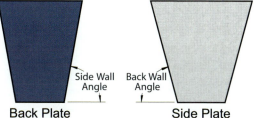

Figure 8.5

In transfer-point design, the shallowest angle is the valley angle between two chutewalls.

Valley Angle

Side Wall Angle

Back Wall Angle

Back Plate

Side Plate

Figure 8.6

A conveyor can have its service-friendly capabilities canceled out by the placement of utility piping and conduits and other components.

maintenance. It is not unusual for the setup and teardown of the scaffold to take longer than the maintenance task. Installing brackets or pockets to accommodate work platforms inside the chute (away from the material flow) is an effective practice that will save a considerable amount of time.

Designing the transfer chute so that maintenance on critical components can be performed without confined-space entry or "hot work" permits will improve maintenance productivity.

A transfer chute that is easy to maintain and clean will be one that is maintained and cleaned, leading to more production and less downtime. *(See Chapter 26: Conveyor Accessibility and Chapter 28: Maintenance for more information.)*

FACTORS IN THE DESIGN OF CHUTES

Conventional Transfer-Chute Design

Conventional transfer-chute design is normally done by an experienced designer or bulk-materials handling engineer using industry-accepted "rules of thumb." Many engineering firms establish their own design rules; many industries have developed consistent approaches to chute design that solve issues particular to their needs. While these various rules may vary, there is general agreement on at least the order of magnitude for many of the design requirements for conventional chute design. Guidelines for the design of conventional transfer chutes have been published in a number of references. The following is a brief summary of some of the more common design rules and approaches.

A conventional transfer chute usually consists of the following basic parts (**Figure 8.7**):

A. Head chute

The area surrounding the head pulley of the feeding conveyor

B. Drop chute

The area where the material is in freefall

C. Loading chute

The area where the material comes in contact with the receiving belt (also called the load zone)

D. Settling zone

While not technically part of the transfer chute, an extension of the chutework attached to the transfer chute to settle airborne dust

System Parameters

The following are the minimum parameters a designer must have before starting to design a transfer chute between two belt conveyors:

A. Rated capacity—tons per hour (st/h)

B. Ambient operating environment ranges

C. Bulk density as conveyed—kilograms per cubic meter (lb_m/ft^3)

D. Loose bulk density—kilograms per cubic meter (lb_m/ft^3)

E. Bulk-material classification—size distribution, material characteristics, and any special conditions

F. Discharge and receiving belt widths, speeds, and trough angles

G. Cross-sectional area of the load on the belt—square meters (ft^2)

H. Process flow sheet showing sequence of conveyors

I. General arrangement drawing showing plan and elevation views, critical dimensions, and the planned relationship between the discharge and receiving conveyors

Figure 8.7

A conventional transfer chute usually consists of the following basic parts: A) Head Chute, B) Drop Chute, C) Loading Chute, and D) Settling Zone.

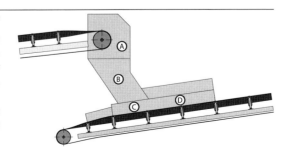

Many times, the listed capacity for conveyors is down-rated 10 to 20 percent from its actual engineered capacity, for several reasons. De-rating the capacity allows for surge loads, reduces spillage, and provides a factor of safety in meeting the specified throughput. When sizing transfer chutes, the conveyor's full load and cross-sectional area should be used.

The material's angle of repose is often used in conventional drop chute design to represent the angle of internal friction and interface friction values of the bulk material. The angle of repose is also used for establishing the minimum slope of chutewalls and the height of the material pile on the inside of the skirtboard. In addition, the angle of repose is often used for calculating the head load or weight of material on a belt that must be started with a full hopper above it. While widely used for these purposes, using the angle of repose for these calculations is often unsatisfactory, because the angle of repose does not represent the ability of the bulk material to adhere to itself or chutewalls.

A better course would be testing the properties of the actual material as it is conveyed through the system. This material testing will establish the range of bulk-material properties that the drop chute must accommodate. It will also help eliminate the most common mistakes made in the design of transfer chutes: the assumptions of maximum lump size and the differences between bulk density as conveyed and loose bulk density. *(See Chapter 25: Material Science for additional information on material properties and testing.)*

Material Trajectory

The path the bulk material takes as it is discharged from the delivery conveyor is called the trajectory. Trajectory is affected by the speed of the belt, the angle of inclination of the discharging belt, and the profile of the material on the belt. In conventional transfer-chute design, the trajectory is plotted and used as a starting point for estimating where the material

stream will first impact the head chutewall. From there, the material stream is assumed to be reflected from the chutewall much like a light beam being bent with a series of mirrors. CEMA's *BELT CONVEYORS for BULK MATERIALS, Sixth Edition*, provides a detailed discussion of calculating and plotting material trajectories.

The most common mistakes made at this stage of design are developing an incorrect initial material trajectory and failing to consider the effects of friction when plotting subsequent reflections of the material stream from the transfer chutewalls.

The current thinking in transfer-chute design is to control the stream of bulk material and not allow it to free fall from the discharge to the receiving belt. With this controlled approach, the designer assumes the material cross section does not fan out or open up significantly. Drop heights are minimized to help reduce material degradation, dust creation, and wear on the receiving belt.

This approach requires some knowledge of the friction values between the bulk material and transfer chute materials. DEM method is being used in conventional chute design as an aid to the designer in assessing the effects of changing properties, such as the coefficient of friction. There are several DEM software packages on the market designed for this purpose.

Distance, Angle, and Overlap between Conveyors

Ideally, all belt-to-belt transfers would be in-line: The discharging and receiving belts would run in the same direction (**Figure 8.8**). This type of transfer allows for sufficient belt overlap in order to avoid loading on the transition area of the receiving belt, where the belt goes from flat at the tail pulley to its full trough angle. Transitioning in this manner also makes it relatively easy to place the material on the receiving belt with the load moving in the direction of the belt, thus reducing unnecessary wear and spillage. In-line transfers are often incorporated into systems in order to reduce the length of the conveyor when insufficient drive power or tension is available for a single belt, to extend the length of the conveyor system, or to accommodate mechanisms to blend, crush, or separate the material.

Figure 8.8

With in-line conveyor transfers, the discharging and receiving belts would run in the same direction.

Figure 8.9

A non-linear transfer may be required to accommodate changes in material flow direction required by site restriction or to allow for material separation or stockpiling.

Figure 8.10

Off-center material loading may push the belt out from under the skirting, allowing the sealing strip to drop down where the belt runs against the seal.

More typically, a change in the direction of the material movement is required as one conveyor loads onto another (**Figure 8.9**). A non-linear transfer may be required to accommodate changes in material flow direction, to allow for diverting the material for stockpiling, or for splitting the material for separation.

Problems associated with non-linear transfer points include: difficulty in maintaining the material's proper speed, trajectory, and angle; problems controlling dust and spillage; and issues of increased wear on (and the resulting higher cost for replacement of) transfer-point components.

If material is loaded on the belt in a direction that is not in line with movement of the receiving belt, wear patterns may become visible on the inside of the head (discharge) chute. These patterns will correspond to the path the material takes as it bounces off the inside of the chute as it tries to attain the direction and speed of the moving belt. Although turbulence may not be visible as the load exits the skirted area, the ricocheting movement of the material within the transfer chute accelerates wear on liners, skirtboard, and sealing systems. The force of the loading material may mistrack the belt and push it out from under the skirting on one side of the belt, allowing the sealing strip to drop down and preventing the belt from returning to its centered position. The belt will attempt to return to its center as material loading changes, forcing the belt into contact with the sealing strip and cutting through the strip, resulting in significant spillage opportunities (**Figure 8.10**).

Fortunately, a number of strategies and components can be employed to guide the flow of material into the desired direction of travel and load it onto the center of the receiving belt.

The most common mistakes made in the transfer chute design stage include not providing enough overlap of the conveyors. This leads to loading on the belt transition and not allowing enough room for installing belt cleaners. Without attention to proper conveyor design, including sufficient overlap, the operation is burdened with a conveyor that plugs often, generates loads of fugitive material, and creates excessive wear problems.

Loading in the transition area of the receiving belt is done in an attempt to reduce costs by saving a few meters of conveyor length. It is recognized that this practice creates numerous problems in loading, sealing, and belt wear and should be avoided.

It should be noted that in order to reduce the load absorption requirements and dust creation opportunities of a conveyor transfer system, drop height should be kept at a minimum; however, engineered hood and spoon designs use gravity to maintain material flow speed (**Figure 8.11**) and often require greater drop heights in order to implement them. Engineered spoons provide many benefits and should be considered as part of the original design or as part of the requirement of a future retrofit. *(See Chapter 22: Engineered Flow Chutes.)*

Design Considerations of the Transfer Chute

The volume of the head (discharge) chute around the discharge pulley is usually dictated by the general arrangement of the conveyors, access requirements for service, and the initial material trajectory.

Head pulley diameter and face width help determine the width and height of the head chute. The space between the chutewall and the pulley rim should be small enough that large lumps are not able to pass from the carrying side to the return and are not caught between the pulley and the chutewall. A typical space is 50 to 75 millimeters (2 to 3 in.) per side. Maintenance of the pulley and pulley lagging as well as access to the shaft bushings should be considered in making this decision.

The head chute should start at the last full transition idler on the delivering conveyor to help contain any fugitive material

that might fall from the belt as the belt changes from troughed to flat on the head pulley. The inlet area of the head chute should be controlled with dust curtains on the carrying sides and barrier seals on the belt return side, because these areas are key factors for controlling the amount of air flowing through the transfer chute (**Figure 8.12**).

Once the bulk-material flow direction has been changed by the first contact with the head chute, material is often channeled into drop (transition) chutes. These drop chutes can be extended with duct-like chutes that place the material stream into proper alignment with the receiving conveyor. All of these drop chutes need to be steep enough to prevent the bulk material from sticking to the walls; they also need to be large enough to prevent plugging.

It is commonly accepted that the drop chute cross-sectional area should be a minimum of four times the cross-sectional area of the material profile. It is also commonly

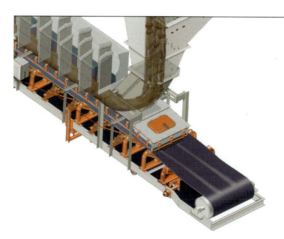

Figure 8.11

Engineered hood and spoon designs use gravity to maintain material flow speed.

Figure 8.12

To control the air flowing through the chute, the inlet area should be controlled with dust curtains on the carrying side and barrier seals on the belt return side.

accepted that the minimum dimensions for width and/or depth should be at least 2.5 times the largest lump expected to pass through the chute. Many designers increase these ratios based on their experience with particular materials. In some cases, where the bulk material is uniform in size and free flowing, these ratios can be reduced, especially when the chute is engineered using the specific properties of the bulk material being conveyed.

The loading (receiving) chute width should be designed to maintain the minimum belt edge necessary for sealing and accommodating mistracking. *(See Chapter 11: Skirtboards.)*

The most common mistake made at this stage of design is making too abrupt a transition between the drop chute and the loading chute, creating chutewall angles that promote buildup leading to plugging. Current design practice is to use valley angles at a minimum of 60 degrees, with 75 degrees preferred (**Figure 8.5**).

Managing Wear and Material Flow

The transfer chute is usually designed for full flow and a consistent material path. However, the flow of a bulk material through the chute will change as the material changes properties, the tonnage changes, the chute wears, or the bulk material builds up on the chutewall.

Deflectors

Deflectors may be used inside a transfer chute to absorb impact and minimize wear, starting at the point where the material trajectory first meets the head chute

(**Figure 8.13**). It is important to provide enough clearance between a deflector and the head pulley of the discharging conveyor to prevent large lumps from blocking the passage or cohesive material from adhering to the plate, which could cause the transfer chute to plug.

Once the material flow leaves the first point of contact with the chute, it may be necessary to fine-tune the flow of the material on start-up of the system. Deflectors, or "kicker plates," are often included in the original plan or installed at start up to steer the material flow.

During the start-up of a new conveyor system, it is common practice to install deflectors within the loading chute to help center the load. The process of getting a desired flow path through the chute is often one of trial and error. These deflector plates should be field-adjustable so they can be repositioned to achieve the desired effect. They should be accessible to allow efficient replacement. Inspection and access points are critical to observing and maintaining the proper direction for deflected materials.

Load placement may be enhanced with deflectors installed on the inside surface of the loading chute to direct lumps of material toward the center of the load zone. Center-loaded lumps are less likely to slip off the edges of the belt or damage the skirtboard seals.

Deflector wear liners inside the bottom of the loading chute next to the belt may reduce the problems associated with off-center loading. One or more deflectors or impact plates may be necessary to retard the forward momentum of the material, redirect it in the proper direction, and center the load on the receiving belt. These liners feature a bend or angle that turns the material toward the center of the belt and away from the belt edges. Deflector wear liners should be used with care, because they may contribute to other problems, such as material entrapment and transfer chute choking.

Figure 8.13

Deflectors may be used inside a chute to absorb impact and minimize wear.

Popular ways to manage the flow of bulk materials through the transfer chute and minimize impact are installation of scalping bars or the use of rock boxes.

Scalping, or Grizzly, Bars

Scalping bars—also called a grizzly or grizzly bars—within the transfer chute allow the fines to pass through first to form a protective bed on the belt. The lumps, which are unable to pass between the bars, slide down the incline and land on the belt on a cushion formed by the previously deposited fines. Plants use grizzlies like a grate at truck dumps or other installations to keep oversize lumps away from conveyor systems (**Figure 8.14**).

Rock Boxes

Rock boxes consist of a ledge inside the drop chute where a pile of the conveyed material accumulates (**Figure 8.15**). Subsequent material moving through the chute flows over or deflects off this pocket of captive material. Abrasive force is shifted from the chutework to the accumulated bed of material, and the overall drop height is reduced and impact force dissipated as material bounces off the material on the ledge (**Figure 8.16**).

Rock ladders, composed of a series of baffles, or "mini" rock boxes, are used to reduce impact and control material velocity over drops of greater distance (**Figure 8.17**). Rock ladder shelves are typically arranged on alternating sides of the chute, so the material never has a free drop of more than 1,5 to 2 meters (5 to 6 ft).

Rock boxes and rock ladders are most appropriate for chutes handling materials such as sand, gravel, or hard rock (**Figure 8.18**). The boxes are most successfully used if physical conditions and flow rates do not change over time, because it is important that the flowing material move consistently across the buildup in the rock box. Care must be taken to accurately judge the cohesive characteristics of the material (under wet conditions, for example)

in order to avoid accumulations that can choke the chute. Rock boxes should not be used in transfer points handling fragile bulk materials that might suffer degradation or materials with large lumps that can block or choke the flow; nor should they be used if a conveyor will carry more than one material.

Figure 8.14

Scalping bars—also called a grizzly or grizzly bars—within the chute allow the fines to pass through first to form a protective bed on the belt. Plants use grizzlies to keep oversize lumps away from the conveyor systems.

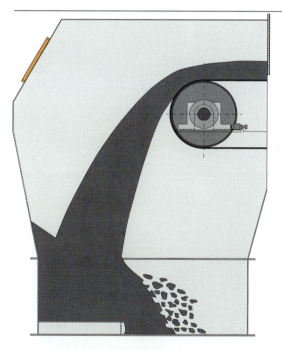

Figure 8.15

A rock box consists of a ledge inside the chute where a pile of the conveyed material accumulates.

Figure 8.16

Rock boxes shift the abrasion from the moving material from the chutework to the bed of material, and impact force is dissipated as material bounces off the material on the ledge.

Impact Plates or Grids

Another method of diverting flow and absorbing impact within the transfer chute is the use of impact plates or grids in the material path (**Figure 8.19**). An impact plate is placed inside the chute to absorb the force of the moving material stream. Impact plates are often used in angular transfers where high belt speeds are present

Figure 8.17

Rock ladders are a series of baffles, or "mini" rock boxes, used to reduce impact and control material velocity over drops of greater distance.

Figure 8.18

Rock boxes and rock ladders are most appropriate for chutes handling materials such as sand, gravel, or hard rock. Note: looking down the chute from the head pulley.

Figure 8.19

Impact plates are placed in the material path inside a loading chute to divert flow and absorb impact.

and circumstances (such as available space and budgets) prevent the engineering of ample chutes.

Some impact grids are designed to catch material to develop a material-on-material impact that preserves the chutewalls. Subsequent material bounces off the captured material without actually hitting the grid or the chutewall. The gap between the head pulley and the impact plate should be carefully considered to minimize problems from oversize rocks or tramp material becoming hung up between the pulley and the plate, or from the buildup of cohesive or high-moisture materials that can choke the transfer chute.

The selection of appropriate materials and careful attention to design and positioning of impact plates and grids may significantly improve the life of these wear components.

Wear Liners

The constant impact and sliding of material against the sides of the transfer chute is the main source of wear in a chute. In addition to the grids, rock boxes, and impact plates discussed above, one way to reduce wear of the chute itself is the use of sacrificial liners inside the chute. Liners may also be installed to reduce wall friction and/or material adhesion. In selecting a material for use as a liner, the goal is to select a material that will both resist abrasion and enhance flow. *(See Chapter 12: Wear Liners for more information.)*

Loading the Receiving Belt

Another phenomenon that occurs at transfer points where material falls vertically onto a high-speed belt is called pooling. Material not yet moving at belt speed piles up on the belt and creates a "pool" of material in the loading zone (**Figure 8.20**). When a lump of material drops onto the belt, it bounces and tumbles, dissipating the energy supplied by the previous conveyor and from its fall until the lump is caught by the motion of the receiving belt. In the meantime, the material can bounce off the

8

pool or pile toward the side or rear of the conveyor, resulting in spillage. The greater the difference between velocity of the material stream and the speed of the receiving belt, the longer and deeper the pool of material. As this body of "pooled" material grows, it becomes increasingly difficult to maintain a sealed, spillage-free transfer point and control belt cover wear.

A speed-up conveyor can be used to remedy this condition (**Figure 8.21**). Another solution is the use of a curved gate, ramp, or spoon to control the speed and direction of the material stream until it reaches the speed and direction of the receiving belt (**Figure 8.22**). These curved loading chutes steer the material flow, "pouring" it onto the center of the receiving belt. The smoother positioning of the load on the receiving conveyor reduces the movement of the material to the edges of the belt and releases less energy and air movement, minimizing dust. The angle at which the chute descends from the unloading structure onto the receiving belt should be flat enough to prevent lumps from bouncing excessively after they land on the belt. A chute with as low as possible valley angle, combined with proper load direction and speed, allows the lumps to strike the belt at a grazing angle (**Figure 8.23**). This allows the material to bounce gently as it is carried in the direction of belt movement rather than rebound back into the face of the incoming material stream. A curved chute reduces the risk of damage to the belt and minimizes material degradation and dust generation.

It should be noted, however, that if the chute angle is too flat, the material stream might slow to the point that it can accumulate on shut down, build up, and eventually plug the chute. Typical valley angles for conventionally designed chutes are between 60 and 75 degrees from the receiving belt line (**Figure 8.5**).

Managing Air Flow

A well-designed and constructed transfer chute can significantly reduce airborne dust by limiting the creation of induced air movement. The skirtboard sections should be large enough to provide a plenum that stills air currents and reduces the positive pressures that can carry airborne particles out of the enclosure. *(See Chapter 7: Air Control and Chapter 11: Skirtboards for more information.)*

The enclosure should be spacious enough to permit a significant reduction in the speed of air currents and, therefore, allow airborne particles to settle back into the load before the conveyor leaves the enclosure.

Chute Structure

The transfer chute is typically fabricated from plates of mild steel or stainless steel, with selection depending on the conveyed material and the conditions in the facility.

Figure 8.20

Pooling occurs when belt cargo that is not yet moving at belt speed piles up in the loading zone.

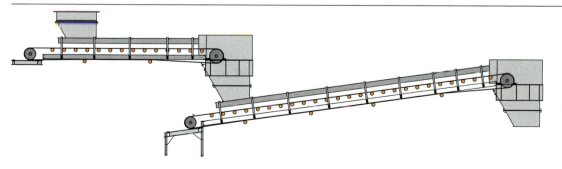

Figure 8.21

A speed-up conveyor can be used to raise the velocity of the material until it reaches the proper speed and direction.

The selection of transfer chute plate thickness depends on the characteristics and volume of material moving through the chute, the structural strength requirements, and the margin for wear if the chute will not be fitted with a replaceable liner

Figure 8.22

A curved gate, ramp, or spoon can place the material stream on the receiving belt with the proper speed and direction.

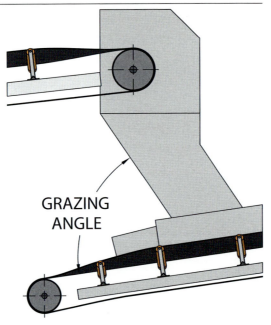

Figure 8.23

The angle at which the chute descends from the unloading structure onto the receiving belt should be flat enough so the material strikes the belt at a grazing angle, to prevent excessive bounce of the lumps.

GRAZING ANGLE

Figure 8.24

Despite the best intentions and practices of chute designers, there are occasions when material will accumulate inside transfer chutes.

system. Local codes usually govern the structural design of chutes, but it is up to the designer to consider all the loads that may be present. Some of the more important loads are the weight of the chute, accumulations of fugitive materials, snow and ice, the weight of a chute full of bulk materials, and wind loads. Work platforms around the chute need to be sturdy enough to handle maintenance activities.

Transfer chutes should be fabricated in sections that are convenient for transport and subsequent erection on site. For retrofit systems, chute sections must also be designed to fit through available openings to reach the construction site.

Care must be exercised in the construction of transfer chutes to avoid imperfections in the surface that might disrupt the material flow and negate the careful engineering that went into the design. Variations of ± 3 millimeters (1/8 in.) may present problems when matching sections of wear liner or truing up the chutework to the belt. The investment of time in a precise chute installation will be returned many times over through improved efficiency, simplified maintenance, and reduced fugitive material.

Despite the best intentions and practices of transfer chute designers, there are occasions when material will accumulate in transfer chutes. Materials with high levels of moisture may adhere to walls or even freeze during winter operations (**Figure 8.24**). Continuous operation may compress the material encrustation more firmly onto the chutewall, allowing for additional material buildup and possibly leading to complete chute blockage. During the chute design process, it is wise to make provisions for future requirements for flow-aid devices, such as vibrators or air cannons. *(See Chapter 9: Flow Aids and Chapter 22: Engineered Flow Chutes.)*

Chute Access

An enclosed transfer chute must have openings to allow for visual inspection and doors for worker entry, and there must be a clear path for workers to reach these openings. Inspection openings, such as hinged access doors, should be positioned away from the flow of material yet located where personnel can observe material movement and inspect for wear (**Figure 8.25**).

Screens or guards should be positioned to protect workers observing material flow from pinch points and rolling components. Covers or doors should be corrosion resistant and provide a dust-tight seal. Safety barriers should be in place to prevent material from escaping the chute and to keep personnel from reaching into the material trajectory.

Often forgotten in the design of transfer chutes is the provision for some method of access to replace liners inside the chute or to maintain belt cleaners.

Consideration of future service requirements is particularly important on transfer chutes too small for personnel to work inside. Fabricating chutes in sections for easy disassembly is one approach to maintenance. *(See Chapter 26: Conveyor Accessibility.)*

TYPICAL SPECIFICATIONS

A. Direction

In general, the transfer chute should be designed to direct the material in the direction of the receiving conveyor and center it on the belt.

B. Drop height

The drop height from the discharge system to the receiving conveyor should be as short as possible while providing

Figure 8.25

Inspection and access doors should be positioned out of the flow of material yet located where personnel can observe material movement and inspect for wear.

⚠ SAFETY CONCERNS

Safety considerations require that access be limited so personnel cannot enter the chute until appropriate safety procedures are followed, including lockout / tagout / blockout / testout procedures of both discharging and receiving conveyors. No one should enter chutes without proper training in confined-space safety procedures.

The structural and liner components of transfer chutes tend to be large and heavy and should be handled with appropriate equipment and due care.

If flow-aid devices (such as air cannons) are installed, proper de-energization and lockout / tagout / blockout / testout procedures must be followed for this equipment prior to service.

Personnel working in, on, or around transfer chutes must be aware of the potential for falling materials, either cargo from the belt above or buildup on the chutewalls. It is recommended that the chute be inspected and thoroughly cleaned before entering for any reason.

It is important to pay attention to safety procedures when working around nuclear devices installed on transfer chutes for level detection or on-line bulk-material analysis.

Chutes and their structures should be grounded to prevent the buildup of static electricity.

adequate space for equipment installation and maintenance.

C. Speed

Material from the discharge should be loaded so it is moving at the same speed as the receiving conveyor is traveling.

D. Slope

The transfer chute should be adequately sloped to prevent material from bouncing excessively after it lands on the receiving conveyor, which can increase dust generation and impact damage.

E. Volume

The volume of the drop chute should be at least four times that of the load stream of the feed conveyor. The transfer sections should be large enough to provide a plenum to minimize air currents.

ADVANCED TOPICS

Chute Width

The belt is 1200 millimeters (48 in.) wide with 30-degree troughing idlers. What is the recommended chute width where the chute matches up with the skirtboards?

The CEMA 2/3 rule results in a chute 800 millimeters (32 in.) wide.

Another method to determine the recommended distance between the skirtboards is based on the amount of belt edge necessary for an effective seal and accomodation of belt wander. The recommended skirtboard width for a belt 1200 millimeters (48 in.) wide with a 30-degree troughing angle is 894 millimeters (35.2 in.). *(See Chapter 11: Skirtboards.)* The difference between the CEMA method and the belt-edge method is more pronounced for very narrow and very wide belts.

Calculating Valley Angles

A new chute with a minimum valley angle of 60 degrees was required. A side wall angle of 75 degrees and a back wall angle of 60 degrees were selected, because these angles were within the recommended range (**Figure 8.26**). The equation can be used to check the design (**Equation 8.1**).

In this example, the valley angle is approximately 57 degrees, so the designer should reconsider the design of the chute to maintain a minimum of 60 degrees as required. If the angles were changed to 65 degrees and 75 degrees, the valley angle would be 61 degrees, which would be steep enough to maintain flow.

Figure 8.26

The valley angle is the angle created by the side wall joining with the back wall.

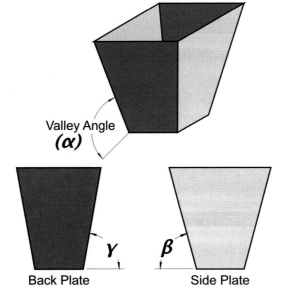

Valley Angle
(α)

Back Plate   Side Plate

Equation 8.1

Calculating Valley Angles

$$\alpha = arc\ cot\left(\sqrt{cot^2\ (\beta) + cot^2\ (\gamma)}\right)$$

Given: *A designer has selected a side wall angle of 75° and a back wall angle of 60°.*
Find: *The valley angle of the chute.*

α	Valley Angle	degrees
β	Back Wall Angle to Horizontal	60°
γ	Side Wall Angle to Horizontal	75°

$\alpha = arc\ cot\left(\sqrt{cot^2\ (60) + cot^2\ (75)}\right) = 57.5$

α	Valley Angle	57.5°

It should be noted that the valley angle will never be greater than the smaller of the other two angles (back wall and side wall).

The design would be an iterative process of selecting wall angles based on geometry and calculating the valley angle. If the valley angle is not appropriate, different wall angles should be selected and the valley angle calculated for the selected angles. This process is repeated until the wall angles fit within the geometry available and the valley angle is in the correct range based on the material.

THE WORK OF CHUTEWORK

In Closing...

Designed correctly, conventional transfer chutes offer an effective method to safely transfer material from one elevation to another, with minimal fugitive material and low maintenance requirements. Incorporating the items discussed in this chapter into the plans will provide both the designer and end user with suitable tools to understand how chutes operate from a practical level and how to design or modify them for improved performance.

Looking Ahead...

This chapter about Conventional Transfer Chutes, the third chapter in the section Loading the Belt, focused on the transfer chute and methods to manage material flow to reduce spillage and dust. The following chapter continues this section with a discussion about Flow Aids.

REFERENCES

8.1   Conveyor Equipment Manufacturers Association (CEMA). (2005). *BELT CONVEYORS for BULK MATERIALS, Sixth Edition*. Naples, Florida.

8.2   Martin Marietta Corporation. *Dust Control Handbook for Minerals Processing*, Contract No. J0235005.

8.3   Morrison, J.N., Jr. (1971). "Environmental Control Applied to Belt Conveyor Transfer Points." In *Bulk Materials Handling: Volume 1*. University of Pittsburgh.

8.4   Taylor, H.J. (1989). *Guide to the Design of Transfer Chutes and Chute Linings for Bulk Materials*. The Mechanical Handling Engineers' Association.

8

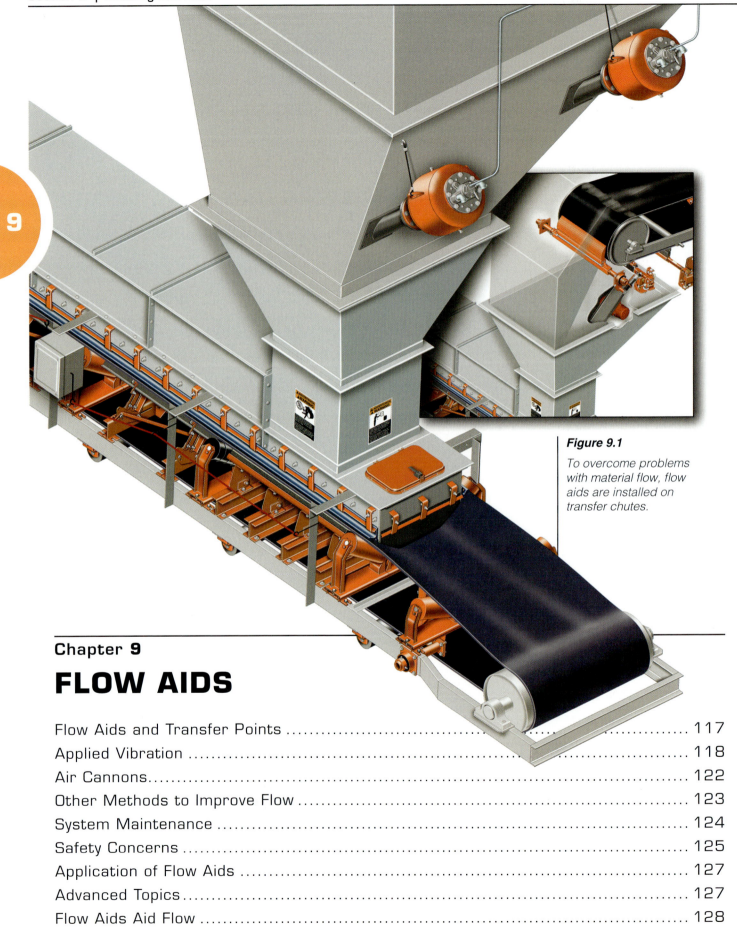

Figure 9.1

To overcome problems with material flow, flow aids are installed on transfer chutes.

Chapter 9

FLOW AIDS

In this Chapter...

In this chapter, we discuss various methods to promote the flow of materials through chutes. These flow aids include both linear and rotary vibrators, air cannons, aeration systems, chute linings, and soft-chute designs. Considerations for selection of the type of flow aid for a particular application, sizing, installation, and maintaining flow aids are offered, along with safety procedures.

Transfer chutes must be designed to accommodate and facilitate the flow of materials they will be handling. But even if the application is ideal and the engineer experienced, changes in material characteristics and/or system demand can create problems with material flow. To overcome these problems, a variety of devices called flow aids are utilized (**Figure 9.1**).

There is a wide range of material characteristics and operating conditions that make the use of flow aids in the original design a practical option. To design a chute that would handle every material situation is virtually impossible. Many times, the most economical solution to sustain the flow with changing material and operating conditions is to include flow-aid devices in the original design. Materials with high moisture-content can adhere to walls or even freeze during winter operations. Continuous operation can serve to compress the material encrustation even more firmly onto the wall. Bulk materials can change in characteristics as the operation progresses through the seam or stockpile. In some cases, the chute can become completely blocked by just a small change in any of these parameters.

Flow aids are installed to promote the flow of materials through a chute or vessel. Because they will affect a conveyor's loading, flow-aid devices can also impact spillage and dust. The accidental, or intended, breakdown of buildup can produce surges, producing overloading, spillage, and mistracking. By designing active flow aids into a chute, the operation gains a level of control over the material flow that is impossible to obtain with static approaches, like low-friction liners, alone.

FLOW AIDS AND TRANSFER POINTS

What is a Flow Aid?

Flow-aid devices are systems used to stimulate or enhance the movement of bulk materials. They can be as simple as an impacting piston vibrator on a chutewall to dislodge material buildup, or as sophisticated as a multiple air-cannon system discharging automatically on a timed cycle to prevent material buildup. Flow-aid devices include rotary or linear vibrators, low-pressure air cannons, and aeration devices, as well as low-friction linings and soft-chute designs. These systems can be combined in any number of ways.

The age-old solution for breaking loose blockages and removing accumulations from chutes and storage vessels was to pound the outside of the walls with a hammer or other heavy object (**Figure 9.2**). However, the more the walls are pounded, the worse the situation becomes, because the bumps and ridges left in the wall from the hammer strikes form ledges that start additional material accumulations (**Figure 9.3**).

A better solution is the application of a flow-aid device to the chute. These devices supply energy precisely where needed to reduce the friction of the walls and break up the material to keep the material moving to the discharge opening.

Figure 9.2

The traditional solution for improving flow from chutes and storage vessels was to pound the outside of the walls with a battering ram, hammer, or other heavy object.

This chapter explores the various methods to promote material flow in a chute. This discussion centers on flow aids applied to conveyor loading and discharge chutes; this information, and these technologies, can also be applied to applications on other material process and storage vessels, including silos, bins, hoppers, bunkers, screens, feeders, cyclones, and heat exchangers.

Flow Aids on Transfer Points

Using material characteristics and process requirements to design a chute to flow efficiently is certainly the best practice. However, materials are unpredictable. The source of the material may change due to economic reasons, or weather conditions can drastically alter its flow characteristics. In these situations, it is a simple and cost-effective approach to apply flow aids to maintain material flow.

In some cases, flow aids are original equipment, incorporated in the design of a system to stabilize flow rates or eliminate anticipated problems. As an example, a flow aid might be designed into a system to move material through a chute that, due to height restrictions, does not have a steep enough angle to maintain consistent material movement. In other cases, flow aids are retrofit components, added to a materials-handling system to overcome problems that were not anticipated in the original design, or that have recently appeared, perhaps due to changes in the condition of the material, the process, or the equipment.

It is wise to incorporate channel mountings for vibrators, or nozzle mounts for air cannons, when a chute is in the fabrication stage. If a problem should arise later, because material characteristics have changed or other unforeseen problems have occurred, it will be a simple matter to install a flow-aid device to remedy the problem.

It is critical that the steel chute and support structure are sound, because the operation of these flow-aid devices can create potentially damaging stress on the structure. A properly designed and maintained chute will not be damaged by the addition of flow aids.

It is important that any flow-aid device be used only when the discharge is open and material can flow from the chute. If used when the discharge is closed, the energy of the flow aid may pack the material more tightly, making flow more problematic when the discharge is opened and causing damage to the bin. The best practice is for the flow aid to be controlled by timers or sensors to prevent any flow-retarding buildup of material. This saves energy, reduces noise, and improves safety, because the flow aid runs only when needed.

Figure 9.3

The more the walls are pounded, the worse the situation becomes, because the bumps and ridges left in the wall from the hammer strikes form ledges that start additional material accumulations.

APPLIED VIBRATION

Vibrators perform the same function as thumping on the outside of a bottle of ketchup: They reduce the cohesion between the material particles and the adhesion between the particles and the wall to increase the flow of material out of the bottom.

The relationship between the bulk material and the frequency of vibration best suited to stimulate that material is proportional to particle size. As a general rule, the smaller the particle, the better it responds to higher vibration frequencies. The relationship between amplitude of vibration and the bulk material is based on cohesive and adhesive forces. As the particle size increases, the amplitude required to cause the bulk material to move increases. Particles that are fine and free flowing (low cohesive) tend to respond well to small amplitudes of vibration; free-flowing particles that are larger respond better to larger amplitudes. Particles that are sticky tend to build up in solid masses that respond well to low-frequency high-amplitude vibration. Generally, the direction of the rotation or the stroke of the vibrator's mass should be in the direction of desired flow of the material.

Linear Vibrators

Linear vibrators activate the material inside a chute or bin by using heavy blows on the outside of the structure's steel walls. In fact, the earliest form of vibration was a hammer. The act of pounding on the chute or bin wall overcomes the adhesive force between the material and the wall surface. However, this hammering on the bin or chutewall often leads to damage to the wall surface (**Figure 9.4**). The marks left on the wall by the hammer, often called "hammer rash," will continue and expand the problem the blow of the hammer was supposed to overcome. In addition, the swinging of the sledgehammer poses the risk of injuries to plant personnel. The piston vibrator was developed to produce this effect without actually swinging a hammer (**Figure 9.5**).

A pneumatic piston (or linear) vibrator uses plant air to move a piston back and forth inside a casing (**Figure 9.6**). In some vibrators, this piston may strike the wall; in other designs, it merely oscillates with enough mass to flex the wall. In both cases, the vibrator provides energy through the wall to the material inside the structure. This force—more controlled than a hammer strike—breaks the adhesion between the material and the wall, so the material will flow out of the structure.

Linear vibration is the best solution for sticky, coarse, high-moisture materials. A convenient test is to take a handful of material and squeeze it into a ball. If the material readily remains in the ball after the fist is opened, linear vibration is probably the best choice.

A piston vibrator would be mounted on the outside of the vessel or chute at the point of the buildup or blockage on the inside. Often these vibrators are attached to a steel channel that is mounted on the chutewall (**Figure 9.7**). This mount spreads the force out over a larger area of the structure to maximize the efficiency

Figure 9.4

Hammering on the bin or chutewall leads to damage to the wall surface, often called "hammer rash."

Figure 9.5

The piston vibrator was developed to produce this pounding effect without actually swinging a hammer.

PISTON MOVEMENT

Figure 9.6

A pneumatic piston (linear) vibrator uses plant air to move a piston back and forth inside a casing to produce a vibratory force.

while preserving the structure from fatigue. Most linear vibrators are driven by plant air and can be controlled remotely, with a solenoid, or locally, with a manual on/off valve.

The vibrator for a particular application is selected according to the weight and characteristics of material in the chute or sloped portion of a bin or hopper (**Table 9.1**). The general "rule of thumb" for

Table 9.1

Typical Vibrator Sizes by Weight of Material Inside the Chute

Maximum Weight of Bulk Material in Chute kg (lb$_m$)	Vibrator Force Required N (lb$_f$)	Diameter of Piston in Linear Vibrator mm (in.)	Bin Wall Thickness Range mm (in.)	Mounting Channel Suggested Length mm (in.)
1315 (2900)	~1300 (~300)	32 (1.25)	1,6 to 3,2 (1/16 to 1/8)	900 (36)
2223 (4900)	~2250 (~500)	50 (2)	4,8 to 6,4 (3/16 to 1/4)	900 (36)
4445 (9800)	~4450 (~1000)	75 (3)	6,4 to 9,5 (1/4 to 3/8)	900 (36)
9979 (22000)	~10000 (~2200)	100 (4)	9,5 to 12,7 (3/8 to 1/2)	1800 (72)

Table 9.2

Vibrator Force Outputs Based on Bulk Density

Bulk Density	Force Ratios
Above 1440 kg/m³ (90 lb$_m$/ft³)	1 newton per 0,7 kg (1 lb$_f$/7 lb$_m$)
Between 640-1440 kg/m³ (40-90 lb$_m$/ft³)	1 newton per 1,0 kg (1 lb$_f$/10 lb$_m$)
Below 640 kg/m³ (40 lb$_m$/ft³)	1 newton per 0,3 kg (1 lb$_f$/3 lb$_m$)

Note: As stated previously, fine, dry materials respond well to high-frequency/low-amplitude vibration, where as larger particles and wet materials respond better to low-frequency/high-amplitude vibration.

Equation 9.1

Linear Vibrator Output Force Requirement

$$LF = \frac{Wt_t}{k_a}$$

Given: 4100 kilograms (9000 lb$_m$) of dry material is plugging a conveyor load chute. **Find:** The linear force required from a vibrator to encourage flow in the given chute.

Variables		Metric Units	Imperial Units
LF	Linear Force Required	newtons	pounds-force
k$_a$	Application Factor *The Application Factor comes from the Rule of Thumb for Vibrator Application.*	1,025 (dry material) 0,82 (wet material)	10 (dry material) 8 (wet material)
Wt$_t$	Weight of Material in Influenced Area	4100 kg	9000 lb$_m$

Metric: $LF = \dfrac{4100}{1,025} = 4000$

Imperial: $LF = \dfrac{9000}{10} = 900$

LF	Linear Force Required	4000 N	900 lb$_f$

typical vibrator applications is to apply 1 newton per 1 kilogram (1 lb$_f$ / 10 lb$_m$) of material weight inside the chute. This general rule assumes that the material is flowable and has a density less than 1440 kilograms per cubic meter (90 lb$_m$/ft^3).

More force will be needed for materials of high density or moisture or of low density. While the ratio above is acceptable for materials between 640-1440 kilograms per cubic meter (40-90 lb$_m$/ft^3), materials with higher or lower bulk densities require different ratios (**Table 9.2**).

The length of the mounting channel and the chutewall thickness best suited for these applications are also dependent on material weight and characteristics; applications outside the parameters given in the table may require specialized engineering.

The calculation of linear force is required for application of a vibrator on a chute (**Equation 9.1**).

After installation, air-powered vibrators must be tuned to the needs of the application by adjusting the air pressure and/or flow rate to maximize the effect on the bulk material.

Rotary Vibrators

In contrast to linear vibrators, other vibrators create a vibratory force through the rotation of an eccentric weight. These rotary vibrators create a powerful vibration much as a household washing machine does when its load is off-center. They supply an energy that is most suited to move fine, dry materials (**Figure 9.8**).

Rotary vibrators can be pneumatically-, hydraulically-, or electrically-powered: The choice for a given application is often determined by the energy supply most readily available at the point of installation.

In rotary pneumatic vibrators, a stream of air drives a mass in a circular orbit to create the vibration; in rotary hydraulic vibrators, it is a stream of hydraulic fluid that moves the mass. In rotary electric vibrators,

eccentric weights are typically mounted on the ends of an electric motor rotor or shaft (**Figure 9.9**).

Rotary vibrators are available in a wide range of sizes and outputs, to match the specifics of each application. In addition, many rotary electric vibrators can be adjusted by altering the overlap of the eccentric weights—increasing or decreasing the amount of unbalance—to provide the desired amount of vibratory force.

Vibration can induce stress into metal structures, and the walls may need to be reinforced at the point(s) of application. Like

Figure 9.7

A piston vibrator should be installed on a steel channel that is mounted on outside of the chute at the point of the buildup.

Figure 9.8

Rotary vibrators create a powerful vibration through the rotation of an eccentric weight or mass around a central axis.

Figure 9.9

In rotary electric vibrators, weights are typically mounted on the ends of an electric motor rotor or shaft.

linear vibrators, rotary vibrators are typically installed on a mount plate or channel that spreads the vibratory energy (and the weight of the device) over a larger surface area (**Figure 9.10**).

Figure 9.10

Rotary vibrators are typically installed on a mount plate or channel that spreads the vibratory energy and the weight of the device over a larger surface area.

Figure 9.11

An air-cannon system is another solution to the buildup of material in transfer chutes.

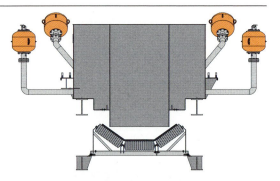

Figure 9.12

Air cannons use plant compressed air to create an eruption of air that will dislodge material buildup in chutes.

Figure 9.13

Some air-cannon systems feed air from one reservoir to a number of discharge nozzles.

Rotary vibrators designed for chutes or hoppers are usually sized based on a 1:10 ratio of output force to the mass of material inside the chute or sloped section of the bin. Generally, the finer the material, the higher the frequency needed to excite the material and make it flow.

On a chute, a rotary vibrator is typically installed in the lower one-fourth to one-third of the structure. If a second vibrator is required, it should be mounted 180 degrees from the first vibrator and halfway up the structure.

Rotary vibrators can be controlled automatically or manually, allowing use only when needed. Once installed, a vibrator must be "tuned" by adjusting its force and/or speed to give the optimum effect for each application.

AIR CANNONS

In addition to vibrators, another solution to buildup of material in transfer chutes is the installation of an air-cannon system (**Figure 9.11**).

Air cannons—sometimes referred to as blasters—use plant compressed air to create an eruption of air that will dislodge material buildup in chutes. Air cannons are simply reservoirs of stored compressed air with fast-acting discharge valves. When the valves are actuated, the air escapes very quickly, creating a wide area of influence (**Figure 9.12**). When strategically located and properly controlled, this blast of air will dislodge or prevent material buildup.

Nozzles and pipes of varying shapes are attached to the chutewall, and air cannons are connected to each nozzle. The nozzles should be positioned to direct the blast toward the outlet or direction of flow. Some systems use independent air reservoirs for each nozzle; other systems use one reservoir for several nozzles (**Figure 9.13**). The nozzles are embedded in the wall, so they can discharge under the layer of accumulated material (**Figure 9.14**). Care must be taken when installing the nozzles to avoid creating additional edges and corners that encourage buildup. The movement of the bulk material may wear the nozzles, and larger lump sizes can deform or destroy the nozzles.

The number of air cannons installed depends on the size and shape of the chute and the nature of the buildup. Typically, one air cannon can keep 1,5 to 2 square meters (15 to 20 ft²) of chutewall free of material. Air cannons with air volume of 50 liters (1.75 ft³) have shown good results in chute applications. Air cannons can be installed at several heights around the vessel of the installation.

Air cannons are available with a variety of sizes of air reservoirs and of discharge diameters, to supply the appropriate amount of force. The firing sequence for an air cannon installation must be adjusted for the specific circumstances of the installation, including the conditions of chute, material, and climate. After satisfactory results are obtained, the cannon(s) can be put on a timer or other automatic control, so the firing cycle will maintain material flow without the attention of plant personnel.

Discharge of the air cannon into the chute can cause an increase of positive pressure within the chute, so it can increase the escape of dust driven out of the chute or the loading zone. In many cases, air cannons are used on sticky materials that require more force than can be supplied by vibration but that will not create high levels of dust. The air generated by air cannons should be included in the generated air calculation. *(See Chapter 7: Air Control, Equation 7.1, Total Airflow Calculation)*

OTHER METHODS TO IMPROVE FLOW

Aeration Systems

Some fine-particle materials, such as flour or cornstarch, will de-aerate when stored—they will become compacted and hard. If they have been stored for too long a period, they will not flow efficiently. Adding low-pressure/high-volume air to the products will allow the materials to flow efficiently again. This is done using a positive displacement blower that supplies air to aeration diffusers, pads, or nozzles mounted inside the vessel (**Figure 9.15**). Some aeration devices rely on the air current alone; some vibrate with the airflow. Air from the pads will break the adhesion between the material and the chutewall for dry material. Wet, sticky materials or lumpy materials do not respond well to this system.

Chute Linings

Lining materials, such as ceramic and engineered plastics, can provide an economical solution to flow problems in a chute. High Density Polyethylene (HDPE), Ultra-High Molecular Weight (UHMW) polyethylene, and ceramics have all shown the ability to promote material flow. The selected material for the lining must be able to handle the levels of impact and/or sliding abrasion seen in the application.

Figure 9.14

The discharge nozzles for an air-cannon system are embedded in the chutewall, so they discharge under the layer of accumulated material.

Figure 9.15

Aeration nozzles can be mounted inside the chute to improve flow.

9

The engineered plastics are usually bolted to the chutewalls with countersunk and covered fasteners. One problem that must be addressed with plastic linings is the difference in their rates of expansion and contraction from the rates of the metal wall. The mounting system must accommodate this difference by allowing the plastic liners to move. If this is not done, the liner will buckle, impeding material flow and quickly wearing out.

Ceramic liners can be installed on metal chutes with glue, welding, or a combination of both techniques.

Proper installation of chute linings is critical to achieving the benefits of lower coefficients of friction. If the sheets or tiles are not installed properly, the ridges where they join may actually increase the effective coefficient of friction over that of steel, worsening the chute's flow properties. Testing of the liner and bulk material is recommended to determine the actual coefficient of friction and predict wear rates.

Soft-Chute Designs

Most chutes are made from rigid metal. However, there are instances in which the chute or its lining can be made of a flexible material. Extremely wet or sticky materials respond well to soft-wall-chute designs.

A soft-chute design uses a space frame made of channel or angle iron. Attached to this frame is a flexible material, such as rubber or conveyor belting. Many times, the natural vibration of the equipment (originating from the conveyor drive or other equipment connected to the system) will prevent material from sticking to the rubber lining.

Vibrators and air cannons can be used to assist a soft chute by activating the flexible lining. One example of using a vibrator to promote flow in a soft chute is a vibrating dribble chute, in which a suspended sheet of plastic becomes a false floor or wall in the chute (**Figure 9.16**). A vibrator is attached to this sheet to keep the material in motion. *(See Chapter 14: Belt Cleaning for information on vibrating dribble chutes.)*

Another technique uses the discharge of air cannons into the back of a flexible rubber blanket installed as a lining on the chutewall (**Figure 9.17**). When the cannons discharge, they give a "kick" to the blanket to dislodge material buildup, like shaking sand out of a towel at the beach. The blanket is secured only at the top. Normally, the blanket is installed on only the flatter, or less "free-flowing," side of the chute. The discharge pipe should be aimed so it inclines down from the cannon to the vessel outlet to prevent material from entering the cannon's discharge opening. This technique works well with wet or sticky materials.

SYSTEM MAINTENANCE

Flow-aid devices are relatively sensitive to proper location and operation. One of the main advantages of using flow aids is that an operation will obtain a level of control over the material flow in a chute

Figure 9.16

A vibrating dribble chute attaches a vibrator to a suspended sheet of plastic to keep the material in motion.

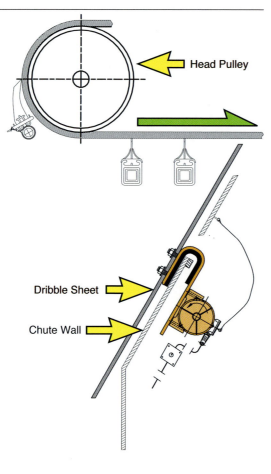

Head Pulley

Dribble Sheet

Chute Wall

⚠ SAFETY CONCERNS

As with all plant equipment, vibrators and air cannons present their own unique safety concerns. Noise and falling or flying material are the primary hazards resulting from the use of flow aids. Noise can be controlled by using flow aids only when needed. Exposure to falling and flying material can be controlled by the location of the flow aid and proper procedures for controlling access to chutes incorporating flow aids. Manufacturer instructions should be followed carefully for the installation, operation, and maintenance of flow-aid systems.

Vibrators should be rigidly mounted to the wall of the structure. Channel mounts should be fastened to the chutewall by stitch welding, in which intervals of weld bead are separated by spaces (**Figure 9.18**). This stitch-welding technique is designed to prevent a failure in the joint from breaking all the way across the mount plate. A monthly inspection of the mount weld area should be made to inspect for cracks in the welds. A safety cable must be installed to prevent the vibrator from falling should its mount fail.

Proper lockout / tagout / blockout / testout procedures should be followed when working on a vibrator or mount.

The mounts and discharge pipes for air cannons should be rigidly attached to the chutewall. The air cannon should be rigidly attached to the mount. It is not recommended that the threaded connection between the mount pipe and the air reservoir be welded completely, as this creates a stressed area, allowing the threads to break.

A safety cable must attach the air reservoir to a structural member to prevent the air cannon from falling in the event of a mount failure (**Figure 9.19**).

Prior to performing any work on the air cannon, the air tank must be totally discharged of air, and the air line supply line shut-off valve must be locked in the closed position to prevent air from filling the tank. It is also wise to pull the pressure relief valve to ensure there is no air left in the air-cannon vessel. Air cannons that fire only in response to a positive pressure signal (and, therefore, cannot discharge accidentally when de-energized) are available.

All inspection and entry doors must be locked to prevent inappropriate entry. Proper vessel entry procedures must be followed, and the air cannons must be properly locked out and discharged before personnel can enter the chute. The chute or vessel must have correct signage, warning of hazards (**Figure 9.20**).

Because flow-aid devices often use compressed air or other energy sources that can create a stored energy hazard, it is critical to follow lockout / tagout / blockout / testout procedures. Even though buildup in a chute may still be in place, its hold on the chutewall might be weakened to the point that a slight disturbance during maintenance can cause it to fall. There is an electrical shock hazard when working on the control systems. The possibility of remote actuation during maintenance and testing must be considered and procedures put in place to prevent unintended actuation.

Areas where vibrators or air cannons are placed may require workers in the vicinity to use hearing protection. The type of vessel and the size of flow-aid system will greatly vary the sound levels. Sound readings should be taken and cautionary signage posted as required (**Figure 9.21**).

If air cannons or aeration devices are used on enclosed bins or chutes, the increase in air pressure must be determined and pressure relief built into the system.

Figure 9.17

Another technique uses the discharge of an air cannon into the back of a flexible rubber blanket to dislodge material accumulations from inside the chute.

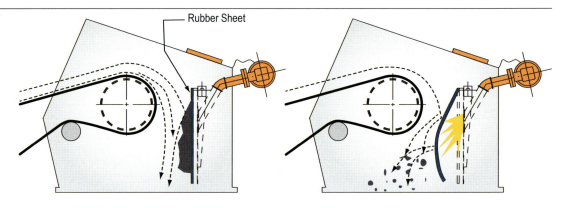

Rubber Sheet

Figure 9.18

Vibrators and mounts should be fastened to the chutewall by stitch welding, in which intervals of weld bead are separated by spaces.

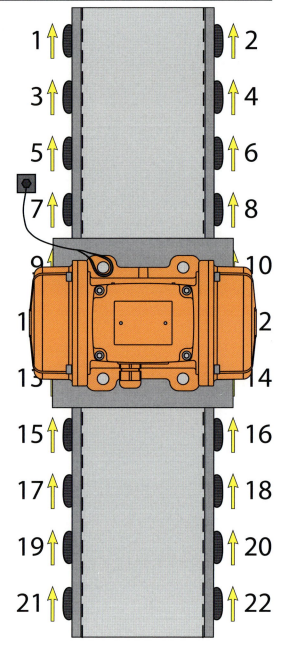

that is not possible any other way. This advantage can also become a problem, because it is very easy to adjust a flow aid out of its optimum operating settings. Often, workers will forget to record settings when doing maintenance, or they will try to adjust the flow aid in response to requests by the operators. This may result in poor performance in material movement, poor energy efficiency, and shorter life for the flow aid. If improperly mounted or adjusted, flow aids may not produce the desired effect and may even make the situation worse. An experienced specialty supplier can usually optimize the initial installation and control settings of a flow-aid system. These settings must be recorded for future reference.

Lack of required air pressure or volume will affect the performance. Keeping dirt and moisture out of the compressed air supply lines is critical for air-powered flow aids. Some pneumatic flow aids require lubrication; others do not. It is important to follow the manufacturers' requirements for air quality and treatment.

Flow aids are often located in areas where they are subject to falling material, impact from moving equipment, the elements, and vibration. Over time, these conditions may deteriorate the flow aid's supply lines and control systems. It is important to follow the manufacturers' recommendations for inspection and routine maintenance of the controls and supply lines.

Flow-aid devices deliver force to the chute and bulk material; over time, components will wear, or even break, under normal conditions. Most flow-aid devices can be rebuilt to extend their useful life. Because clearances and fits are critical to the proper operation of flow aids, it is recommended that flow-aid devices be rebuilt and repaired by the manufacturer—or that the manufacturer train plant maintenance personnel to properly rebuild the equipment.

Since flow aids usually operate intermittently, they may appear to be ready to work, when, in fact, they are not operating at optimum levels. The flow-aid device should be tested periodically, according to the manufacturer's suggestions, to make sure it is operating properly. An experienced specialty supplier can often tell from the sound or effect of the flow-aid device if it is in need of repair or adjustment.

APPLICATION OF FLOW AIDS

The typical characteristics and applications for various flow-aid systems can be compared (**Tables 9.3** and **9.4**).

In many cases, it is advantageous to "oversize" the flow-aid device—especially vibrators—by one model or unit size, so that the device can be turned down for its regular duties. If the requirements increase, a new vibrator does not need to be purchased.

There are generalized rules for sizing and positioning flow aids, but experience in diagnosing the problem and adjusting the flow aid for maximum effect is more art than science. The selection, installation, and control of flow aids are best done by a specialty supplier drawing on accumulated knowledge from numerous installations.

After a review of specific characteristics of any potential application, including the nature of the problem and the characteristics of the material, the choice of flow-aid type is often related to the available source of power at the point of application.

ADVANCED TOPICS

Sizing a Vibrator as a Flow Aid

Most vibrator manufacturers will provide the force output for their various units. The user is charged with determining the force required for a given application (**Equation 9.1**). From that force requirement the proper vibrator can be selected, using the manufacturers' technical data.

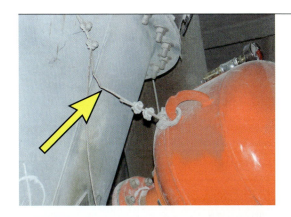

Figure 9.19

A safety cable must attach the air cannon's air reservoir to a structural member of the vessel.

Figure 9.20

Air-cannon systems require suitable safety signage and vessel entry procedures.

Figure 9.21

Many flow-aid devices require that noise protection signage be posted.

Table 9.3

Characteristics of Common Flow-Aid Devices

	Linear Vibrator	Rotary Vibrator	Air Cannon	Aerator
Electric		Yes		
Pneumatic	Yes	Yes	Yes	Yes
Hydraulic		Yes		
Variable Speed	Yes	Yes	Can discharge as often as air reservoir can be filled	Yes
Variable Force	Yes	Yes	Yes	Yes
Vibration Range (vpm)	1400 to 5500	Pneumatic Vibrators 3000 to 25000 vpm Electric Vibrators 600 to 3600 rpm		

Table 9.4

Suitable Applications for Flow Aids by Material Characteristics

Material Characteristic	Linear Vibrator	Rotary Vibrator	Air Cannon	Aerator
Light Fluffy		X		X
Small Particles		X		X
Large Particles	X		X	
Sticky Fines	X		X	
Explosive Materials*	X	X	X	
Interlocking Particles			X	
Thick Buildup of Material	X		X	

* Contact the manufacturer for products and accessories specifically designed for hazardous duty rating and locations.

FLOW AIDS AID FLOW

In Closing...

Any kind of flow-aid system has to be properly engineered to provide a benefit to an operation. Material specifications; process characteristics; and the number, size, and location of devices are all critical elements in an efficient flow-aid system. If not properly engineered and applied for an application, flow-aid devices can create additional problems.

Looking Ahead...

This chapter about Flow Aids, the fourth chapter in the section Loading the Belt, presents flow aids as a means to improve flow. The following chapter continues this section and focuses on Belt Support.

REFERENCES

9.1 Conveyor Equipment Manufacturers Association (CEMA). (2005). *BELT CONVEYORS for BULK MATERIALS, Sixth Edition*. Naples, Florida.

9.2 Any manufacturer and most distributors of conveyor products can provide a variety of materials on the construction and use of their specific products.

10

Figure 10.1

For an effective, minimum-spillage transfer point, the belt's line of travel must be stabilized with proper belt support in the conveyor's loading zone.

Chapter 10

BELT SUPPORT

In this Chapter...

This chapter focuses on belt support in the conveyor load zone to prevent the escape of fugitive materials and to prevent damage to the belt and other components. Topics covered include idlers, slider beds, and impact cradles, as well as several alternative methods for maintaining a stable belt line. Equations to calculate power requirements needed for belt support are provided.

The building of an efficient conveyor load zone is like the construction of a house: It starts with a good foundation. For a house, the foundation consists of the footings and/or walls of the basement; in a conveyor belt system, the foundation is a stable, sag-free belt line.

For a conveyor to control dust and spillage, the design engineer must do whatever is practical to keep the belt's line of travel consistently steady and straight. While there are many factors that influence the belt's running line both inside and outside the loading zone, a key ingredient is the provision of proper belt support.

For an effective, minimum-spillage transfer point, it is essential that the belt's line of travel be stabilized with proper belt support in the loading zone (**Figure 10.1**).

BENEFIT OF STABILITY

A flat, sag-free belt line in the skirted area is essential to successfully sealing the load zone (**Figure 10.2**). Ideally, the belting should be kept flat, as if it were running over a table that prevented movement in any direction except in the direction the cargo needed to travel; it would eliminate sag and be easier to seal.

Belt sag, when viewed from the side of the transfer point, is the vertical deflection of the belt from a straight line as drawn across the top of the two adjacent idlers (**Figure 10.3**). The shape of the sagging belt is assumed to be a catenary curve, a natural curve formed when a cable is suspended by its endpoints.

If the belt sags between idlers below the loading zone or flexes under the stress of loading, fines and lumps will work their way out the sides of the conveyor, dropping onto the floor as spillage or becoming airborne as a cloud of dust. Worse, these materials can wedge into entrapment points where they can gouge the belt or damage the sealing system and other components, worsening the spillage problem. A small amount of belt sag—sag that is barely apparent to the naked eye—is enough to permit fines to become entrapped, leading to abrasive wear on the skirtboard-sealing system and the belt surface. A groove cut into the belt cover along the entire length of the belt in the skirted area can usually be attributed to material captured in entrapment points (**Figure 10.4**). When belt sag is prevented, the number and size of

Figure 10.2

A flat, sag-free belt line in the skirted area is essential to successfully sealing the load zone.

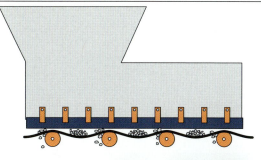

Figure 10.3

Belt sag is a vertical deflection of the belt from a straight line as drawn across the top of the two adjacent idlers.

Figure 10.4

A groove cut into the belt cover along the entire length of the belt in the skirted area can usually be attributed to material captured in entrapment points.

10

entrapment points are reduced, therefore reducing the possibility of belt damage.

In order to prevent spillage and reduce the escape of dust particles, belt sag must be eliminated wherever practical to the extent possible. It is particularly important to control sag in the conveyor's loading zone, where the cargo constantly undergoes changes in weight. These changes in load carry fines and dust out of the sealing system and push particles into entrapment points between the wear liner or skirt seal and the belt.

Methods to Control Sag

One method for reducing belt sag along the entire length of the conveyor is to increase the belt tension. There are drawbacks to this, however, such as increased drive power consumption and additional stress on the belt, splice(s), and other components. When utilizing additional tension to reduce sag, the maximum rated tension of the belting should never be exceeded.

After achieving the belt tension required by the conveyor belt and the load on the system, the recommended method for reducing belt sag is to improve the conveyor's belt-support system (**Figure 10.5**).

Proper Belt Support

The key to a stable, sag-free line of belt travel is proper support. The amount of support needed is determined by the unique characteristics of each individual conveyor, its loading zone(s), and its material load. The factors to be assessed include the trough angle and speed of travel of the

conveyor being loaded, the weight of the material, the largest lump size, the material drop height, and the angle and speed of material movement during loading.

It is essential that the belt be stabilized throughout the entire length of the load zone. Support systems extended beyond what is minimally required will provide little harm other than an incidental increase in conveyor power requirements. A belt-support system that is left shorter than required can lead to fluctuations in the belt's stability at the end of the support system, potentially creating spillage problems that will render the installed belt-support system almost pointless. Belt support is like money: It is much better to have a little extra than to fall a little short.

Basics of Building Belt Support

It is essential that the stringers—the conveyor's support structure upon which all other components are installed—are straight and parallel for proper belt support. If not, they should be straightened or replaced. Laser surveying is the preferred method for checking stringer alignment. *(See Chapter 16: Belt Alignment.)*

Footings must provide a rigid support structure to prevent stringer deflection. The amount of material being loaded and the level of impact forces must be considered to prevent excessive deflection under load. Properly spaced stringers tied to rigid footings ensure a good base for the remaining structure.

Conveyor Equipment Manufacturers Association (CEMA) provides a valuable resource for construction standards for conveyors and loading zones: "Conveyor Installation Standards for Belt Conveyors Handling Bulk Materials" *(Reference 10.1)*.

There are a number of techniques and components that can be used, independently or in combination, to control belt sag by improving belt support in the loading zone. They include idlers, belt-support cradles, and impact cradles.

Figure 10.5

To reduce belt sag, improve the conveyor's belt-support system.

BELT SUPPORT WITH IDLERS

The basic means of support for a conveyor belt is idlers. An idler consists of one or more rollers—with each roller containing one or more bearings to ensure it is free rolling. The rollers are supported by, or suspended from, a framework installed across the conveyor stringers (**Figure 10.6**). Idlers are the most numerous of conveyor components, in terms of both the number used on a particular conveyor and the number of styles and choices available. There are many types, but they all share the same responsibilities: to shape and support the belt and cargo, while minimizing the power needed to transport the materials.

The Idler Family

Idlers are classified according to roll diameter, type of service, operating condition, belt load, and belt speed; they are rated on their load-carrying capacity based on calculated bearing life. CEMA uses a two-character code that expresses the idler classification and implied load rating, with a letter-based code followed by idler diameter in inches, resulting in classes from B4 to F8 (**Table 10.1**). Other regions may have different classification systems.

Regardless of the codes and classifications, the key is to make sure each conveyor is consistent throughout—that all idlers on a given conveyor conform to the same standards and, ideally, are supplied by the same manufacturer.

There is a wide variety of general categories of idlers, depending on their intended application.

Carrying Idlers

Carrying idlers provide support for the belt while it carries the material. They are available in flat or troughed designs. The flat design usually consists of a single horizontal roll for use on flat belts, such as belt feeders.

Figure 10.6

An idler consists of one or more rollers, each with one or more bearings. The rollers are supported by, or suspended from, a framework installed across the conveyor stringers.

Idler Classifications (Based on CEMA Standards)					
CEMA Idler Classification	**Roll Diameter**		**Belt Width**		**Description**
	mm	**in.**	**mm**	**in.**	
B4	102	4	450-1200	18-48	Light Duty
B5	127	5	450-1200	18-48	
C4	102	4	450-1500	18-60	Medium Duty
C5	127	5	450-1500	18-60	
C6	152	6	600-1500	24-60	
D5	127	5	600-1800	24-72	
D6	152	6	600-1800	24-72	
E6	152	6	900-2400	36-96	Heavy Duty
E7	178	7	900-2400	36-96	
F6	152	6	1500-2400	60-96	
F7	178	7	1500-2400	60-96	
F8	203	8	1500-2400	60-96	

Table 10.1

Metric dimensions are conversions by Martin Engineering; belt widths may not be actual metric belt sizes.

Figure 10.7

The troughed idler set usually consists of three rolls—one horizontal roll in the center with inclined (or wing) rolls on either side.

Figure 10.8

"Picking" idlers incorporate a longer center roll and shorter inclined rollers to supply a large cargo area.

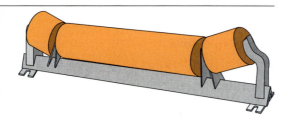

Figure 10.9

With in-line idlers, the centerlines of the three rolls are aligned.

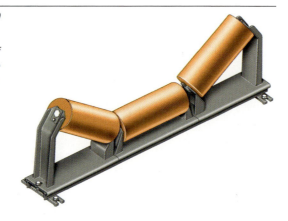

Figure 10.10

Offset idlers, with the center roller placed on a centerline different from the wing rolls, can reduce the overall height of the idler set.

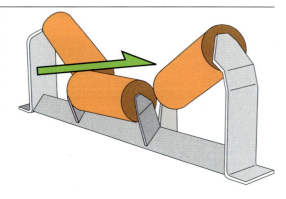

Figure 10.11

Used to support the belt on its way back to the loading zone, return idlers normally consist of a single horizontal roll hung from the underside of the conveyor stringers.

The troughed idler usually consists of three rolls—one horizontal roll in the center with inclined (or wing) rolls on either side (**Figure 10.7**). The angle of the inclined rollers from horizontal is called the trough angle. Typically, all three rolls are the same length, although there are sets that incorporate a longer center roll and shorter inclined rollers called "picking" idlers. This design supplies a larger flat area to carry material while allowing inspection or "picking" of the cargo (**Figure 10.8**).

Troughed idler sets are available as in-line idlers (**Figure 10.9**)—the centerlines of the three rolls are aligned—and offset idlers—the center roll has a centerline different from the wing rollers, usually with the belt passing over the center roller in advance of the wing rollers (**Figure 10.10**). Offsetting the idlers can reduce the overall height of the idler set and, accordingly, is popular in underground mining applications, where headroom is at a premium. Offset idlers eliminate the gap between the rollers, reducing the chance of a type of belt damage called junction-joint failure.

Return Idlers

Return idlers provide support for the belt on its way back to the loading zone after unloading the cargo. These idlers normally consist of a single horizontal roll hung from the underside of the conveyor stringers (**Figure 10.11**). V-return idlers, incorporating two smaller rolls, are sometimes installed to improve belt tracking (**Figure 10.12**).

Training Idlers

There are a number of designs for training idlers that work to keep the belt running in the center of the conveyor structure. Typically, these idlers are self-aligning: They react to any mistracking of the belt to move into a position that will attempt to steer the belt back into the center (**Figure 10.13**). They are available for both carrying side and return side application. *(See Chapter 16: Belt Alignment.)*

Belt-training idlers should never be installed under the carrying side of the belt in the load zone, as they sit higher than the adjacent regular carrying idlers and raise the belt as they swivel.

Impact Idlers

Rubber-cushioned impact idlers are one solution for absorbing impact in the belt's loading zone (**Figure 10.14**). These idlers use rollers composed of resilient rubber disks to cushion the force of loading. Impact idlers typically have the same load rating as standard idlers, because they utilize the same shafts and bearings. The rubber covers absorb some of the energy to provide the benefit of shock absorption.

One disadvantage of using impact rollers in the load zone is that each idler supports the belt only at the top of the roller. No matter how closely spaced, the rounded shape of the roller and the ability of the rubber to deflect under the load will allow the conveyor belt to oscillate or sag away from the ideal flat profile (**Figure 10.15**). This sag allows and encourages the escape or entrapment of fugitive material. The space interval between impact rollers offers little protection from tramp materials dropping from above and penetrating the belt.

Even impact idlers are subject to impact damage, suffering damaged bearings and rollers from "too large" lumps or unusual impacts (**Figure 10.16**). Idlers with worn or seized bearings cause the belt to run erratically, allowing mistracking and spillage over the sides of the belt. Idlers damaged from severe impact or seized due to fugitive material increase the conveyor's power consumption significantly. In many cases, it becomes more effective to absorb impact with impact cradles, as discussed below.

Idler Spacing

The spacing between the rolling components has a dramatic effect on the idlers' support and shaping missions. Idlers placed too far apart will not properly support the belt nor enable it to maintain the desired profile. Placing idlers too close together will

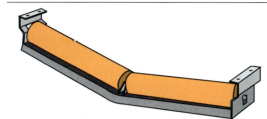

Figure 10.12

V-return idlers, incorporating two smaller rolls, are sometimes installed to improve belt tracking.

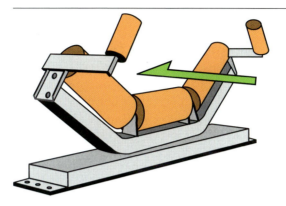

Figure 10.13

Available for both the carrying side and return side of the conveyor, training idlers self-align to steer the belt back into the proper path.

Figure 10.14

Impact idlers use rollers composed of resilient rubber disks to cushion the force of loading.

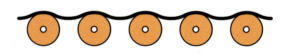

Figure 10.15

The rounded shape of the roller and the ability of the rubber to deflect under the load will allow the conveyor belt to oscillate or sag away from the ideal flat profile.

Figure 10.16

Although designed to cushion loading forces, impact idlers are subject to impact damage, suffering damaged bearings, rollers, or bent frames from "too large" lumps or unusual impacts.

10

improve belt support and profile, but will increase conveyor construction costs and may lead to an increase in the conveyor's power consumption.

Normally, idlers are placed close enough to support a fully loaded belt so it will not sag excessively between them. If the belt is allowed too much sag, the load shifts as it is carried up and over each idler and down into the valley between. This shifting of the load increases belt wear and power consumption. The sag also encourages material spillage. CEMA has published tables of recommended idler spacing for applications outside the loading zone (**Table 10.2**).

The spacing of return idlers is determined by belt weight, because no other load is supported by these idlers and sag-

related spillage is not a problem on this side of the conveyor. Typical return idler spacing is 3 meters (10 ft).

Idlers in the Skirted Area

The basic and traditional way to improve belt support, and so reduce belt sag under a loading zone or anywhere else along the conveyor, is to increase the number of idlers. By increasing the number of idlers in a given space—and consequently decreasing the space between the idlers—the potential for belt sag is reduced (**Figure 10.17**). Idlers can usually be positioned so that their rolls are within 25 millimeters (1 in.) of each other (**Figure 10.18**).

However, this method is not without drawbacks. As the idlers are packed more tightly, it becomes more difficult to service them. Idler sets are typically maintained by laying the framework over on its side to allow the rolls to be lubricated or replaced. If the idlers are closely spaced, there is no room available for the idler set to be laid on its side to allow the maintenance to be performed (**Figure 10.19**). To reach one set of idlers, one or more adjacent sets must be removed, creating a "falling domino" chain reaction.

Figure 10.17

The traditional method to reduce belt sag under a loading zone is to increase the number of idlers in a given space, consequently decreasing the space between the idlers.

Table 10.2

Recommended Idler Spacing for Applications Outside the Loading Zone as Published by CEMA								
		Carrying Side Idler Spacing Outside the Loading Zone						
		Weight of Material Handled in Kilograms per Cubic Meter (lb$_m$/ft³)						
Return Idler Spacing	**Belt Width**	**480 (30)**	**800 (50)**	**1200 (75)**	**1600 (100)**	**2400 (150)**	**3200 (200)**	
m (ft)	**m (in.)**	**m (ft)**	**m (ft)**	**m (ft)**	**m (ft)**	**m (ft)**	**m (ft)**	
3,0 (10.0)	457 (18)	1,7 (5.5)	1,5 (5.0)	1,5 (5.0)	1,5 (5.0)	1,4 (4.5)	1,4 (4.5)	
3,0 (10.0)	610 (24)	1,5 (5.0)	1,4 (4.5)	1,4 (4.5)	1,2 (4.0)	1,2 (4.0)	1,2 (4.0)	
3,0 (10.0)	762 (30)	1,5 (5.0)	1,4 (4.5)	1,4 (4.5)	1,2 (4.0)	1,2 (4.0)	1,2 (4.0)	
3,0 (10.0)	914 (36)	1,5 (5.0)	1,4 (4.5)	1,2 (4.0)	1,2 (4.0)	1,1 (3.5)	1,1 (3.5)	
3,0 (10.0)	1067 (42)	1,4 (4.5)	1,4 (4.5)	1,2 (4.0)	1,1 (3.5)	0,9 (3.0)	0,9 (3.0)	
3,0 (10.0)	1219 (48)	1,4 (4.5)	1,2 (4.0)	1,2 (4.0)	1,1 (3.5)	0,9 (3.0)	0,9 (3.0)	
3,0 (10.0)	1372 (54)	1,4 (4.5)	1,2 (4.0)	1,1 (3.5)	1,1 (3.5)	0,9 (3.0)	0,9 (3.0)	
3,0 (10.0)	1524 (60)	1,2 (4.0)	1,2 (4.0)	1,1 (3.5)	0,9 (3.0)	0,9 (3.0)	0,9 (3.0)	
2,4 (8.0)	1829 (72)	1,2 (4.0)	1,1 (3.5)	1,1 (3.5)	0,9 (3.0)	0,8 (2.5)	0,8 (2.5)	
2,4 (8.0)	2134 (84)	1,1 (3.5)	1,1 (3.5)	0,9 (3.0)	0,8 (2.5)	0,8 (2.5)	0,6 (2.0)	
2,4 (8.0)	2438 (96)	1,1 (3.5)	1,1 (3.5)	0,9 (3.0)	0,8 (2.5)	0,6 (2.0)	0,6 (2.0)	

Metric conversions added by Martin Engineering; belt widths may not be actual metric belt sizes.

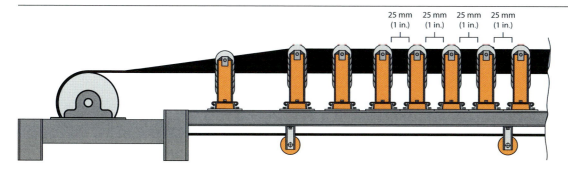

Figure 10.18

Idlers can usually be positioned so that their rolls are within 25 millimeters (1 in.) of each other.

25 mm (1 in.)   25 mm (1 in.)   25 mm (1 in.)   25 mm (1 in.)

10

Track-Mounted Idlers

Track-mounted idlers that slide into position are a solution to the problems in servicing closely-spaced idlers. These idlers are mounted on a steel beam that forms a track, allowing the individual rollers to be installed or removed with a slide-in/slide-out movement perpendicular to the path of the conveyor (**Figure 10.17** and **Figure 10.20**). Idlers used in track-mounted configurations can be steel rollers or rubber ring impact-style rollers. With track-mounted idlers, each individual roller, or each set, can be serviced without laying the frame on its side or raising the belt.

The track upon which idlers (and/or other belt-support components) slide provides a supplement to the conveyor structure. This track could be incorporated into the design of the conveyor as part of the structure (**Figure 10.21**). Incorporating a slide-in-place system during the conveyor's design stage allows the use of modular belt-support structures, idlers, cradles, or combination units and simplifies component installation. This is particularly beneficial on wide belts, where large components might otherwise require cranes or other heavy equipment for installation.

Tips for Idler Installation

When installing idlers in a transfer point, they should be square with the stringers and aligned horizontally and vertically across the conveyor. Variations will cause entrapment points, capturing material that will lead to belt damage and spillage. Laser surveying can be used to ensure the alignment of all rolling components. *(See Chapter 16: Belt Alignment.)*

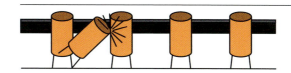

Figure 10.19

If the idlers are closely spaced, there is no room available for the idler set to be laid on its side for maintenance.

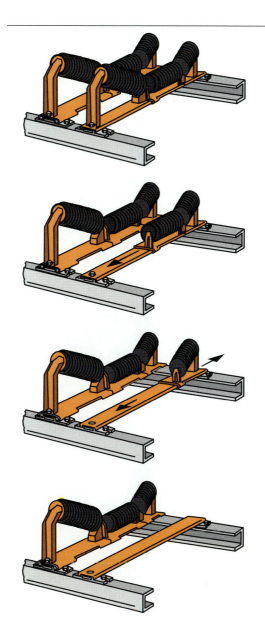

Figure 10.20

Track-mounted idlers solve the problems in servicing closely-spaced idlers by allowing the individual rollers to be installed or removed with a slide-in/slide-out movement.

Idler standards have tolerances for roll diameter, roundness (or "run out"), center roll height, and trough angle. Even a slight difference in an idler's dimensions—the difference from one manufacturer to another—can create highs and lows in the belt line, making it impossible to provide effective sealing. Idlers must be aligned with care and matched so as not to produce humps or valleys in the belt. Idlers should be checked for concentricity; the more they are out of round, the greater is the ten-

dency for the belt to flap or bounce. Only idlers supplied by the same manufacturer and of the same roll diameter, class, and trough angle should be used in the skirted area of a conveyor.

BELT-SUPPORT CRADLES

So important is the "flat table" concept to good sealing that many designers now use cradles in place of idlers under conveyor loading zones (**Figure 10.22**). Instead of using an idler's rolling "cans," cradles use some variety of low-friction bars to support the belt profile.

In this discussion of belt-support systems, the terms cradle, bed, or saddle should be considered synonymous.

All belt-support cradles perform two functions—controlling belt sag in the load zone to curtail spillage and providing a slick surface upon which the belt can ride. In addition, impact cradles reduce belt damage by absorbing the forces from the landing of material on the belt. Other benefits of the use of cradles under the transfer point include a reduction in moving parts and elimination of required lubrication. The modular design of the typical cradle system allows the belt support to be extended as far as the circumstances require.

Edge-Seal-Support Cradles

Edge-sealing-support systems are designed to provide continuous support of the belt and maintain a straight belt profile at the belt edges.

One form of edge-seal support is a "side rail" configuration. This system places one or more low-friction bars on both sides of the conveyor directly under the skirtboard seal (**Figure 10.23**). The bars function to support the sides of the belt, allowing effective sealing of the belt edge.

Each edge-seal cradle installation may be one or more cradles long, depending on the length of the transfer point, the speed of the belt, and other conveyor characteristics.

Figure 10.21

The track upon which idlers slide can be incorporated into the design of the conveyor as part of the structure.

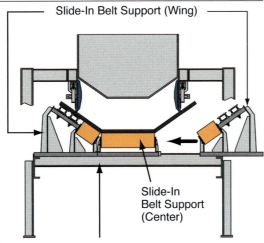

Slide-In Belt Support (Wing)

Slide-In Belt Support (Center)

Conveyor Cross Member (I-Beam)

Figure 10.22

To provide the flat table that allows effective sealing, many conveyors use cradles instead of idlers in the conveyor loading zone.

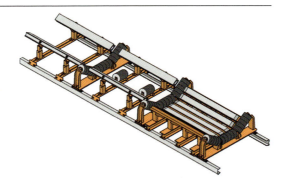

Figure 10.23

"Side rail" edge support places one or more low-friction bars on both sides of the conveyor directly under the skirtboard seal.

10

The top of these bars should be installed in line with the top of the entry and exit idlers, to avoid the creation of entrapment points (**Figure 10.24**). When multiple edge-sealing cradles are used, idlers should be placed between the cradles.

On faster, wider, more heavily loaded belts, the edge-seal cradles may need more than one bar on each side to support the belt edge. On wider belts, it is often necessary to add a center support roll or an additional low-friction bar under the middle of the belt (**Figure 10.25**).

Edge-support slider bars can be manufactured from low-friction plastics such as Ultra-High Molecular Weight (UHMW) polyethylene. These materials provide a low-drag, self-lubricating surface that reduces heat accumulation and undue wear on either the belt or the bar. One proprietary design features bars formed in an "H" or "box" configuration, allowing for the use of both the top and bottom surfaces (**Figure 10.26**).

At conveyor speeds above 3,8 meters per second (750 ft/min), the heat created by the friction of the belt can reduce the performance of the plastic bars. Consequently, the use of stainless steel support bars has found acceptance in these applications. Stainless steel bars should also be incorporated in applications with service temperatures above 82 degrees Celsius (180° F).

Safety regulations may limit the choice of materials used in bar support systems. Most countries have regulations requiring anti-static and/or fire–resistant materials used in contact with the belt in underground applications. Other regional or plant requirements may govern materials to be used.

The low-friction bars should be supported in a mounting frame that is adjustable, to allow easy installation, alignment, and maintenance. This frame should accommodate various idler combinations and chutewall widths and allow for adjustment due to wear.

The bars should be held in the support position without the risk of the mounting hardware and fasteners coming into contact with the belt. For example, the bolts holding the bars in place should be installed parallel rather than perpendicular to the belt (**Figure 10.27**).

An edge-support cradle may add incrementally to the friction of the belt and to the conveyor's power requirements. However, this marginal increase in energy consumption is more than offset by the elimination of the expenses for cleanup of skirt leakage, entrapment-point damage to the belt, and unexpected downtime necessary for idler maintenance or belt replacement.

Figure 10.24

To avoid the creation of entrapment points, the top of the belt-support bars should be installed in line with the top of the entry and exit idlers.

Figure 10.25

On wider belts, it may be necessary to add additional support rolls or low-friction bars in the center of the support cradle.

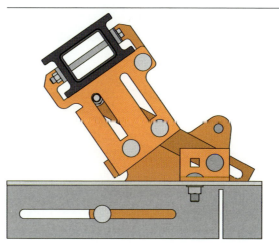

Figure 10.26

One proprietary cradle features bars formed in an "H" or "box" configuration, allowing the use of both the top and bottom surfaces.

10

Impact Cradles

Nothing can damage a conveyor's belting and transfer-point components and create material leakage as rapidly and dramati-

cally as impact in the loading zone from heavy objects or lumps with sharp edges (**Figure 10.28**). Whether arising from long material drops or large lumps—or boulders, timber, or scrap metal—these impacts will damage components like idlers and sealing strips. Impact can also create a "ripple" effect on the belt, de-stabilizing its line of travel and increasing the spillage of material. Heavy or repeated impacts can also damage the belt cover and weaken its carcass. Consequently, system engineers do a variety of things to reduce impact levels in loading zones, including the inclusion of engineered chutes, rock boxes, or designs that load fines before lumps.

However, in many cases, it is not possible to totally eliminate impact, so it becomes necessary to install some sort of energy-absorbing system under the loading zone. If one were to lay a belt on a concrete floor and strike it with an ax or a hammer, the belt would be damaged. However, if one would place layers of foam between the belt and the floor, the belt would be somewhat protected. This is the way an impact belt-support system protects the belt under severe impact loading conditions.

Impact cradles are installed directly under the material-drop zone to bear the brunt of the shock of the material hitting the belt as it loads (**Figure 10.29**). These cradles are usually composed of a set of individual impact-absorbing bars assembled into a steel support framework. The bars are composed of durable elastomeric materials that combine a slick top surface—allowing the belt to skim over it to minimize friction—and one or more sponge-like secondary layers to absorb the energy of impact (**Figure 10.30**).

Some manufacturers align a group of long bars—typically 1,2 meters (4 ft) in length—into a cradle, with the bars running parallel to the direction of belt travel. Other manufacturers use shorter modular segments that align to form a saddle that is perpendicular to belt travel. These saddles are typically 300 millimeters (12 in.) in width. The number of cradles and saddles

Figure 10.27

The bars should be held in the support position without the risk of the mounting hardware and fasteners coming into contact with the belt.

Figure 10.28

Impact in the loading zone from long material drops or large lumps can damage components and create spillage.

Figure 10.29

Impact cradles are installed directly under the material drop zone to bear the brunt of the shock of the material hitting the belt.

Figure 10.30

Impact cradles are composed of a steel framework holding a set of impact-absorbing bars. The bars combine a slick top surface and one or more sponge-like secondary layers to absorb the energy of impact.

required is determined by the length of the impact zone. The number of bars required in a given cradle or saddle is determined by the width of the conveyor belt.

Some systems feature a slick top surface and a cushioned lower layer permanently attached; others feature separate components that are put together at the application. Impact cradles are available in a track-mounted design, which simplifies replacement of bars when required (**Figure 10.31**).

The limit to the amount of impact that can be absorbed by the belt in combination with an impact cradle is based on the belt's ability to resist crushing energy. For loading zones with the highest levels of impact, the entire impact-cradle installation can be mounted on a shock-absorbing structure, such as springs or air cushions. While this does reduce the stiffness of the entire loading zone and so absorbs impact force, it has the drawback of allowing some vertical deflection of the belt in the skirted area, making it harder to seal the load zone.

Standard for Impact Cradles

CEMA STANDARD 575–2000 provides an easy-to-use rating system for impact cradles utilized in bulk-materials handling applications. This system gives manufacturers and users a common rating system to reduce the chance for misapplication.

The cradle-classification system is based on the impact energy created by the bulk material to establish a duty rating for the given application. The impact-force requirement is determined for each application by calculating the worst-case impact. For a given application, the impact from both the single largest lump (**Equation 10.1**) (**Figure 10.32**) and a continuous homogeneous flow (**Equation 10.2**) (**Figure 10.33**) should be calculated. Most applications will use the larger of these two forces. The reference numbers for impact force are then used to select one of the three ratings from a chart (**Table 10.3**).

The equations used by CEMA are generally accepted as reasonable approximations of impact forces. The CEMA

Figure 10.31

Impact cradles are available in a track-mounted design, which simplifies replacement of bars when required.

$$F_I = W + \sqrt{2 \cdot k \cdot W \cdot h_d}$$

Equation 10.1

Calculating Impact Force from a Single Lump of Material (CEMA STANDARD 575-2000)

Given: A lump of material with a weight (force) of 475 newtons (100 lb$_f$) drops 4 meters (13 ft) onto an impact cradle with an overall spring constant of 1000000 newtons per meter (70000 lb$_f$/ft). **Find:** The impact force created by the lump of material.

	Variables	Metric Units	Imperial Units
F_I	Impact Force	newtons	pounds-force
k	Spring Constant of System that is Absorbing the Impact	1000000 N/m	70000 lb$_f$/ft
W	Weight (Force) of the Largest Lump of Material	475 N	100 lb$_f$
h_d	Drop Height	4 m	13 ft
Metric: $F_I = 475 + \sqrt{2 \cdot 1000000 \cdot 475 \cdot 4} = 62119$			
Imperial: $F_I = 100 + \sqrt{2 \cdot 70000 \cdot 100 \cdot 13} = 13591$			
F_I	Impact Force	62119 N	13591 lb$_f$

10

STANDARD notes that the impact from a maximum lump size almost always yields the highest impact force and, therefore, should govern the impact rating specified for a given application. A completely thorough analysis would involve adding the force absorbed by the lump with the force absorbed by a stream and cross referencing the force value.

The dimensions for cradle construction are based on CEMA's long-established idler-classification system. They include the ratings: B, C, D, E, or F, followed by the nominal idler diameter as measured in inches (e.g., 5, 6, or 7).

Cradles with Bars and Rollers

A number of "combination cradle" designs are available, which use bars for a continuous seal at the belt edge but also incorporate rollers under the center of the belt (**Figure 10.34**). These hybrid designs are popular as a way of combining the low power consumption of rollers with the flat sealing surface of impact or slider bars. With a hybrid design, the running friction is kept low by supporting the center of the belt with conventional rollers. This reduces the power consumption of the conveyor. The belt edge is continuously supported, eliminating belt sag between the idlers. This reduces spillage to a mini-

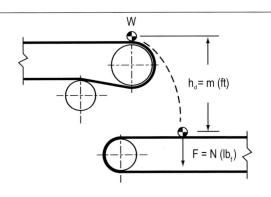

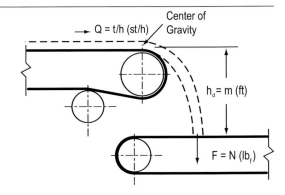

Figure 10.32 (left)

Impact Calculation from the Single Largest Lump

Figure 10.33 (right)

Impact Calculation from a Continuous Homogeneous Flow

Equation 10.2

Calculating Impact Force from a Stream of Material (CEMA STANDARD 575-2000)

$$F_s = k \cdot Q \cdot \sqrt{h_d}$$

Given: A stream of material drops 4 meters (13 ft) onto an impact cradle at the rate of 2100 tons per hour (2300 st/h). **Find:** The impact force created by the stream of material.

	Variables	Metric Units	Imperial Units
F_s	Impact Force	newtons	pounds-force
Q	Material Flow	2100 t/h	2300 st/h
h_d	Drop Height	4 m	13 ft
k	Conversion Factor	1,234	0.1389

Metric: $F_s = 1,234 \cdot 2100 \cdot \sqrt{4} = 5183$

Imperial: $F_s = 0.1389 \cdot 2300 \cdot \sqrt{13} = 1152$

F_s	Impact Force	5183 N	1152 lb$_f$

Table 10.3

CEMA STANDARD 575-2000 Impact Bed/Cradle Rating System			
Code	Rating	Impact Force (N)	Impact Force (lb$_f$)
L	Light Duty	<37800	<8500
M	Medium Duty	37800-53400	8500-12000
H	Heavy Duty	53400-75600	12000-17000

Metric conversions added by Martin Engineering.

mum. Since the central rollers operate in a virtually dust-free environment, the life of idler bearings and seals is extended, thus reducing long-term maintenance costs. These designs are most commonly seen on high-speed conveyors operating above 3,8 meters per second (750 ft/min) or applications where there is a heavy material load that would create high levels of friction in the center of the conveyor.

Another possibility is to use cradles incorporating impact bars in the center with short picking idlers closely spaced on the wings. Here the design intention is to provide superior impact cushioning in the center of the belt, while reducing friction on the belt edges.

CRADLE INSTALLATION

Multiple-Cradle Systems

It is often appropriate to install combination systems, incorporating both impact-absorbing cradles and seal-support cradles (**Figure 10.35**). As many impact cradles as necessary should be installed to support the belt to the end of the impact zone. Side-seal-support cradles then complete the system over the distance required to stabilize the load in the skirted area.

These systems provide an efficient way to combine optimum belt support with maximum cost-efficiency in system construction and power consumption.

Cradle Alignment

The impact cradle is usually installed so that the bars in the center of the cradle are set slightly—12 to 25 millimeters (0.5 to 1 in.)— below the normal unloaded line of the belt (**Figure 10.36**). This allows the belt to absorb some of the force of impact when the material loading deflects it down onto the cradle, while avoiding continuous friction and wear on the bars. The wing bars—the bars on the sides of the cradle—should be installed in line with the entry, exit, and intermediate idlers to prevent belt sag and entrapment points (**Figure 10.37**).

It is important that the bar directly under the steel chute or skirtboard wall be precisely aligned with the wing idlers.

Cradles can be welded or bolted to the stringers; it may be better to bolt the systems in place, as this will allow more efficient maintenance. Some impact cradles are available in a track-mounted design, which simplifies cradle installation or the replacement of bars when required.

Figure 10.34

"Combination cradles" use bars for a continuous seal at the belt edge and rollers under the center of the belt.

Figure 10.35

Combination systems, incorporating both impact-absorbing cradles and seal-support cradles, can be installed to provide a stable belt line.

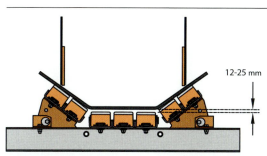

Figure 10.36

Impact cradles are usually installed so that the bars in the center of the cradle are set 12 to 25 millimeters (0.5 to 1 in.) below the normal unloaded line of the belt.

The wing bars of an impact cradle should be installed in line with the entry, exit, and intermediate idlers to prevent belt sag and entrapment points.

12-25 mm

10

Figure 10.38

A well-designed impact cradle will simplify installation with the use of adjustable wing supports, which allow the cradle to be placed under the belt before its sides are raised to the appropriate trough angle. Note: adjustment range for a 35° troughing angle is ± 2.5°.

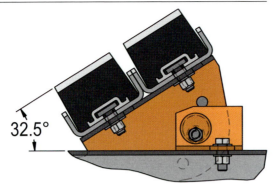

Adjustment Detail 32.5°

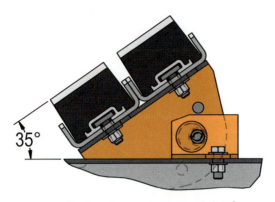

Adjustment Detail 35°

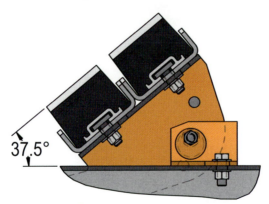

Adjustment Detail 37.5°

Figure 10.39

When two or more cradles are installed, the use of intermediate idlers—idlers placed between the adjacent cradles—is recommended.

Installation of impact cradles is simplified through the use of adjustable wing supports, which allow the cradle to be slid under the belt in a flat form; the sides are then raised to the appropriate trough angle (**Figure 10.38**). It is important that the cradle be designed to allow some simple means of adjustment of bar height and angle. This will enable the cradle to work with idlers of varying manufacturers and allow compensation for wear.

Idlers Between Cradles

When two or more cradles are installed, the use of intermediate idlers—that is, idlers placed between the adjacent cradles—is recommended (**Figure 10.39**). Installing an idler set between two cradles (or putting each cradle between two idlers) will reduce the drag of the conveyor belt over the bars. This reduces the conveyor's power consumption. In addition, the heat buildup in the bars will be reduced, giving the bars and belt longer life expectancies.

Idlers should be specified before and after each 1200 millimeter (4 ft) cradle; the number of idler sets required for a given transfer point is the same as the number of cradles required plus one. To ensure uniformity for a stable belt line, all of these idlers should be of the same manufacturer with the same size roller. Impact idlers should be used between cradles under the loading zone; conventional idlers can be used outside the impact area. Track-mounted idlers should be used between cradles to allow for ease of maintenance.

In some impact areas, it may be acceptable to go as far as 2,4 meters (8 ft) between intermediate idlers. These applications might include long loading zones where it is difficult to predict the location of the impact and where rollers might be damaged by point-impact loading. These would also include transfer points under quarry and mine dump hoppers, at pulp and paper mills where logs are dropped onto belts, or at recycling facilities that see heavy objects ranging from car batteries to truck engines dropped on conveyors.

ALTERNATIVE METHODS OF BELT SUPPORT

This book discusses several methods of alternative conveying systems. *(See Chapter 33: Considerations for Specialty Conveyors.)* In addition, there are other methods of supporting conventional belting on more-or-less conventional conveyor structures.

Catenary Idlers

Catenary idlers, sometimes called garland idlers, are sets of rollers—typically, three or five—linked together on a cable, chain, or other flexible connection and suspended from the conveyor structure below the belt (**Figure 10.40**). These idler sets swing freely under the forces of loading material, acting to absorb impact and centralize the load. Their flexible mounting allows the idlers to be quickly moved or serviced and provides some amount of self-centering.

Catenary idlers are typically seen in very heavy-duty applications such as conveyors seeing high impact levels and large volumes of material. Typical installations would include conveyors under the discharge of bucket wheel excavators and under the loading zones of long overland conveyors carrying run-of-mine material (**Figure 10.41**). Catenary idlers are also commonly used in the metalcasting industry.

However, the "bounce" and swing of catenary idlers and the changes this motion can add to the belt path, particularly when the material is loaded off-center, must be considered when engineering a conveyor system (**Figure 10.42**). As the catenary idler swings, the belt moves from side to side. This allows the escape of fugitive material out the sides of the loading zone and creates mistracking that exposes belt edges to damage from the conveyor structure (**Figure 10.43**). Consequently, greater edge distance must be left outside the skirtboard to allow for sealing.

Air-Supported Conveyors

Another concept for stabilizing the belt path is the air-supported belt conveyor.

These conveyors replace carrying-side idlers and cradles with a trough-shaped plenum below the belt. The belt is supported by a film of air that is released from the plenum (**Figure 10.44**). *(See Chapter 23: Air-Supported Conveyors.)*

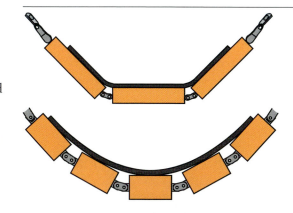

Figure 10.40

Catenary idlers, sometimes called garland idlers, are sets of rollers—typically, three or five—linked together on a cable, chain, or other flexible connection and suspended from the conveyor structure below the belt.

Figure 10.41

Typical installation of catenary idlers include conveyors under the discharge of bucket wheel excavators and under the loading zones of long overland conveyors carrying run-of-mine material.

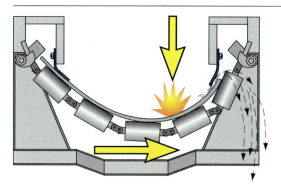

Figure 10.42

Their suspension allows catenary idlers to bounce and swing under loading impact, changing the belt path and making sealing difficult.

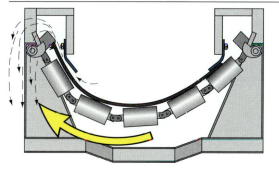

Figure 10.43

As the catenary idler swings, the belt moves from side to side, allowing the escape of fugitive material out the sides of the loading zone and creating mistracking.

10

SYSTEM MAINTENANCE

A key to providing the proper line and stability for a conveyor is the maintenance on the belt-support systems. Proper maintenance of these components will keep the belt from developing unwanted dynamic action that would defeat the belt-support system's ability to control fugitive materials.

The maintenance procedures required for a conveyor belt-support system will vary by the construction and components of the particular system, but should include the following:

A. Inspection of rolling components—including pulleys and idler "cans" (rollers)—for wear and operation (Do they still roll?)

B. Replacement of "stalled," "seized," damaged, or worn rollers

C. Lubrication of bearings in rolling components as appropriate—some idlers are manufactured as "sealed for life," so no lubrication would be required

D. Inspection of belt-support cradles

E. Adjustment of cradles to compensate for wear

F. Realignment and/or replacement of bars showing abuse or wear

G. Removal of material accumulations from rollers, frames, cradle structure, and support bars as required

It is important to refer to the manufacturer's instructions for the required maintenance for any specific component.

Idlers should not be over-lubricated. This can damage the bearing seals, allowing fugitive materials to enter the bearing, increasing the friction while decreasing life. Excess oil and grease can spill onto the belt where it can attach to the cover, decreasing life. Excess grease can also escape onto handrails, walkways, or floors, making them slippery or hazardous. Idlers equipped with sealed "greased for life" bearings should not be lubricated.

It may be best to select the components of the belt-support system with ease of maintenance in mind. Otherwise, the time required to perform maintenance and/ or the difficulty in executing these chores will reduce the likelihood that this essential maintenance will actually be performed.

TYPICAL SPECIFICATIONS

A perfectly flat and straight belt line in the skirted area is essential to successfully seal a transfer point. Belt sag should be minimized to no more than 3 millimeters (0.125 or 1/8 in.) through the load zone. Specifications include the following:

A. Impact cradles in load area

To absorb the shock of loading impact and to stabilize the belt line, full impact cradles should be used under the belt in the direct load area. The impact cradle section should be no longer than 1,2 meters (4 ft.) with an idler installed a minimum of every 1,2 meters.

B. Cradle installation in load area

The cradles should be designed to match the profile of the troughed belt and should be installed so that the bars in the center of the cradle are 12 to 25 millimeters (0.5 to 1 in.) below the normal unloaded track of the belt.

C. Track-mounted cradles

The bars should be installed in a cradle form designed for ease of installation and service without requiring the raising of the belt or the removal of adjacent idlers or the cradle itself. The cradle

Figure 10.44

Air-supported conveyors stabilize the belt line by supporting the belt with a film of air rising from a trough-shaped pan.

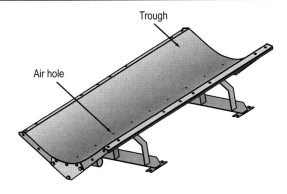

Trough

Air hole

should be constructed in three track-mounted sections for ease of access and maintenance.

D. Edge-support bars and center-support rollers

In the skirted stabilization area directly after the loading point, seal-support cradles with low friction edge-support bars and center-support rollers should be used.

E. In line with idlers

The cradles should be designed in line with the entry and exit idlers, as well as any intermediate idlers.

F. Method for adjustment

The design should include a method for vertical and radial adjustment of the bar to the belt.

In all cases, the equipment selected must not only provide adequate belt support but must also be able to maintain the belt in constant contact with the skirting system to assure sealing efficiency.

ADVANCED TOPICS

Idler Spacing and Belt Sag

In *BELT CONVEYORS for BULK MATERIALS, Sixth Edition*, CEMA recommends that conveyor belt sag between idlers be limited to 2 percent for 35-degree idlers and 3 percent for 20-degree idlers *(Reference 10.2)*. The CEMA method refers to limiting sag outside the load zone to prevent spillage.

In the load zone, the sag must be much less than that recommended by CEMA to

SAFETY CONCERNS

Workers must be aware of the following hazards specific to the loading zone and trained to perform inspections, cleaning, and maintenance in a safe manner:

A. Pinch Points

A moving belt creates pinch or nip point hazards between the rotating and stationary components of the load zone.

B. Heavy Components

Many of the belt-support and load-zone components are heavy, creating lifting hazards.

C. Tight Quarters

Load zones are often in tight quarters with limited access, areas that are sometimes considered confined spaces.

D. Water, Snow, or Ice

Load zones are often in locations exposed to weather, so they are subject

to accumulations of water, snow, or ice, creating additional slip, trip, and fall hazards.

E. Storage Area

The open area around the tail end of the conveyor and the load zone often becomes a storage area for spare and replaced equipment. This practice creates trip and fall hazards around load zones.

F. Auxiliary Equipment

Auxiliary equipment is often automated and can start without warning, creating potentially dangerous situations.

Established lockout /tagout / blockout / testout procedures must be followed before adjusting or maintaining any belt-support system. It is important to ensure the area is clear of obstructions and to follow all confined-space entry requirements.

prevent spillage; dusting; and wear of the belt, wear liner, and skirt seal. For example (**Equation 10.3**), using the CEMA method results in recommended maximum sag between idlers of 12,5 millimeters (0.5 in.) for 35-degree idlers and 19 millimeters (0.75 in.) for 20-degree idlers. This is clearly unacceptable sag for control of fugitive materials in the load zone.

Sag (ΔY_s) is proportional to the weight (force) of the belt and bulk material ($W_b + W_m$) [newtons (lb$_f$)] and the idler spacing (S_i) [millimeters (in.)], and it is inversely proportional to the minimum belt tension in the load zone (T_m) [newtons (lb$_f$)] (**Equation 10.3**). To control fugitive materials, it is recommended that the designer manage the belt tension and idler spacing in the load zone to keep belt sag at no more than 3 millimeters (0.12 in.) and preferably 0.0. Even with very little sag, if belt support is not continuous, fugitive materials can escape and cause wear.

The example (**Equation 10.3**) shows that with idler spacing of 600 millimeters (24 in.), there is 3,37 millimeter (0.135 in.)

of sag. If the idler spacing in the example is reduced to 178 millimeters (7 in.), belt sag drops to 1,0 millimeter (0.039 in.).

If a belt-support system such as an impact cradle or air-supported conveyor section is used, idler spacing (S_i) can be assumed to be 0.0; the calculation then yields belt sag of 0.0, because there should be no sag when the belt is a continuous, flat surface.

Cradles and Power Requirements

Belt-support systems have a significant effect on the power requirements of a conveyor. Changes in belt support will have a particularly noticeable effect on short or under-powered systems. It is recommended that the theoretical power requirements of proposed changes in belt-support systems be calculated to ensure there is adequate conveyor drive power available to compensate for the additional friction placed on the conveyor.

Added kilowatts (hp) consumption can be calculated by determining the added belt tension, using the standard methods

---

Equation 10.3

Calculating Belt Sag

$$\Delta Y_s = \frac{(W_b + W_m) \cdot S_i \cdot k}{T_m}$$

Given: A belt that weighs 550 newtons per meter (38 lb$_f$/ft) is carrying 3000 newtons per meter (205 lb$_f$/ft) of material. The idlers are spaced at 600 millimeters (2 ft) and the tension in the area is 24000 newtons (5400 lb$_f$). Find: The belt sag.

Variables		Metric Units	Imperial Units
ΔY_s	Belt Sag	millimeters	inches
W_b	Weight (Force) of the Belt per Length of Belt	550 N/m	38 lb$_f$/ft
W_m	Weight (Force) of the Material per Length of Belt	3000 N/m	205 lb$_f$/ft
S_i	Idler Spacing	600 mm	2 ft
T_m	Belt Tension	24000 N	5400 lb$_f$
k	Conversion Factor	0,038	1.5

Metric: $\Delta Y_s = \dfrac{(550 + 3000) \cdot 600 \cdot 0,038}{24000} = 3,37$

Imperial: $\Delta Y_s = \dfrac{(38 + 205) \cdot 2 \cdot 1.5}{5400} = 0.135$

ΔY_s	Belt Sag	3,37 mm	0.135 in.

---

recommended by CEMA. The coefficient of friction of the new (or proposed) support systems, multiplied by the load placed on the belt support from belt weight, material load, and sealing system, equals the tension. There is no need to allow for the removal of idlers, the incline of the conveyor, or other possible factors, as estimates provided by this method will in most cases produce results higher than the power consumption experienced in actual use. In applications where there is a lubricant, such as water, consistently present, the actual power requirements may be one-half, or even less, of the amount estimated through these calculations.

The tension added by a skirtboard sealing-support system can be calculated (**Equation 10.4**).

The tension added by an impact bed can be calculated (**Equation 10.5**).

The tensions due to the impact bed and the support bed can be related to the power requirements added to the drive on a conveyor belt (**Equation 10.6**).

PAY NOW, OR PAY (MORE) LATER

In Closing...

Seemingly simple changes in a conveyor system such as changing the belt specifi-

cation or adding belt support can result in dramatic changes in the drive power required.

In its sixth edition of *BELT CONVEYORS for BULK MATERIALS*, CEMA details a relatively complex formula for determining conveyor belt tension and power requirements. Current conveyor engineering computer software offers similar equations and, given the proper data, will perform the calculation.

The installation of improved belt-support systems can increase the conveyor's drive power requirements. However, the true implications of improved belt-support systems are seen when they are compared to the power consumption of a conveyor where idler bearings drag or the idlers themselves build up with material due to transfer-point spillage induced by belt sag.

As demonstrated by R. Todd Swinderman in the paper "The Conveyor Drive Power Consumption of Belt Cleaners" *(Reference 10.3)*, "Fugitive material can also impair the operation of conveyor systems, increasing power consumption significantly." For example, Swinderman calculated that a single frozen impact idler set would require approximately 1,5 kilowatt additional power (2.1 hp), while a seized steel idler set can demand as much as 0,27

$$\Delta T_s = (W_b \cdot L_b \cdot 0.1) + (F_{ss} \cdot 2 \cdot L_b)$$

Given: A conveyor belt weighing 130 newtons per meter (9 lb$_f$/ft) is supported under the seal for 6 meters (20 ft). The seal presses on the belt with a force of 45 newtons per meter (3 lb$_f$/ft).
Find: Tension added to the belt due to the sealing support.

Variables		Metric Units	Imperial Units
ΔT_s	Tension Added to the Belt due to the Sealing Support	newtons	pounds-force
W_b	Weight (Force) of the Belt per Length of Belt	130 N/m	9 lb$_f$/ft
F_{ss}	Rubber Strip Sealing Load	45 N/m	3 lb$_f$/ft
L_b	Length Belt Support	6 m	20 ft
Metric: $\Delta T_s = (130 \cdot 6 \cdot 0,1) + (45 \cdot 2 \cdot 6) = 618$			
Imperial: $\Delta T_s = (9 \cdot 20 \cdot 0.1) + (3 \cdot 2 \cdot 20) = 138$			
ΔT_s	Tension Added to the Belt due to the Sealing Support	618 N	138 lb$_f$

Equation 10.4

Calculating the Tension Added to the Belt due to Sealing Support

10

Equation 10.5

Calculating Tension Added to the Belt due to the Impact Bed

$$\Delta T_{IB} = \left[(W_b \cdot L_b) + (F_{ss} \cdot 2 \cdot L_b) + \left(\frac{Q \cdot L_b \cdot k}{V} \right) \right] \cdot \mu_{ss}$$

Given: *A conveyor belt weighing 130 newtons per meter (9 lb$_f$/ft) is supported by an impact bed for 1,5 meters (5 ft). The seal presses on the belt with a force of 45 newtons per meter (3 lb$_f$/ft). The belt carries 275 tons per hour (300 st/h) and travels at 1,25 meters per second (250 ft/min). The support systems use a UHMW sliding surface.* **Find:** *Tension added to the belt due to the impact bed.*

	Variables	Metric	Imperial
ΔT_{IB}	Tension Added to the Belt due to the Impact Bed	newtons	pounds-force
W_b	Weight (Force) of the Belt per Length of Belt	130 N/m	9 lb$_f$/ft
L_b	Length Belt Support	1,5 m	5 ft
F_{ss}	Rubber Strip Sealing Load	45 N/m	3 lb$_f$/ft
Q	Material Flow	275 t/h	300 st/h
V	Belt Speed	1,25 m/s	250 ft/min
μ_{ss}	Friction Coefficient Per CEMA 575-2000	0,5 – UHMW 1,0 – Polyurethane 1,0 – Rubber	0.5 – UHMW 1.0 – Polyurethane 1.0 – Rubber
k	Conversion Factor	2,725	33.33

Metric: $\Delta T_{IB} = \left[(130 \cdot 1,5) + (45 \cdot 2 \cdot 1,5) + \left(\dfrac{275 \cdot 1,5 \cdot 2,725}{1,25} \right) \right] \cdot 0,5 = 615$

Imperial: $\Delta T_{IB} = \left[(9 \cdot 5) + (3 \cdot 2 \cdot 5) + \left(\dfrac{300 \cdot 5 \cdot 33.33}{250} \right) \right] \cdot 0.5 = 137.5$

ΔT_{IB}	Tension Added to the Belt due to the Impact Bed	615 N	137.5 lb$_f$

Equation 10.6

Calculating the Power Consumption Added to the Belt Drive due to Sealing and Impact Support

$$P = (\Delta T_S + \Delta T_{IB}) \cdot V \cdot k$$

Given: *A conveyor belt traveling 1,25 meters per second (250 ft/min) is supported by an impact bed and a seal-support system that add 615 newtons (137.5 lb$_f$) and 618 newtons (138 lb$_f$) respectively.* **Find:** *The power consumption added to the drive due to the sealing and impact support.*

	Variables	Metric Units	Imperial Units
P	Power Consumption Added to Belt Drive	kilowatts	horsepower
ΔT_S	Tension Added to the Belt due to the Sealing Support (*Calculated in Equation 10.4*)	618 N	138 lb$_f$
ΔT_{IB}	Tension Added to the Belt due to the Impact Bed (*Calculated in Equation 10.5*)	615 N	137.5 lb$_f$
V	Belt Speed	1,25 m/s	250 ft/min
k	Conversion Factor	1/1000	1/33000

Metric: $P = \dfrac{(618 + 615) \cdot 1,25}{1000} = 1,54$

Imperial: $P = \dfrac{(138 + 137.5) \cdot 250}{33000} = 2.09$

P	Power Consumption Added to Belt Drive	1,54 kW	2.09 hp

kilowatts (0.36 hp). One idler with a 25 millimeter (1 in.) accumulation of material would add 0,32 kilowatt additional power (0.43 hp) to the conveyor's drive requirements. These additional requirements would be multiplied by the number of idlers affected.

The use of improved belt support and sealing techniques places additional requirements on conveyor drive systems. However, these additional requirements and costs will seem minor when compared to the power consumed by operating with one "frozen" idler or several idlers operating with a material accumulation. By implementing the proper belt-support systems, a plant can prevent the many and more costly problems that arise from the escape of fugitive material.

It would be better to design a system that incorporates the slightly elevated power consumption required to prevent spillage, rather than suffer the much higher power consumption and greater consequences that arise from fugitive material. The costs for installation and operation of proper belt-support systems represent an investment in efficiency.

Looking Ahead...

This chapter about Belt Support, the fifth chapter in the section Loading the Belt, discussed the importance of proper belt-support systems to maintain a stable belt line to prevent fugitive material and dust. The following three chapters continue this section and discuss additional ways to prevent spillage, focusing on Skirtboards, Wear Liners, and Edge-Sealing Systems.

REFERENCES

10.1 Conveyor Equipment Manufacturers Association (CEMA). (2005). "Conveyor Installation Standards for Belt Conveyors Handling Bulk Materials." In *BELT CONVEYORS for BULK MATERIALS, Sixth Edition*, Appendix D, pp. 575–587. Naples, Florida.

10.2 Conveyor Equipment Manufacturers Association (CEMA). (2005). *BELT CONVEYORS for BULK MATERIALS, Sixth Edition*, p. 133. Naples, Florida.

10.3 Swinderman, R. Todd, Martin Engineering. (May 1991). "The Conveyor Drive Power Consumption of Belt Cleaners," *Bulk Solids Handling*, pp. 487–490. Clausthal-Zellerfeld, Germany: Trans Tech Publications.

10

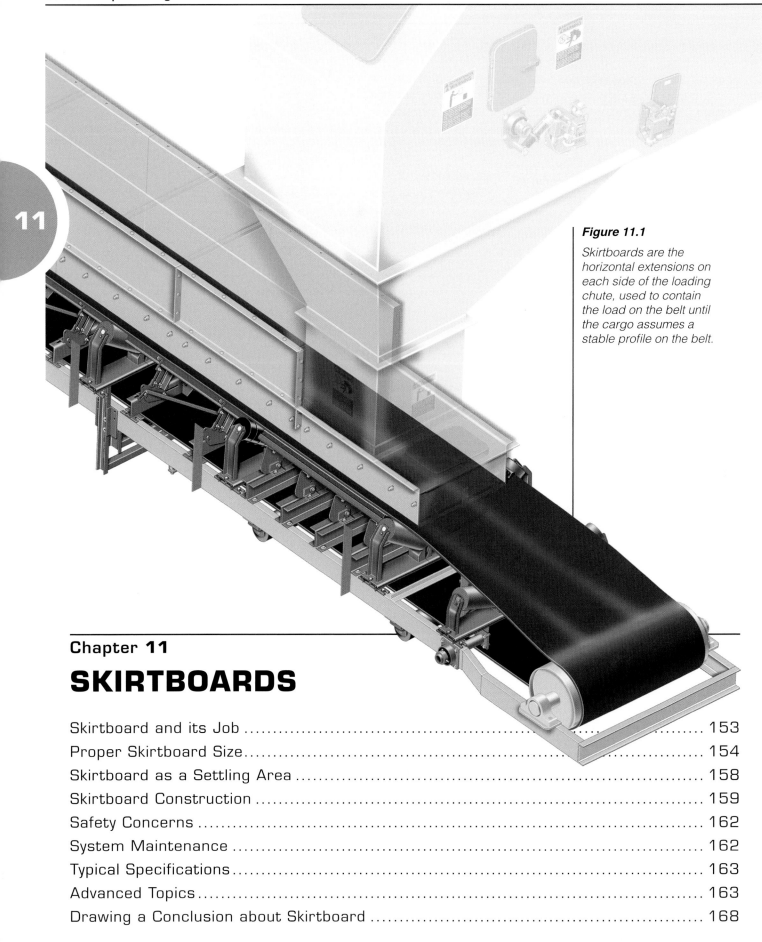

Figure 11.1

Skirtboards are the horizontal extensions on each side of the loading chute, used to contain the load on the belt until the cargo assumes a stable profile on the belt.

Chapter 11

SKIRTBOARDS

Figure 11.2

Skirtboard usually extends out from the load point in the direction of material travel on both sides of the belt, confining and shaping the cargo until it has settled into the desired profile.

In this Chapter...

In this chapter, we discuss skirtboards and the functions they serve for reducing spillage and dust. We provide equations for determining proper length and width of skirtboards and give examples for both. We also include information about skirtboard construction.

Skirtboards are used to contain the load as the material is placed onto the belt until it assumes a stable profile (**Figure 11.1**). Skirtboards—which may be referred to in the industry as skirt plates, steel skirting, or sometimes merely chute or chutework—are almost always constructed of steel plate. In this book, the term skirtboard is used to define the construction material that extends out from the load point in the direction of material travel on both sides of the belt (**Figure 11.2**). The terms "rubber skirting," "skirtboard seal," "side wipers," "dust seal," "seal strips," "sealing," and "edge seal" refer to the elastomer strip installed on the metal skirtboard to prevent the escape of fines. *(See Chapter 13: Edge-Sealing Systems.)*

The primary purpose of skirtboard is to keep the load on the conveyor, preventing the material from spilling over the belt edge while the load is settled into the belt trough and has reached belt speed. The skirtboard of each transfer point must be engineered to match the characteristics of the material being transferred, the receiving conveyor, the drop height between the conveyors, and the way the transfer point is loaded and used.

Best practices in chute and skirtboard design now provide the opportunity for a much cleaner and more efficient material-handling system. This chapter discusses what has become the best practice for the design and application of transfer-point skirtboard systems.

SKIRTBOARD AND ITS JOB

The skirtboard and the wear liner placed inside the skirtboard combine with the elastomer sealing system to form a multiple-layer seal (**Figure 11.3**). The elastomer seal strips cannot and should not be expected to withstand significant material side pressures or contact with pieces of material larger than small fines. The skirtboard and wear liner form the first lines of defense and are intended to contain fugitive material and prevent any head pressure present in the system from placing the material in contact with, and thus prematurely wearing, the sealing system.

In addition, the skirtboard and its covering form a settling zone, which is used for effective dust management (**Figure 11.4**). In the settling zone, the current of air traveling with the moving stream of material is slowed and controlled, allowing airborne particles to fall back into the main material body (**Figure 11.5**).

Figure 11.3

Skirtboard combines with the transfer point's wear liner and sealing system to form an effective multiple-layer seal.

Figure 11.4

The settling zone is usually an enlarged portion of the covered skirtboard area at the transfer point.

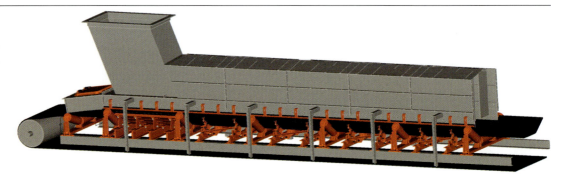

11

Inadequately sized skirtboard always leads to poor conveyor performance in the forms of material spillage, fugitive material fines, excessive dust, and much higher operating costs for the end user. It is imperative that the skirtboard be designed properly in length and height so material can be contained and fugitive material controlled.

PROPER SKIRTBOARD SIZE

Skirtboard Length

Skirtboard length refers to the additional length of steel wall beyond the impact zone. The impact zone is the area of the loading chute if it was extended down to the belt.

Skirtboard should extend in the belt's direction of travel past the point where the material load has fully settled into the profile that it should maintain for the remainder of its journey on the conveyor.

Sometimes the load never becomes completely stable, and, consequently, skirtboard is required for the entire length of the conveyor. This is most common with very fine materials that are easily made airborne,

materials that tend to roll, or conveyors with multiple load points. Belt feeders, typically short in length and loaded to nearly the full width of the belt, are commonly skirted for their full length.

The minimum length for the skirtboard should be based on the total air movement and the speed of the belt, using the following guidelines:

A. Metric measurements

If airflow is under 0,5 cubic meters per second, the length of skirtboard is 0,6 meter for every 0,5 meter per second of belt speed. If airflow is greater than 0,5 cubic meters per second, the length of skirtboard is 0,9 meter for every 0,5 meter per second of belt speed.

B. Imperial measurements

If airflow is less than 1000 cubic feet per minute, the length of skirtboard is 2 feet of skirtboard for every 100 feet per minute of belt speed. If airflow is over 1000 cubic feet per minute, the length of skirtboard is 3 feet for every 100 feet per minute of belt speed. *(See Advanced Topics: Equation 11.1.)*

To prevent spillage or damage to the belt, skirtboard should end above an idler rather than between idlers (**Figure 11.6**). This in itself may increase the overall length of the skirtboard.

What may provide a more telling answer for skirtboard length is the need for enclosing the dust-suppression and/or dust-collection systems, as discussed elsewhere in this book. *(See Chapter 19: Dust Suppression and Chapter 20: Dust Collection.)* The walls

Figure 11.5

This settling zone slows the air and allows the conveyed product to settle and cleaner air to escape.

of a dust-control enclosure can effectively serve as skirtboard, with the length necessary for effective dust-control systems generally providing more than what is required for load stabilization.

Penalties for increasing the length of skirtboard are the additional maintenance cost for longer liners and seals, a minimal increase in the cost of the steel for the walls, and a slight increase in the conveyor's power requirements. The extra power consumption results from the added friction created by the longer steel wall and the additional length of sealing strip installed. This is usually a modest increase that provides long-term benefits that greatly outweigh the minimal up-front cost. *(For more information about the power consumption of edge-sealing systems, see Chapter 13: Edge-Sealing Systems.)*

There are also times when conditions such as the incline of the belt, the shape of the conveyed product, or the depth of the material bed require the length of the skirted area to be increased substantially to prevent material spillage.

When in doubt, it is always better to have the skirtboard a little longer than is minimally required by the above equation. An extra 25 percent in additional settling area length is a recommendation that will improve dust control with only a minimal increase in power requirements and expense for steel.

When the conveyor incorporates multiple load zones, the calculation should use the total airflow from all loading points to establish the minimum dimensions of the settling zone after the last loading point.

Skirtboard Width

The distance between the two sides of the skirtboard is usually determined by belt capacity requirements, the space needed to establish an effective seal on the outside of the skirtboard is too often ignored.

The importance of designing the system so there is enough "free-belt" distance

(between the outside of the skirtboard and outer belt edge) should not be understated. A conveyor designer must always consider the effect of possible belt wander on the ability to effectively seal between the stationary skirtboard system and the moving belt. By maintaining the largest possible "free-belt" distance, the designer can help eliminate a great deal of the common spillage and dusting problems often associated with the transfer of bulk materials from one conveyor to another. The benefits realized by maintaining the correct "free-belt" distance to allow a sealable skirtboard system will be further enhanced by incorporating the proper belt support under the skirtboard and by installing a highly effective skirtboard-sealing system. *(See Chapter 10: Belt Support and Chapter 13: Edge-Sealing Systems.)*

Because the act of troughing a belt works to diminish a belt's width, the phrase Effective Belt Width is used to represent the width of a troughed belt. This is not the carrying width (the distance between the skirtboard) but rather the measurement of the horizontal width of a troughed conveyor belt that is measured across the width parallel to the bottom roller (**Figure 11.7**).

Several standards are available for setting the distance between the skirtboards and thereby establishing the "free-belt" edge distance. The Conveyor Equipment Manufacturers Association (CEMA) and the Deutsches Institut für Normung (DIN 22101) have both established formulas that can be referenced.

Best practice indicates that in order to ensure adequate "free-belt" edge to properly apply edge seals and provide a tolerance for belt mistracking, the skirtboards should

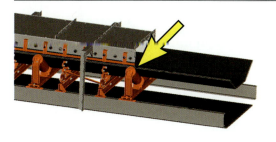

Figure 11.6

Steel skirtboards should end above an idler to prevent spillage or damage to the belt.

Recommended Loading-Zone Design — Table 11.1

Columns: I) Belt Width · II) Trough Angle · III) Effective Belt Width (See Figure 11.7 "A") · IV) Recommended Chute Width (See Figure 11.7 "B")

Metric (mm)

Belt Width	Trough Angle	Effective Belt Width	Recommended Chute Width
300	0°	300	NR
	20°	288	NR
	30°	273	NR
	35°	264	NR
	40°	253	NR
	45°	241	NR
500	0°	500	270
	20°	480	264
	30°	455	NR
	35°	440	NR
	40°	422	NR
	45°	402	NR
650	0°	650	420
	20°	624	408
	30°	592	393
	35°	572	383
	40°	549	372
	45°	523	360
800	0°	800	570
	20°	768	552
	30°	729	529
	35°	704	515
	40°	675	499
	45°	644	481
1000	0°	1000	770
	20°	960	744
	30°	911	711
	35°	879	691
	40°	844	668
	45°	805	642
1200	0°	1200	970
	20°	1152	936
	30°	1093	894
	35°	1055	867
	40°	1013	837
	45°	966	803
1400	0°	1400	1170
	20°	1344	1128
	30°	1275	1076
	35°	1231	1043
	40°	1182	1005
	45°	1127	964
1600	0°	1600	1370
	20°	1536	1320
	30°	1457	1258
	35°	1407	1219
	40°	1350	1174
	45°	1288	1125
1800	0°	1800	1570
	20°	1728	1512
	30°	1639	1440
	35°	1583	1395
	40°	1519	1343
	45°	1449	1286
2000	0°	2000	1770
	20°	1920	1703
	30°	1821	1622
	35°	1759	1570
	40°	1688	1512
	45°	1609	1447
2200	0°	2200	1970
	20°	2112	1895
	30°	2004	1804
	35°	1935	1746
	40°	1857	1681
	45°	1770	1608
2400	0°	2400	2170
	20°	2304	2087
	30°	2186	1986
	35°	2111	1922
	40°	2026	1849
	45°	1931	1769
2600	0°	2600	2370
	20°	2495	2279
	30°	2368	2169
	35°	2287	2098
	40°	2194	2018
	45°	2092	1930
2800	0°	2800	2570
	20°	2687	2471
	30°	2550	2351
	35°	2462	2274
	40°	2363	2187
	45°	2253	2091
3000	0°	3000	2770
	20°	2879	2663
	30°	2732	2533
	35°	2638	2450
	40°	2532	2356
	45°	2414	2252
3200	0°	3200	2970
	20°	3071	2855
	30°	2914	2715
	35°	2814	2626
	40°	2701	2525
	45°	2575	2413

Imperial (in.)

Belt Width	Trough Angle	Effective Belt Width	Recommended Chute Width
18	0°	18.0	9.0
	20°	17.3	8.8
	35°	15.8	NR
	45°	14.5	NR
24	0°	24.0	15.0
	20°	23.0	14.6
	35°	21.1	NR
	45°	19.3	NR
30	0°	30.0	21.0
	20°	28.8	20.3
	35°	26.4	19.0
	45°	24.1	17.8
36	0°	36.0	27.0
	20°	34.6	26.1
	35°	31.7	24.3
	45°	29.0	22.6
42	0°	42.0	33.0
	20°	40.3	31.9
	35°	36.9	29.6
	45°	33.8	27.4
48	0°	48.0	39.0
	20°	46.1	37.6
	35°	42.2	34.8
	45°	38.6	32.3
54	0°	54.0	45.0
	20°	51.8	43.4
	35°	47.5	40.1
	45°	43.5	37.1
60	0°	60.0	51.0
	20°	57.6	49.1
	35°	52.8	45.4
	45°	48.3	41.9
72	0°	72.0	63.0
	20°	69.1	60.6
	35°	63.3	55.9
	45°	57.9	51.6
84	0°	84.0	75.0
	20°	80.6	72.2
	35°	73.9	66.5
	45°	67.6	61.2
96	0°	96.0	87.0
	20°	92.1	83.7
	35°	84.4	77.1
	45°	77.3	70.9
108	0°	108.0	99.0
	20°	103.7	95.2
	35°	95.0	87.6
	45°	86.9	80.5
120	0°	120.0	111.0
	20°	115.2	106.7
	35°	105.5	98.2
	45°	96.6	90.2

Notes: Dimensions were determined by calculation rather than field measurements. Metric measurements are rounded to the nearest millimeter. Imperial measurements are rounded to one decimal. Thickness of steel in chute or skirtboard is not considered. Three-piece troughing idlers of equal length are assumed. Belt edge distances in metric consider 90 mm for the side sealing + 25 mm for margin for belt misalignment. Belt edge distances in Imperial consider 3.5 in. for the side sealing + 1 in. for margin for belt misalignment. Particle size of the bulk material is not considered.

be located with a minimum of 115 millimeters (4.5 in.) of actual belt width on each side of the conveyor belt (**Table 11.1**).

The edge distance should be increased to a minimum of 150 millimeters (6.0 in.) when five roll catenary idlers are used in the load zone to compensate for the extra belt misalignment that is a characteristic of belts using catenary idlers.

The width of the skirtboard should be checked to ensure that the height of the material bed exiting the skirting area along with the material's angle of repose do not combine to create a spillage condition.

Skirtboard Height

Belt width and speed, material lump size, and air speed at the discharge must be considered when determining the height of the skirtboard required for a given transfer point.

The skirtboard should be tall enough to contain the material load when the belt is operating at normal capacity and to pass lumps without jamming. As the size of lumps included in the load goes up, so must skirtboard height; at minimum, the height must be sufficient to contain the largest pieces. The skirtboard should be tall enough to contain two of the largest pieces stacked on top of each other.

In the sixth edition of *BELT CONVEYORS for BULK MATERIALS*, CEMA has published tables specifying the minimum height for uncovered skirtboard on conveyors with 20-, 35-, and 45-degree idlers, carrying cargos of various sized lumps. In summary, it specifies approximately 300 millimeters (12 in.) is tall enough for particles of 50 millimeters (2 in.) or smaller, carried on flat or 20 degrees troughed belts up to 1800 millimeters (72 in.) wide, or for 35 degrees and 45 degrees troughed belts up to 1200 millimeters (48 in.) wide. The table specifies skirtboards with a minimum height of 825 millimeters (32.5 in.) for belts as wide as 2400 millimeters (96 in.) with lumps up to 450 millimeters (18 in.).

For materials that may create a dust problem, it is good practice to increase the height of the skirted area to serve as a plenum, a space that will reduce the positive air pressure. This area will serve to "still" the dust-laden air so that the particles can settle back onto the cargo of the conveyor.

For maximum dust control, the chutewalls (skirtboards) must be high enough to furnish a cross-sectional area in the load zone that provides a maximum air velocity above the product bed of less than 1 meter per second (200 ft/min). *(See Advanced Topics: Equation 11.2.)* This larger volume combined with multiple dust curtains provides an ample chamber to accommodate the positive pressures of air movement without most common bulk materials blowing out of the enclosure (**Figure 11.8**). In many circumstances, to achieve this limited airspeed, the skirtboard height must be increased to 600 millimeters (24 in.) or more. Very light materials with extremely small particles or dusty materials may require exit velocities as low as 0,25 meters per second (50 ft/min).

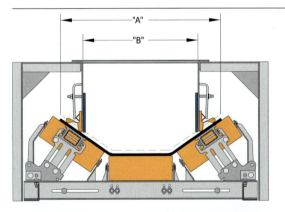

Figure 11.7

Effective Belt Width (A) is the width of the conveyor belt when troughed. The actual load-carrying distance (B) is then reduced by the requirement for edge sealing outside the skirtboard.

Figure 11.8

Skirtboard height can be increased to reduce airspeed, minimizing the positive pressures of air movement without most common bulk materials blowing out of the enclosure.

The skirtboard must still be tall enough to contain the cargo's largest lump if it is placed at the top of the material load profile. If the calculation does not yield a sufficient height, the calculated height should be replaced with a height of 2.5 times the largest lump.

Obviously, there is a practical limit to the height that a skirtboard can be made. If the settling zone/skirtboard height requirement becomes excessive, a dust-collection system that can account for the transfer point's total air movement and an active dust-control system (e.g., dust-suppression system or baghouse dust-collection system) will need to be installed.

Chutewalls need to be high enough and located so that dust exhaust pickups do not pull fines off the pile. The collectors may pull in so much material they quickly plug. If the skirtboard walls are not high enough, energy will be wasted removing dust that would have shortly settled on its own, and the dust-collection system will be larger and more expensive than necessary. *(See the discussion of settling zones in Chapter 18: Passive Dust Control.)*

SKIRTBOARD AS A SETTLING AREA

Covering the skirtboard with a steel or fabric system is recommended for dust control (**Figure 11.9**). Unless a specific reason exists not to cover it, the skirtboard should be enclosed with a cover, lid, or roof. The covering of the skirtboard is required to create the plenum needed to allow dust to settle and air movement to be stilled. A large plenum is useful in controlling the clouds of dust driven off by the forces of transferring the stream of material. *(See Chapter 7: Air Control and Chapter 18: Passive Dust Control.)*

The incorporation of properly-placed dust curtains within the skirtboard cover system will help slow airflow and significantly decrease the release of airborne dust from the exit end of the transfer point. *(See Chapter 18: Passive Dust Control for more information about dust curtains.)*

In addition, placing a "roof" over the skirtboard will contain the occasional lump of material that through some random circumstance comes through the loading chute onto the belt with force sufficient to "bounce" it completely off the belt.

CEMA recommends that these skirtboard covers should be slanted down from the loading chute to the skirtboard, allowing for material that is not yet moving at belt speed, to avoid material jams. In *BELT CONVEYORS for BULK MATERIALS, Sixth Edition*, CEMA provides tables for minimum heights of uncovered skirtboards and minimum belt width based on lump size. Generally accepted practice is to keep the skirt width and height at least 2.5 times the largest lump size. By keeping the skirtboards tall, two things are accomplished: avoiding material jams and providing a large area to dissipate air velocity and let dust settle.

Skirtboard coverings are not the semi-circular "hoods" commonly seen along the run of a conveyor but are typically a flat roof between the two skirtboards. In most cases, a covering of steel is best. These covers can be clamped in place, allowing for inspection or maintenance. Fabric or rubber is most often applied to connect vibrating equipment to stationary chutes or skirtboards.

The covering should be designed so that it will support the weight of a worker, or it should be guarded and marked with "No Step" warnings, so nobody falls through the covering.

Figure 11.9

Covering the skirtboard with a steel or fabric system is recommended to create the plenum needed to allow dust to settle and air movement to be stilled.

Openings in the skirtboard or covers should be provided for service and inspection; these openings should be provided with doors to prevent material escape and minimize the outflow of air.

SKIRTBOARD CONSTRUCTION

Clearance Above the Belt

Even under the most ideal conditions, steel skirtboards can be hazardous to the belt. Fluctuations in the belt's line of travel may allow the belt to move up against the steel, where it can be gouged or cut. In addition, material can wedge under the skirtboard to abrade the belt's surface.

It is critical to raise the bottom edge(s) of the skirtboard far enough above the conveyor so they never come in contact with the belt cover. As the distance above the belt increases, so does the difficulty in providing an effective seal. Skirtboard is sometimes installed with a clearance of several inches above the belt to facilitate belt replacement. When the steel is placed this far above the belt surface, it is virtually impossible to provide an effective seal on the outside of the skirtboards when there is side pressure.

An ineffective seal perpetuates itself. Material leaks out, accumulating on idlers, leading to mistracking and other problems that result in an unstable belt line. The belt flexes up and down and wanders from side to side. Plant engineers and maintenance staffs, mindful of the need to prevent the belt from coming into contact with the chutework, increase the belt-to-skirting clearance. This dramatically increases the difficulty in sealing the transfer point, resulting in increased spillage. This increased spillage results in a continuing vicious cycle of belt wander, rolling component failures, and increased operating costs.

The closer the steel and belt are placed together, the easier it is to maintain a seal between them. It is critical to provide relief in the direction of belt travel. The gap under the steel should form a wedge-shaped opening that allows conveyed material to ride along the steel skirting and sealing rubber, rather than become wedged into an opening by the ceaseless force of belt motion. The skirtboard should open gradually, both horizontally and vertically, from the loading point in the direction of belt travel to permit entrapped material to free itself (**Figure 11.10**).

It is recommended that the lower edges of the skirt plates be positioned 6 millimeters (1/4 or 0.25 in.) above the belt at the belt's entry into the loading zone. This dimension should be uniformly increased in the direction of belt travel to 9 to 12 millimeters (3/8 or 0.38 to 0.5 in.) as the belt exits the skirtboard (**Figure 11.11**). This close clearance cannot be accomplished unless the belt travel is stabilized within a plus-or-minus tolerance of 1,5 millimeters (1/16 or 0.063 in.) at the entry (tail pulley) end of the chute.

It is critical that the centerline of the skirtboard construction be in line with the centerline of the belt to prevent belt mistracking. If the two are not in line, the unequal forces from the cargo's center of gravity and the friction against the skirtboard will cause a chronic mistracking of the belt and accelerated wear on the wear liners and skirt seal. With the steel positioned close to the belt line, it is critical to the safety of the belt that the belt be prevented from rising up off the idlers during conveyor startup. This is one reason why the elevation of the tail pulley, commonly known as the half-trough arrangement, is not a good idea, as this practice encourages

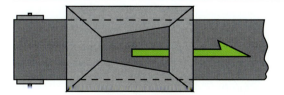

Figure 11.10

To reduce the risk entrapped material might gouge the belt, the skirtboard should open (or self-relieve) both horizontally and vertically in the direction of belt travel. (Illustration is exaggerated to show effect.)

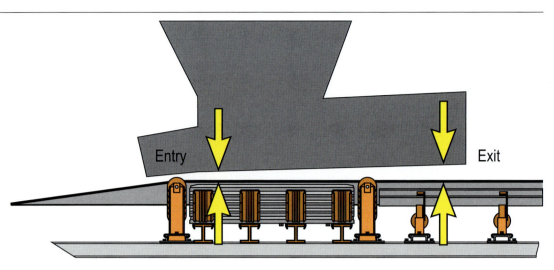

Figure 11.11

The lower edges of the skirt plates should be positioned 6 millimeters (1/4 or 0.25 inch) above the belt at the belt's entry into the loading zone. This dimension should be uniformly increased in the direction of belt travel to 9 to 12 millimeters (3/8 or 0.38 to 0.5 in.) as the belt exits the skirtboard.

the belt to rise. The employment of using a half-trough arrangement is generally done in the interest of shortening the transition distance. *(See Chapter 6: Before the Loading Zone for additional information about half-trough transitions.)* It is important that the belting specifications and tension be calculated correctly to minimize the risk of the belting lifting off the idlers. Hold-down rollers can be installed to keep the belt on the idlers.

Rough bottom edges or warped steel can create difficult conditions, capturing material to increase the drag on the conveyor drive and/or abrade the belt surface. Ceramic blocks or wear plates must be carefully installed to avoid jagged or saw-toothed edges that can trap material or damage the belt (**Figure 11.12**). The rule is to maintain a smooth flow surface on the bottom edge of the skirtboard and eliminate all entrapment points. Skirtboard steel and chute liners must be installed very carefully, with all seams well matched.

The gap left between the skirt and the belt surface should be sealed by a flexible, replaceable elastomer sealing system ap-

plied to the outside of the skirtboard. *(See Chapter 13: Edge-Sealing Systems.)*

Skirtboard Construction

The strength and stability of the skirtboard are very important to its success. Many times conveyor skirtboard is supported by cantilever brackets that are not rigid enough to withstand the impact of material or the vibration of equipment. This risks a structural failure that endangers the belt and the transfer point itself.

The thickness of the skirtboard must be sufficient to withstand side pressures that may occur when the chute becomes plugged or the belt rolls backward. As it is located close to the belt, any movement of the skirtboard must be prevented to minimize the risks of damage.

Except in very light applications, the minimum thickness of mild steel used for skirtboard construction should be 6 millimeters (0.25 in.). On belts moving over 3,7 meters per second (750 ft/min) or 1300 millimeters (54 in.) or more wide, the minimum thickness should be 10 millimeters (3/8 in.). For applications with belts moving over 5 meters per second (1000 ft/min) or 1800 millimeters (72 in.) wide, the minimum thickness should be 12 millimeters (0.5 in.).

Skirtboards should be installed on structural steel supports on approximately 1,2-meter (48-in.) centers so the supports

Figure 11.12

Skirtboard should be installed so it maintains a smooth flow surface on the bottom edge of the skirtboard to eliminate all material entrapment points.

11

do not interfere with the spacing of, or access to, belt-support cradles and idlers. The most common support design is an angle iron "A-frame," installed on approximately the same spacing as the carrying idlers. These "A-frames" should be rigid and well-gusseted, and they should be installed far enough above the belt to allow easy access for the adjustment or replacement of the skirtboard seal (**Figure 11.13**).

At least one "A-frame" should be positioned at the beginning of the skirtboard and another at the end. Closer spacing should be considered in the conveyor's impact zone to the extent of doubling the support structure.

There are minimum sizes of angle iron that need to be used to construct these "A-frames" (**Table 11.2**). These specifications are best suited for low-density, free-flowing materials. For belt feeders, or for handling high-density materials like ore or concentrate, heavier steel and closer spacing is required.

It is important that proper clearance be provided between the bottom of the skirtboard supports and the belt to allow room for the installation and maintenance of a skirtboard seal and clamp system. The minimum clearance between the horizontal support and the belt at the skirtboard wall should be 230 millimeters (9 in.).

If there is dynamic vibration in the system caused from either belt movement or other operating machinery such as breakers, crushers, or screens, the skirtboard may need to be isolated from it.

These recommendations are for standard conveyor installations and normal duty ratings where the belt is approximately waist high and the idlers are standard width. For other applications, such as "double height" skirtboard or severe conditions, additional support structure may be required. A conveyor or structural engineer should be consulted for advice on skirtboard thicknesses and required supporting structures.

Skirtboard for Conveyors with Multiple Loading Points

Where a belt is loaded at more than one point along the conveyor's length, care must be used in the positioning of the skirtboard at the loading points. The skirtboard at the subsequent load points must be designed to allow the previously loaded material to pass freely, without being "plowed" off the belt by the skirtboard or chute steel of the following loading point.

Because it is impossible to rely on material being loaded evenly and centrally onto a conveyor with multiple load points, a certain amount of plowing and spillage of the material is probably inevitable. This fugitive material contributes to higher

Figure 11.13

Skirtboard must be properly supported, and the support structure must be far enough above the belt to allow access to the sealing system.

Recommended Angle Iron Sizes for Skirtboard Supports		Table 11.2
Conveyor Specifications	**A-Frame Angle Iron Size**	
Below 3,7 m/s (750 ft/min) or 1400 mm (54 in.) wide	50 x 50 x 5 mm (2 x 2 x 3/16 in.)	
3,7 m/s to 5 m/s (750 ft/min to 1000 ft/min) or 1400 mm to 1800 mm (54 to 72 in.) wide	75 x 75 x 6,4 mm (3 x 3 x 1/4 in.)	
Over 5 m/s (1000 ft/min) or 1800 mm (72 in.) wide	75 x 75 x 9,5 mm (3 x 3 x 3/8 in.)	

11

operating and cleanup costs as well as premature equipment failures. Therefore, it is a sound practice to incorporate continuous skirtboards.

When loading points are relatively close together, it is usually better to provide a continuous skirtboard between the two loading points and a deeper trough angle than would be normally used rather than use individual skirtboards at each loading point (**Figure 11.14**).

Another excellent approach for situations with multiple load zones would be the installation of an air-supported conveyor. Air-supported conveyors are uniquely suited for multiple load zones, as they require only a centered load, rather than conventional skirtboard or skirt seals (**Figure 11.15**). *(See Chapter 23: Air-Supported Conveyors.)*

Figure 11.14

When a conveyor has multiple loading points relatively close together, it may be best to provide a continuous skirtboard between the loading points.

Figure 11.15

Air-supported conveyors are uniquely suited for multiple load zones, as they require only a centered load, rather than conventional skirtboard or skirt seals.

SYSTEM MAINTENANCE

As skirtboard is basically a steel wall without moving parts, there is little preventive maintenance to be performed. If the skirtboard also functions as the wear liner, wear can be an issue. The skirtboard may be subject to corrosion and require periodic replacement. If the conveyor is subject to frequent jamming, the skirtboard may be deformed, increasing the chance of belt damage. Skirtboard covers should be secured in place and access doors closed after inspections and maintenance. Periodic checks should be performed to make sure the skirtboards are structurally capable of containing the bulk materials and properly positioned above the belt.

SAFETY CONCERNS

Even when properly installed with proper relief and without sawtooth edges, skirtboard still represents an unforgiving edge in close proximity to the moving belt. Care must be taken when working in the vicinity of the skirtboard to avoid becoming pinched or caught between moving components and the steel structure.

Even when constructed of steel, conveyor skirtboard covers are not intended as walkways and should not be used as work platforms. They should be guarded and marked with "No Step" warnings.

Cutting and welding are common maintenance procedures in the load zone. Established "hot work" and fire watch procedures should be followed. Chutes and covered skirtboard sections are often considered confined space and require workers to follow special precautions. *(See Chapter 2: Safety for information about confined space.)* Chutes and feeding equipment can contain large amounts of buildup or accumulation of loose bulk materials that can fall during maintenance. Established lockout / tagout / blockout / testout procedures must be followed.

TYPICAL SPECIFICATIONS

The following specifications are for conveyors that handle free-flowing and relatively uniform bulk materials, such as coal and crushed rock.

A. Transfer point

The transfer point will be equipped with steel skirtboards on either side of the belt as an extension of the chute.

B. Load settling

The skirtboards will be long enough to allow the load to settle into the profile to be carried.

C. Reduced air velocity

The skirtboard cross-sectional area will be sufficient to reduce the air velocity to 1,0 meter per second (200 ft/min) to allow dust to settle before the load exits the skirted area.

D. Covers

The skirtboard system will be fitted with covers to allow it to act as a plenum that allows dust to settle out of the air.

E. Free-belt area

The skirtboard will be designed to allow sufficient free-belt area on each side to allow for effective sealing.

F. Clearance above the belt

The bottom edge of the skirtboard should be installed 6 millimeters (0.25 in.) above the belt at the exit of the load zone, opening slightly to 9 to 12 millimeters (0.38 to 0.5 in.) at the exit.

G. No rising

The belt should be prevented from rising off its support structure, even when operating unloaded.

H. Construction

The skirtboard system will be solidly constructed with proper supports that do not interfere with the ability to install or maintain conveyor components, including belt-support cradles, idlers, or skirtboard-sealing systems.

ADVANCED TOPICS

Sample Problems: Calculating Dimensions for Skirtboard / Settling Zone

There are equations to determine the appropriate minimum dimensions for the skirtboard at the transfer points in the four problems below (**Equation 11.1** Skirtboard Length and **Equation 11.2** Skirtboard Height). *(For Total Airflow in the sample problems, see Equation 7.1 or measure the total airflow.)*

$l_{sb} = \dfrac{V \cdot CF}{k}$		Units	
		Metric	Imperial
l_{sb}	Skirtboard Length (from loading zone to end of chute)	meters	feet
V	Belt Speed	m/s	ft/min
CF	Chute Factor	If $Q_{tot} < 0{,}5$ m³/s = 0,6	If $Q_{tot} < 1000$ ft³/min = 2
		If $Q_{tot} > 0{,}5$ m³/s = 0,9	If $Q_{tot} > 1000$ ft³/min = 3
k	Conversion Factor	0,5	100

Equation 11.1

Skirtboard Length

$h_{sb} = \dfrac{Q_{tot}}{CW \cdot v}$		Units	
		Metric	Imperial
h_{sb}	Skirtboard Height	meters	feet
Q_{tot}	Total Airflow	m³/s	ft³/min
CW	Chute (Skirtboard) Width	m	ft
v	Target Air Speed	m/s	ft/min

Equation 11.2

Skirtboard Height

Figure 11.16

Sample Problem #1

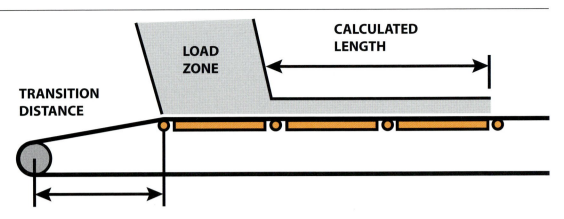

11

Table 11.3

(Figure 11.16)

Skirtboard Sample Problem #1				
Given:	Material	Sub-bituminous Coal	**Find:**	Minimum Skirtboard Length (**Equation 11.1.1**)
	Belt Width	1 m (36 in.)		
	Belt Speed	3 m/s (600 ft/min)		Minimum Skirtboard Height (**Equation 11.2.1**)
	Width of Skirtboard	0,6 m (2 ft)		
	Measured Airflow	0,56 m³/s (1200 ft³/min)		

Equation 11.1.1

Skirtboard Length Sample Problem #1

$$l_{sb} = \frac{V \cdot CF}{k}$$

Given: *Belt speed of 3 meters per second (600 ft/min) and an airflow of 0,56 cubic meters per second (1200 ft³/min).* **Find:** *Minimum Skirtboard Length.*

Variables		Metric Units	Imperial Units
V	Belt Speed	3 m/s	600 ft/min
CF	Chute Factor	0,9	3
k	Conversion Factor	0,5	100

Metric: $l_{sb} = \dfrac{3 \cdot 0,9}{0,5} = 5,4$ Imperial: $l_{sb} = \dfrac{600 \cdot 3}{100} = 18$

l_{sb}	Minimum Skirtboard Length (from loading zone to end of chute)	5,4 m	18 ft

Equation 11.2.1

Skirtboard Height Sample Problem #1

$$h_{sb} = \frac{Q_{tot}}{CW \cdot v}$$

Given: *An airflow of 0,56 cubic meters per second (1200 ft³/min), a chute (skirtboard) width of 0,6 meters (2 ft), and a target air velocity of 1 meter per second (200 ft/min).* **Find:** *Minimum Skirtboard Height.*

Variables		Metric Units	Imperial Units
Q_{tot}	Total Airflow	0,56 m³/s	1200 ft³/m
CW	Chute (Skirtboard) Width	0,6 m	2 ft
v	Target Air Speed	1 m/s	200 ft/min

Metric: $h_{sb} = \dfrac{0,56}{0,6 \cdot 1} = 0,93$ Imperial: $h_{sb} = \dfrac{1200}{2 \cdot 200} = 3.0$

h_{sb}	Minimum Skirtboard Height	0,93 m	3 ft

Note: In some real applications, calculated results may be impractical, so engineering judgment must be exercised.

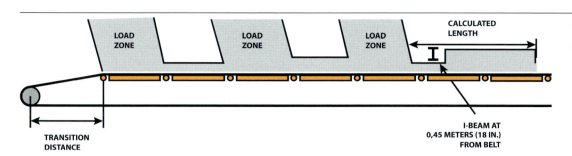

Figure 11.17

Sample Problem #2

TRANSITION DISTANCE

CALCULATED LENGTH

I-BEAM AT 0,45 METERS (18 IN.) FROM BELT

Skirtboard Sample Problem #2				
Given:	Material	Wood Chips	**Find:**	Minimum Skirtboard Length **(Equation 11.1.2)**
	Belt Width	1,27 m (48 in.)		
	Belt Speed	3,5 m/s (700 ft/min)		Minimum Skirtboard Height **(Equation 11.2.2)**
	Width of Skirtboard	1,0 m (3 ft)		
	Measured Airflow	0,28 m³/s (600 ft³/min) from each load zone; all load zones operate at once		

Table 11.4

(Figure 11.17)

$$l_{sb} = \frac{V \cdot CF}{k}$$

Equation 11.1.2

Skirtboard Length Sample Problem #2

Given: *Belt speed of 3,5 meters per second (700 ft/min) and an airflow of 0,84 cubic meters per second (1800 ft³/min).* **Find:** *Minimum Skirtboard Length.*

	Variables	Metric Units	Imperial Units
V	Belt Speed	3,5 m/s	700 ft/min
CF	Chute Factor	0,9	3
k	Conversion Factor	0,5	100

Metric: $l_{sb} = \dfrac{3,5 \cdot 0,9}{0,5} = 6,3$	Imperial: $l_{sb} = \dfrac{700 \cdot 3}{100} = 21$	
l_{sb}	Minimum Skirtboard Length (from loading zone to end of chute)	6,3 m	21 ft

$$h_{sb} = \frac{Q_{tot}}{CW \cdot v}$$

Equation 11.2.2

Skirtboard Height Sample Problem #2

Given: *An airflow of 0,84 cubic meters per second (1800 ft³/min), a chute (skirtboard) width of 1 meter (3 ft), and a target air velocity of 1 meter per second (200 ft/min).* **Find:** *Minimum Skirtboard Height.*

	Variables	Metric Units	Imperial Units
Q_{tot}	Total Airflow	0,84 m³/s	1800 ft³/m
CW	Chute (Skirtboard) Width	1,0 m	3 ft
v	Target Air Speed	1 m/s	200 ft/min

Metric: $h_{sb} = \dfrac{0,84}{1,0 \cdot 1} = 0,84$	Imperial: $h_{sb} = \dfrac{1800}{3 \cdot 200} = 3$	
h_{sb}	Minimum Skirtboard Height	0,84 m	3 ft

Note: This extended height skirtboard **must** begin immediately downstream of the "I-beam" above the belt.

Note: In some real applications, calculated results may be impractical, so engineering judgment must be exercised.

Figure 11.18

Sample Problem #3

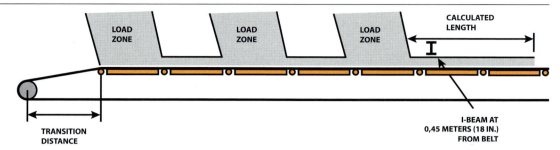

Table 11.5

(Figure 11.18)

Skirtboard Sample Problem #3

Given:	Material	Anthricite Coal	Find:	Minimum Skirtboard Length (*Equation 11.1.3*)
	Belt Width	1,27 m (48 in.)		
	Belt Speed	3,5 m/s (700 ft/min)		Minimum Skirtboard Height (*Equation 11.2.3*)
	Width of Skirtboard	1,0 m (3 ft)		
	Measured Airflow	0,28 m³/s (600 ft³/min) from each load zone; each load zone operates one at a time		

Equation 11.1.3

Skirtboard Length Sample Problem #3

$$l_{sb} = \frac{V \cdot CF}{k}$$

Given: *Belt speed of 3,5 meters per second (700 ft/min) and an airflow of 0,28 cubic meters per second (600 ft³/min).* **Find:** *Minimum Skirtboard Length.*

	Variables	Metric Units	Imperial Units
V	Belt Speed	3,5 m/s	700 ft/min
CF	Chute Factor	0,6	2
k	Conversion Factor	0,5	100

Metric: $l_{sb} = \dfrac{3,5 \cdot 0,6}{0,5} = 4,2$ Imperial: $l_{sb} = \dfrac{700 \cdot 2}{100} = 14$

l_{sb}	Minimum Skirtboard Length (from loading zone to end of chute)	4,2 m	14 ft

Equation 11.2.3

Skirtboard Height Sample Problem #3

$$h_{sb} = \frac{Q_{tot}}{CW \cdot v}$$

Given: *An airflow of 0,28 cubic meters per second (600 ft³/min), a chute (skirtboard) width of 1 meter (3 ft), and a target air velocity of 1 meter per second (200 ft/min).* **Find:** *Minimum Skirtboard Height.*

	Variables	Metric Units	Imperial Units
Q_{tot}	Total Airflow	0,28 m³/s	600 ft³/m
CW	Chute (Skirtboard) Width	1,0 m	3 ft
v	Target Air Speed	1 m/s	200 ft/min

Metric: $h_{sb} = \dfrac{0,28}{1,0 \cdot 1} = 0,28$ Imperial: $h_{sb} = \dfrac{600}{3 \cdot 200} = 1$

h_{sb}	Minimum Skirtboard Height	0,28 m	1 ft

Note: The skirtboard can fit below the "I-beam" above the belt.

Note: In some real applications, calculated results may be impractical, so engineering judgment must be exercised.

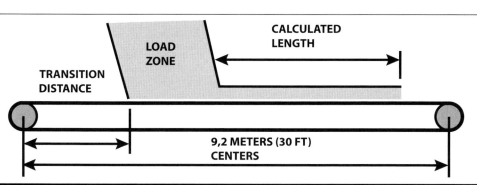

Figure 11.19

Sample Problem #4

11

Skirtboard Sample Problem #4					
Given:	Material	Anthracite Coal	**Find:**	Minimum Skirtboard Length **(Equation 11.1.4)**	
	Belt Width	1,8 m (72 in.) (feeder belt)			
	Belt Speed	0,5 m/s (100 ft/min)			
	Material Depth	0,3 m (1 ft)		Minimum Skirtboard Height **(Equation 11.2.4)**	
	Width of Skirtboard	1,5 m (5 ft)			
	Flat Belt				
	Measured Airflow	0,047 m³/s (100 ft³/min)			

Table 11.6

(Figure 11.19)

$$l_{sb} = \frac{V \cdot CF}{k}$$

Equation 11.1.4

Skirtboard Length Sample Problem #4

Given: Belt speed of 0,5 meters per second (100 ft/min) and an airflow of 0,047 cubic meters per second (100 ft³/min). **Find:** Minimum Skirtboard Length.

	Variables	Metric Units	Imperial Units
V	Belt Speed	0,5 m/s	100 ft/min
CF	Chute Factor	0,6	2
k	Conversion Factor	0,5	100

Metric: $l_{sb} = \dfrac{0,5 \cdot 0,6}{0,5} = 0,6$	Imperial: $l_{sb} = \dfrac{100 \cdot 2}{100} = 2$		
l_{sb}	Minimum Skirtboard Length (from loading zone to end of chute)	0,6 m	2 ft

Note: The skirtboard **must** extend the entire loaded length of the belt, because it also functions to keep the material on the belt.

$$h_{sb} = \frac{Q_{tot}}{CW \cdot v}$$

Equation 11.2.4

Skirtboard Height Sample Problem #4

Given: An airflow of 0,047 cubic meters per second (100 ft³/min), a chute (skirtboard) width of 1,5 meters (5 ft), and a target air velocity of 1 meter per second (200 ft/min). **Find:** Minimum Skirtboard Height.

	Variables	Metric Units	Imperial Units
Q_{tot}	Total Airflow	0,047 m³/s	100 ft³/m
CW	Chute (Skirtboard) Width	1,5 m	5 ft
v	Target Air Speed	1 m/s	200 ft/min

Metric: $h_{sb} = \dfrac{0,047}{1,5 \cdot 1} = 0,03$	Imperial: $h_{sb} = \dfrac{100}{5 \cdot 200} = 0.1$		
h_{sb}	Minimum Skirtboard Height	0,03 m	0.1 ft

Note: The height **must** be at least 0,3 meters (1 ft), because that is the height of the material on the belt.

Note: In some real applications, calculated results may be impractical, so engineering judgment must be exercised.

DRAWING A CONCLUSION ABOUT SKIRTBOARD

In Closing...

Skirtboard plays a key role in the control of both dust and spillage. By centering the cargo, properly designed skirtboard systems will reduce spillage; by forming a plenum for materials to settle out of the air, skirtboard helps prevent the escape of dust. Both are essential steps in the struggle to improve conveyor efficiency by total material control.

Looking Ahead...

This chapter about Skirtboards, the sixth chapter in the section Loading the Belt, focused on the use of skirtboards to reduce spillage and dust. Two chapters remain in this section: Chapter 12: Wear Liners and Chapter 13: Edge-Sealing Systems.

REFERENCES

11.1   Conveyor Equipment Manufacturers Association (CEMA). (2005). *BELT CONVEYORS for BULK MATERIALS, Sixth Edition*. Naples, Florida.

11.2   The website http://www.conveyor-beltguide.com is a valuable and non-commercial resource covering many aspects of belting.

11.3   Any manufacturer and most distributors of conveyor products can provide a variety of materials on the construction and use of their specific products.

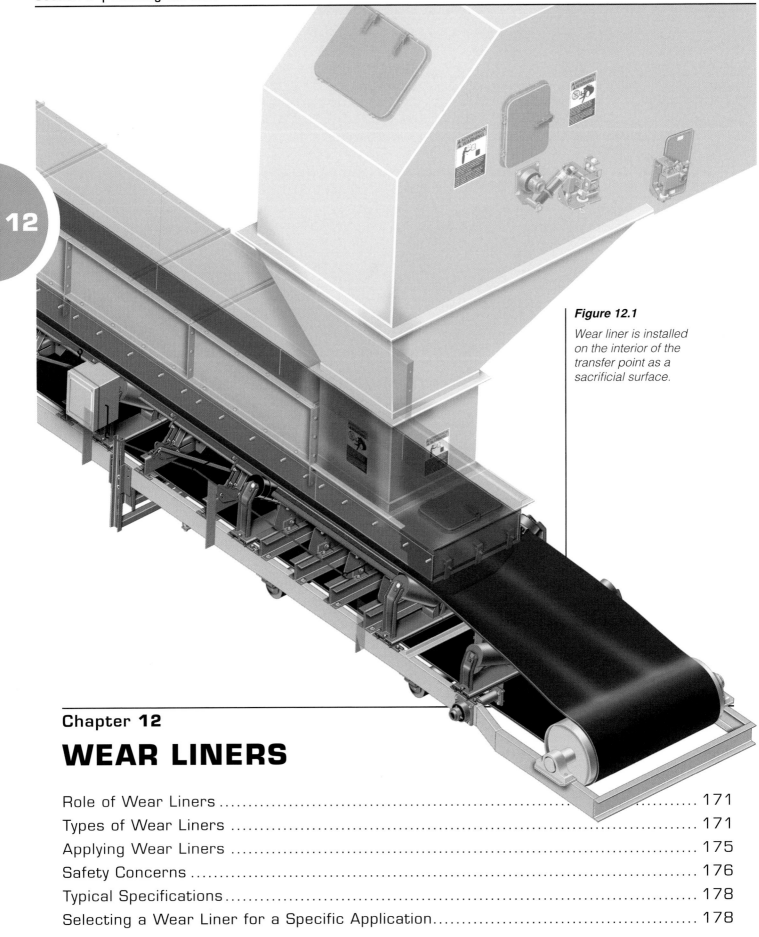

Figure 12.1

Wear liner is installed on the interior of the transfer point as a sacrificial surface.

Chapter 12

WEAR LINERS

In this Chapter...

This chapter will cover the reasons for the installation of wear liner, the three styles of wear liners commonly used, and the various materials used as wear liners. We also discuss selection and correct installation techniques for wear liner in the skirtboard section of a transfer point.

Wear liner is a material installed on the interior of the transfer point as a sacrificial surface to be worn away by contact with the moving bed of material (**Figure 12.1**).

ROLE OF WEAR LINERS

In plans for a low-spillage transfer point, wear liner serves multiple purposes:

A. It provides a sacrificial, easily-replaceable wear surface to protect the walls of the chute and the skirtboard.

B. It helps center the material load.

C. It prevents the material load from applying high side forces onto the sealing strips, thereby improving the seal's service-life.

D. Some types of wear liners can reduce friction, impact, noise, and degradation of the bulk material.

The forces of material flowing through a transfer point and dropping onto a belt inside a traditional transfer area create tremendous outward pressures. Wear liner is installed to control this side pressure and protect the components of the transfer point. The wear liner is a key component in the containment of the conveyed material in the skirted area (**Figure 12.2**). This side pressure of the material, if uncontrolled, will push material fines and dust away from the center of the material pile and under the skirtboard, resulting in spillage.

Wear liners are installed on the inside of the skirtboard(s) to protect the skirting seal. They have the mission of separating the job of sealing from the function of load placement. By creating a dam between the material pile and the edge-sealing strips,

the wear liners greatly reduce the side-loading forces that reach the sealing strips. With wear liners installed, the sealing strips do not have to act as a wall to contain the material load but rather act only as a seal, a purpose for which they are much more suited. This arrangement improves the effectiveness and life expectancy of the sealing system while reducing the risk of damage by material entrapment.

There are only a few instances where the installation of a wear liner will not greatly enhance the sealability of a transfer point and the life expectancy of its components. These would be very lightly-loaded belts or belts handling non-abrasive, low-density materials. In all other circumstances, properly installed and maintained wear liners will reduce the material side-loading forces to increase sealing efficiency and sealing-strip life.

TYPES OF WEAR LINERS

Configurations of Wear Liner

Four styles of wear liner are commonly seen today: straight, spaced, deflector, and tapered (**Figure 12.3**).

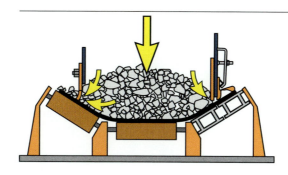

Figure 12.2

Without the protection offered by a wear liner, the elastomer sealing system is not strong enough to contain the forces that push material over the edge of the belt.

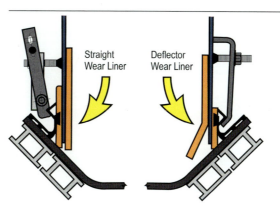

Straight Wear Liner

Deflector Wear Liner

Figure 12.3

Left: Straight Wear Liner. Right: Deflector Wear Liner.

Straight Wear Liner

Straight wear liner has the capability of preventing side-loading forces on the skirting seals without choking the chute and constricting material flow. Straight wear liner has been used on all sizes of belts (**Figure 12.4**). The real benefit of straight wear liner is that it provides improved life and improved sealing effectiveness without closing down the effective load area. In an era when more and more production is asked of fewer and fewer resources, it is important to maximize system capacity by utilizing the full width of the loading chute and conveyor belt. Straight wear liner is a good choice to meet both current and future requirements for flow of most bulk materials.

Straight wear liner is also best for belts with multiple loading points, whether installed in one long transfer point or through several loading zones.

Spaced Wear Liner

A variation on straight wear liner installation technique is spaced wear liner (**Figure 12.5**). This hybrid technique can be used in applications where mechanical dust collection is present. To assist in sealing, the liners are not installed directly onto the wall of the skirtboard but rather separated slightly—25 to 50 millimeters (1 to 2 in.)—from it. The space between the skirtboard and the wear liner is used as negative pressure area. Fines and airborne dust in this area can be pulled from this space by the conveyor's dust-collection system.

This technique is better suited for use on new conveyor systems, so the requirement for the "free belt edge distance" can be engineered into the dimensions of the loading zone from the beginning without reducing the carrying capacity of the conveyor. While the dimensions of this space are not large, typically 25 to 50 millimeters (1 to 2 in.) of free space on each side of the conveyor, it is important in a spaced wear-liner installation that the liner be installed so its top edge is well above the height of the pile of material in the loading zone.

Deflector Wear Liner

Deflector wear liner incorporates a bend so the bottom half of the liner is bent inward, toward the middle of the belt (**Figure 12.6**). This angle provides a "free" area between the rubber skirting and the wear liner. This area is useful, as fines that have worked their way under the bottom edge of the wear liner still have an area on the belt on which to travel; they are not au-

Figure 12.4

Straight wear liner improves sealing without choking the effective load area.

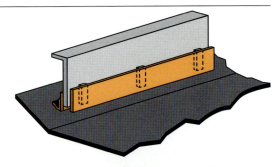

Figure 12.5

Spaced wear liner incorporates an open area behind the liner where dust collection can be applied.

Figure 12.6

Deflector wear liner creates a free area between the liner and the sealing system.

12

tomatically ejected from the system. These particles are contained by the sealing strip and have a path to travel down the belt to the exit area of the transfer point. The fines that work their way out to challenge the rubber seal are relatively free of applied forces; they are isolated from the downward and outward force of the material load.

The drawback of deflector wear liner is that it reduces the effective cross-sectional area of the skirtboard area. This, in turn, reduces the volume of material that can pass through the transfer point, and consequently may require adjustments in the chute dimensions or in a system's operating schedule to maintain a specified capacity. This consideration is particularly important on smaller belts—less than 750 millimeters (30 in.) wide—or belts running near capacity. By reducing the loading zone's cross section, deflector liner may also reduce the maximum allowable lump size, leading to material jams.

In addition, deflector wear liner should not be used in loading zones that see impact. In such applications, the liner faces higher wear and the opportunity for pieces of material to rebound off the belt and wedge into the deflector liner's open bottom area, creating the risk of belt abrasion. Deflector liner also concentrates the abrasive wear from material impact onto the "bent" area and "lip" of the liner. If the wear is concentrated in one spot, the wear can create an opening where lumps of material can build up, increasing the chances of belt abrasion (**Figure 12.7**).

Tapered Wear Liner

Tapered wear liner is usually cast from molybdenum (moly) steel for use in heavy-duty applications. The cross section of the casting is trapezoidal to reduce the gap at the concurrence of the belt, liner, and skirting seal while presenting sufficient wear thickness where the material impacts or slides along the skirted area. To keep the weight of individual castings to a reasonable weight for handling, the tapered wear liners are usually made 300 to 400 millime-

ters (12 to 16 in.) wide. Because cast wear liners are heavy and are supplied in short lengths, it is difficult to install them so the bottom edge is in a smooth, straight line. Poor installation can create pockets where bulk materials can be trapped and wear the belt.

Wear-Liner Materials

Straight and deflector wear liners are typically supplied as sheets of material, often 1200 millimeters (48 in.) long, 200 millimeters (8 in.) high, and 12 millimeters (1/2 or 0.5 in.) thick. Cast liners are usually supplied in pieces that are 300 to 400 millimeters (12 to 16 in.) wide, 200 to 500 millimeters (8 to 20 in.) high, and 25 to 75 millimeters (1 to 3 in.) thick. The liners can be supplied with pre-drilled holes to simplify field installation.

There are a number of materials suitable for use as wear liners (**Table 12.1**).

Mild Steel Wear Liner

Mild steel wear liner is commonly used on materials with very low abrasion or on belts with light loads or low operating hours. Materials such as sawdust, wood chips, and garbage would be good examples of material suitable for mild steel wear liners. In addition, projects with demands for low initial costs but which require good short-term results are also candidates for mild steel wear liner.

If the environment is damp or otherwise corrosive, the higher corrosion rate of mild steel may add additional friction to the material body in the loading zone.

Figure 12.7

Deflector liner also concentrates the abrasive wear from material impact onto the "bent" area and "lip" of the liner.

Mild steel wear liner can be supplied in either the straight or deflector pattern.

Abrasion-Resistant Plate Wear Liner

Abrasion-resistant plate wear liner (AR plate) provides a much longer life than wear liner fabricated from mild steel. AR plate is a good, all-around wear liner, capable of handling more-abrasive materials such as sand, hard rock mining ores, and coal. The wear life may extend five to seven times longer than mild steel. AR plate is available in either straight or deflector styles.

Ceramic-Faced Wear Liner

Ceramic-faced wear liner is a good, long-term wear liner for continuously-operating belts carrying highly-abrasive material where impact is minimal. A mild steel backing plate faced with ceramic blocks is a good choice in these circumstances. These ceramic blocks are glued and/or plug welded to the mild steel backing, usually on the bottom 100 millimeters (4 in.) of the plate. On more heavily loaded belts, the ceramic blocks can also be applied higher up the backing plate to reduce wear.

Ceramic-faced wear liner has been shown to work well with coal and wood chips. Ceramic-faced wear liner can be supplied in both straight and deflector styles.

Any time liners are faced with castable materials, whether ceramic or alloys such as magnesium steel, extreme care must be taken to align the blocks during their installation on the steel plate. The bottom edge of the installation must be positioned with care to avoid pinch points and "stair steps" that can trap material.

Stainless Steel Wear Liner

Stainless steel wear liner is a choice that falls between mild steel and AR plate in abrasion resistance. The chemical resistance of stainless steel is often required for applications where the possibility of corrosion on mild steel or AR plate exists. The coefficient of friction between the bulk material and stainless steel varies significantly, and power requirements should be reviewed if retrofitting with stainless steel liners. Stainless steel wear liner can be supplied in both straight and deflector styles.

Table 12.1

Wear-Liner Materials					
Lining Material	Initial Cost	Sliding Abrasion Resistance	Impact Resistance	Temperature Resistance	Low-Friction Quality
Mild Steel	Low	G	G	VG	NR
Abrasion-Resistant Plate	Medium	VG	G	VG	NR
Stainless Steel	High	G	G	E	VG
Chromium Carbide Overlay	Medium	E	G	VG	VG
Rubber	High	G	E	NR	NR
Polyurethane	High	E	E	NR	G
UHMW	Medium	G	NR	NR	E
Ceramic Tile					
Quarry Tiles	Low	G	NR	G	G
Vitrified Tiles	Low	VG	NR	VG	VG
Basalt Tiles	Medium	VG	G	VG	G
Alumina Tiles	High	E	G	E	G

Note: Performance Comparison of Possible Wear-Liner Materials. Ratings: E–Excellent; VG–Very Good; G–Good; NR–Not Recommended.

Chromium Carbide Overlay

Chromium carbide overlay is a very hard material suitable for conveyors seeing very high levels of abrasion. Alone, chromium carbide is very brittle, so it is overlaid onto a backing plate for installation. The backing plate can be of mild or stainless steel, depending on application requirements. The hard facing will rank between 53 and 65 Rockwell "C" hardness; some overlay materials "work harden" under contact with the cargo, so they score a 75 hardness on the Rockwell "C" scale. Also referred to as "clad plate," these materials are available in two designations: single-weld or double-weld pass. For wear liner applications, the double-weld pass grade is typically used. This material is not suited for high impact and, as a result, is used only in the straight style of wear liner.

Plastic Wear Liner

Plastic wear liners are a more recent development. Recently, wear liners composed of Ultra-High Molecular Weight (UHMW) polyethylene, or urethane have been installed. In many of these installations, the liner sits directly on the belt to control extremely fine, dusty materials. Slotted holes in the liner panels allow adjustment to keep the wear liner in contact with the belt.

Applications of UHMW as a wear liner show success with fine, powdery products such as sand, fly ash, and electric-arc furnace (EAF) dust. In addition, as UHMW is accepted by the U.S. Food and Drug Administration, it is suitable for use with powdery foodstuffs. Urethane liners are now being used successfully in gold-mining and ore-processing applications for their light weight and ease of replacement.

Plastic materials have been applied only as straight wear liners; the abrasion that would be seen in applications as a deflector design would dramatically reduce the service-life. Care should be taken to not install plastic liners in conditions that exceed the material's service temperature or that see a high belt speed; doing so could raise the liner's temperature to a softening point, shortening the material life.

Wear Liner for Curved Chutes

Many of these materials are also suitable for lining curved chutes for applications where there is a need for wear resistance or reduced friction. Examples would include ceramic tiles or AR plate used as a liner in a curved chute handling coal or UHMW used in a chute for wood chips.

Wear Liner Cost vs. Value

While the initial cost of a wear liner should be a significant consideration, it is more important that the material be selected on the basis of its performance and service-life. Factors that should also be considered include:

A. Friction coefficient

B. Resistance to material adhesion

C. Resistance to sliding-abrasion wear

D. Resistance to impact-abrasion wear

E. Resistance to corrosion

F. Attachment method

G. Installation cost

H. Maintenance cost

Choosing the correct material for use as a wear liner could increase the initial cost of the transfer point. However, the use of a liner material specifically tailored to a given application should produce a better return on investment when considering the labor of replacing prematurely worn liners and the increased time for cleanup of spilled material.

APPLYING WEAR LINERS

Application of Wear Liner

With the exception of UHMW and urethane wear liners, all wear-liner systems must be installed with a relieving angle from the entry area opening toward the exit area of the transfer point. The distance above the belt will vary with the product size. As with the skirtboard steel, the desire is for the liner to create a larger opening toward the exit of the loading zone to prevent material entrapment.

As stated earlier, UHMW and urethane wear liners are typically installed so the bottom edge will touch, or lie, on the belt.

At the entry area, the space between the belt and the bottom edge of the wear liner is generally in the range of 3 to 10 millimeters (1/8 to 3/8 in.), with the closer dimension specified for materials with smaller particle sizes. At the exit end, the distance

Figure 12.8

Poorly-aligned wear liner creates material entrapment points.

Figure 12.9

Wear liner should be installed in a straight line that gradually opens toward the exit of the transfer point.

will typically be 10 to 20 millimeters (3/8 to 3/4 in.). Again, the smaller distance is for finer materials, whereas the larger dimension is for materials containing larger lumps. Proper belt support to eliminate belt sag and vibration is essential to preserving the belt in the face of this narrow spacing.

A word of caution: When joining pieces of wear liner, it is imperative the bottom edges line up smoothly without creating a jagged, "saw-toothed" pattern (**Figure 12.8**). If the bottom edges are not precisely aligned, entrapment points will be created. The conveyed material will then create exceptionally-high pressure points in these areas that will lead to material spillage or, even worse, wedges of material will accumulate into "teeth" that will abrade the belt. To prevent these buildups, the bottom edge of the wear liner should be straight, as if a string line were stretched from the entry to the exit of the transfer point (**Figure 12.9**). Again, relief from entrapment should be provided with a slight increase in the distance from the belt surface to the bottom edge as the belt moves toward the exit of the transfer point.

Installing Wear Liner

Wear plate can be applied by methods including bolts, welding, or a combination of both.

Wear liners are commonly installed with countersunk bolts that provide smooth surfaces on the interior face of the skirtboard. These bolts also allow the simple replacement of the liner. Liners can be welded

⚠ SAFETY CONCERNS

When installing wear liners, it is important to bear in mind that these are large panels, usually of steel, and may have sharp edges. They are heavy and usually awkward to handle, particularly when trying to maneuver and install them in the close quarters between the skirtboards on narrow conveyors. Proper lifting tackle and equipment to restrain the liners should be used while completing the installation.

The operation's established lockout / tagout / blockout / testout procedures, confined-space regulations, and other appropriate safety policies should be followed.

into position, with the obvious drawback being the difficulty in replacing worn liners. Should the installation require that the wear liner be welded into position, care must be taken to use the correct welding materials and techniques to match the liner material.

Another installation technique calls for the wear liner to be plug welded from outside the transfer point (**Figure 12.10**). With this technique, holes are drilled or cut through the plated steel wall. Then the back of the liner is welded to the chute wall. This system provides installation without bolt heads or holes protruding into the loading zone to act as targets for material abrasion. The liner provides its full thickness for wear-life. At the end of the liner's life, replacement can be performed by cutting out the plug welds and installing new liners using the same holes.

When welding the liner in place, care must be taken to control the stress introduced into the lining metal. Abrasion-resistant plate, when applied as a liner, must be applied the same way a person would apply wallpaper. If a sheet of wallpaper is applied by securing the four outside edges first, big air bubbles will be trapped in the center of the sheet. A similar situation occurs if AR plate is installed in the same fashion, but instead of trapped air bubbles, residual stress will be created in the plate that will try to escape. When the structure starts to flex under normal operation, these stresses can introduce cracks in the wear liner. If not caught in time, a large section of liner can break off or the chute wall will bow.

To avoid this stress, it is important to use proper welding technique. The accepted "best practice" is called backstep or "back welding." It calls for stitch welding on the top of the plate (**Figure 12.11**). At each weld, the bead is drawn back toward the welded end. Correct welding rod selection is imperative to assure the strength and durability of the weld joint.

Careful attention must be paid to the strength of the conveyor structure when installing liners for the first time. Unless properly reinforced, the support structure could be too weak to support the added weight of liners, risking costly damage and downtime.

Chute Design for the Purpose of Maintenance

As with any enclosure, it is important to provide a simple way to inspect the interior. Doors through the chute wall or skirtboard must be included as a mechanism to inspect the condition of liners.

Ideally, the chute will be large enough to allow personnel inside to perform the work of installation and replacement. If the conveyor size is not such that there is room for personnel to work inside the transfer-point enclosure, the skirtboard and liner system should be designed so that the entire assembly can be opened, laid over on its back, or lifted off the structure. This would allow the replacement of the liner and the reinstallation of the wall assembly with minimum downtime, inconvenience, and cost.

It would be valuable to permanently install anchors for rigging at suitable locations above the skirtboard to allow easier lifting and placement of wear liner sheets.

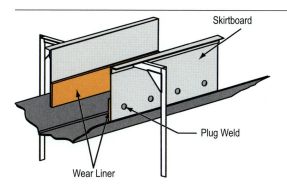

Figure 12.10

Wear liners can be installed from the outside of the skirtboard by plug welding.

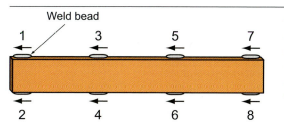

Figure 12.11

Backstep welding is "best practice" for the installation of wear liners to prevent warping.

TYPICAL SPECIFICATIONS

A. Wear liner

The skirtboard will be equipped with a wear liner. Affixed to the interior of the skirtboard, this liner will protect the edge-sealing system from the side-loading forces of the material load.

B. Position

The wear liner will be positioned within 3 to 10 millimeters (1/8 to 3/8 in.) of the belt at the entry area and within 10 to 20 millimeters (3/8 to 3/4 in.) of the belt at the exit end of the transfer-point skirtboard.

C. Alignment

The bottom of liners will be precisely aligned to eliminate any jagged or "saw-toothed" edges, which can capture material.

D. Plug-welding technique

The wear liner will be installed using a plug-welding technique to prevent the intrusion of bolt holes into the liner material.

SELECTING A WEAR LINER FOR A SPECIFIC APPLICATION

In Closing...

The choice of the "best" liner material for a given application is usually specific to that given application, driven by the material conveyed and the system.

Sometimes the material choice is driven from higher up in the organization, where a corporate engineer has some experience—positive or negative—with a particular liner choice and so mandates (or forbids) the use of a specific material. Other times, a choice will be made to keep all materials uniform within a plant—to use existing material, or make future orders simpler—even in an application where the material specified is less than ideal.

There are many references, articles, and specialty suppliers of wear-liner materials to assist in choosing a liner for a specific application. Plant personnel will usually know what has been tried in an application, and they will know its track record. They will know what has worked, and perhaps more importantly, what has failed—or what was deemed too short-lived—in a specific application.

This institutional memory is a valuable tool in liner choice. But it should not be relied upon exclusively. This background knowledge should be compared against accurate records showing the dates of installation and the tonnage conveyed over a specific liner material in a given location. The record-keeping is the key to validating the wear-liner material choice made.

Looking Ahead...

This chapter about Wear Liners, the seventh chapter in the section Loading the Belt, discussed their use in a low-spillage transfer point. The following chapter, Edge-Sealing Systems, concludes this section.

REFERENCES

12.1 Conveyor Equipment Manufacturers Association (CEMA). (2005). *BELT CONVEYORS for BULK MATERIALS, Sixth Edition*. Naples, Florida.

12.2 The website http://www.conveyor-beltguide.com is a valuable and non-commercial resource covering many aspects of belting.

12.3 Any manufacturer and most distributors of conveyor products can provide a variety of materials on the construction and use of their specific products.

12

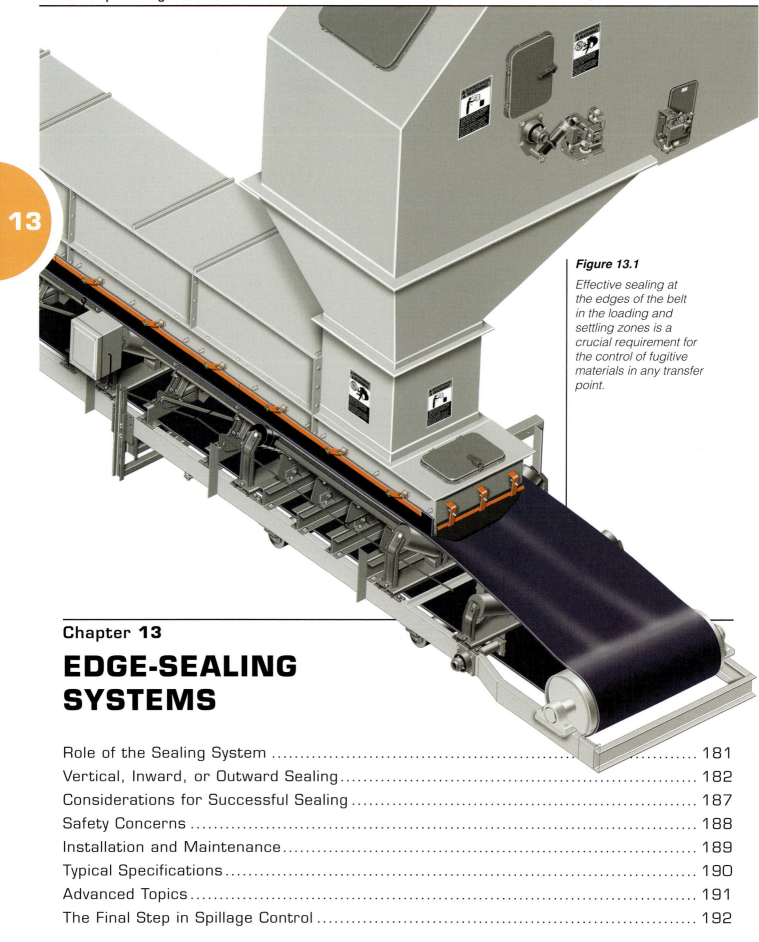

13

Figure 13.1
Effective sealing at the edges of the belt in the loading and settling zones is a crucial requirement for the control of fugitive materials in any transfer point.

Chapter **13**

EDGE-SEALING SYSTEMS

In this Chapter...

In this chapter, we conclude the discussion about sealing the loading zone of a transfer point with a focus on edge-sealing systems. Three main types of sealing systems are described, with advantages and disadvantages of each, along with various engineered systems. Guidelines for selection, installation, use, and maintenance of edge-sealing systems are also discussed. The chapter closes with equations for calculating additional power consumption required.

A crucial requirement in any transfer point designed for reduced spillage and high efficiency is an effective sealing system at the edges of the belt (**Figure 13.1**). The seal should start in the loading area and continue to the end of the settling zone. An edge-sealing system, typically a flexible elastomer strip, is installed on the outside of the skirtboard on both sides of the belt to bridge the gap between the steel structures and the moving belt.

Instead of the first step in preventing conveyor spillage, the skirtboard seal is the last chance to control fugitive material and prevent its release. Stabilizing the belt line, with properly installed belt support, and controlling the amount of material leakage, with a wear-liner system installed close to the belt, improve the performance of the belt's edge-sealing system. A flexible, multiple-layer system incorporating a level of self-adjustment provides effective material containment for a transfer point and improves the operation of the belt conveyor.

A functional edge-sealing system requires the use of belt support, skirtboards, wear liners, and an edge seal (**Figure 13.2**). Belt-support systems are discussed in Chapter 10, skirtboards in Chapter 11, and wear liners in Chapter 12. This chapter will focus primarily on the skirtboard, or edge, seal as the important final component in sealing the loading zone of a transfer point for the prevention of the escape of fine particles of fugitive material.

ROLE OF THE SEALING SYSTEM

What Sealing Systems Can and Cannot Do

In the past, a typical edge-sealing system was a vertical strip of elastomer clamped to the outside of the chute or skirtboard steel. The rubber sealing strip would bridge the gap from the steel to the belt, which was typically 25 to 50 millimeters (1 to 2 in.) or more.

This elastomer sealing strip was expected to perform an almost miraculous function. With the belt not properly supported, and/or no wear liner or worn out wear liner in place, the elastomer edge seal was required to contain the full weight of the material load while attempting to adjust for an undulating belt path. To ask flexible sealing strips alone to do more than contain light material or dust on the belt is asking for the unattainable. The material load will quickly abrade the sealing strip, or push it away from the skirtboard, and allow for the resumption of spillage (**Figure 13.3**).

In an attempt to stop the escape of particles, plant personnel continually adjust the sealing strips down to the belt, thereby increasing the sealing pressure and leading to several undesirable results. The increase in sealing pressure raises the conveyor's power requirements, sometimes to the

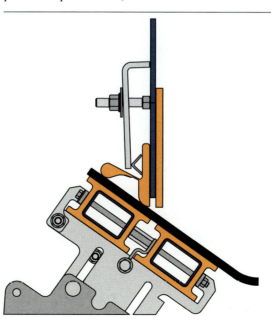

Figure 13.2

Effective sealing at the edge of the belt requires belt support, wear liners, skirtboards, and edge seal.

13

point in which it is possible to stall the belt. The increased friction causes heat buildup that will soften the elastomer sealing strip and shorten the elastomer's life, sometimes to the point in which the seal will virtually melt away. This increased wear is most obvious at the points where there is the highest pressure, typically directly above the idlers (**Figure 13.4**).

On a belt that is not properly supported in the load zone, the belt will sag between the idlers and allow material to become entrapped between the wear liner or the edge seal and the belt. The entrapped material greatly accelerates belt and edge-sealing

system wear and increases the drive power requirement.

Goal of a Sealing System

The goal of any sealing system is to contain conveyed material fines and dust on the belt. Desirable attributes include minimal seal-to-belt contact area and minimal seal-downward pressure, and reasonable life. Minimizing these items reduces drag against the conveyor, minimizes wear on the belt and the seal, and minimizes additional power required to drive the belt.

The job of sealing the edges of a conveyor load zone is a challenging one. Even in the best transfer points with the belt tracking properly on a stable line, the sealing system faces a certain amount of sideways pressure and vibration due to variations in loading and other conditions. The sealing system must be designed to conform to these fluctuations in belt travel to form an effective seal. The seal must be rugged enough to stand up to belt abrasion and splice impact without undue wear and without catching the splice. The sealing system should offer a simple adjustment mechanism to compensate for wear.

No sealing system can stand up for long in the face of abuse from the material load. If the seals are not sheltered from the material flow, both the effectiveness and the life of the sealing strips will be diminished. With the impact of the loading material onto a sealing system, the material forces the sealing strips down onto the belt, accelerating wear in both seal and belt. The transfer point should be constructed to avoid both loading impacts on the seals and material flow against the seals.

Figure 13.3

Without an effective sealing system, material will spill over the sides of the belt, shortening component life and increasing maintenance and cleanup chores.

Figure 13.4

With the higher pressure, the seal wears faster and unevenly. The uneven wear, in the form of scallops, will be most obvious directly above the idlers where pressure is the highest.

Figure 13.5

The first skirtboard seals were fabricated in-house from readily-available materials such as used belting or large "barge" ropes. These primitive sealing systems were pushed down onto the belt edges or held in position by gravity.

VERTICAL, INWARD, OR OUTWARD SEALING

Engineered Sealing Systems

The first skirtboard seals were fabricated in-house from readily available materials such as used belting or large "barge" ropes (**Figure 13.5**). These primitive sealing systems were pushed down onto the belt edges or held in position by gravity. While they were inexpensive, these systems were not very successful. They became impregnated with material that abraded the belt, and they lacked an easy method to adjust for wear. Eventually, the disappointing results of these homemade techniques led to the desire for, and design of, more effective systems.

Now, the state of the art in engineered transfer-point components has progressed from sealing strips that barely contained lumps of material to the current systems that prevent the escape of fines and even dust. A number of engineered sealing systems are now commercially available. In general, these systems consist of a long strip of elastomer, held against the lower edge of the skirtboard by an arrangement of clamps.

For effective sealing, it is critical that there be adequate free-belt distance. Free-belt distance—the amount of belt outside the skirtboard on both sides of the conveyor—provides the space available for the sealing system (**Figure 13.6**). Too often, in the interest of putting the greatest load on the narrowest belt, the free-belt distance is reduced. This invariably comes at the cost of sealing system effectiveness. *(For more information about free-belt distance and effective belt width, see Chapter 11: Skirtboards.)*

There are a number of different approaches to skirtboard sealing. A simple way to classify these systems is to describe where each contacts the belt: Some drop straight down from the skirtboard, some extend back inside the skirtboard, and some seal on the outside of the skirtboard.

Vertical Sealing

Vertical sealing systems typically use only a single rubber sealing strip (**Figure 13.7**). Often, one supplier offers a system of clamps, and another provides the rubber strip. Sometimes, a specially-shaped elastomer strip is installed; other times, strips of rubber or, worse, used belting are applied (**Figure 13.8**). The sealing system selected

should always be less abrasion resistant than the top cover of the belt it is sealing.

A specific caution must be given against the use of any previously used or leftover belting as a skirting seal. Used belting is normally loaded with abrasive materials, such as sand, cinders, or fines, from its years of service. All belting, new or used, contain fabric reinforcement or steel cords that will grind away at the moving belt, wearing away the protective top cover and

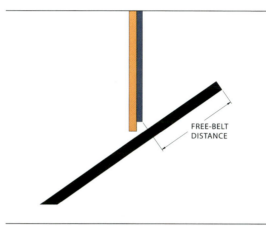

Figure 13.6

Free-belt distance—the amount of belt outside the skirtboard on both sides of the conveyor—provides the space available for the sealing system.

FREE-BELT DISTANCE

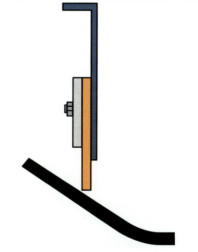

Figure 13.7

The vertical sealing systems typically use a rubber or elastomer strip which might have a special proprietary shape. A system of clamps is used to hold the skirtboard rubber in place.

Figure 13.8

Strips of rubber or, worse, used belting are sometimes seen. Regardless of the material used, the sealing system should always be less abrasion resistant than the top cover of the belt.

13

leading to premature failure and costly replacement.

Another type of vertical sealing system uses a series of interlocking sealing blocks installed outside the skirtboard on a special mounting plate. The interlocking blocks can be moved downward (toward the belt) but resist upward movement (**Figure 13.9**). These blocks can be easily adjusted down to the belt using only a hammer; however, each block must be individually adjusted,

and over-adjustment is a common problem. When over-adjusted, these blocks can easily stall a belt.

Main advantages of straight-up-and-down, or vertical, skirtboard seals are:

A. Low in cost

B. Narrow belt-edge distance (free-belt distance) requirements

C. Can be self adjusting

Main disadvantages of straight-up-and-down skirtboard seals are:

A. Often difficult to adjust accurately

B. Easily over-adjusted, causing premature wear

C. Prone to entrapping material, leading to belt damage

D. Susceptible to leakage of dust and fines

A third type of vertical edge-sealing system, designed to overcome many of the disadvantages associated with straight-up-and-down sealing, is the floating-sealing system. This system uses sealing strips mounted to the steel skirtboard on independent, freely-rotating link arms (**Figure 13.10**). The links allow the sealing strip to float on the belt, reacting to changes in the belt line while remaining in sealing contact with the belt (**Figure 13.11**). This design allows the sealing system to self-adjust, using its own weight to compensate for wear. The self-adjusting function allows this type of sealing system to overcome obstacles provided by inconsistencies in the belt line due to improper belt support or surges in the material loading.

Inward Sealing

Some sealing systems are clamped on the outside of the chute with the elastomer strip curled back under the steel. With these types of systems, the seal is formed on the inside of the skirtboard. Because the seal lies inward, the wear liner must be spaced far enough above the belt to allow for some free vertical movement of the seal (**Figure 13.12**). These inward systems have had

Figure 13.9

One class of sealing systems is easily adjustable because it is made in segments. A series of modular sealing blocks are installed outside the skirtboard on a special mounting plate that allows the interlocking blocks to be moved downward (toward the belt) but resist upward movement.

Figure 13.10

One edge-sealing system that would be classified as straight-up-and-down is the floating-sealing system. This system uses sealing strips mounted to the steel skirtboard on independent, freely-rotating link arms.

Figure 13.11

This system allows the sealing system to self-adjust, using its own weight to compensate for wear or variations in travel.

some success on conveyors carrying light, fluffy materials and fine, non-abrasive materials, such as carbon black. Inward sealing systems are also useful as a temporary solution on belts with limited edge-sealing distance, where a lack of belting outside the skirtboard steel limits the space available for the application of a sealing system. These systems are sometimes useful in areas of high internal-chute pressure—under a rail car dump, for example—where, when properly applied, the material loaded on the seal would tend to assist in the sealing effort. It should be noted that the seal may wear quickly, and material trapped under the seal would tend to prematurely wear the belt's top cover.

Sealing inward is sometimes applied due to a belt with a severe mistracking problem, because the belt would be least likely to travel out from under this type of system. In this situation, it would be better to solve the mistracking issues rather than apply a seal to overcome the problem. *(See Chapter 16: Belt Alignment.)*

The protective benefit from the installation of wear liner can be neutralized when the sealing system reaches back under the skirtboard, placing the sealing strip inside the wear liner (**Figure 13.13**). The sealing strip is abraded by the material load, and material can more easily be entrapped against the belt.

Main advantages of inward skirtboard-sealing systems are:

A. Self-adjusting

B. Handle light, fluffy materials and fine, non-abrasive materials

C. Require limited edge distance (free-belt distance)

D. Handle high internal chute pressure

E. Handle severely mistracking belts

Principal disadvantages of inward skirtboard seals are:

A. Shorter life of the seal due to being in the material flow

B. Prone to material entrapment under the sealing strip leading to premature belt wear

C. Reduced load-carrying area of the belt due to the "rainbow" effect (**Figure 13.14**)

A hybrid of these systems combines the floating vertical-sealing system, described in

Figure 13.12

Some sealing systems are clamped on the outside of the chute with the sealing strip curled back under the steel to form a seal on the inside of the chutewall. Because the seal lies inward, the wear liner must be spaced far enough above the belt to allow for some free vertical movement of the seal.

Figure 13.13

The protective benefit from the installation of wear liner is somewhat neutralized when the sealing system reaches back under the chutewall to place the sealing strip on the inside of the steel enclosure. Inside the wear liner, the sealing strip can easily be damaged by the forces of the material load.

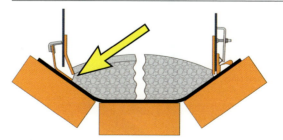

Figure 13.14

Inward skirtboard seals reduce the usable area of the belt by taking up space where the load could be carried. Due to a "rainbow" effect in the edges of the carrying area, capacity is reduced all the way across the belt.

13

Vertical Sealing above, with an "L"-shaped rubber strip (**Figure 13.15**). The foot of the rubber "L" extends inward, under the skirtboard steel and wear liner, toward the bulk materials. This one-piece sealing system serves as both the material (lump) seal and a dust seal. The rubber "foot" improves the seal's resistance to side pressure from the belt cargo and increases the range of belt mistracking the sealing system can tolerate.

Figure 13.15

One hybrid of these systems combines the floating vertical-sealing system, described in Vertical Sealing above, with an "L"-shaped rubber strip. The foot of the rubber "L" extends inward, under the skirtboard steel and wear liner, toward the bulk materials.

Figure 13.16

The final variation in sealing systems are those systems that seal on the outside of the skirtboard steel.

Figure 13.17

Multiple-layer sealing systems incorporate two layers: a primary strip that is pushed gently down to the belt to contain most particles, and a secondary strip that lies on the belt's outer edge to contain any fines or dust that push under the wear liner and primary strip.

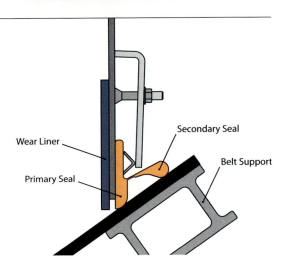

Wear Liner

Primary Seal

Secondary Seal

Belt Support

This type of system is particularly useful for transfer points where the wear liner is installed 25 millimeters (1 in.) or more above the belt. This spacing is used to preserve the belt from being damaged by running up against the wear liner and is a common practice in some industries where conveyors and conditions are less than ideal, such as coal mining and aggregate production.

This hybrid system allows the seal to float on the belt, rising and falling with any belt movement, including splice passage. The low application pressure reduces wear in both the belt and sealing strip.

Outward Sealing

The final variations of edge-sealing systems are systems that seal on the outside of the skirtboard steel (**Figure 13.16**). The most effective design combines multiple-layer seal effectiveness with the simplicity of single-strip systems.

When conveyor maintenance workers prepare for cold weather activities, they dress in layers. They know it is better to put on multiple layers of clothing—undershirt, shirt, sweatshirt, and jacket—than to wear one thick layer. The same concept can be used for transfer-point sealing: It is better to work with several thin layers than with one thick, general-purpose layer. In sealing, the first layer is provided by the wear liner installed inside the chute. Extending down close to the belt, the wear liner keeps the large particles of material well away from the belt edge. *(See Chapter 12: Wear Liners.)* The next layer is provided by the sealing system.

Multiple-layer seal designs feature rugged single-strip elastomers manufactured with a molded-in flap that serves as a secondary seal (**Figure 13.17**). This outrigger, or secondary strip, typically forms one or more channels that would capture the fines and gently carry them along the belt before depositing them back into the main body of material.

13

The system's primary seal is clamped against the outside of the chutework, extending vertically down and lightly touching the belt. It is applied with light pressure onto the belt, and the clamp applies force horizontally toward the chute, rather than down onto the belt. Because the clamping force is horizontal, the primary strip contains the material without the application of high pressure on the belt that would increase wear and conveyor power consumption. This primary strip will contain most of the material that has escaped past the wear liner.

As an outrigger, the secondary seal requires only the force of its own elasticity to provide sealing pressure, and, consequently, will wear a long time without the need for adjustment.

Installing the preferred one-piece design is a simple procedure: Unroll the seal to the proper length, cut it, and attach it to the skirtboard using a clamping system. A one-piece seal avoids any unnecessary joints and the handling of multiple pieces. Sealing systems should be provided in different thicknesses, to handle different duty applications, and in different materials, to handle different needs such as food grade, high temperature, and underground applications.

Advantages of outward sealing are:

A. Long lasting, because they are positioned away from material flow and sheltered by the skirtboard and wear liner

B. Can be self-adjusting

C. Low required sealing pressure due to "labyrinth" or multiple-layer seal design

D. Adapt to existing clamp systems

Disadvantages of outward sealing are:

A. Require greater belt-edge distance (or free-belt distance)

B. Vulnerable to damage if belt mistracks from underneath the seal

CONSIDERATIONS FOR SUCCESSFUL SEALING

Edge Distance Requirements

Care must be taken in selecting an edge-sealing system that fits into the available area between the belt's edge and the steel skirtboard. Sealing strips should not extend out to the belt's edge, as it increases the risk of damage to the seal or the belt in the event of belt mistracking.

In general, 115 millimeters (4.5 in.) outside the skirtboard steel on each side of the loading zone is recommended as the minimum distance required to establish an effective seal. *(For additional information about edge distance, see Chapter 11: Skirtboards, especially Table 11.1).*

Edge Seals and Belt Wander

All skirt-sealing systems are vulnerable to damage from a mistracking belt. If the belt wanders out from underneath one side of the skirtboard, the unsupported sealing strip hangs down below the line of the belt (**Figure 13.18**). When the belt moves back into a centered position, the seal will be abraded from contact with the edge of the moving belt or bent backward into an unnatural position and torn or worn away. Either outcome risks a significant increase in spillage. The keys to avoiding edge-seal damage are to provide adequate edge distance and to control belt tracking. *(See Chapter 16: Belt Alignment.)*

Figure 13.18

If the belt wanders out from underneath one side of the skirtboard, the unsupported sealing strip (green) hangs down below the line of the belt, leading to seal damage and spillage.

Sealing System and Belt Cover Wear

A research project, published in the journal *Bulk Solids Handling* in 1995, examined to what extent engineered-belt cleaning and sealing systems increased or decreased belt wear *(Reference 13.1)*. This study tested the abrasion of several edge-sealing systems against typical conveyor belting. The conclusions of the study reported the use of more sophisticated belt cleaning and sealing systems with adequate maintenance can extend the life of the conveyor belt. Although belt wear is introduced by these devices, the amount of wear is approximately one-half the rate expected when the belt runs through accumulations of fugitive material resulting from the lack of, or failure of, cleaning and sealing systems.

Avoiding Grooves in the Belt

It is a common misconception that the sealing system must be "softer" than the belt cover to ensure the seal wears before the belt. It is possible to make seals from materials with a wide range of hardness and wear resistance that are appropriate for edge seals. To prevent the sealing strips from wearing the belt, the sealing strips should be composed of materials with lower abrasion resistance than the belt's top cover, ensuring the seal will wear before the belt cover. Abrasion resistance is not measured by durometer, a rating of hardness, but rather by an abrasion index, such as Pico, Deutches Instit für Normung (DIN), or Taber ratings.

Many belts suffer wear along the skirted area at a set distance in from either edge of the belt (**Figure 13.19**). Often the sealing system is blamed for this wear; however, most often this wear is not caused by abrasion from the sealing strip. Rather, these grooves are started by the entrapment of fines and small lumps of material between the liner and the belt. This entrapped material begins scratching the belt's surface and gradually wears through the top cover. This entrapped material is most often seen on belts without proper belt support or on conveyors where material loading begins while the belt is in transition *(see Chapter 6: Before the Loading Zone)*, as the changes in belt shape makes it easy for material to be trapped under the skirtboard. Whenever determining the cause of a belt groove, it is

Figure 13.19

Many belts suffer wear along the skirted area at a set distance in from either edge of the belt.

⚠ SAFETY CONCERNS

It is important to follow established lockout / tagout / blockout / testout procedures before any work is done to the edge-sealing strips or any other item on or near the conveyor belt. Manufacturer guidelines for inspection intervals and maintenance procedures should be followed.

Sealing strips should never be raised while the belt is running, because this will place the worker in close proximity to the moving belt. This action may also allow material lumps, fines, or dust to be ejected from the transfer point at the worker.

It is recommended that the area of the transfer point be guarded on both sides of the conveyor. The guards should prevent untrained personnel from adjusting the sealing strips or becoming trapped in the numerous pinch points of the transfer zone.

important to inspect conveyors and chutes located above the affected conveyor for leakage and spillage.

INSTALLATION AND MAINTENANCE

Installation Guidelines

A sealing system must form a continuous strip along the sides of the steel skirtboard. If simple, end-to-end butt joints are employed to splice lengths of sealing strip together, material eventually pushes between the adjoining surfaces and leaks out. An interlocking or overlapping joint is best to prevent this spillage. A better solution is to use sealing strips available as one continuous strip, without need for a seam or joint.

With all edge-sealing systems, it is a good idea to round off the seal's leading edge at the tail end of the conveyor where the belt enters the back of the loading zone (**Figure 13.20**). Presenting a rounded edge to the moving belt reduces the chance a mechanical belt fastener can catch the sealing strip, and either rip it or pull it off the chute.

Maintenance of the Sealing System

When specifying a skirt-sealing system, it is wise to consider its mechanism for adjusting and replacing the wearable rubber. As the conveyor runs, the heat generated by the friction of the belt against the skirting seal combines with the abrasive nature of the material fines to erode the sealing strip. To counter this wear, the sealing strip must periodically be adjusted down against the belt.

Applying too much downward pressure to the sealing system leads to additional power requirements to move the belt; it also leads to extra wear in both the belt and the seal.

If the procedures for the service of skirting rubber are cumbersome or complicated, three detrimental consequences are likely:

A. No adjustment

Adjustment does not happen at all, so the skirting sealing strips wear, gaps open, and leakage resumes.

B. Infrequent adjustment

Adjustment is made too infrequently, so spillage occurs intermittently.

C. Over-adjustment

The maintenance person or conveyor operator, to compensate for not making regular adjustments, will over-adjust the seal. Applying too much force down onto the belt risks damaging the belt or catching a splice and ripping out the entire section of sealing strip.

To prevent these problems, skirtboard-seal maintenance procedures should be as free of complications, tools, and downtime as possible.

Sealing systems that rest gently on the belt, using little more than the pressure of their own weight or the tension built into the design, can minimize the need for maintenance adjustment.

Some multiple-layer sealing systems provide a self-adjusting function, as the elastomeric memory maintains the sealing pressure. As the legs of the secondary strip wear, the natural resilience of the elastomer strip keeps it down on the belt, maintaining seal effectiveness.

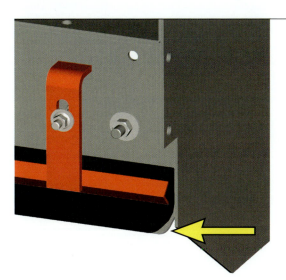

Figure 13.20

With all edge-sealing systems, it is a good idea to round off the seal's leading edge at the tail end of the conveyor where the belt enters the back of the loading zone.

TYPICAL SPECIFICATIONS

A. Low-maintenance design

A skirtboard-sealing system should be installed on the conveyor transfer point to eliminate material spillage and provide a positive dust seal at the sides of the loading zone. This sealing system should be a low-maintenance design providing an effective seal without the application of downward pressure onto the belt.

Figure 13.21

Seal-contact distance is the width of belt edge in contact with the edge seal.

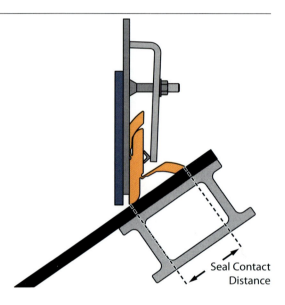

Seal Contact
Distance

B. Wear liners

It is recommended that appropriate wear liners be installed on the inside of the chute and skirtboard to protect the sealing system from material side-loading forces.

C. Self-adjustment function

To reduce maintenance intervals, the sealing system should be designed to self-adjust, with the system flexing to maintain sealing effectiveness in response to wear or changes in the belt's line of travel.

D. Single extended-length strip

The sealing strip should be supplied in a single extended-length strip (on each side of the conveyor) to provide a continuous seal without a seam.

E. Severity of the application

When selecting a skirtboard-sealing system, it is important to match the severity of the application. Factors such as belt speed, material load, and free-belt distance should be reviewed to make sure the application receives a suitable system (**Table 13.1**).

Table 13.1

Typical Comparative Selection Guide for Skirtboard Seal			
Basic Characteristics of Sealing Systems	**Sealing System Type**		
	Inward Sealing	Vertical Sealing	Outward Sealing
May Sacrifice Cargo Capacity	Yes	No	No
Recommended Free-Belt Distance (Outside the Skirtboard)	≥ 115 mm (4.5 in.)	≥ 115 mm (4.5 in.)	≥ 115 mm (4.5 in.)
Seal-Contact Distance (Figure 13.21)	≥ 38 mm (1.5 in.)	≤ 51 mm (2 in.)	≈ 75 mm (3 in.)
Wear Liner Required	Yes	Yes	Yes
Wear-Liner Distance Above Belt	≥ 25 mm (1 in.)	≤ 25 mm (1 in.)	≤ 25 mm (1 in.)

ADVANCED TOPICS

Special Designs of Sealing Systems

Some special circumstances require combination or hybrid sealing systems. In the case of a flat feeder belt where there is very fine material under high pressure, a variation of sealing outward in the conventional upright manner is commonly used.

Power Requirements of Sealing Systems

In order to be effective in keeping material on the belt, sealing systems must exert some pressure against the belt. This pressure will increase the drag against the belt and, therefore, increase the power consumption of the conveyor. The additional power requirement is directly dependent on the length and width of the seal and the pressure applied to the seal to keep it in contact with the belt. It is independent of the width of the belt.

The Conveyor Manufacturers Association (CEMA) *(Reference 13.2)* provides a formula for calculation of skirtboard-seal drag of generic rubber edge seals installed along both sides of a transfer point (**Equation 13.1**).

The tension (**Equation 13.1**) is related to the additional power required to drive the conveyor belt (**Equation 13.2**).

Note: this resistance is in addition to the drag of the material load against the skirtboards/wear liners.

In test facilities and field situations, it has been found that many skirtboard sealing-systems can be adjusted down onto the belt with very high levels of force. In these cases, actual tension should be measured to avoid underestimating the actual drag on the belt. Reasonable approximations for start-up requirements and the resulting forces have been measured in the field (**Table 13.2**) (**Equations 13.1** and **13.2**).

Operating power requirements are typically one half to two-thirds of start-up power requirements. If actual conditions are known, actual power requirements or tension should be measured or calculated and used in these equations.

Force Between Belt and Sealing Strip with Various Sealing Systems	F_{ss}	
	Units	
Effective Normal Force (F_{ss}) between Belt and Seal	Metric	Imperial
Type of Skirting	N/m	lb$_f$/ft
Rubber Slab: SBR Rubber—60 to 70 Shore D	45	3
Self-Adjusting Flat Rubber: *Martin® Self-Adjusting Exterior Skirting—Heavy Duty* or similar	78	5.25
1-Piece Multi-Barrier: *Martin® ApronSeal™ Single Skirting—Performance Duty* or similar	30	2
1-Piece Multi-Barrier Heavy Duty: *Martin® ApronSeal™ Single Skirting—Heavy Duty* or similar	50	3.3

Table 13.2

THE FINAL STEP IN SPILLAGE CONTROL

In Closing...

Rather than being the first step in solving conveyor spillage, the skirtboard seal is the last chance to contain fugitive material and prevent its release. The better the job done with belt support and wear-liner systems to contain material and keep it away from the edge, the better the performance will be of the belt's edge-sealing system. A flexible, multiple-layer system incorporating some level of self-adjustment will provide effective material containment for a transfer point and improve the operations of the belt conveyor. Periodic inspection and maintenance will extend life, reduce damage, improve performance, and boost satisfaction. This will ensure that an operation receives optimum value for its investment in an engineered sealing system.

Equation 13.1

Calculation for Tension Added to Belt due to Skirtboard Seal

$$\Delta T_{ss} = n \cdot \mu_{ss} \cdot F_{ss} \cdot L$$

Given: Rubber slab skirting installed on both sides over 6 meters (20 ft) of belt. **Find:** Tension added to the belt due to the skirtboard seal.

	Variables	Metric Units	Imperial Units
ΔT_{ss}	Tension Added to the Belt due to the Skirtboard Seal	newtons	pounds-force
n	Number of Skirtboard Seals	2	2
μ_{ss}	Friction Coefficient *(Per CEMA 575-2000)*	0,5 – UHMW 1,0 – Polyurethane 1,0 – Rubber	0.5 – UHMW 1.0 – Polyurethane 1.0 – Rubber
F_{ss}	Normal Force between Belt and Seal per Length	45 N/m *(Table 13.2)*	3 lb$_f$/ft *(Table 13.2)*
L	Length of Skirted Section of Conveyor	6 m	20 ft
Metric: $\Delta T_{ss} = 2 \cdot 1 \cdot 45 \cdot 6 = 540$			
Imperial: $\Delta T_{ss} = 2 \cdot 1 \cdot 3 \cdot 20 = 120$			
ΔT_{ss}	Tension Added to the Belt due to the Skirtboard Seal	540 N	120 lb$_f$

Equation 13.2

Calculation for Additional Power Required to Drive the Belt

$$P = \Delta T_{ss} \cdot V \cdot k$$

Given: Rubber slab skirting installed on both sides over 6 meters (20 ft) of belt. Belt is traveling at 3 meters per second (600 ft/min). **Find:** The power added to the drive due to the skirtboard seal.

	Variables	Metric Units	Imperial Units
P	Power Added to Belt Drive	kilowatts	horsepower
ΔT_{ss}	Tension Added to the Belt due to the Skirtboard Seal *(Calculated in **Equation 13.1**)*	540 N	120 lb$_f$
V	Belt Speed	3 m/s	600 ft/min
k	Conversion Factor	1/1000	1/33000
Metric: $P = \dfrac{540 \cdot 3}{1000} = 1{,}6$			
Imperial: $P = \dfrac{120 \cdot 600}{33000} = 2.2$			
P	Power Added to Belt Drive	1,6 kW	2.2 hp

Looking Ahead...

This chapter about Edge-Sealing Systems concludes the section Loading the Belt and the discussion about sealing the loading zone of a transfer point to prevent the escape of fine particles and fugitive material. The following chapter begins the section related to the Return Run of the Belt with a discussion about Belt Cleaning; the following two chapters continue that section with information about Pulley-Protection Plows and Belt Alignment.

13

REFERENCES

13.1 Swinderman, R. Todd, Martin Engineering. (October–December 1995). "Belt Cleaners, Skirting and Belt Top Cover Wear," *Bulk Solids Handling*. Clausthal-Zellerfeld, Germany: Trans Tech Publications.

13.2 Conveyor Equipment Manufacturers Association (CEMA). (2005). "'Universal Method' for Belt Tension Calculation." In *BELT CONVEYORS for BULK MATERIALS*, Sixth Edition, pp. 104–129. Naples, Florida.

SECTION 3
RETURN RUN OF THE BELT

Figure 14.1

Properly installed and maintained engineered belt-cleaning systems remove carryback and return it to the main material flow.

Chapter **14**

BELT CLEANING

In this Chapter...

In this chapter, we focus on different types of belt cleaners: their design, applications, installation, and maintenance. We discuss the nature and cost of carryback, belt cleaners' use in reducing carryback, and methods to assess their performance in doing so. In Advanced Topics, we provide equations to determine additional power needed when using belt cleaners.

The normal process of carrying bulk material on a belt conveyor results in a separation of the load into a layer of moist fines resting on the belt, with coarser, dryer material above the fines, and the largest lumps on top. The lumps, most of the coarse material and some of the fines discharge in the normal trajectory; a portion of the coarse grit and fines will cling to the belt. Known as carryback, this residual material is carried back on the belt as it returns to the tail pulley. As the particles dry, they lose cohesive and adhesive strength and are then dislodged by the return rollers and bend pulleys. Most of the material eventually falls from the belt, accumulating in piles under the belt; building up on the return idlers, pulleys, and conveyor structure; or becoming airborne dust.

The cleanup of fugitive materials, such as carryback, is an activity associated with many serious accidents. Removing carryback that accumulates on equipment and the ground can be labor- and equipment-intensive, often requiring crews of people and expensive equipment or services such as loaders and vacuum trucks.

Properly installed and maintained engineered belt-cleaning systems can be economically justified to mitigate this problem by removing carryback and returning it to the main material flow (**Figure 14.1**).

CONVEYOR BELT CARRYBACK

Carryback as Fugitive Material

Carryback, material that adheres to the belt past the discharge point of the head pulley, accounts for much of the fugitive material present in any conveyor system. Like transfer-point spillage, carryback presents serious problems for conveyor systems, creating consequences in maintenance, downtime, and plant efficiency. These problems present themselves as accumulations of fugitive material, leading to belt mistracking, shortened belt-life, premature component failures, and other problems that create conveyor downtime and expensive repairs. Accumulations of fugitive material on the ground or as clouds of dust in the air present fire and explosion hazards; slip, trip, and fall hazards; and health hazards. They also may become a magnet for unwanted attention from neighbors and regulatory agencies. No matter where it lands, carryback is a leading cause and indicator of conveyor inefficiencies.

To control the damage and expense that carryback creates, conveyor belt-cleaning systems are installed. Typically, belt-cleaning systems are one or more scrapers mounted at or near the discharge (head) pulley to remove residual fines adhering to the belt as it passes around the head pulley (**Figure 14.2**).

The Nature of Carryback

Carryback is typically the belt's cargo in its worst state. Carryback particles are normally smaller in size with higher moisture content than the conveyed material in general. Vibration of the belt as it rolls over the idlers creates a settling action. The smallest

Figure 14.2

Typical belt-cleaning systems feature one or more scrapers mounted at or near the discharge (head) pulley to remove residual fines.

fines, along with excess moisture, sift to the bottom of the pile where they can create an adhesive mixture that clings to the belt. Removed from the belt, this mixture will attach to other surfaces, including the belt-cleaning system and the vertical walls of the chute (**Figure 14.3**). When allowed to dry, these accumulations can become concrete-hard and reduce belt-cleaning efficiency, damage the belt or other equipment, and lead to chute plugging problems (**Figure 14.4**).

14

Figure 14.3

Removed from the belt, carryback can attach to other surfaces, including the belt-cleaning system and the vertical walls of the chute.

Figure 14.4

Allowed to dry, carryback accumulations can become concrete-hard and reduce belt-cleaning efficiency, damage the belt or other equipment, and lead to chute plugging problems.

Figure 14.5

Released carryback material builds up on idlers, creating out-of-round components that lead to belt mistracking.

The Cost of Carryback

Carryback is often a more cumbersome and costly problem than transfer-point spillage. Because it can drop off at any point along the conveyor return, fugitive carryback will require cleaning crews to work along the full length of the conveyor. This makes its removal more extensive and expensive than the more localized cleanup of spillage.

Released carryback material builds up on idlers, creating out-of-round components that contribute to belt mistracking (**Figure 14.5**). It works its way into bearings, increasing friction and leading to bearing failures. A large German lignite-mining firm calculated that it replaced 12 percent of the return idlers in its operation every year. Roughly 30 percent of those replacements were caused by wear in the support ring or face of the idler, wear that originated due to the release of material along the belt line.

Because carryback accumulates unevenly on the rolling components of the conveyor, it causes belt mistracking, leading to material spillage and to off-center loading of conveyors. Conveyor belt mistracking leads to shortened belt-life, excessive labor costs for maintenance and cleanup, unscheduled downtime, and safety hazards.

Measuring the Amount of Carryback

To better understand the carryback problem, the amount of material that clings to the belt and is carried back should be quantified. When the problem can be quantified, it is easier to determine how effective the installed belt-cleaning systems are, and what efforts, in the form of additional cleaning systems and improved cleaner maintenance schedules, are necessary. Carryback is measured as the dry weight of material in grams per square meter (oz/ft^2) of the portion of the belt surface carrying material. The bulk-materials handling industry in Australia has been a leader in designing and using systems to accurately assess the amount of carryback on a given belt. These systems have been used to evaluate the success of belt-cleaning equipment, to aid in the design of new equipment, and to monitor belt-cleaning performance contracts.

There are several ways to determine the amount of carryback. A small sample can be collected from a section of the belt, using a putty knife to manually scrape carryback material from the stopped (locked out) belt. By scraping a measured area, collecting the material in a pan or on a plastic sheet, and then weighing the material, a value for carryback per unit area can be calculated. One drawback to this method is that in the process of stopping the belt, the amount and character of the carryback will change.

A method developed by belt-cleaning pioneer Dick Stahura utilized a carryback collection pan affixed with scraper blades (**Figure 14.6**). This pan was designed to be held against the return side of the moving belt to remove residual material and collect it in the pan. Due to safety concerns, this method has been automated, resulting in a carryback gauge as developed by Australia's University of Newcastle Research Associates, Ltd, (commonly known as the TUNRA Group) for the International Conveyor Technology (ICT) group. This ICT Carryback Gauge has the ability to sample the entire width of the belt using a moving sampler that traverses the belt at a constant rate for a predetermined amount of time (**Figure 14.7**).

These testing methods provide a snapshot of carryback levels and cleaning performance for a given interval of time. Changes in material specifications, throughput, or climate conditions can dramatically alter the amount of material remaining on the belt; therefore, a program to periodically determine carryback should be implemented.

Carryback Calculations

By accepting this collected sample as an average amount of carryback or by taking several samples and averaging them for more accurate results, it is possible to determine the mean amount of carryback present on the belt and the expected variance as the standard deviation (**Figure 14.8**).

A seemingly modest amount of one gram (0.035 oz) of carryback removed from a section of belting indicates there would be a substantial amount of carryback remaining on this belt, material that could be

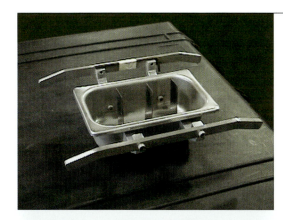

Figure 14.6

Belt-cleaning pioneer Dick Stahura developed a carryback collection pan affixed with scraper blades.

Figure 14.7

The ICT Carryback Gauge uses a moving sampler that traverses the belt at a constant rate for a predetermined amount of time.

Figure 14.8

By accepting this collected sample as an average amount of carryback or by taking several samples and averaging them for more accurate results, it is possible to determine the mean amount of carryback present on the belt and the expected variance as the standard deviation.

released from the belt and then accumulate under this single conveyor (**Figure 14.9**).

This one gram of material collected from the conveyor would be the same amount of material as found in the small packets of artificial sweetener found on tables in many restaurants. However, with the belt length, width, and speed seen in modern conveyors, this small amount of carryback becomes tons of material left on the belt per year. This will then drop from the belt all along the conveyor return, fouling equipment and creating a mess that takes time, effort, and money to fix.

The one gram of material used in the above example represents a small quantity of carryback and would indicate to some this belt is already clean. More typically, the actual measurement of material on conveyor belts shows carryback in the range of 7 to 250 grams per square meter (0.2 to 7.3 oz/1.2 yd²). This material becomes airborne dust and/or accumulates in piles under the conveyor or in buildups on and around rolling components.

A more "scientific" measurement would be to sample the belt right after the discharge point and then sample the belt again right before it enters the load zone on the conveyor. In addition to showing how much material adheres to the belt past the discharge, this would yield a measurement of how much carryback drops from the belt during the return run. Not only is this material lost to the process, but this carryback collects on return idlers and pulleys and creates future maintenance problems.

Analyzing the Material

Testing the bulk material is useful in determining how it will behave on the belt. This behavior depends on a number of parameters: bulk density, interface friction, interface cohesion, interface adhesion, and the condition of the belt itself. Testing has shown that for most (if not all) materials, cohesion (sticking to itself) and adhesion (sticking to the belt) increase as moisture increases. This behavior occurs to a critical point, as shown in a bell-shaped curve, until enough moisture has been applied to begin fluidizing the material and reducing the cohesion (**Figure 14.10**). The exact variation in adhesion and cohesion with moisture content will vary from material to material and from site to site.

Testing to determine the adhesion of a material to the belt can be omitted, as it can be found mathematically from the values of friction and cohesion. An optimal cleaning pressure can be determined using these results and belt-cleaner specifics (cleaning angle, blade profile, and blade composition). Advanced modeling techniques are used to predict the number of belt cleaners at a given blade-to-belt pressure necessary to remove the carryback layer from the conveyor belt. *(See Chapter 25: Material Science, for more information on bulk-material testing.)*

Figure 14.9

Even a modest amount of carryback left on the belt can accumulate to significant piles under the conveyor.

Figure 14.10

Testing has shown that cohesion and adhesion increase as moisture increases until enough moisture has been applied to begin fluidizing the material and reducing the cohesion. The exact variation in adhesion and cohesion with moisture content will vary from material to material and from site to site. Note: Moisture content is the % of weight loss between the wet material and the material after it has dried.

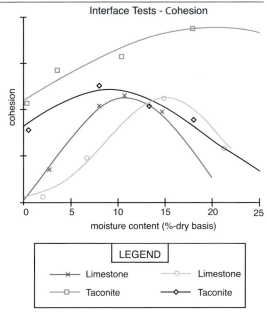

Interface Tests - Cohesion

cohesion

moisture content (%-dry basis)

LEGEND
- Limestone
- Limestone
- Taconite
- Taconite

The important thing to remember is that eventually the conveyor will see the "worst-case material" during the life of the belt.

CONVEYORS AND BELT CLEANING

The Rise of Engineered Belt-Cleaning Systems

For many years, belt cleaners were home-made affairs, often a slab of rubber, a left-over piece of used belting, or a discarded piece of lumber wedged into the structure or held against the belt by a counterweight (**Figure 14.11**). These systems proved to be unwieldy, cumbersome, and generally ineffective. Plant operating requirements necessitated the use of wider, faster, more deeply troughed, and more heavily-loaded conveyors. These requirements led to the development of engineered belt-cleaning systems to protect the plant's investment by extending the service-life of expensive belts and other conveyor components. These systems usually consist of a structural support (mainframe), cleaning element (blade), and tensioner.

Engineered cleaning systems are designed to reduce space requirements by enabling the cleaner to be installed in the discharge chute. By simplifying blade replacement, maintenance time and labor requirements are minimized. By incorporating advanced materials such as plastics, ceramics, and tungsten carbide, service-life of the blade is extended further, reducing maintenance requirements. By engineering the blade's edge and improving the tensioning devices that hold the cleaning edge against the belt, cleaning performance is enhanced. Through the use of engineered cleaning systems, the adhesive mass of fines and moisture traveling on the return side of the belt can be almost completely removed.

Monitoring Cleaner Performance

The ability to measure carryback allows for the development of a belt-cleaning performance specification for a given material-handling facility and bulk material. A complete specification will detail the performance required in terms of the aver-age carryback for the facility. The supplier must be required to design, supply, install, commission, and maintain the belt-cleaning system, guaranteeing average carryback levels are not exceeded.

After the belt-cleaning system is installed, carryback tests should be conducted to assess the performance of the belt-cleaning system. An ongoing testing and record-keeping program will yield information on periodic maintenance requirements and provide payback data for cleaning system maintenance and upgrade opportunities.

By monitoring belt-cleaning performance through carryback testing, the facility can assess the savings of possibly upgrading to more efficient cleaning systems.

Performance analysis and maintenance programs, implemented by in-house departments, are seldom seen as a priority, due to the overwhelming challenges otherwise presented to the facility. The easiest way to get results from an investment in belt-cleaning equipment is by awarding a service contract to a specialist in the supply, installation, maintenance, and analysis of belt-cleaning systems.

Designing Conveyors for Effective Cleaning

When considering the construction of new conveyors, it is desirable to include in the design requirements a specification for belt-cleaner performance. This specification should include an allowance for the amount of carryback measured in grams

Figure 14.11

For many years, belt-cleaning systems were homemade.

per square meter (oz/ft^2) passing the cleaning system. Plant owners should demand, and engineers should design, conveyor systems with adequate cleaner systems to ensure carryback is maintained below the level specified in the contract. It would encourage the conveyor system designer to include adequate space for installation and maintenance of the belt-cleaning system and to include components on the conveyor that are compatible with the goal of effective belt cleaning.

A common problem in the employment of belt-cleaning devices occurs when insufficient space is provided in the design of the head frame and housing for an adequate multiple belt-cleaner system. This commonly occurs because conveyor engineers have not taken into account the true nature of the conveyed material, particularly when it is in its worst condition. Conveyor engineers should design for clearance and access according to the Conveyor Equipment Manufacturers Association's (CEMA) recommendations, and the design should allow the belt-cleaner manufacturer to fit and mount the appropriate system, including cutting the holes in the chute after erection and belt installation.

A key to effective belt cleaning is mating the blade to the belt. It stands to reason that the better the blade matches the belt profile, the better it will clean. Anything making it more difficult for the blade to stay in contact with the belt as it moves past the belt cleaner must be avoided in the design of a conveyor system. These

undesirable factors include wing pulleys, out-of-round pulleys, and poorly-selected or poorly-installed lagging. Any pulsation or vibration of the belt's surface lowers cleaning efficiency and adversely affects the life of the belt.

Vulcanized splices are the preferred method of splicing the belt in order to provide optimal performance of the belt-cleaning system. Improperly installed mechanical fasteners can catch on belt cleaners and cause them to jump and vibrate, or "chatter." Mechanical splices should always be recessed according to the manufacturer's recommendations in order to avoid unnecessary damage to the cleaner and the splice.

After installing a belt cleaner, periodic inspection, adjustment, and maintenance are required to maintain effective performance. Just as cleaners must be designed for durability and simple maintenance, conveyors themselves must be designed to enable easy service, including required clearances for access.

Belt Turnovers

To eliminate problems caused by a dirty belt contacting the return idlers, a conveyor belt can be twisted 180 degrees after it passes the discharge point. This "turnover" puts the belt's cargo-carrying side up and brings the clean surface of the belt into contact with the return idlers (**Figure 14.12**). Theoretically, the carryback should remain on the belt as it moves along the return run. The belt must be turned back 180 degrees when it enters the tail section in order to properly position the belt top cover as it enters the load zone. The distance required to accomplish the 180-degree turnover of the belt is from 12 to 20 times the belt width at both ends of the conveyor.

Turnovers require a special structure and rollers using additional vertical space under the conveyor. Consequently, turnovers are usually justifiable on only long overland conveyors.

Figure 14.12

To eliminate problems caused by a dirty belt contacting the return idlers, a conveyor "turnover" brings the clean surface of the belt into contact with the return idlers.

Another concern is that carryback will dry out and could become airborne all along the return run.

The act of twisting the belt causes the dirty side of the belt to contact the turnover rollers. Because this takes place at the point where idler alignment and cleanliness are most crucial, fugitive material here is particularly problematic. To minimize the carryback released when the belt is twisted, an effective belt-cleaning system must be installed at the conveyor's discharge. It is often more cost effective to install an advanced cleaning system such as a wash box than to install a turnover system.

Belting and Belt Cleaners

The condition of the belt will have a dramatic influence on the performance of the belt-cleaning system. It is difficult to clean a belt that has become cracked, frayed, delaminated, or gouged due to exposure to the elements, chemical attack, or belt misalignment. Belt cleaning can be made difficult by patterns on the belt surface such as those seen in polyvinyl chloride (PVC) belting. In both of these cases, the only effective method to remove residual material is by washing the belt.

Some belting manufacturers continue the ill-advised embossing of identification numbers and corporate logos into the top covers of the belts (**Figure 14.13**). It is easy to appreciate the marketing value of this practice. However, it is just as easy to recognize the difficulties that these emblems in the top covers of the belts can create in effectively cleaning and sealing the conveyor systems. A better practice is to brand the belts on the bottom cover side.

Steel-cable belts often show on their surface the pattern of the cables hidden inside the rubber. This gives the effect of hard and soft streaks in the belt. To remove the "streaks," cleaning pressure is increased beyond what is necessary, so wear in the belt cover and the cleaning blades accelerates.

All methods of belt cleaning and all blade materials will wear the belt's top cover to some degree. At least one manufacturer of belting incorporates a factor for this wear into the design of the top cover. It is generally accepted that it is better to wear the belt slowly by installing belt-cleaning systems than to wear it rapidly by dragging it through dirt and across the non-rotating or damaged idlers that arise when the belt is not cleaned. In reality, a well-designed cleaning system has much less negative effect on belt cover-life than does the loading of the material onto the belt. Top cover selection should be governed by the considerations of material loading rather than worries over belt cleaning.

Sometimes a conveyor belt in generally "good condition" will suffer from a longitudinal groove damage that renders conventional cleaning methods less effective. This longitudinal damage could arise from many sources, including a lump of material or tramp iron wedged against the belt. To improve cleaning efficiency, the groove could be dressed by applying a urethane patch to fill the groove. This patching compound may need to be reapplied several times during the life of the belt. Localized cleaning with an air knife, brush cleaner, or alternate device may also be a solution for keeping the belt in service longer.

The Impact of Cleaning on Belt-Life

The question of how much the use of an engineered belt-cleaning system against the moving belt will shorten the service-life of a belt is worthy of some consideration.

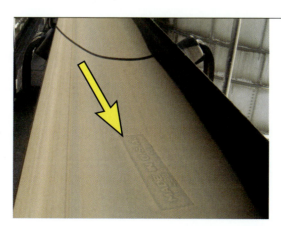

Figure 14.13

Some belting manufacturers emboss corporate logos into the belts' top covers, creating problems in cleaning and sealing conveyor systems.

14

The mechanisms of wear depend a great deal on the amount of heat generated in the blade and top cover. Field observations indicate, particularly for elastomeric blade materials, the highest wear rate to both blade and belt occurs when there is no material on the belt.

"Belt Cleaners and Belt Top Cover Wear," a research study by R. Todd Swinderman and Douglas Lindstrom, has examined the issue of whether belt cleaners adversely affect the life of the belting *(Reference 14.1)*. The results of this study showed a belt cleaner can induce wear of the belt, but the rate of wear was still less than allowing the belt to run through abrasive carryback without benefit of cleaners.

Similar results were reported in a study that compared belt-life and belt-failures in facilities using engineered cleaning and sealing systems to those in facilities that did not use these systems. Performed in India, this survey reviewed over 300,000 hours of operation on 213 belts in facilities handling lignite, limestone, and iron ore. This study showed that facilities using the engineered cleaning and sealing systems had belts that lasted an average of 150 percent longer (and required only 50 percent of the cleanup labor) than belts in the facilities not equipped with engineered cleaning and sealing systems. This survey of operating facilities echoes the laboratory research indicating that while belt-cleaning systems do introduce some wear to the belt cover, the end result is "the cleaner the belt, the longer it will last."

CLEANING SYSTEM DESIGN

The Basics of Effective Cleaning Systems

The basics for installing an effective belt-cleaning system include the following criteria: far forward, out of the material flow, and with minimal risk to the belt.

Far Forward

To minimize the release of carryback into the plant, belt cleaning should take place as far forward on the conveyor's return run as possible (**Figure 14.14**). Typically, at least one cleaner is installed at a point where the belt is still in contact with the head pulley. Normally, this cleaner is installed on the face of the head pulley, just below the point where the material leaves the belt. This position, called the primary position, provides a significant advantage in that the carryback is immediately returned to the main material flow. This reduces the potential for release onto rolling components and into the plant environment. With the cleaners tensioned against the belt and the belt still against the head pulley, control of the blade-to-belt pressure is more precise. The head pulley provides a firm surface against which to mount the cleaner.

Utilizing the space available and mounting the first cleaner in what is considered the primary position creates more space available for the installation of one or more cleaners in the secondary and tertiary positions. As with the primary cleaner, the farther forward each added cleaner is installed, the less chance there is for carryback to escape and the less need there is for devices like dribble chutes or scavenger conveyors to return recovered material to the flow.

Out of the Material Flow

It is important that cleaners are installed out of the flow of the material and that the material cleaned from the belt does not adhere and build up on the blades or structure (**Figure 14.15**).

Figure 14.14

To minimize the release of carryback into the plant, belt cleaning should take place as far forward on the conveyor's return run as possible.

A cleaner installed in the trajectory of the material may experience premature wear on the support frame and the back of the blades, making it necessary to change the blades before the cleaning edge is worn out. Preferred placement of a cleaner in the primary position involves installing the cleaner so the blade tip is below the horizontal centerline of the pulley.

A cleaner installed outside the material trajectory can still acquire a buildup of material that adheres to its outside surfaces. Cleaners should be designed to minimize the chance for material adhesion. This is accomplished by avoiding flat surfaces and pockets that can capture material and by utilizing non-stick materials for cleaner construction. In the proper environment, water sprayed on the surface of the belt— or on the cleaners—assists in softening the material and minimizing material buildup. *(See Chapter 24: Belt-Washing Systems.)*

With Minimal Risk to the Belt

An essential consideration in the selection of a belt-cleaning device is minimizing any risk that the cleaner could damage the belt or a splice, the very systems it was installed to protect. Belt-cleaning systems must be designed so the blade is capable of moving away from the belt when a splice, damaged section of belt, or other obstruction moves past the cleaner with the belt. The cleaner's tensioning systems, particularly on the primary cleaner where the angle of attack is more acute, should include a mechanism to provide relief from the shock of the splice impact.

An aggressive primary cleaner with a lot of cleaning pressure will have a tendency to more quickly wear away the top cover of the belt. These cleaners inherently provide an increased risk of catching on a protruding splice or flap of belt.

Care should be taken in choosing an appropriate material to put in contact with the belt. Materials such as strips of used belting should never be applied as a belt-cleaning or sealing material, because they may include steel cables or abrasive fines. These embedded materials cause excessive wear of the belt's top cover.

Belt-Cleaner Design Principles

Although other belt-cleaning systems— most notably brush and pneumatic systems—are available, most belt cleaners are blade cleaners: They use a blade to scrape material away from the belt's surface. These devices require an energy source— such as a spring, a compressed air reservoir, or a twisted elastomeric element—to hold the cleaning edge against the belt. The blade that directly contacts the belt is subject to abrasive wear; it must be periodically readjusted and eventually replaced to maintain cleaning performance.

Coverage of the Belt

Typically, the blades of a cleaner do not cover the full width of the belt, because the full width of the belt is not typically used to carry material. CEMA specifies the minimum blade coverage based on belt width (**Table 14.1**).

Various belt-cleaner manufacturers have their own standard or typical blade coverage. Many manufacturers allow for more than the minimum coverage, but rarely does the blade width need to be equal to or greater than the belt width.

For improved cleaning, the carrying width of the material on the belt should be observed or calculated and matched by the cleaner's width.

In some cases, providing a blade width that is wider than the material load on the

Figure 14.15

Belt cleaners should be installed out of the flow of the material.

Table 14.1

Minimum Belt-Cleaner Blade Coverage			
Metric Standard Belt Sizes* (mm)		Imperial Standard Belt Sizes (in.)	
Belt Width	Minimum Cleaner Coverage	Belt Width	Minimum Cleaner Coverage
300	200	18	12
500	330	24	16
650	430		
800	530	30	20
1000	670	36	24
1200	800	42	28
1400	930	48	32
1600	1070		
1800	1200	54	36
2000	1330	60	40
2200	1470	72	48
2400	1600	84	56
2600	1730		
2800	1870	96	64
3000	2000	108	72
3200	2130	120	80

*Modeled after CEMA; *Metric measurements are based on standard metric belt sizes, not conversions of Imperial sizes.*

Figure 14.16

Top: In a situation where the blade is wider than the material flow, the center of the blade may wear quicker than the outer edges.

Bottom: As the center of the blade wears, it allows material to pass inside the tip; the outer edges will be held away from the material and so do not wear.

belt can lead to undesirable wear patterns. The center section of the blade will wear faster than the portion of the blade on the outside section of the belt, because there is more abrasive cargo material in the center. The outside portion of the cleaning blade will then hold the center section of the blade away from the belt. Carryback can then flow between the belt and the blade, accelerating wear on this center section of the blade (**Figure 14.16**).

The material on the belt also provides a lubrication and cooling effect for the blade; therefore, care should be taken to prevent covering too much of the belt. Without this lubrication effect, a buildup of heat on the outside edge can cause the blade to fail and/or damage the belt.

Reducing blade coverage on the belt can help alleviate the problem of heat. However, care should be taken when reducing blade coverage, especially on a cupped belt. If the belt curls over the edge of a cleaner blade, it is exposed to the sharp edge of the blade. Some cleaners use a more flexible, nonmetallic blade on their outer edges to avoid this problem. Another solution is to flatten the belt with the use of hold-down rolls (**Figure 14.17**).

In some applications, the blade must be as wide as or wider than the belt. A cleaner used as a squeegee to dry a belt may need to be the full width of the belt to catch all the wet areas. Some materials like fly ash tend to spread on the belt or flow horizontally across the belt cleaner. In this case, if the blade does not extend to the full width, material can build up between the belt and the cleaner support shaft, where it can harden and damage the belt.

Single or Multiple Blade Segments

A multiple-blade design with individual spring or elastomer support on each individual cleaning blade will keep each blade in proper cleaning tension against the belt, yet allow each blade to yield to a lower pressure than the tensioning device's total applied force. In other words, narrow

14

blades can match up better against the belt, follow changes in surface contour, bounce away from the belt for splice passage, and return to cleaning position more easily than a single, monolithic blade. This means a multiple-blade design will be more efficient and safer for the cleaner and the belt.

New developments in urethane have improved the ability of single-blade primary cleaners to maintain contact with the belt.

There are a number of materials used for cleaner blades, ranging from rubber and urethane to mild and stainless steels. Blades are available with inserts of tungsten carbide or fillers such as glass beads to enhance abrasion resistance and cleaning performance.

Cleaner manufacturers have extended the range of urethane materials available to provide improved performance for specific conditions, including improved resistance to abrasion, heat, chemicals, or humidity. In some cases, the unique combination of characteristics in a specific application requires a comparative testing program to determine the best material for that application.

Angle of Attack

The angle of attack for the cleaning blades against the belt is a subject of importance. Generally speaking, there are two alternatives: positive-rake (or peeling) blades and negative-rake (or scraping) blades (**Figure 14.18**). With a positive rake, the blades are opposed to the direction of belt travel (**Figure 14.19**); with a negative-rake cleaning angle, the blades are inclined in the direction of travel, typically at an angle of 3 to 15 degrees from the vertical depending on the type of splice (**Figure 14.20**). Blades installed in a position that is vertical, or perpendicular to the belt, at the point of installation are said to have a zero-rake angle.

Metal blades in a positive-rake position are quickly honed to razor sharpness by the moving belt and can cause expensive dam-

Figure 14.17

One solution to a belt curling over the edge of the cleaner is to flatten the belt with the use of hold-down rolls.

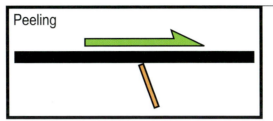

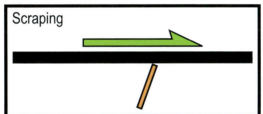

Figure 14.18

The angle of attack for the cleaning blades against the belt is a subject of importance. Generally speaking, there are two angles of attack for belt cleaning: positive-rake (or peeling) blades and negative-rake (or scraping) blades.

14

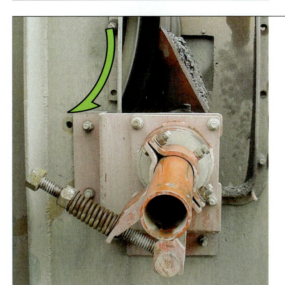

Figure 14.19

Blades with a positive rake are opposed to the direction of belt travel.

Figure 14.20

Blades with a negative-rake cleaning angle are inclined in the direction of travel.

age if they are knocked out of alignment. Positive-rake blades are occasionally subject to high-frequency vibration that causes the blades to "chatter," repeatedly jabbing their sharpened edges into the belt cover.

Negative-rake blades allow material to build up on the inclined cleaning edge, which can force the blade away from effective cleaning contact; however, all belt cleaners, regardless of the angle of attack, are subject to buildup absent regular cleaning and service (**Figure 14.21**). With a negative-rake blade, the upstream edges of the cleaning blade will not bite into the belt surface, even if held against the belt with excess pressure.

A general opinion is that a positive rake is acceptable for primary cleaners, which are applied at very low pressures against the belt. However, it is advisable to use negative-rake blades in secondary and tertiary locations where higher blade-to-belt cleaning pressures and use of metal blades present more risk to the belt, splice, and cleaner itself.

SELECTING A BELT-CLEANING SYSTEM

Selecting a belt cleaner for a given application requires the assessment of a number of factors. The following is the basic information that a supplier would need in order to recommend a suitable belt-cleaning system:

A. Belt width and speed

B. Width of the cargo on the belt

C. Pulley diameter

D. Material characteristics (including lump size, moisture content, temperature, abrasiveness, and corrosiveness)

E. Conveyor length

Conveyor length is a significant variable, as the undulating action of the belt as it moves over the idlers causes the fines to settle through the material and become compacted on the belt. On long overland conveyors, this effect can be significant. For this reason, longer conveyors are almost always harder to clean than shorter belts.

Short belts or belts that are allowed to run empty for long periods of time can suffer problems with the heat generated by the belt cleaner. A blade in contact with the belt will create heat due to the friction of the belt against the blade. Belts that are allowed to run for long periods without a cargo can cause heat to build up in the blade and the blade-holding mechanism, reducing blade-life or damaging the holder. If the belt is short, the top cover of the belt may not dissipate the heat and will degrade. A belt cleaner provided with high tension against the belt will aggravate this problem to the point in which the blade may stick to the belt when the belt stops.

Additional variables that may affect the ultimate performance of the selected system and so should be reviewed in the selection of a cleaning system include:

A. Space available for installation and service

B. The possibility of changes in material characteristics (e.g., from wet and sticky to dry and dusty)

C. Severe temperature extremes

D. Cuts, gouges, rips, or cracks in the belt surface from age or abuse

E. Numerous, non-recessed, or damaged mechanical splices

F. Belt vibration, from material buildup on

Figure 14.21

Negative-rake blades allow material to build up on the inclined cleaning edge, which can force the blade away from effective cleaning contact; however, all belt cleaners, regardless of the angle of attack, are subject to buildup absent regular cleaning and service.

14

head pulleys and other rolling components, making it difficult to keep a cleaner in contact with the belt

G. Material that will adhere to or entangle the cleaning device

H. Material accumulation in the dribble chute

Other considerations in the development of a supplier proposal and the evaluation of that proposal include:

A. Level of cleaning performance desired/required

B. Level of maintenance required/available

C. Level of installation expertise required/available

D. Initial price versus cost of ownership

E. Manufacturer's record (including service capabilities and performance guarantees)

There are a number of "shortcuts" available to help match a belt-cleaning system to an application, including on-line selector systems that analyze material characteristics and conveyor specifications to provide a belt-cleaner recommendation.

Information Required for Specifying a Cleaning System

In a paper presented to the 2004 Annual Meeting of the Society for Mining, Metallurgy, and Exploration (SME), R. Todd Swinderman listed the basic information to be supplied by an end-user requesting a proposal for a belt-cleaning system and the information that should be provided by a cleaning-system manufacturer in a proposal to the end-user customer.

The user's request for proposal shall include:

A. Hours of operation of the conveyor system

B. Rated tons per hour of the system

C. Percentage of time the conveyor is conveying material while running

D. Belt brand and description: belt width, age, condition, and tracking

E. Belt speed and whether it is one directional or reversing; if reversing, the percent of use in each direction

F. Temperature, humidity, and other environmental or operating conditions that may effect the operation or life of belt-cleaning equipment

G. Material specification according to the CEMA STANDARD 550 *Classification of Bulk Solids Standard*

H. Level of belt cleaning that is to be obtained in grams per square meter (oz/ft^2) on the carrying portion of the return run of the conveyor belt

I. Pulley diameter and the pulley lagging condition, type, and thickness

The equipment supplier's proposal shall include:

A. Recommendation for the appropriate system to meet the user's required level of cleaning performance

B. Statement of the expected average life and cost of wear parts

C. An allowance for changes in the cleaning performance over time, as the conditions of the belt or the bulk material change

D. Test method, test equipment, and the reference location to be used to measure performance; test equipment designed to produce repeatable results that are representative of the entire carrying width and length of the belt; test method and test equipment documented so that the tests can be repeated

E. Installation instructions

F. Required maintenance procedures and intervals

G. Guarantee of performance

H. Terms and conditions of payment

A Cleaner Suitable for All Conveyors?

A review of the marketplace shows there

are many competing designs for cleaning systems. This poses the question: Why has the industry not settled on one successful design that can provide acceptable results in all applications?

An engineer for a major mining company wrote in a summary of the various belt cleaners installed at his operations:

> Due to the diverse materials with their greatly differing physical properties as well as the diverse environmental conditions in open cast mining, there is currently no universally applicable belt cleaner, which would fulfill the requirements of all situations without problems.

This writer was discussing only one company's open cast lignite mines in Germany. This is a challenging application to be sure, but it is only one of the myriad of environments belt cleaners are required to endure. The universe of belt-cleaner applications is so much broader—including great varieties of materials, conditions, and conveying systems—that it requires a number of different choices.

The problem is to provide a cleaning system that is adjustable to fit most of these situations. Indeed, one of the problems in developing a universal cleaning system is that each equipment manufacturer, each conveyor engineer, and plenty of plant maintenance personnel have their own ideas on how to properly clean a conveyor belt. With so many different variables in applications, each of these designs has found some level of success.

Plant engineers or maintenance workers who design and install a belt cleaner of their own design would stop by every day to ensure it was working properly. By paying close attention to the cleaner, periodically adjusting the blade-to-belt tension and knocking off any material buildup, the designer would ensure its performance would at least equal many of the commercial cleaning systems presently available. The key element allowing this system to achieve acceptable cleaning performance

was probably the heightened level of service. Regardless of a belt cleaner's design, all belt-cleaning systems will function better given regular maintenance attention.

Cleaning Systems for "Worst-Case" Materials

Conveyor cleaning systems should not be designed to match only the limited challenges of "normal" operating conditions. Rather, they should be designed to stand up to the worst possible applications that may be encountered. If 99 percent of the time the conveyed material is dry, the day will surely come when the material becomes wet and sticky, and the cleaning system may not be adequate for the challenge. This single event could cost several times more than what was "saved" by designing for the "normal" operating conditions. With a cleaning system designed for "worst-case" conditions, the over-design will likely provide the benefit of improved wear-life and reduced service requirements when operating conditions are normal. When the conditions change for the worse, the cleaning system can stand up to the challenge.

It takes some work to analyze material in its worst state condition, but any sophisticated belt-cleaning-system manufacturer will have completed such tests to understand material behaviors and provide the best product to handle many different operating conditions.

SYSTEMS APPROACH TO BELT CLEANING

The Need for a System

State-of-the-art in belt-cleaning technology recommends that more than one "pass" at the belt should be made in order to safely and effectively remove carryback material. Like shaving, it is safer and more effective to make a series of gentle strokes over the surface than one "all at once," high-pressure, aggressive-angle assault.

It is usually most effective to install a multiple-cleaner system, composed of a

14

primary cleaner and at least one secondary cleaner (**Figure 14.22**). The primary cleaner is installed on the head pulley using low blade-to-belt pressure to remove the top layer and majority of the carryback. This allows the secondary cleaner, tensioned at optimum belt-cleaning pressure, to perform final, precision removal of adhering fines without being overloaded with a mass of carryback. The two styles of cleaners are given different responsibilities in the task of cleaning the belt and so are designed and constructed differently. Many conveyors can be addressed with the dual-cleaning system (**Figure 14.23**); however, there are applications that will need additional belt cleaners to achieve maximum carryback removal under all conditions (**Figure 14.24**).

The phrase "multiple-cleaning systems" can refer to any combination, ranging from the commonly seen primary-cleaner and secondary-cleaner system to more sophisticated systems that include one or more pre-cleaners and one or more secondary and/or tertiary cleaners. A conveyor system with a large head pulley allows the installation of more than one pre-cleaner in the primary cleaner position. Where it is necessary to get a very clean belt, the primary cleaner and secondary cleaners could be supplemented with a belt-washing system that incorporates water sprays, belt cleaners, and belt wipers that "squeegee" the belt dry. (*See Chapter 24: Belt-Washing Systems.*)

In addition to providing improved cleaning, a multiple-cleaner system increases the time between required service sessions. Two cleaners, each applied with a somewhat lower blade-to-belt pressure, should extend cleaner blade-life longer than a cleaner applied with higher pressure to provide the sole cleaning benefit.

A successful multiple-cleaning system is one that fits into the conveyor structure, achieves the desired level of cleaning, and minimizes maintenance requirements.

Primary Cleaners

The primary cleaner, sometimes called the pre-cleaner or the doctor blade, is installed on the face of the head pulley just below the trajectory of the material discharging from the belt (**Figure 14.25**). This position allows the material removed from the belt to fall in with the main cargo as it leaves the belt, minimizing any overloading of the dribble chute or other reclamation system.

Figure 14.22

A multiple-cleaner system, composed of a primary cleaner and at least one secondary cleaner, is often the most effective system.

Figure 14.23

Many conveyors can be satisfactorily cleaned with a dual-cleaning system.

Figure 14.24

Some applications will need additional belt cleaners to achieve maximum carryback removal.

Pre-cleaners, with low blade pressure but an aggressive angle of attack, are installed to shear the coarse grit off the carryback layer (**Figure 14.26**). This clears away most of the carryback and enables the secondary cleaner to remove additional, underlying material.

The primary-cleaner blade should be installed in a positive-rake position, inclined against the movement of the belt and pulley, typically at an angle between 30 to 45 degrees to a line tangent to the belt's surface at the point of contact. Rather than blocking the path of the material, a pre-cleaner diverts the material away from the belt so it can return to the main material flow or move down the back of the cleaner into the discharge (**Figure 14.27**). Using this low angle of attack, in combination with elastomer pre-cleaner blades applied with light pressure against the belt, results in low wear rates for both the blade and the belt surface. If the angle of attack were greater (i.e., a zero-rake angle or negative-rake position), more pressure would be required to hold the blade in position against the onslaught of material. Increasing the pressure increases the risk of damage to the belt.

To minimize the risk to the belt, the splice, and the cleaner from even a lightly-tensioned blade held in a positive-rake position, pre-cleaners usually use blades of resilient elastomer, such as urethane or rubber, rather than metal. A blade-to-belt pressure of approximately 14 kilopascals (2 lb_f/in.²) combines cleaning performance with safety for the belt. This low blade-to-belt pressure means that the system will be able to relieve—that is, to bounce the blade away from the belt—when an obstruction such as a mechanical splice moves past the cleaning edge, thus reducing the risk of damage. Properly applied cleaning pressure improves blade-life and reduces belt wear. Too little pressure allows material to slide between the blade and the belt, where it can become trapped and impart wear on both. Too much pressure accelerates wear and increases energy required to move the belt.

In order to achieve an acceptable level of cleaning with only a single pre-cleaner (unsupported by any secondary or additional cleaners), it would typically be necessary to tension the blade against the belt with higher blade-to-belt pressure than should be recommended for the preservation of belt-life.

Pre-cleaner characteristics include wear area, constant-angle cleaning, and constant-area cleaning.

14

Figure 14.25

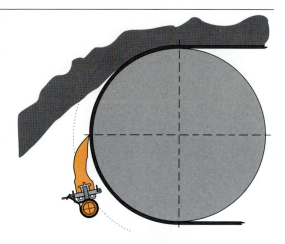

The primary cleaner, sometimes called the pre-cleaner or the doctor blade, is installed on the face of the head pulley just below the trajectory of the material discharging from the belt.

Figure 14.26

Pre-cleaners, with an aggressive angle of attack but low blade pressure, shear off the top of the carryback layer.

Figure 14.27

Rather than blocking the path of the material, a pre-cleaner diverts the material away from the belt to return to the main material flow.

Wear Area

One property that should be defined in any pre-cleaner is the amount of blade material that can be worn away by the belt. This is called wear area. This area can be found by laying out the blade and head pulley in a drafting software package. The mounting distance, center of blade rotation, and head-pulley diameter should be based on manufacturer specifications. The blade is rotated into the head pulley to the 100 percent wear-life and the interfering area calculated (**Figure 14.28**). This wear area allows for comparison of blade designs, because blade-life is neither a function of belt coverage nor individual blade widths.

Constant-Angle Cleaning

To overcome the problem of the blade angle changing as the blade wears, a radially-adjusted belt cleaner can incorporate a specifically designed curved blade. This design has been termed "CARP" for Constant Angle Radial Pressure. With a "CARP" blade design, the cleaning angle remains the same throughout the life of the blade. This "constant angle" design has the obvious advantage of maintaining cleaner efficiency throughout the service-life of the blade.

Constant-Area Cleaning

Many new cleaner blades are designed with a tip that has a small contact area against the belt. This tip or "point" allows the blade to "wear in" quickly, to achieve a good fit to the belt, regardless of the diameter of the head pulley.

As cleaner blades wear, the surface area of the blade touching the belt increases. This causes a reduction in blade-to-belt pressure and a corresponding decline in cleaner efficiency. Therefore, the system's tensioner requires adjustment (retensioning) to provide the additional pressure for consistent cleaning performance. It would be better to design cleaners that do not suffer from this gradual increase of blade-to-belt area (**Figure 14.29**). When combined with tensioner design, the CARP principle described above has proven capable of minimizing the change in area throughout a blade's wear-life.

Secondary Cleaners

Secondary cleaners are defined as any cleaner located in the secondary position on the return run of the belt (**Figure 14.30**). The secondary position is the area from just past the point where the belt leaves contact with the head pulley to just before the belt contacts the snub pulley. This location is still within the discharge or dribble chute, allowing the removed carryback to return to the main material flow by gravity.

As the pre-cleaner performs the initial rough cleaning, the secondary cleaner is

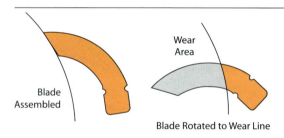

Figure 14.28

The wear area of an elastomer pre-cleaner blade is an indicator of the life of the blade.

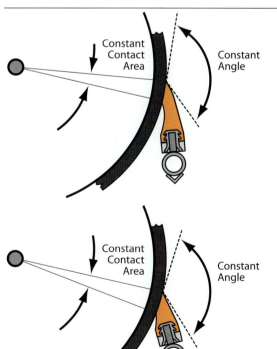

Figure 14.29

A Constant Angle Radial Pressure (CARP) blade is designed to maintain the cleaning angle through all stages of its blade-life. As cleaner blades wear, the surface area of the blade touching the belt should remain constant.

dedicated to performing the fine cleaning of the material that has passed by the primary cleaner. More than one secondary cleaner may be required to achieve the desired level of cleanliness.

The positioning of additional cleaner(s) is important. The closer the removal of carryback is performed to the conveyor's discharge point, the smaller the risk of problems with the buildup of fines in the dribble chute. On a belt with a vulcanized splice, the best location for an additional

cleaner is to place it in contact with the belt while the belt is still against the head pulley (**Figure 14.31**). This allows the secondary cleaner to scrape against a firm surface for more effective material removal.

If conveyor structural members, space limitations, or poor mechanical splices make it impossible to install a cleaner in the preferred position, secondary cleaners should be mounted where the material will be returned to the material flow: that is, where the cleaned material will fall within the chute. But if a secondary cleaner is installed in a position where its pressure against the belt changes the belt's line of travel, cleaning performance will be less effective (**Figure 14.32**). In this case, increasing applied pressure serves only to alter the belt line more, without improving cleaning performance. Hold-down rolls or other devices should be installed to maintain a stable belt line.

The angle of attack of the secondary cleaner blade to the belt is an important consideration. Metal- or ceramic-tipped blades in a positive-rake position are quickly honed to extreme sharpness by the belt's movement. These sharpened blades raise the risk that an adjustment made by an untrained operator may result in too much pressure or the incorrect angle being applied for the cleaner to quickly release from belt obstructions like mechanical splices. The result could be damage to the belt, splice, or the belt cleaner itself. Consequently, when obstructions or mechanical splices are present or anticipated, it is recommended that secondary blades be angled in the direction of the belt's travel—a negative-rake angle—rather than opposing the belt in a positive-rake position. Testing has indicated that an angle of 7 to 15 degrees in the direction of belt travel maintains cleaning efficiency while allowing easier passage of obstructions (**Figure 14.33**).

A moving belt does not present a consistent and uniform surface. Narrow, independent blades that are individually suspended

Figure 14.30

Secondary cleaners are defined as any cleaner located in the area from just past the point where the belt leaves contact with the head pulley to just before the belt contacts the snub pulley.

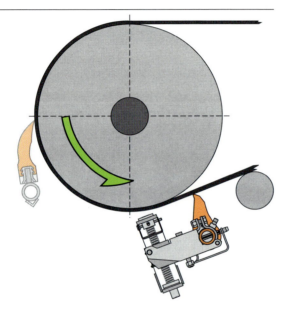

Figure 14.31

The best location for a secondary cleaner is to place it in contact with the belt while the belt is still against the head pulley, allowing it to scrape against a firm surface.

Figure 14.32

Cleaning performance will be less effective if a secondary cleaner is installed in a position where its pressure changes the belt's line of travel.

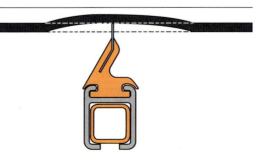

14

have the best potential to remain in precise contact as the changing belt surface passes across the cleaning edge. It is also beneficial if these individual blades can pivot, or rock, from side to side to instantly adjust to the changing contours of the belt surface. Research indicates that a cleaner formed from a line of individual, independent blades each 75 to 200 millimeters (3 to 8 in.) wide is well suited for effective secondary cleaning.

The Bureau of Mines' study, *Basic Parameters of Conveyor Belt Cleaning*, points out that the initial blade-wear occurs at the edges where individual blades adjoin *(Reference 14.2)*. The testing showed material would pass through the spaces between the adjacent blades and slowly enlarge these spaces. This passage, in turn, accelerated blade wear and allowed more material to pass. To minimize this erosive wear, an overlapping blade pattern can be used, created by an alternating long-arm/short-arm pattern (**Figure 14.34**). This prevents "stripes" of carryback down the belt created by gaps between the blades. Alternately, two cleaners with blades in a line can be installed with the gaps offset.

Secondary-cleaner blades themselves can be of a hard material—ceramic or tungsten carbide, for example—that resists the buildup of heat stemming from the friction against the belt's surface. Some operations prefer to avoid application of a metal blade against the belt, so a variety of urethane formulations have been developed for secondary cleaners.

One secondary-cleaner design applies the natural resilience, or "spring," of the elastomer used in the blades to reduce the need for cleaner adjustment. These urethane blades are forced against the belt, so they "flex" in the direction of belt travel (**Figure 14.35**). Over time, the blade's resilience continues to push the blade tip against the belt, even as the blade is being worn away by the movement of belt and material. This makes the blades self-adjusting and so reduces the need for follow-up service for adjustment (**Figure 14.36**).

Tertiary Cleaners

Tertiary cleaners are sometimes applied for final cleaning. The tertiary location for cleaners is normally considered as the area past the snub pulley and outside the discharge chute (**Figure 14.37**). This location is outside the area that allows the easy return of material to the main material flow, which requires the use of an auxiliary chute or scavenger conveyor. There may be multiple cleaners applied in this location to achieve the results required by the operation (**Figure 14.38**).

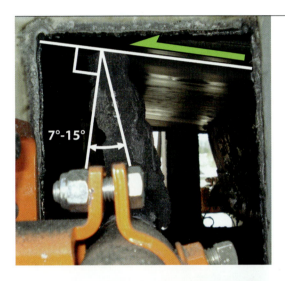

Figure 14.33

An angle of 7 to 15 degrees rake in the direction of belt travel maintains cleaning efficiency while allowing easier passage of obstructions.

Figure 14.34

To avoid "stripes" of carryback created by gaps between the blades, an overlapping blade pattern, created by an alternating long-arm/short-arm pattern, can be used.

Figure 14.35

This cleaner incorporates urethane blades that are forced against the belt, so they "flex" in the direction of belt travel.

14

Figure 14.36

Over time, the resilience of the urethane blade continues to push the blade tip against the belt, even as the blade is being worn by the movement of the belt, making the blade self-adjusting.

Figure 14.37

The location for tertiary cleaners is the area past the snub pulley and outside the discharge chute.

Figure 14.38

There may be multiple cleaners installed outside the area that allows the easy return of material to the main material flow, which requires the use of an auxiliary chute or scavenger conveyor.

Figure 14.39

Specialty cleaners and belt-washing systems are often located in the tertiary position.

14

The tertiary cleaner is normally used to clean off water and small particles that pass around or between the secondary-cleaner blades. Specialty cleaners and wash boxes or additional secondary-style cleaners are often located in the tertiary position (**Figure 14.39**). Squeegee blades to remove moisture—from the material or applied in the course of belt cleaning—are typically installed in the tertiary position (**Figure 14.40**).

One problem that can be seen with a tertiary cleaner is heat buildup. If the belt is clean or dry before it reaches the tertiary cleaner, the blade(s) in the tertiary cleaner may accumulate heat, allowing the material in the blade or blade holder to break down or damage the surface of the belt. A multiple-cleaning system should be carefully checked to eliminate this possibility. The use of a fine water spray to lubricate the belt and reduce the strength of the carry-back is very effective in increasing tertiary cleaner efficiency and maintenance intervals. *(See Chapter 24: Belt-Washing Systems.)*

Matching Cleaner to Application

The increasing sophistication of belt-cleaner design has allowed the development of cleaning systems to conform to the needs of each specific or specialized application. These alternative designs include specialized materials for cleaner construction, such as urethanes, intended for high-temperature, high-moisture, or high-abrasion conditions. In addition, there are a variety of cleaning systems engineered to match special challenges, ranging from light-duty food-grade conveyors to heavy-duty mine applications.

Mine-Grade Belt Cleaners

The high material volumes, fast speeds, wide belts, and large-diameter pulleys seen in many mining operations pose special challenges for belt-cleaning systems. The removal of overburden in some German lignite mines features conveyors with belt widths up to 3200 millimeters (126 in.) wide, operating at speeds of 10,5 meters per second (2067 ft/min). To withstand these abusive conditions, extra-heavy-duty mine-grade belt-cleaner systems have been developed (**Figure 14.41**). These systems are marked by massive mainframes to withstand large lumps and high volumes of material, massive blades to provide extended wear-life, and durable tensioning systems to reduce the need for system maintenance (**Figure 14.42**).

The high speed of these conveyors makes it difficult to use secondary cleaners effectively. The higher operating speeds and resultant higher vibration of these belts, combined with higher blade-to-belt pressure typical of a secondary cleaner, produce both higher wear and added risk to both the belt and cleaner. As a result, these applications may feature two pre-cleaners on the head pulley; the pulleys are normally large enough to allow this practice (**Figure 14.43**).

Crust Breakers

In applications such as the handling of crushed ore in copper and other hardrock mines, or in the conveying of overburden in lignite mining, moist material particles can sift down to the bottom of the conveyed load and stick to the belt with such strong adhesion that this material will refuse to be thrown from the conveyor at the discharge. Instead, this paste-like material will adhere to the belt as it passes the pulley in a layer 75 to 100 millimeters (3 to 4 in.) or more thick. This crust of material can quickly overwhelm a conventional cleaning system, leading to poor cleaning performance and shortened cleaner-life and risking the productivity of the entire material-handling system.

To overcome this problem, some operations will install a "crust breaker." This cleaning edge is installed on the head pulley just below the material trajectory. Here it serves as a doctor blade to limit the amount of material that gets through to the conventional pre-cleaner installed just below (**Figure 14.44**). Fabricated from ceramic-covered metal plates, the crust

Figure 14.40

Squeegee blades to remove moisture from the belt are typically installed in the tertiary position.

Figure 14.41

To withstand abusive conditions, extra-heavy-duty mine-grade systems have been developed.

Figure 14.42

Mine-grade belt cleaners are marked by heavy-duty mainframes, massive blades to provide extended wear-life, and durable tensioning systems.

Figure 14.43

To avoid using secondary cleaners on high-speed/high-vibration belts, some operations may use two pre-cleaners on large head pulleys.

14

Figure 14.44

A crust breaker serves as a doctor blade to limit the amount of material that gets through to the conventional pre-cleaner installed just below.

Figure 14.45

Specialized reversing belt cleaners have been developed for conveyors that run in two directions or have substantial rollback.

14

Figure 14.46

Typically, reversing belt cleaners are designed with a blade that is installed perpendicular to the belt and capable of deflecting a modest amount—7 to 15 degrees—in both directions of belt motion.

Figure 14.47

Their vertical installation and tensioning allows reversing cleaners to be installed on one-direction belts in narrow spaces where secondary cleaners with a trailing-arm design would not fit.

breaker's cleaning edge is installed so it is close to, but does not touch, the belt. That way, it reduces the amount of material that reaches the pre-cleaner and that could pass behind the pre-cleaner to damage (or overload) the secondary cleaner. With the "crust breaker" installed in advance of the conveyor's pre-cleaner, the conventional cleaners can provide improved cleaning and longer blade-life.

Cleaners for Reversing Belts

Some conveyors run in two directions or have substantial rollback, so it is critical that cleaners installed on these systems work well in either direction of operation, or at least are not damaged by belt reversal. Specialized cleaners have been developed for reversing belts (**Figure 14.45**). These cleaners are usually installed perpendicular to the belt and so are tensioned vertically into the belt surface. Typically, the cleaners are designed with a blade that is capable of deflecting a modest amount—7 to 15 degrees—in either direction of belt motion (**Figure 14.46**).

Of course, reversing cleaners could be installed on one-direction belts. A feature of reversing belt cleaners that encourages their use on non-reversing belts is their vertical installation and tensioning. This feature allows the reversing cleaner to be installed in narrow spaces where secondary cleaners with a trailing-arm design would not fit (**Figure 14.47**).

Food-Grade-Cleaning Systems

Some cleaning systems are engineered especially for the small pulleys and slower belt speeds common to food-processing plants. Constructed of food-grade materials and able to withstand the operation's frequent washdown cycle and cleaning chemicals, these systems are a match for food-processing applications (**Figure 14.48**).

Cleaners for Chevron Belts

Belts with ribs, cleats, or chevrons are used to convey materials that would slide back as the load moves up an incline.

These raised elements pose problems when attempting to remove carryback. Belt cleaners that utilize blades with "fingers" that have the ability to walk over the obstructions are required for cleaning chevron-style belts (**Figure 14.49**). This design can effectively clean chevrons/ribs/cleats up to 20 millimeters (0.75 in.) tall (**Figure 14.50**).

Cleaners for Pocket Belts

Belts with very deep cleats and/or sidewalls handling sticky materials are difficult to clean. The most common way of cleaning these belts is to beat the belt with a linear vibrator or rotating beater-bar-style cleaner when the belt is upside down and horizontal on the return run. These systems require frequent maintenance and are only partially effective.

Pulley Cleaners

It is possible for fugitive material to fall onto the clean side of the belt during its return run and for this fugitive material to then accumulate on the conveyor's snub or bend pulleys. To keep the belt tracking true, it is then necessary to install pulley-cleaning devices. To clean adhering material from a pulley, a scraper with an elastomer blade is mounted slightly below the pulley's horizontal centerline on the side of the pulley away from the belt (**Figure 14.51**). This will allow removed material to fall into a receptacle or accessible area for removal.

Rotary-Brush Cleaners

Cleaning systems composed of a rotating brush applied against the belt can be used effectively on dry materials (**Figure 14.52**). These systems can be free-wheeling (turned by the motion of the belt), but they are more effective when driven by an electric motor. Brush cleaners often encounter problems with sticky or moist materials that build up in the bristles of the brush (**Figure 13.53**). A beater bar or comb can be installed to assist in clearing the buildup from the bristles.

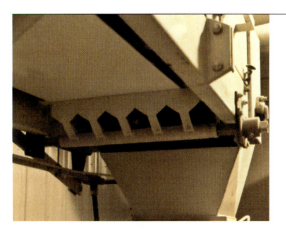

Figure 14.48

Designed for the pulley sizes and belt speeds typical of food-processing industries, food-grade belt cleaners are a match for food-processing applications.

Figure 14.49

Cleaners utilizing blades with "fingers" that have the ability to walk over obstructions are required for cleaning chevron-style belts.

14

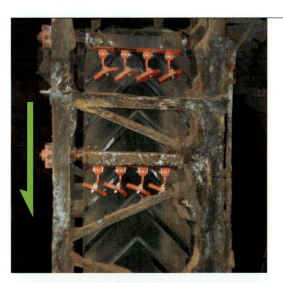

Figure 14.50

The "fingers" allow the belt-cleaner blades to effectively clean chevrons/ribs/cleats. Note: photo taken from below conveyor.

Figure 14.51

To remove carryback from a pulley, a belt cleaner with an elastomer blade is mounted slightly below the pulley's horizontal centerline on the side of the pulley away from the belt.

Pneumatic (Air-Knife) Cleaners

An air-knife belt-cleaning system directs a stream of compressed or fan-produced pressurized air to shear off carryback (**Figure 14.54**). Air-knife cleaners can be mounted in the primary, secondary, or tertiary positions. These cleaners are of interest because they do not contact the belt.

Figure 14.52

Cleaning systems composed of a rotating brush applied against the belt can be used effectively on dry materials.

14

Figure 14.53

Brush cleaners often encounter problems with sticky or moist materials that build up in the bristles of the brush.

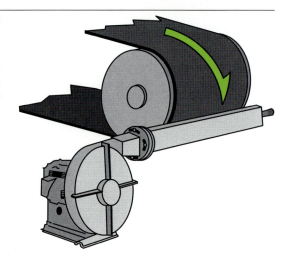

Figure 14.54

An air-knife belt-cleaning system directs a stream of compressed or fan-produced pressurized air to shear off carryback.

An air-knife system can be effective in removing dry materials and is sometimes used on very wet materials with low adhesion in applications like coal cleaning plants. When used on dry materials, such as alumina, these systems are part of a dust pick-up station, with the dust blown by the air knife into a dust-collector hood. Air-knife systems can be used to dry belts that have been wetted from material moisture or from water added to improve belt-cleaning efficiency, as in a wash-box system. *(See Chapter 24: Belt-Washing Systems for a more thorough, detailed discussion about wash-box systems.)*

The disadvantages of air-knife systems include the continuing expense of providing air to the knife and problems with the plugging of the air outlet(s). On dry materials, they can create additional airborne dust. With wet materials, splatter will build up on chute walls.

The Use of Water in Belt Cleaning

In many applications, an increase in moisture level of the carryback material is directly related to increased adhesion to the belt, so it increases the difficulty of handling and removing the material. This effect is seen with increases in moisture content up to a specific level—unique to each material—at which point the adhesion then drops off. Therefore, the use of water is a major advantage in cleaning conveyor belts handling almost any material.

The use of a simple spray bar just behind the primary cleaner or in front of the secondary cleaner will do many things to improve the cleaning process (**Figure 14.55**). A small amount of water sprayed against the belt immediately after the pre-cleaner aimed at the underside of the blade will act as a release agent, moistening the belt and material and reducing adhesion to most surfaces. It also serves as a coolant to the secondary-cleaner blades to prevent the "baking on" of the material. As a bonus, the life of cleaning blades can be extended by the lubricating action of the water.

On belts that can run empty for long periods of time, belt-cleaner blades will generate heat due to the friction between the belt surface and the tip of the blade. The faster the belt speed, the quicker the heat is generated. The use of a water spray to lubricate the belt will eliminate this problem by reducing the friction and cooling the blade.

These spray bars might not need to apply much more than a mist to the belt surface. Any excess water left on the belt after the secondary cleaners can be removed by the use of a soft urethane "water-squeegee blade" as a tertiary-cleaning system (**Figure 14.56**).

The results gained by the correct application of water in belt-cleaning systems more than justify its consideration in most material-handling operations. In a paper presented to the 1990 International Coal Engineering Conference in Australia, J.H. Planner reported that adding a water spray to various conventional cleaning systems raised cleaning efficiency from the 85 percent range to the 95 percent range *(Reference 14.3). (See Chapter 24: Belt-Washing Systems.)*

BLADE-TO-BELT PRESSURE

Optimal Cleaning Pressure

A key factor in the performance of any cleaning system is its ability to sustain the force required to keep the cleaning edge against the belt. Blade-to-belt pressure must be controlled to achieve optimal cleaning with a minimal rate of blade wear.

There is a popular misconception that the harder the cleaner is tensioned against the belt, the better it will clean. Research has demonstrated this is not true. A 1989 study by the Twin Cities Research Center of the U.S. Bureau of Mines examined the issue of optimizing the blade-to-belt pressure to offer the best level of cleaning without increasing blade wear, belt damage, and/or conveyor power requirements. This research is published in *Basic Parameters of Conveyor Belt Cleaning (Reference 14.2)*. The study evaluated the cleaning effectiveness and wear characteristics of various steel blades by holding them perpendicular to a running belt with measured amounts of pressure to remove a moistened sand/lime mixture. The study found that the amount of carryback and blade wear both decrease as blade pressure is increased up to an optimum blade pressure. The study established this optimum secondary blade-to-belt pressure at 76 to 97 kilopascals ($11–14$ lb_f /in.^{2}) of blade-to-belt pressure (**Figure 14.57**). Increasing pressure beyond this range raises blade-to-belt friction, thus shortening blade life, increasing belt wear, and increasing power consumption without improving cleaning performance. An over-tensioned blade normally exhibits accelerated, but even, wear; some discoloration or "scorch" marks; and belt top-cover particles in carryback baked on the blade.

Operating a belt cleaner below this optimum pressure provides less effective cleaning and can cause rapid blade wear. A belt cleaner barely touching the belt may appear to be in working order from a distance; whereas in reality, material is being forced between the blade and the belt at high velocity. This passage of

Figure 14.55

The use of a simple spray bar just behind the primary cleaner or in front of the secondary cleaner will improve the cleaning process.

Figure 14.56

Any excess water on the belt after the secondary cleaners can be removed by the use of a soft urethane "water-squeegee blade" as a tertiary-cleaning system.

material between the belt and the blade creates channels of uneven wear on the face of the blade. As material continues to pass between the blade and the belt, these channels increase in size, rapidly wearing the blade. A blade that has been under-tensioned will normally exhibit a jagged edge with wear lines on the wear surface.

The Bureau of Mines study also reported cleaning effectiveness decreased over time because of uneven blade wear. Grooves worn into the blade edge allow the passage of carryback that cannot be eliminated by increasing the blade-to-belt contact pressure. The report noted: "Once a cleaner blade's surface is damaged, no realistic

level of blade-to-belt pressure will allow the blade to conform to the belt's surface for proper cleaning."

Tensioning Systems/Devices

Blade-to-belt cleaning pressure is maintained by a tensioning device. These tensioners range in sophistication from concrete block counterweights and locking collars to torque storage couplings and engineered air-spring systems plumbed to the plant's compressed air supply (**Figure 14.58**). The reason for choosing a specific tensioner depends on the conveyor specification as well as plant preferences.

All tensioning systems should be designed to allow the cleaning edge to relieve itself away from the belt in order to allow passage of mechanical splices and other obstructions. Tensioners should be self-relieving to minimize risk of injury to personnel or equipment if the blades are "pulled through" by obstructions or holes in the belt (**Figure 14.59**).

A tensioning device should be designed to be compatible with the cleaner in order to provide a consistent blade-to-belt pressure throughout the life of the blade. When cleaner adjustment and re-tensioning are required, the tensioner should allow this maintenance to be performed simply, without requiring tools or more than one service worker.

Some cleaners use the resilience of a urethane blade, when compressed and locked into position, to supply the cleaning pressure (**Figure 14.60**). When installed, these blades are deflected by being forced against the belt. As the blade wears, it "stands taller" to maintain cleaning pressure. Because the blade itself supplies both cleaning pressure and shock-absorbing capacity, the cleaner does not need a conventional tensioner. Instead, the blade assembly is forced against the belt, and the mainframe is locked into position, slightly compressing the blades to set the initial blade-to-belt pressure.

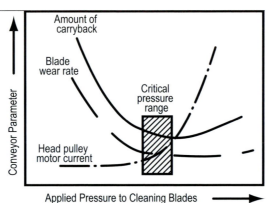

Figure 14.57

The study found that the amount of carryback and blade wear both decrease as blade pressure is increased, up to an optimum blade-to-belt contact pressure of 76 to 97 kilopascals (11–14 lb$_f$/in.^{2}).

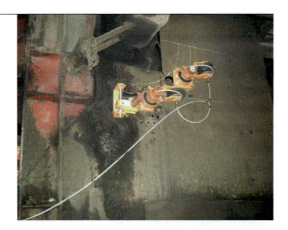

Figure 14.58

More sophisticated tensioning systems include engineered air-spring systems plumbed to the plant's compressed-air supply.

Figure 14.59

Tensioning systems should allow the cleaning edge to relieve away from the belt in order to allow passage of mechanical splices and other obstructions.

14

Linear or Radial Adjustment

There are competing theories for belt-cleaner adjustment. There are linearly-adjusted cleaners that are pushed up (in a line) against the belt, and radially-adjusted cleaners that are installed with a mainframe as an axis and rotated into position (**Figure 14.61**).

Radially-adjusted cleaners have several practical advantages over the linear design. They are easier to install, can be adjusted from one side of the belt, and can more readily rotate away from the belt to absorb the shock inherent in belt motion and splice passage.

Linearly-adjusted cleaners generally require access to both sides to provide even adjustment (**Figure 14.62**). Because of this, the tensioners for these cleaners often have some form of powered adjustment, such as an air bag, that can be remotely controlled. Linear tensioners maintain a constant-cleaning angle as the blade wears and can be designed to allow for easy withdrawal of the cleaner for maintenance without removing the tensioner.

In addition, some hybrid systems incorporate vertical tensioning with a radial relief mechanism (**Figure 14.63**).

Maintaining the angle of the blades against the belt is important for ensuring effective cleaning. If the angle of contact is altered by blade wear, cleaner performance will similarly "decay." A well-designed belt cleaner must control the cleaning angle across its wear-life.

CLEANER INSTALLATION

A critical ingredient in the performance of any belt-cleaning system is its installation. Improper installation will have an adverse effect on how the cleaner performs; it will reduce blade-life and cleaning efficiency. The installation instructions from the manufacturer should be closely followed.

Figure 14.60

Some cleaners use the resilience of a urethane blade to supply the cleaning pressure.

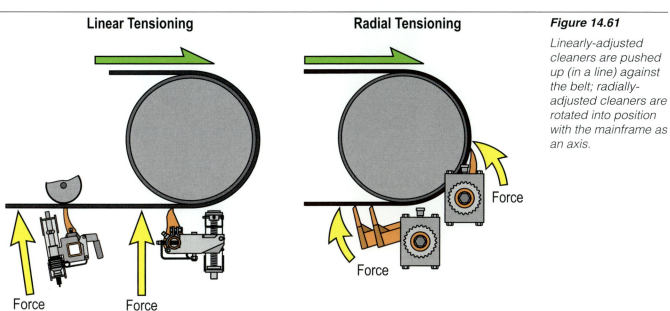

Linear Tensioning

Force   Force

Radial Tensioning

Force

Force

Figure 14.61

Linearly-adjusted cleaners are pushed up (in a line) against the belt; radially-adjusted cleaners are rotated into position with the mainframe as an axis.

Considerations affecting the installation position of a belt cleaner include:

A. Cleaner design

B. Tensioner and mounting requirements

C. Bolting or welding the cleaner in place

D. Installation on chutewall or hung from stringer

E. Position of conveyor structural beams, bearings, and drives

Figure 14.62

Linearly-adjusted cleaners generally require access to both sides of the belt to provide even adjustment.

Figure 14.63

Some hybrid systems incorporate vertical tensioning with a radial relief mechanism.

Figure 14.64

Placing the cleaner at the proper distance from the belt helps avoid "pull through," in which the belt pulls the cleaner all the way around into an inverted position, which usually results in a bent mainframe.

Regardless of the brand of belt cleaner, the critical factor in cleaner installation is that the cleaner support frame be installed at the correct distance from the surface of the belt. Placing the cleaner at the proper distance from the belt helps avoid "pull through" problems, in which the belt pulls the cleaner into the belt and all the way around into an inverted position, which usually results in a bent mainframe (**Figure 14.64**). Maintaining the proper dimension places the blades at the correct angle of attack against the belt for the best cleaning, proper blade wear, and longest life. The correct distance will be different from cleaner style to cleaner style.

It is strongly recommended that the manufacturer install and maintain the belt cleaners on both new and retrofit applications, because most performance problems with new belt-cleaner systems are due first to improper installation and second to lack of maintenance. Using the manufacturer (or manufacturer-approved contractors) for installation makes certain of proper installation and continued performance.

Troubleshooting Cleaner Installation

If a cleaning system is performing poorly, but the blades do not show excessive wear and the tensioner is set correctly, other problems may be present. These problems might include:

A. The support frame is not parallel to the pulley.

B. The cleaner is not installed the proper distance from the belt surface.

C. The pressure applied to the cleaner is changing the belt line.

D. The blades are not centered on the belt.

Any of these factors will impair a cleaner's ability to remove carryback. The cleaner's operators or installation manual should be reviewed to determine the appropriate corrective actions.

Belt Flap and Belt Cleaning

Belt flap can create problems in belt cleaning. Belt flap is an oscillation of the belt and is most often seen on the low-tension (return) side of the conveyor. This vibration has been measured with amplitudes as large as 25 millimeters (1 in.). The movement can be so strong that it destroys belt cleaners or plows and shortens the bearing-life of return idlers. The amplitude of the vibration can make it difficult to keep the blades in contact with the belt, thus reducing the cleaning effects. To control belt flap, the spacing of return idlers can be varied or a hold-down roller can be used to try to quiet the belt.

Pre-Cleaner "Heeling"

Pre-cleaners are designed to have the tip of the blade contact the belt first. As the tip wears, the primary cleaner normally rotates into the belt to maintain contact between the blade and the belt. However, problems can arise when an elastomer pre-cleaner blade is mounted too close to the belt. A primary cleaner installed in this manner, regardless of the blade's design, will have the heel of the blade tip contact the belt first. This "heeling" creates a gap between the belt and the blade tip (**Figure 14.65**). Conveyed material collects in this gap, and the accumulation forces the blade away from the belt. Once the blade is pushed away from the belt, larger amounts of material then pass between the belt and the blade, greatly increasing wear on the blade and belt and decreasing cleaning efficiency. The solution is to maintain the proper installation distance, so that the leading edge of the blade first contacts the belt.

The Problem with Over-Tensioning

Optimal cleaning efficiency results from the combination of the right scraping angle and adequate tension against the belt. As noted in the Bureau of Mines paper *Basic Parameters of Conveyor Belt Cleaning*, increasing the blade-to-belt pressure of a cleaner does not necessarily improve cleaning performance *(Reference 14.2)*. Increasing

the pressure can reduce cleaning efficiency and shorten wear-life. Even when properly installed, if an elastomeric pre-cleaner blade is over-tensioned, the force is shifted from the entire contact area toward the heel of the blade. This results in a "mini heeling" situation and frequently can cause the blade to wear to a thin flap at the tip that can reduce cleaning efficiency (**Figure 14.66**).

If a secondary cleaner is over-tensioned, its cleaning angle may be changed to the point in which carryback is trapped in the wedge-shaped region between the blade and belt (**Figure 14.67**). This will create a

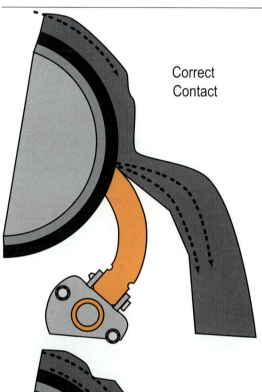

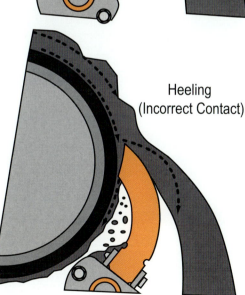

Correct Contact

Heeling (Incorrect Contact)

Figure 14.65

"Heeling"—when the heel of the pre-cleaner blade contacts the belt first—creates a gap between the belt and the blade tip; material will collect in this gap and force the blade away from the belt.

14

Figure 14.66

When an elastomeric pre-cleaner blade is over-tensioned, the force is shifted from the entire contact area toward the heel of the blade. This can cause the blade to wear to a thin flap at the tip, reducing cleaning efficiency.

Figure 14.67

If a secondary cleaner is over-tensioned, its cleaning angle may be changed to the point in which carryback becomes trapped between the blade and belt, pushing the belt up and reducing the effective cleaning pressure.

Figure 14.68

It is not unusual for carryback material to adhere to the surface of a vertical, low-friction liner.

Figure 14.69

A cleaner encapsulated with a sticky or dried accumulation of material cannot function properly.

buildup of material that will push the belt up and reduce the effective cleaning pressure. Material will pass between the blade and belt, again resulting in a poor cleaning efficiency and increased blade and belt wear.

Handling Material Cleaned from the Belt

The fact that carryback clings to the belt past the discharge point indicates that it has different characteristics from the rest of the conveyor cargo. The particles are finer and have higher moisture content, so they have different flow qualities from those typical of the main body of material. It is not unusual for carryback material to adhere to the surface of a vertical, low-friction liner (**Figure 14.68**). Even after its removal from the belt, carryback presents problems in capture, handling, and disposal.

Because of the characteristics of carryback material, it is usually best to locate the belt cleaners as close to the discharge point as possible. Returning as much carryback as possible back to the main material flow reduces the need to handle this difficult material outside of the process. The sticky fugitive material that carries back farther along the belt may build up within the chute or need to be handled by dribble chutes or scavenger conveyors, thus increasing the costs and complexity of the material-handling system. To ensure effective cleaning performance, the buildup of material on the cleaner or dribble chute must be prevented. A cleaner that becomes encapsulated with a sticky or dried accumulation of material cannot function properly (**Figure 14.69**).

Collecting and returning the carryback to the main material body can present a serious complication in the design of discharge chutes. Ideally, the conveyor's main discharge chute is sufficiently large so the material cleaned from the belt can fall through the same chute, where it is reunited with the main stream of material. But in many cases, auxiliary chutes or systems need to be added.

14

Dribble Chutes

On conveyors where the cleaning systems are positioned so the material removed from the belt does not freely return to the main material body, a dribble chute, or fines chute, is usually required. This is a separate part of the discharge chute that directs the removed carryback back into the main material flow. This auxiliary chute must be large enough and designed with a steep enough wall angle to ensure the material falls away from the cleaning system and prevents the encapsulation of the cleaner in these sticky materials. It is advisable to install a dribble chute with an angle as near vertical as possible and to line it with a low-friction material such as Ultra-High Molecular Weight (UHMW) polyethylene. It may be useful to incorporate flow aids to help in moving the carryback material away from the cleaners.

One way to solve the problem of buildup on a dribble chute is to create a dynamic sub-floor inside the chute. This can be accomplished by mounting a sheet of smooth, low-friction, abrasion-resistant plastic such as UHMW polyethylene, parallel to the chute floor with one end unsecured, so the sheet is free to move. A vibrator is mounted to this sheet, providing a dynamic action to prevent material buildup (**Figure 14.70**). Because this vibrating sheet is isolated from the steel chutework by a rubber cushion, there is very little force applied to the structure to cause metal fatigue (**Figure 14.71**).

An alternative system could include a flexible curtain or sheet of rubber used as a chute liner, which is periodically "kicked" with the discharge from an air cannon. This flexes the liner, causing any adhered material to drop away. *(See Chapter 9: Flow Aids.)*

Figure 14.70

An electric vibrator can be mounted to a sheet of low-friction plastic to create a vibrating dribble chute.

Figure 14.71

By isolating the vibrating sheet from the steel chutework with a rubber cushion, there is very little force applied to the structure to cause metal fatigue.

Figure 14.72

A powered plow or push conveyor can be used to move material cleaned from the belt back to the main material stream.

14

Access into the dribble chute should be provided to allow personnel to clear build-ups and provide a periodic wash down to prevent blockages.

Scavenger Conveyors

When a separate dribble chute is impractical, it may be beneficial to provide a scavenger conveyor (**Figure 14.72**). This is a smaller auxiliary conveyor, installed below the main system, that returns the cleanings to the main material flow. Small screw conveyors, scraper-chain conveyors, electrical or hydraulic plow or push conveyors, and vibratory conveyors are commonly used as scavenger systems (**Figure 14.73**).

An advantage of scavenger conveyors is that they allow the placement of several tertiary belt cleaners in a location more convenient for service. The material cleaned from the belt can be transported, even uphill, back into the main chute. The main disadvantage of these systems is they represent the addition of another piece of mechanical equipment that must be periodically cleaned and maintained.

Figure 14.73

Vibratory conveyors are often seen as scavenger systems.

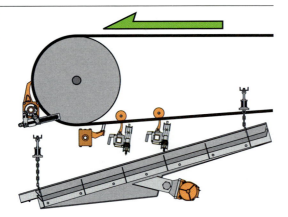

Figure 14.74

After installing a belt cleaner, periodic inspection, adjustment, and maintenance are required.

SYSTEM MAINTENANCE

The Importance of Maintenance

Even the best-designed and most efficient belt-cleaning system requires maintenance and adjustment on a regular basis, or its performance will deteriorate over time. Proper maintenance of belt-cleaning systems reduces wear on the belt and cleaner blades, prevents damage, and ensures efficient cleaning action. Lack of maintenance on belt-cleaner systems not only produces a failure to clean effectively but also adds considerable risk to the conveyor system.

Conveyor designers and cleaner manufacturers must both design their equipment to simplify these vital maintenance activities. Maintenance requirements and procedures should be reviewed during the selection process for a cleaning system. Advance planning for cleaner service will allow maintenance activities to be performed expeditiously and will translate into improved belt cleaning and minimum downtime. Service chores that are simple and "worker-friendly" are more likely to be performed on a consistent basis.

After installing a belt cleaner, periodic inspection, adjustment, and maintenance are required (**Figure 14.74**). Just as cleaners must be designed for durability and simple maintenance, conveyors must be designed to enable easy service, including required clearances for access. Elements that can be incorporated into a conveyor belt-cleaning system to improve maintenance procedures include:

A. Adequate service access with ample clearances and work spaces, as recommended by CEMA

B. Access windows with easy-to-operate doors installed on both sides of the pulley, in line with the axis of the belt cleaners

C. Cleaning elements that slide in and out for service without requiring removal of the mainframe

D. Components including blades and

mainframe that resist corrosion and abuse

E. Components that allow the quick performance of required service—that can be adjusted or replaced with simple hand tools without waiting for a maintenance crew with power tools to perform the work

Mandrel-mounting systems that allow slide-in/slide-out positioning of a cleaner assembly offer an opportunity for faster service (**Figure 14.75**). Some facilities have made arrangements allowing this service to be performed while the belt is running, assuming the appropriate regulatory and safety committee approvals are granted and proper personnel training is supplied.

All applicable safety rules should be followed when performing any cleaning system maintenance procedure. Only trained and certified personnel following appropriate lockout / tagout / blockout / testout procedures should be considered for these practices.

Tips for Belt-Cleaner Maintenance

If plant management and personnel devote energy and attention to maintaining the cleaner's performance, they will be repaid with more efficient cleaning. It is usually best to assign belt-cleaning system maintenance to dedicated plant personnel or specialty contractors, because they will then have a commitment to maintaining the system.

The problem is that most in-house inspections never happen, and when they do occur, they are "walk by" inspections by people who are not trained in what to look for or how to maintain the cleaners. Most managers will feel this is a simple task that should be done in-house; the truth is that cleaner maintenance is never a priority, so rarely is it done. The use of outside specialty contractors ensures that belt-cleaner maintenance is done properly. Specialty contractors often notice other developing conveyor system or component problems that can be avoided.

While specific maintenance instructions are provided for each cleaner and tensioner by its manufacturer, there are regular and routine procedures that should be performed at specific intervals.

Manufacturer recommended and required local safety procedures must be followed when performing maintenance on belt-cleaning systems.

Daily: Remove Accumulated Material from the Cleaner

With the belt stopped, clean off any material that has gathered between the cleaner blades and the belt or that has built up on the arms of secondary cleaners. Often, a rotation of the cleaner away from the belt followed by a few sharp raps of the blades back into the stopped belt will dislodge this material. In other conditions, a quick rinse with a water hose or high-pressure spray will remove the buildup and allow inspection of the blades.

Weekly: Check Cleaner Performance

Check the work of the cleaning system. Carryback remaining on the belt could indicate worn out blades or improper tension.

Weekly: Check Blade Wear

Inspect the cleaning elements for wear. Some brands of blades incorporate a vis-

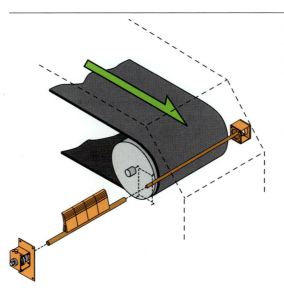

Figure 14.75

Mandrel-mounting systems that allow slide-in/slide-out positioning of a cleaner assembly offer an opportunity for faster service.

ible wear line; for others, a check of the manual will be needed to ascertain the limits of safe and effective wear.

Weekly: Check Tensioner Adjustment

The most critical element in maintaining cleaner performance is keeping the cleaning edge tensioned against the belt. As blades wear, the tensioner may need to be adjusted to accommodate the blade's shorter length. The manual for each tensioner should provide specific instructions for re-tensioning.

ASSESSING BELT-CLEANER PERFORMANCE

Improving Belt-Cleaning Performance

There are a number of tactics that can be applied to upgrade the performance of a plant's belt-cleaning system. A general program for improving belt-cleaner performance might include:

A. Follow the manufacturer's instructions

Ensure systems are installed and maintained to the manufacturer's recommendations. Observe the recommended maintenance interval.

B. Standardize and systemize

Standardizing to one brand or style of cleaner on all conveyors in a plant and/or all plants in a company will simplify procedures and minimize spares. If there are different belts and materials in the operation, consider adopting one cleaner platform plant-wide that allows for altering the cleaner by application or as bulk-material properties change. Some manufacturers offer cartridge-style cleaner platforms that facilitate customization while minimizing the number of mainframe and tensioner systems. In addition, cleaner maintenance practices can be systemized—either by managing them inside the plant or by outsourcing maintenance to a specialized contractor—to improve execution and accountability.

C. Raise the bar

Continuously upgrade performance requirements. Expect a clean plant and demand performance. Find a supplier that will guarantee their product and work with them to understand the cleaning process. Implement cleaning performance measurements, such as blade-life testing to improve blade selection or carryback testing to check performance. Consider installing additional cleaners on problem conveyors.

Another strategy to optimize a cleaner installation is to perform research. It is now possible to analyze cleaning systems to identify the blade-to-belt pressure that optimizes both cleaning efficiency and blade-life. To make optimization easier, some belt cleaners are marked with wear percentages molded into the blades; tensioning systems have been developed that provide continuous and constant spring pressure.

The optimization process consists of setting the cleaning pressure at a given level and recording the length of time and/or the total amount of conveyed material it takes for the blade to reach the 25-percent wear indicator. The pressure is then adjusted and the cleaner used until it wears to the next 25-percent wear indicator. Cleaning efficiency must also be measured, using the quantitative methods discussed earlier in this chapter or visual qualitative measurements. In this way, the plant can determine what pressure provides the longest blade-life while maintaining acceptable cleaning efficiency.

The results will vary from application to application, and even from conveyor to conveyor within the same plant. One German lignite mine found that a higher cleaning pressure resulted in a longer blade-life; a similar operation reported that lower pressures resulted in longer blade-life while still maintaining acceptable levels of cleaning.

Goals for Belt Cleaning

Belt cleaning is a process, and like any other process, the results follow a curve (**Figure 14.76**). The amount of material removed is proportional to the amount of "effort" put into it. This "effort" could be money, cleaning pressure, number of cleaners, or a combination of these.

This cleaning process curve can be shown on a graph where the effort of removing carryback for a certain cleaning percentage is measured against the cost of that amount of carryback (including cleanup, maintenance, idler replacement, and the value of lost material). At some point along the curve, the expense for installing, operating, and maintaining additional cleaning systems is greater than the cost of leaving the remaining material on the belt (**Figure 14.77**).

The costs for a belt-cleaning system that could achieve a cleanliness level of "100 percent" would probably outweigh the system's benefits. A significant amount of the carryback that remains on the belt past cleaning systems with a high level of cleaning will stay on the belt throughout the conveyor's return run. This material will still be on the belt when the conveyor reenters the load zone. Consequently, it might not be worth the expense of cleaning it all off.

Even more important, to clean "100 percent" by mechanical scraping alone, the cleaner(s) would be applied with so much pressure the cleaners would endanger the belt cover. And regardless of the efficiency of the cleaning system, some carryback will remain on the belt trapped in small cracks and grooves in the belt's surface. Therefore, it is impossible to reach "100 percent" cleaning.

It is best to have goals for belt cleaning that are reasonable and obtainable for the operation of the plant. With proper equipment selection and continuing maintenance, reasonable goals will let a facility show suitable improvements and return on investment. Cleaning systems should be designed, installed, and maintained to meet actual operating requirements with an acceptable level of carryback.

Developing a Standard for Belt-Cleaning System Performance

The amount of carryback remaining on a belt is more dependent upon the characteristics of the bulk material and the physical parameters of the conveyor system than it is on tonnage of material conveyed or other factors. R. Todd Swinderman proposed a performance-based standard for belt cleaning in a paper presented to the 2004 Annual Meeting of the Society for Mining, Metallurgy, and Exploration (SME) *(Reference 14.4)*. The purpose of his proposal was to propose a standard method for specifying belt-cleaner systems based on user expectations and equipment performance over time. His paper proposing three levels of cleaning performance is detailed below. *(See Chapter 31: Performance Measurements for the Swinderman Scale.)*

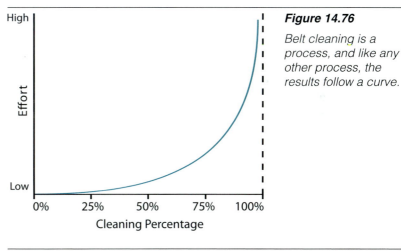

Figure 14.76

Belt cleaning is a process, and like any other process, the results follow a curve.

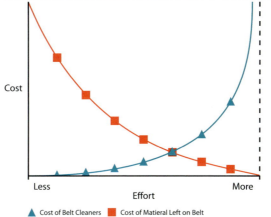

Figure 14.77

At some point along the declining curve representing the amount of material removed and the cost of that carryback, the cost of installing and maintaining additional cleaning systems is greater than the cost of leaving the remaining material on the belt.

Level I cleaning is generally specified when concerns about carryback are not critical. Cleaning systems normally associated with achieving Level I performance are single- or double-cleaner systems with one-piece blades receiving poor to average maintenance. Frequent cleanup of the carryback that falls from the conveyor return will likely be required. Level I would be specified for materials that are easy to remove from the belt, in operations that transport low tonnages of material, in plants that are intermittent in operation, or where carryback can be easily cleaned up and returned to the process.

Level II cleaning is generally specified when carryback is of concern but does not create a significant safety or environmental problem. Cleaning systems normally associated with Level II performance would be multiple-cleaner systems with segmented blades undergoing maintenance at the manufacturer's specified interval. Level II cleanliness would be specified for high-volume operations, for operations where the spilled material has moderate value, or where manual cleanup under the conveyers once per week is acceptable.

Level III cleaning is generally specified when concerns about carryback are critical. Concerns range from safety to environmental to product contamination. Cleaning systems normally associated with achieving a Level III performance are multiple-

cleaner systems in conjunction with at least one low-volume water spray. Difficult-to-clean bulk materials may require the use of a wash-box system with multiple cleaners using a combination of low-volume water sprays to lubricate the cleaners and high-volume sprays to keep the wash box and discharge piping freely flowing. Level III performance would be specified where prevention of spillage is required, contamination of cargo on the belt is a concern, the bulk material has a high value per ton, or where cleanup under the conveyors once a month is acceptable.

The Swinderman paper noted the higher the level of cleanliness desired, the more sophisticated the cleaning system and the better its performance will need to be (**Table 14.2**).

TYPICAL SPECIFICATIONS

A. In general

   a. It is important to design conveyor belt-cleaning systems for the problems presented by the "worst-case" material conditions, rather than for "normal" operating conditions. This will allow cleaning systems to better handle changes in the materials.

   b. Belt cleaners should be installed as close to the material discharge point as possible, ensuring effective cleaning by supporting the cleaning ele-

Table 14.2	**The Mean Level of Carryback Allowed to Remain on the Cleaned Portion of the Belt**			
	Level of Cleaning	**Level I**	**Level II**	**Level III**
	Mean Carryback Level (Dry Weight g/m²)	250	100	10
	Mean Carryback Level (Dry Weight oz/ft²)	0.82	0.33	0.03

Notes:

1. Because environmental and operating conditions vary, the level of cleaning is based on the mean of a standard distribution curve that will be unique to each conveyor and bulk material. Therefore, the mean of the measurements from all similar systems conveying similar bulk materials in a facility is to be used in measuring the total belt-cleaner performance.

2. Carryback factors (Cb_f), the percentage of material that ultimately falls from the belt and accumulates below the conveyor, has been measured to average 75% of the measured carryback at Level I, 50% at Level II, and 25% at Level III (Reference 14.5).

ments against a firm surface.

c. Belt cleaners should be installed outside the material trajectory and positioned so that cleaned materials cannot build up on the blades and structures.

d. Belt-cleaning systems should be designed to provide less than full-belt-width coverage to allow for minor variations in belt tracking and provide optimum blade-to-belt contact.

e. Each head pulley should have a belt-cleaning system consisting of (at a minimum) a primary cleaner and a secondary cleaner with provisions for the addition of tertiary cleaners.

f. The cleaning systems shall be designed to allow simple maintenance and blade-replacement procedures. Periodic maintenance should be performed as per manufacturers' recommendations to keep the belt cleaners operating at peak performance.

B. Primary cleaners

a. Primary cleaners perform the initial rough cleaning. They should be designed with flexible (elastomer) blades and radial-adjustment tensioning devices.

b. Primary cleaners should be installed on the face of the head pulley just below the trajectory of the material, utilizing a positive-rake cleaning angle.

c. Primary-cleaner blades should incorporate a constant-cleaning angle and area design.

d. Pre-cleaners should be designed for use on one-directional and reversing belts.

e. Reversing belts should have a pre-cleaner installed on each discharge pulley.

C. Secondary cleaners

a. Secondary cleaners remove the majority of the material that passes by the pre-cleaner's blades. Secondary blades should contact the belt just past the point where the belt has left the head pulley. Alternatively, the cleaners can be located behind the head pulley with a hold-down roll above the blades. The hold-down roll should be a minimum of 100 millimeters (4 in.) in diameter.

b. Secondary-cleaner blades should be designed to contact the belt in a negative-rake position.

c. The blades should be constructed of tungsten carbide or similar abrasion-resistant material.

d. On one-directional belts, the cleaners should be adjusted with a radial-adjustment tensioning device, and on reversing belts with a vertical-spring tensioner.

e. Reversing belts should have a reversing secondary cleaner installed as close as possible to each terminal (discharge) pulley.

D. Tertiary cleaners

a. Space should be planned in the design of conveyor load zones for the possible addition of tertiary cleaners.

b. Tertiary cleaners should utilize a separate dribble chute or scavenger conveyor to return the carryback to the main material flow.

E. Other

a. Necessary utilities (water, electricity, compressed air) should be available at points convenient to the belt-cleaner installation.

b. Clearances and access in accordance with CEMA recommendations should be provided in the conveyor design.

ADVANCED TOPICS

Belt Cleaners and Power Requirements

Applying belt cleaners increases the drag against the belt and raises a conveyor's power consumption (**Equations 14.1** and **14.2**).

A study by R. Todd Swinderman published in *Bulk Solids Handling* has examined how much power the application of a belt cleaner consumes from a conveyor's overall drive power *(Reference 14.6)*. The power requirement is calculated for the belt width actually contacted by the cleaner. In most cases, cleaning blades do not contact the full width of the conveyor belt.

The paper considers a belt 900 millimeters (36 in.) wide moving at speeds of 0,5; 2,0; 3,5; and 5,0 meters per second (100, 400, 700, and 1000 ft/min). Blade coverage of the cleaners against the belt is 762

millimeters (30 in.). The power consumption added to the conveyor's drive by the tensioning of various types of belt cleaners ranges from 0,14 to 3,8 kilowatts (0.2 to 5.1 hp) (**Table 14.3**).

An application is calculated using a commercially available conveyor engineering computer software program. The specifications used in the program are: a belt 1200 millimeters (48 in.) wide, operating at 3,0 meters per second (600 ft/min), conveying 1350 tons per hour (1500 st/h) of coal for a distance of 90 meters (300 ft) at an incline of 14 degrees. The weight of the

---

14

Equation 14.1

Calculating Tension Added to the Belt due to the Belt Cleaner

$$\Delta T_{BC} = I_{BC} \cdot \mu_{BC} \cdot F_{BC}$$

Given: A belt 900 millimeters (36 in.) wide has a cleaner on the head pulley. The cleaner exerts a force of 0,88 newtons per millimeter (5.0 lb$_f$/in.) on the belt, and the coefficient of friction is 0,6.
Find: Tension added to the belt due to the cleaner.

Variables		Metric Units	Imperial Units
ΔT_{BC}	Tension Added to the Belt due to the Cleaner	newtons	pounds-force
μ_{BC}	Friction Coefficient	0,6	0.6
F_{BC}	Normal Force Between Belt and Cleaner per Length of Cleaner	0,88 N/mm	5.0 lb$_f$/in.
I_{BC}	Length of Cleaner Blade	900 mm	36 in.
Metric: $\Delta T_{BC} = 900 \cdot 0,6 \cdot 0,088 = 47,5$			
Imperial: $\Delta T_{BC} = 36 \cdot 0.6 \cdot 0.5 = 10.8$			
ΔT_{BC}	Tension Added to the Belt due to the Cleaner	475 N	108 lb$_f$

Equation 14.2

Calculating Power Consumption Added to the Belt Drive

$$P = \Delta T_{BC} \cdot V \cdot k$$

Given: A belt cleaner adds 475 newtons (108 lb$_f$) tension to a belt traveling 3 meters per second (600 ft/min). **Find:** The power consumption added to the drive due to the cleaner.

Variables		Metric Units	Imperial Units
P	Power Consumption Added to Belt Drive	kilowatts	horsepower
ΔT_{BC}	Tension Added to the Belt due to the Cleaner *(Calculated in **Equation 14.1**)*	475 N	108 lb$_f$
V	Belt Speed	3 m/s	600 ft/min
k	Conversion Factor	1/1000	1/33000
Metric: $P = \dfrac{47,5 \cdot 3}{1000} = 0,14$			
Imperial: $P = \dfrac{10.8 \cdot 600}{33000} = 0.2$			
P	Power Consumption Added to Belt Drive	1,4 kW	2.0 hp

---

belt is specified as 22,3 kilograms per meter (15 lb_m/ft), and idlers are spaced every 600 millimeters (24 in.). This conveyor would require a total drive power of 107 kilowatts (143 hp).

If 1,2 kilograms per square meter (0.25 lb_m/ft²) of carryback were present on the belt, this would amount to 10,9 additional tons per hour (12 st/h) of load. By itself this additional load would require very little additional power to carry: 1 kilowatt (1.3 hp) additional power, for a total of 108 kilowatts (144 hp). Conveyor problems do not arise from the power consumed by the weight of the carryback, but rather from the impact on the conveyor hardware of this carryback as it is released into the environment.

A single frozen impact idler set would require approximately 1,2 kilowatts (1.6 hp) of additional power. One seized steel idler set can demand as much as 0,27 kilowatts (0.36 hp) additional power. This study also notes that a 25 millimeter (1 in.) layer of carryback on a single return roller can add as much as 0,32 kilowatt (0.43 hp) to the conveyor's drive requirements.

These additional power requirements for the problems arising from fugitive material should be compared to the power requirements of a typical dual-cleaning system. Continuing the example above, for a 1200-millimeter (48-in.) belt traveling 3 meters per second (600 ft/min) and incorporating a dual-cleaning system, the power

requirement would be 1,3 kilowatts (1.7 hp) for the pre-cleaner and 2,1 kilowatts (2.8 hp) for the secondary cleaner.

The combined additional power consumption of 3,4 kilowatts (4.5 hp) required for the use of an effective multiple-cleaner system represents an increase of only three percent over the 107 kilowatts (143 hp) required for the conveyor without any cleaners. This "conveyor power penalty" applied by the belt-cleaning system is only a little more than the power consumed at the rate of 0,27 kilowatts (0.36 hp) for a single seized idler set or 0,32 kilowatts (0.43 hp) required by 25 millimeters (1 in.) of accumulation of material on a single return roller.

As noted by Swinderman, the consequences of not installing and properly maintaining effective belt cleaners proves a more serious drain of conveyor drive power through the added friction caused by idlers with material accumulations or seized bearings.

Belt Cleaning and Dust Control

Carryback is a major source of dust in conveying. This dust is created when the dirty side of the belt interacts with idlers and belt pulleys. Belt cleaners greatly reduce the overall generation of dust in the conveying of bulk materials, because they reduce the total amount of material carried back on the return run of the belt. As belt-cleaner designs have improved, the overall

14

Table 14.3

Power Consumption Added to Conveyor Drive Requirement by Various Types of Belt Cleaners								
	Belt Speed, m/s (ft/min)							
	0,5 (100)		**2 (400)**		**3,5 (700)**		**5 (1000)**	
Blade Type	**kW**	**(hp)**	**kW**	**(hp)**	**kW**	**(hp)**	**kW**	**(hp)**
Urethane-Bladed Pre-Cleaner	0,14	(0.2)	0,52	(0.7)	0,97	(1.3)	1,34	(1.8)
Metal-Bladed Secondary Cleaner	0,22	(0.3)	0,89	(1.2)	1,57	(2.1)	2,24	(3.0)
Urethane-Bladed Secondary Cleaner	0,37	(0.5)	1,49	(2.0)	2,68	(3.6)	3,80	(5.1)

Note: All testing performed utilizing tension supplied by Martin Engineering to recommended cleaning pressure

level of dust originating as carryback has been reduced to the point in which belt cleaners have become a critical passive method of dust control. This advanced topic will offer a method to estimate the reduction in dust that can potentially be released from the belt into the environment.

Moisture Content and Particle Size

The dustiness of a bulk material is related to air velocity, particle size, and cohesiveness (**Figure 14.78**).

Several factors influence the dustiness of a given bulk material. The factors most important to this discussion are moisture content and particle size. Generally, additional moisture content increases the material's particle size and cohesiveness. As the particle size and cohesiveness increase, the Dustiness Index decreases. A decrease in dustiness is seen with moisture levels as low as 2.5 percent; most materials having a moisture content of 16 percent or more will have a Dustiness Index that is effectively zero. Particle size is also a significant variable; generally, bulk materials with particles over 100 microns (0.0039 in.) have a very low Dustiness Index *(Reference 14.7)*. In general, if the carryback has, on average, more than 16 percent moisture, and the particle size is greater than 100 microns (0.0039 in.), very little dust reduction can be expected from belt cleaning.

Carryback is usually high in moisture content, because the undulating action of the belt causes the fines and moisture to migrate toward the belt surface. The typical moisture content of carryback is from 15 percent to 50 percent. The particle-size distribution of carryback is dependent upon where it is measured and the type of belt cleaner used. If there are multiple belt cleaners, the size of carryback particles that pass each successive cleaner will become smaller (**Table 14.4**). For a typical dual system, the average particle size passing a pre-cleaner with elastomeric blades is on the order of 1000 microns (0.039 in.), with a range from 1 micron (3.93 x 10^{-5} in.) to 5 millimeters (0.2 in.). The average size of particles passing a secondary cleaner with hard metal blades will be on the order of 50 microns (0.00196 in.), with a range from 1 micron (3.93 x 10^{-5} in.) to 250 microns (0.0098 in.).

Belts that Run Empty

It is evident that a set of conditions of low moisture and small particle size must exist for dust to be released from carryback. Most of the time, carryback is of sufficient moisture content to agglomerate the smaller particles, sticking to the belt and preventing the release of dust. Much of the dust generated by belt cleaners and downstream components, such as idlers and bend pulleys, occurs when the belt is

---

Figure 14.78

Relationship in Creating Airborne Dust

$$\text{Dust Generated} \quad \alpha \quad \frac{\text{Air Velocity}}{\text{Particle Size} \bullet \text{Cohesiveness}}$$

Table 14.4

Characteristics of Carryback that Passes the Cleaning System			
Belt-Cleaning System	Belt Cleaners Disengaged	Urethane-Bladed Pre-Cleaner Only Engaged	Pre-Cleaner (with Urethane Blade) and Metal Blade Secondary Cleaner Engaged
Size of Particles	Large, Small, and Fine Particles (10 mm to 1 micron)	Small and Fine Particles (5 mm to 1 micron)	Fine Particles (250 microns to 1 micron)
Moisture Content	Lowest Moisture Content (~15%)	Increased Moisture Content (~30%)	Very Wet (~50%)

Notes: Conveyed Material: Crushed limestone, 200 mm (8 in.) minus. Belt Speed: 2 m/sec (394 ft/min). Collection Time for Each Sample: 30 seconds

allowed to run for prolonged periods of time without materials being conveyed. If the belt runs long enough without cargo, the carryback on the belt dries, and its particle size is reduced by contact with rolling components. Under these conditions, the majority of the carryback will be released into the environment in the form of dust. Of course, there are exceptions, such as when handling very dry materials like alumina; however, for coals and most minerals, belt cleaning greatly reduces the dust generated, by removing most of the carryback from the belt.

A case in point is demonstrated by the bunker-room conveyor system at a coal-fired power plant. The plant would leave these conveyors running when it was not actively loading coal. When coal was not being loaded, the airborne dust would diminish and the air clear. However, after running unloaded long enough for the material on the belt to dry, the level of dust in the air would climb again. When the belt was empty, the carryback dried out and was dislodged from the belt into the air by the return rollers, belt cleaners, and other components. When the loading of the conveyors began again, the dust level would actually decrease, because the cargo

contained incidental moisture and dust-suppression chemicals (**Figure 14.79**).

While this example does not represent results from before and after installation of belt cleaners, it does show that the periods of highest dust generation correspond to the times when the conveyor was running without coal on the belt. One of the things that can be deduced from these results is that when the belt runs empty, the carryback eventually dries out. The dry carryback is then available to become airborne dust, with most of this dust released in a short period of time when the moisture content of the carryback presumably reaches a critical low value. It is evident that a significant gain in dust control would have been achieved if the plant simply turned off the conveyor when not conveying coal. When the belt is allowed to run empty, the carryback dries out, and the belt cleaner takes the fine dry material from the belt, as do the return idlers.

Estimating the Dust Avoided by Belt Cleaning

Presumably, the dust created within the chute is dealt with by the conveyor's passive or active dust-control system(s); the amount of dust generated by the belt on the return

14

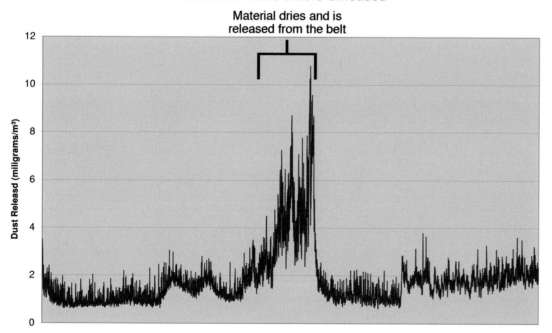

Figure 14.79

Dust emissions from a power plant tripper belt increased for an interval of time after the belt ran unloaded for enough time to allow the carryback material to dry out and become dislodged from the belt.

run is directly proportional to the amount of carryback on the belt. Conditions change continuously in bulk-materials handling, making it difficult to calculate a precise value for dust created by the conveying process or avoided by belt cleaning. Calculating the amount of dust that could be released requires a number of assumptions and estimates; without actual application-specific data, it is basically a theoretical exercise. Using the same assumptions and estimates, it can be shown that installing a sufficient belt-cleaning system, maintaining it properly, and not allowing the belt to run empty for prolonged periods will result in a significant reduction in dust potentially escaping from the conveyor.

A Sample Problem

A typical conveyor in a mining or power-generation application with no belt-cleaning system installed will suffer carryback at an average of 500 grams per square meter (1.6 oz/ft²) on the belt surface that is carrying cargo. Adding a properly installed and maintained dual-cleaning system will reduce the carryback to less than 100 grams per square meter (0.3 oz/ft²). More advanced systems—using more cleaners or a belt-washing station—can reduce carryback further to 10 grams per square meter (0.03 oz/ft²). Whereas 10 grams per square meter (0.03 oz/ft²) may seem like a high value, it must be considered that a single scratch in the belt surface that is 0,14 millimeter (0.006 in.) wide and 0,14 millimeter (0.006 in.) deep can contain 10 grams (0.4

oz) of material (with a specific gravity of 1.0) per meter (3 ft) of belt length. It can be assumed, based on the level of cleaning, that from 25 percent to 75 percent of the carryback left on the belt after belt cleaning is "hiding" in the belt in cracks, holes, and general surface roughness (*Reference 14.4*).

The other primary factors needed to make an estimate of the dust generation of belt cleaning are belt speed and operating hours. To reduce the number of calculations and charts required, a standard belt speed of 1 meter per second (200 ft/min) is used; other belt speeds can be extrapolated linearly—for example, a belt speed of 3 meters per second (600 ft/min) means three times the values at 1 meter per second (200 ft/min).

Sample Calculations

Assumptions

The assumptions (**Table 14.5**) are based on typical values for carryback and experience in measuring performance of belt-cleaning systems under a wide variety of conditions in coal and hard rock mining. To be conservative, it is assumed that the Dustiness Index is 100 percent and that all carryback in this example is 100 microns (0.0039 in.) or less; therefore, all carryback is potential airborne dust. Levels I, II and III represent standard categories for carryback that roughly correspond to 1 cleaner, 2 cleaners, and 3 cleaners or a belt-washing system, respectively.

Table 14.5	**Example Assumptions**			

	Level of Belt Cleaning			
Assumptions	**No Cleaners**	**Level I**	**Level II**	**Level III**
Carryback g/m² (oz/ft²)	500 (1.6)	250 (0.8)	100 (0.3)	10 (0.03)
Carryback Factor	88%	75%	50%	25%
Particle Size µm (in.)	100 (0.004) minus	100 (0.004) minus	100 (0.004) minus	100 (0.004) minus
Width of Belt Cleaned	67%	67%	67%	67%
Belt Speed m/s (ft/min)	1,0 (200)	1,0 (200)	1,0 (200)	1,0 (200)

Definitions

The following definitions are used to provide explanation for the assumptions:

A. Carryback

Carryback is the dry weight of material adhering to the belt after the belt discharges its cargo. The amount of carryback on the belt can be measured using a carryback gage and subsequent laboratory procedures. If the adhesion properties of the bulk material are known, a better estimate of the amount of carryback on the belt can be made *(Reference 14.8)*. With no cleaners engaged, the assumption for this example is that the carryback will be 500 grams per square meter (1.639 oz/ft²) of belt surface cleaned.

B. Carryback factor

The carryback factor is the estimated percentage of carryback that will be dislodged from the belt downstream from the belt cleaners by components such as return idlers and bend pulleys. The cleaner the belt is, the lower will be percentage of carryback that will fall from the belt, because either the remaining carryback is captured in cracks and damages in the belt or the particles have sufficient adhesive strength to remain on the belt surface.

C. Particle size

For this example, it is assumed that 100 percent of the carryback is small enough to become airborne dust. In actual practice, a sieve analysis can be performed to determine the percentage of the carryback particles that are small enough to become potential airborne dust and the results of the calculations multiplied by the percentage of particles less than 100 microns (0.0039 in.).

D. Width of belt cleaned

Only a portion of the belt is in contact with the bulk material, and this is the width that must be cleaned of carryback. The assumption is that the CEMA "rule of thumb" of two/thirds belt width for skirtboard spacing is a reasonable estimate for this variable. The actual width can be measured and used in place of this assumption.

E. Belt speed

Belt speed is the speed of the belt in meters per second (ft/min). A value of 1.0 meters per second (200 ft/min) is used for this example. Results for other belt speeds can be extrapolated by multiplying by the actual belt speed in meters per second (ft/min). The same extrapolation is possible with the time period (minutes to hours, days, weeks, etc.) and belt width, because the relationship to carryback quantities generated is linear.

Equation

Calculation can determine the amount of potential dust generated from carryback (**Equation 14.3**). Additional calculations can be made for various belt widths and levels of cleaning performance, with the rest of the variables remaining the same (**Figure 14.80**). As can be seen for the example above, the reduction in potential dust is 89 percent by installing a system that will achieve Level II cleaning. Typically, a system that will achieve a Level II cleaning performance consists of at least one precleaner and one secondary cleaner, both properly selected and sized for the application. The dust loading can be determined by calculating the amount of air flowing through the transfer point per minute and adding the dust generated per minute *(Reference 14.9)*.

Conclusion

A major source of dust is conveyors that are allowed to run empty for long periods of time. As the belt runs empty, the carryback dries out and is more easily released into the environment by contact with components such as return idlers and bend pulleys. By reducing carryback, engineered belt-cleaning systems reduce the potential amount of dust released into the conveyor system and the environment significantly.

14

The proper installation and maintenance of belt cleaners inside the containment housing is critical in maintaining cleaning effectiveness and mitigating dust genera- tion. If actual measurements are available for the critical variables, a reasonable engi- neering estimate of the dust released by the belt after belt cleaning can be calculated.

14

Equation 14.3

Calculating Potential Dust Generated

$DG = BW \cdot Cb_f \cdot DI \cdot BS \cdot WC \cdot Cb \cdot k$		
Given: A belt 1500 millimeters (60 in.) wide is carrying material at a speed of 1,0 meter per second (200 ft/min). The carryback factor is 88 percent, the dust index is 100 percent, and the width cleaned is 67 percent. The carryback level for no cleaners is 500 grams per square meter (1.639 oz/ft²). *Find:* The potential dust generated.		

	Variables	Metric Units	Imperial Units
DG	Dust Generated	kilograms per minute	pounds-mass per minute
BW	Belt Width	1500 mm	60 in.
Cb_f	Carryback Factor	0,88 (88%)	0.88 (88%)
DI	Dustiness Index	1,0 (100%)	1.0 (100%)
BS	Belt Speed	1,0 m/s	200 ft/min
WC	Width Cleaned	0,67 (67%)	0.67 (67%)
Cb	Carryback	500 g/m²	1.639 oz/ft²
k	Conversion Factor	0,00006	0.00521

Metric: $DG = 1500 \cdot 0,88 \cdot 1 \cdot 1 \cdot 0,67 \cdot 500 \cdot 0,00006 = 26.5$		
Imperial: $DG = 60 \cdot 0.88 \cdot 1 \cdot 200 \cdot 0.67 \cdot 1.639 \cdot 0.00521 = 60.4$		

DG	Dust Generated	26,5 kg/min	60.4 lb$_m$/min

Figure 14.80

Potential dust generated for different belt widths and cleaning levels.

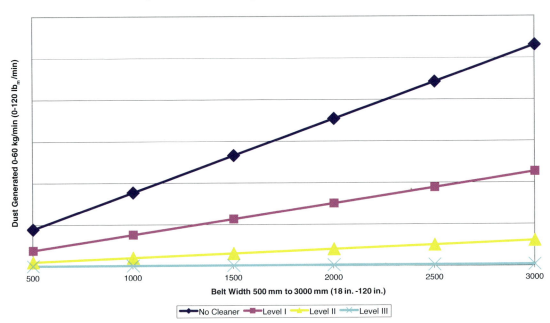

Potential Dust Generated
Assuming a dust index of 1, a belt speed of 1 m/s (200 ft/min), a width cleaned of 67%, and carryback factors and carryback amounts taken from Table 14.5

SAFETY CONCERNS

Failure to install appropriate belt-cleaning systems or maintain them is frequently the root cause of many of the accidents that occur when personnel are cleaning around moving conveyors.

Great care must be taken when observing or inspecting belt-cleaner systems. It is recommended that only trained and qualified personnel install and maintain belt-cleaning and their related systems. The manufacturer's manual usually provides important information, and industry groups such as CEMA provide safety information and standardized warning signage.

A pre-job safety analysis, commonly called a JSA, should be completed prior to the installation or maintenance of belt cleaners or the cleanup of any accumulation of fugitive material. Some topics to cover in a JSA are:

A. Lockout / tagout / blockout / testout procedures must be followed.

B. No maintenance or adjustment procedures should be attempted while the conveyor is running without strict compliance with national, state, local, and in-house safety regulations.

C. Belt cleaners installed in elevated positions or over hazardous areas, such as a barge load-out over a river, may require special precautions such as fall-protection systems.

D. Maintenance of belt cleaners in enclosed areas requires that confined-space procedures be followed.

E. Chutes often contain equipment, such as flow-aid devices and samplers, which can start automatically. All devices must be locked out / tagged out / blocked out /tested out separately from the main conveyor drive to prevent injury to employees.

F. Belt cleaners are often located in areas subject to accumulations of fugitive materials and accumulations of grease, water, or trash that can create slip and trip hazards. These accumulations should be removed before beginning maintenance procedures.

G. Warning stickers that are no longer readable should be replaced.

H. Manufacturer recommended and required local safety procedures must be followed when performing maintenance on belt-cleaning systems.

14

BENEFITS OF CARRYBACK CONTROL

In Closing...

Belt cleaners come in a variety of types and materials and must be selected to fit the application and material conditions. For effective cleaning, selection should be accomplished by specialists experienced with their design and the characteristics of the material to be conveyed.

When properly selected, installed, and maintained, engineered belt-cleaning systems can be effective in reducing carryback. Reducing spillage and dust, in turn, reduces maintenance and accidents, many of which occur when workers are cleaning up fugitive materials. Reducing carryback also protects the belt and conveyor components from damage, extending their service-life and preventing further spillage caused by belt wander.

Looking Ahead...

This chapter about Belt Cleaning, the first chapter in the section Return Run of the Belt, discussed ways to remove carryback to prevent fugitive material from falling from the belt on the return run. The following two chapters, Pulley-Protection Plows and Belt Alignment, continue this section and describe additional methods to reduce spillage.

REFERENCES

14.1 Swinderman, R. Todd and Lindstrom, Douglas, Martin Engineering. (1993). "Belt Cleaners and Belt Top Cover Wear," *National Conference Publication No. 93/8*, pp. 609–611. Paper presented at The Institution of Engineers, Australia, 1993 Bulk Materials Handling National Conference.

14.2 Rhoades, C.A.; Hebble, T.L.; and Grannes, S.G. (1989). *Basic Parameters of Conveyor Belt Cleaning*, Report of Investigations 9221. Washington, D.C: Bureau of Mines, US Department of the Interior.

14.3 Planner, J.H. (1990). "Water as a means of spillage control in coal handling facilities." In *Proceedings of the Coal Handling and Utilization Conference: Sydney, Australia*, pp. 264–270. Barton, Australian Capital Territory, Australia: Institution of Engineers, Australia.

14.4 Swinderman, R. Todd, Martin Engineering. (2004). "Standard for the Specification of Belt Cleaning Systems Based on Performance." *Bulk Material Handling by Conveyor Belt 5*, pp. 3–8. Edited by Reicks, A. and Myers, M., Littleton, Colorado: Society for Mining, Metallurgy, and Exploration (SME).

14.5 Martin Supra Engineering. (2008) *CarrybackTtest/Sum/SBM-001-SBW-05-2008*. Unpublished report for P.T. Martin Supra Engineering: Newmont, Indonesia.

14.6 Swinderman, R. Todd, Martin Engineering. (May 1991). "The Conveyor Drive Power Consumption of Belt Cleaners," *Bulk Solids Handling*, pp. 487–490. Clausthal-Zellerfeld, Germany: Trans Tech Publications.

14.7 Wood, J. P. (2000). *Containment in the Pharmaceutical Industry*. Informa Health Care.

14.8 Roberts, A.W.; Ooms, M.; and Bennett, D. *Conveyor Belt Cleaning – A Bulk Solid/Belt Surface Interaction Problem*. University of Newcastle, Australia: Department of Mechanical Engineering.

14.9 Swinderman, R. Todd; Goldbeck, Larry J.; and Marti, Andrew D. (2002). *Foundations 3: The Practical Resource for Total Dust & Material Control*. Neponset, Illinois: Martin Engineering.

14

14

15

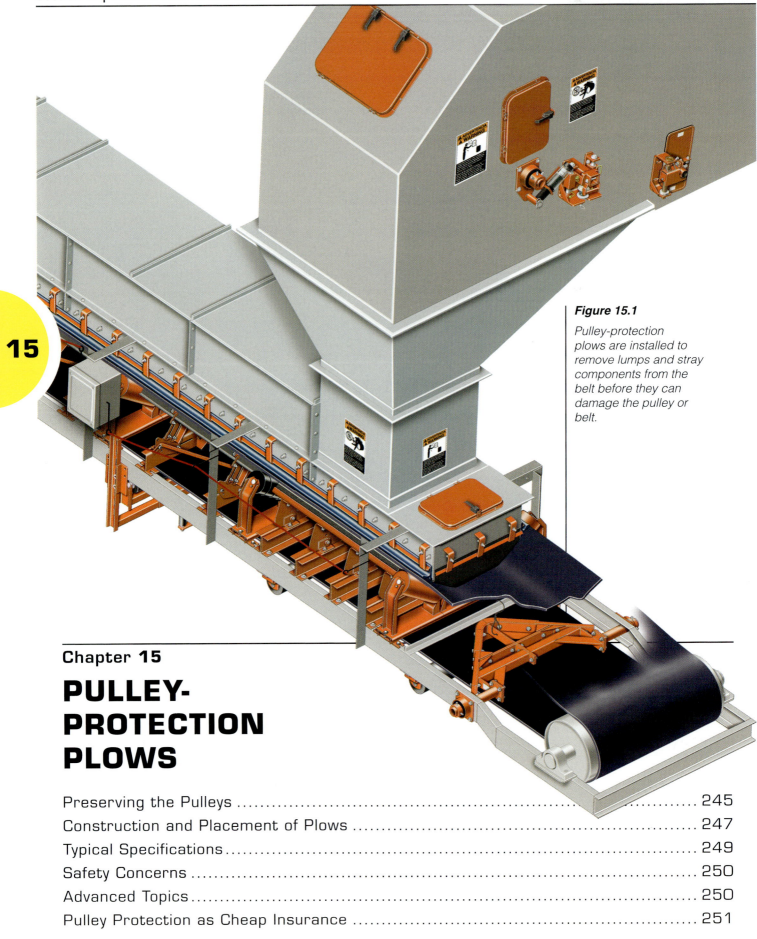

Figure 15.1

Pulley-protection plows are installed to remove lumps and stray components from the belt before they can damage the pulley or belt.

Chapter 15

PULLEY-PROTECTION PLOWS

In this Chapter...

This chapter examines the use of pulley-protection plows as a form of low-cost "insurance" against damage to the belt and pulley. It looks at the need for such plows and the damage that can occur without them along with considerations to keep in mind when selecting and placing pulley-protection plows.

Pulley-protection plows are devices that block any large lumps or stray conveyor components, such as idler rollers, belt-cleaner blades, or other tramp iron, from becoming trapped between the belt and the tail or other pulley(s) where they can damage the pulley or belt (**Figure 15.1**). Protection plows, while not designed as belt cleaners, can remove fugitive materials with a simple, low-pressure scraping that directs the material off the return belt much like a snowplow cleans a road.

As the conveyor belt returns from its discharge point (typically the head pulley) to its loading zone, it will pass over a number of pulleys. These return-side rolling components include the take-up pulley, the snub pulley(s), and, right before the belt reaches the loading zone, the tail pulley. Occasionally along its return run, the belt will collect and carry a lump of spilled material, tramp iron, or even a stray conveyor component to the tail pulley on the non-carrying side of the belt. If these objects are not removed from the belt, they can become trapped between and damage the pulley and the belt. This is why pulley-protection plows—commonly called tail plows—are installed near the tail of the conveyor, their most prevalent mounting location (**Figure 15.2**).

PRESERVING THE PULLEYS

Threats to the Pulley and the Conveyor

The entrapment of anything between the belt and the pulley can do significant damage to a conveyor system (**Figure 15.3**). When fugitive material is trapped between

the belt and the pulley, one or more failures are likely to occur:

A. Degradation of the fugitive material

If the material fails, it will break up into fines and be carried between the belt and the pulley. Material trapped in this location can allow the belt to slip against the pulley, causing the non-carrying underside of the belt to wear. Even small particles and fines can wear and grind away on the less durable, more easily-damaged inside surface of the belt. Furthermore, material that builds up on tail pulleys will cause belt wander that in turn can damage the belt edge and/or the conveyor structure.

B. Failure of the belt

Any material entrapped between the pulley and the belt has the potential for forcing its way out through the top cover of the belt, particularly if the material is a lump with sharp edges. This material creates an uneven belt surface and can be a starting point for longitudinal and profile rips, holes, or edge gouges along the length of the belt.

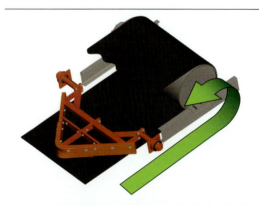

Figure 15.2

The most common location for pulley-protection plows is at the tail of the conveyor.

Figure 15.3

Entrapment of material between the pulley and the belt can damage the pulley and/or the belt.

15

15

C. Failure of the pulley

> If the material and the belt do not fail, the face of the tail pulley is likely to be damaged. A damaged pulley will lead to belt misalignment or damage and pulley slippage.

The most damaging problem arising from the entrapment of material between the belt and tail pulley is the fact that it can become a repeating phenomenon. Once a piece of material reaches the pulley, it can be pinched between the belt and pulley, carried around the pulley's rotation, and then ejected back onto the return side of the belt. Once there, it will again travel toward the pulley to be consumed again (**Figure 15.4**). In essence, if it initially fails to break something, the lump will keep trying until a failure occurs or the lump is removed from the belt. If the material is strong enough, it could destroy the entire tail-pulley section of a conveyor and damage the belt.

Avoiding Pulley Damage

In a conveyor system, where stability is a key to the control of fugitive material, any damage to the belt or pulley can adversely affect the system's performance. By eliminating possible sources of damage to the conveyor, the entire system is improved and the risks of dust and spillage are drastically reduced.

The basic protection against this trapping of material between the belt and pulley is control of loading. The correct trajectory and drop height for material, along with the relationship between the speed of the loading material and the speed of the moving belt, are factors that help settle the load, reduce agitation, and minimize material spillage. Maintaining proper alignment of the belt is also required to reduce spillage that can drop material onto the return side of the belt.

An additional method available for preventing cargo from falling onto the belt return is to enclose or cover the return belt with decking. On long conveyors, this can turn into an expensive proposition; therefore, decking is rarely applied at locations other than near the loading zone. Even with decking applied to the full run of a conveyor, material could accumulate on top of the decking and eventually spill onto the return run of the conveyor, creating the need for a pulley-protection device.

With ideal installations, and regardless of other precautions, there is still the possibility that lost components or conveyed material will spill onto the inside of the belt. Consequently, there is a need for a system to prevent these items from damaging the conveyor's rolling components. These pulley-protection plows are most commonly installed at the tail pulley, but depending on the characteristics of the specific material and the individual conveyor, they may also be useful to protect the take-up or other pulleys (**Figure 15.5**).

Figure 15.4

Top: A lump captured between the pulley and the belt can be carried around the pulley and ejected back onto the belt to become captured again.

Bottom: A pulley-protection plow will remove lumps from the belt to prevent their capture.

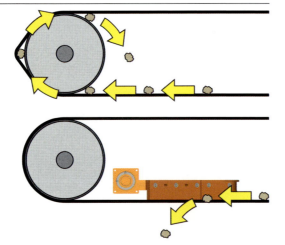

Figure 15.5

Usually installed at the tail pulley, plows can also be installed to protect the take-up or other pulleys.

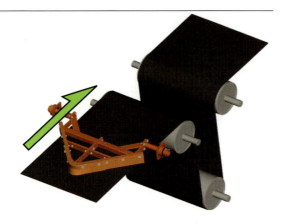

This is a Job for a Plow

A pulley-protection plow removes fugitive materials with a simple, low-pressure scraping that directs the material off the belt much like a snowplow cleans a road. Instead of cleaning fines off the belt, the primary mission of a plow is to block any large lumps or stray conveyor components, such as idler rollers, belt-cleaner blades, or other tramp iron, from entering the tail pulley where they can damage the belt (**Figure 15.6**).

A plow that is installed slightly above the belt has the potential to capture a lump of material against the belt and, therefore, risks surface abrasion and damage with the possibility of ripping the belt. Pulley-protection plows are usually designed to float on the belt's surface, using either the weight of the plow or a tensioning mechanism to hold the plow with a slight pressure, 13 to 20 kilopascals (2 to 3 lb_f/in.^{2}), against the belt. These plows are built of heavy-duty construction and tall enough to keep fast moving materials from going over the top of the plow.

More than Lumps?

If the belt is carrying significant quantities of fines or slime on its inside surface, an additional step of providing a return belt-cleaning system should be taken. At the expense of consuming additional conveyor drive power, this system would provide effective removal of material, reducing the risk of belt slip and material accumulation on the pulley.

A pulley-protection plow utilized for cleaning fines from the belt should be located in a position, such as right under the load point, to scrape the spillage off the belt and discharge the material where it can be easily collected. Care should be taken when discharging fugitive materials in the area of the tail pulley, as it can create a number of other problems, including material buildup below the conveyor. As with any belt cleaner, removed material that piles up under the conveyor can lead to premature wear of the belt top cover.

CONSTRUCTION AND PLACEMENT OF PLOWS

Plow Construction

Pulley-protection devices are usually designed as a linear or V-shaped plow using a steel frame with a rubber, urethane, or plastic blade that directs any fugitive material off the belt. To prevent large lumps from "jumping" the plow and becoming hung up in the plow's suspension, in the conveyor structure, or between the belt and pulley, plows should stand as tall as the largest lump conveyed, with a minimum height of 100 millimeters (4 in.). In the case of high-speed belts, it may be advantageous to increase the height of the plow to one-half the total height of the pulley it is protecting. It is beneficial to cover the "interior space" of the plow to prevent material from becoming caught in the plow itself.

The plow should include a safety cable, which would attach to a point above and in front of its leading edge. In the event of a mounting failure, this cable would prevent the plow from traveling into the pulley and causing the damage the plow is trying to prevent.

On belt conveyors that travel in one direction only, the return belt cleaner is usually a "V"-plow (**Figure 15.7**). The point of the "V" is toward the head pulley so that any loose material carried on the belt's inside surface would be deflected off the conveyor by the wings of the plow.

If the belt has a reversing operation or suffers significant rollback, the installed device should be a diagonal plow that pro-

Figure 15.6

A V-plow is installed so the point of the "V" is toward the head pulley, deflecting belt-borne material off the belt with a low-pressure scraping action.

vides cleaning protection in both directions (**Figure 15.8**). Diagonal plows are normally installed across the belt at an angle of 45 degrees to the direction of travel (**Figure 15.9**). If the belt operates in two directions, where either pulley can serve as the tail pulley, then a diagonal plow should be installed at each end of the conveyor.

Placement of Pulley Protection

Plows should be carefully located so that the material removed from the belt does not create a hazard as it falls or where it accumulates.

Just as it is important to have a roller above a secondary belt cleaner that provides downward pressure to keep the cleaner from pushing the belt up, it is important to have one or two pressure rollers below

Figure 15.7

A V-plow is used on belts that travel in only one direction.

Figure 15.8

On reversing conveyors, a diagonal plow will provide cleaning protection in both directions of operation.

Figure 15.9

Diagonal plows are installed across the belt at an angle of 45 degrees to the direction of travel.

the pulley-protection plow installation. In this case, the mission is to prevent the plow from changing the belt line by pushing the belt down so that material can pass underneath the blade. Depending on the space available, this can be a single idler roller placed directly under the plow or a pair of return idlers, one installed before the plow and one after.

Like any other conveyor component that will be in contact with the belt, the installation of a pulley-protection device will increase the friction against the moving belt. Consequently, this drag will increase the conveyor's drive power requirements.

In the sixth edition of *BELT CONVEYORS for BULK MATERIALS*, the Conveyor Equipment Manufacturers Association (CEMA) offers a recommended setting of 2 pounds-force per inch of belt width as the normal force for plow-to-belt pressure. (The metric equivalent is 0,35 newtons per millimeter of belt width.) This pressure can be converted into power consumption using formulas (**Equation 15.1**).

Selection Considerations for Pulley Protection

When specifying a pulley-protection device, there are a number of factors that should be considered. A plow should:

A. Provide firm but flexible pressure

Firm but flexible pressure will allow the device to clean the belt surface. The intent of the device is to remove material effectively and efficiently yet adjust automatically to accommodate for the wear of its blade and fluctuations in belt movement, speed, and path.

B. Be securely mounted

The plow must be firmly mounted in order to minimize the risk of it breaking away from its installation to endanger the conveyor components it was installed to protect. The installation should include a safety cable to protect the conveyor system should the plow installation fail (**Figure 15.10**).

$P = BW \cdot f_c \cdot V \cdot f \cdot k$

Equation 15.1

Power Consumption of a Pulley-Protection Plow

Given: *A urethane plow on a 900-millimeter (36-in.) belt traveling at 3 meters per second (600 ft/min).* **Find:** *The power added to the drive due to the plow.*

	Variables	Metric Units	Imperial Units
P	Power Consumption Added to Belt Drive	kilowatts	horsepower
BW	Belt Width	900 mm	36 in.
f_c	Load per Belt Width (Per CEMA)	0,35 N/mm	2 lb_f/in.
V	Belt Speed	3 m/s	600 ft/min
f	Friction Coefficient *(Per CEMA STANDARD 575-2000)*	0,5 (UHMW) 1,0 (Urethane) 1,0 (Rubber)	0.5 (UHMW) 1.0 (Urethane) 1.0 (Rubber)
k	Conversion Factor	1/1000	1/33000

$$\text{Metric: } P = \frac{900 \cdot 0,35 \cdot 3 \cdot 1}{1000} = 0,945$$

$$\text{Imperial: } P = \frac{36 \cdot 2 \cdot 600 \cdot 1}{33,000} = 1.3$$

P	Power Consumption Added to Belt Drive	0,945 kW	1.3 hp

C. Be designed for ease of installation

The plow should be easy to install to minimize downtime of the system during the installation procedure. For example, the device should fit within the conveyor structure without requiring extensive modifications to the device or the structure.

D. Be designed with a durable, easily replaceable blade

In order to provide a long service-life and allow fast maintenance, the blade should be fabricated of a material suited to withstand application conditions, and it should be attached so it can be easily removed and replaced when worn.

E. Be readily accessible

The plow should be installed in an area where it can be observed during operation and easily maintained.

TYPICAL SPECIFICATIONS

One or more low-pressure pulley-protection plows should be positioned on the return side of the belt to remove fugitive material before it can become entrapped between the belt and a rolling component according to the following specifications:

A. Flexible pressure

If the device is designed to contact the belt surface, the design should allow the plow or cleaner to "float" across the belt surface with a firm but flexible pressure.

B. Safety cable

The device must be fitted with a safety cable to protect the belt and pulley should an unexpected mount failure occur.

Figure 15.10

A safety chain provides protection in the event of a failure of the plow mount.

C. Replaceable blade

The design should incorporate an easily replaceable blade of rubber, plastic, or urethane.

D. Full belt coverage

The plow blade should provide full coverage of the belt to prevent material lumps from "slipping around" the plow on the outside.

E. Location

The plow should be located so that the material cleaned from the belt can be safely ejected from the conveyor without hitting the stringer, other components, or walkways; it should be in a location that is safe and convenient for cleanup.

F. Unidirectional conveyors

On unidirectional conveyors, a "V"-plow should be installed between the last return idler and the tail pulley. Additional devices might be required to protect other pulleys or clean the bottom cover of the belt.

G. Reversing conveyors

On reversing conveyors, diagonal return plows should be installed on both ends of the conveyor and mounted across the belt at a 45-degree angle.

ADVANCED TOPICS

The impact caused by a lump of material or other objects carried on the belt can be quite large if conditions such as high belt speeds and large lump sizes are present. These large impact forces must be considered when selecting equipment, especially in view of the continuing demand to increase belt speeds.

The only variable controllable by the designer of a plow is the spring constant (k) in that component. This variable is the ability of the plow blade to absorb, cushion, or deflect a moving lump without damage. In the same way that dropping an egg onto a mattress (rather than a concrete floor) reduces impact force, the use of "softer" materials in blades and the incorporation of springs or other flexible elements in the plow mounting increases the chance the plow will withstand the lump's impact forces. As an example, a lump of material weighing 2,25 kilograms (5 lb_m) traveling on a belt moving 3 meters per second (600 ft/min) will strike a plow with a force of 815 newtons (183 lb_f). However, if that lump traveling at the same speed hits a plow equipped with a blade that has twice the impact absorption properties, it will strike with a force of only 199 newtons (45 lb_f). This reduced force of impact results in

SAFETY CONCERNS

Because tail-protection plows are positioned on the return run or non-carrying side of the belt and near the tail pulley, they are often in enclosed and nearly inaccessible locations. This makes them difficult to service and even poses a safety risk for personnel performing an inspection.

Safety concerns are paramount when the conveyor is in operation. It is important that great care be taken to avoid entanglement in rotating equipment when performing an inspection. No service procedures should be attempted while the conveyor is running. Proper lockout / tagout / blockout / testout procedures should be practiced prior to working on conveyors and/or their components to ensure the belt does not move.

Warning signs should be placed at all plow locations indicating pinch points. Care should also be taken, as objects can be thrown from the belt by the plow.

15

lower requirements for the strength of the plow, which in turn reduces the cost of the equipment.

An understanding of these impact forces, combined with new developments in design, allows an operation to more closely match the application to the design of the pulley-protection plow. This allows a more cost-effective selection of a protection system that meets performance requirements. A plow manufacturer should be able to calculate the impact forces for applications and determine the appropriate pulley-protection plow. There may be heavier objects that can strike the plow, such as teeth from loader buckets and fallen conveyor rollers, but the vast majority of impacts would be within the strength capabilities of the plow.

PULLEY PROTECTION AS CHEAP INSURANCE

In Closing...

While most pulley-protection devices are fairly simple devices, some innovations demonstrate the advantages of using engineered systems rather than homemade scrapers. Through innovations in design and construction, pulley-protection plows are available to provide the benefits of an engineered system while minimizing the initial investment (**Figure 15.11**). These engineered systems can provide a long-term solution that provides savings through improved performance, extended service-life, and reduced maintenance expenses rather than the false economies of the homemade unit. With today's use of computer-aided design systems to develop new conveyors, engineered plows can be positioned during an early phase of the conveyor's engineering process. Engineered systems ensure there is space to install, operate, inspect, and maintain a pulley-protection plow.

Figure 15.11

Engineered pulley-protection plows provide savings through improved performance, extended service-life, and reduced maintenance expenses.

Installed between a pulley and the nearest return idler, pulley-protection devices represent a form of low-cost "insurance" when weighed against the out-of-pocket costs of conveyor maintenance, damage, and possible premature replacement of the belt and/or pulley.

Looking Ahead...

In looking at methods to control dust and spillage, two topics have been addressed that relate to the Return Run of the Belt: Belt Cleaning and, in this chapter, Pulley-Protection Plows. The third and final chapter in this section will address Belt Alignment.

REFERENCES

15.1 Conveyor Equipment Manufacturers Association (CEMA). (2005). *BELT CONVEYORS for BULK MATERIALS, Sixth Edition*. Naples, Florida.

15.2 The website http://www.conveyorbeltguide.com is a valuable and non-commercial resource covering many aspects of belting.

15.3 Any manufacturer and most distributors of conveyor products can provide a variety of materials on the construction and use of their specific products.

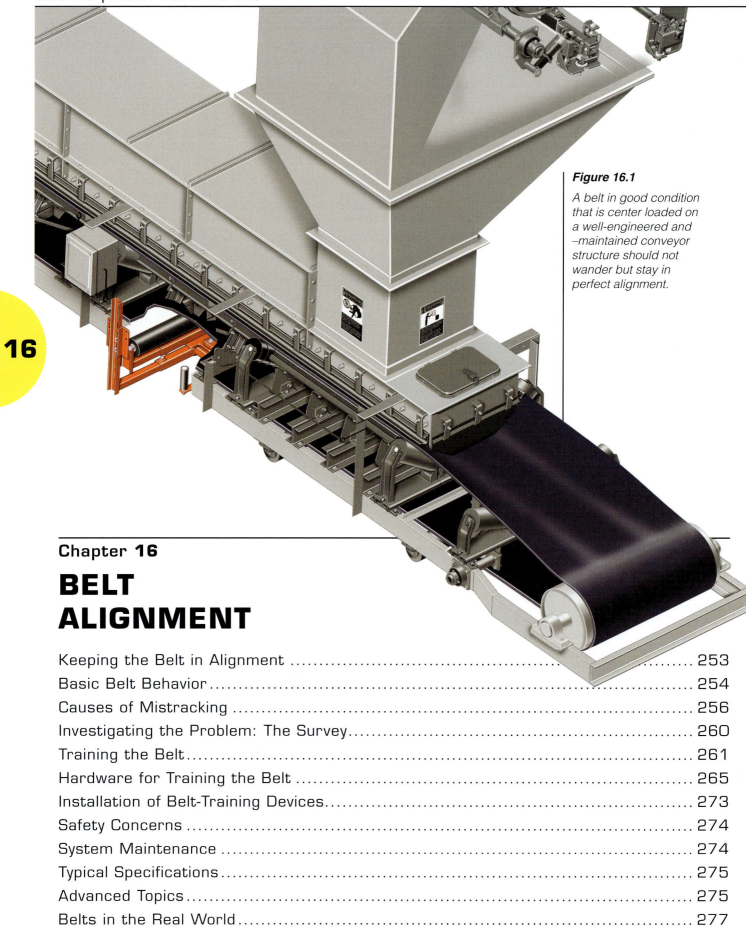

Figure 16.1

A belt in good condition that is center loaded on a well-engineered and –maintained conveyor structure should not wander but stay in perfect alignment.

16

Chapter **16**

BELT ALIGNMENT

In this Chapter...

In this chapter, we focus on belt alignment and its relationship to fugitive materials: causes of mistracking as well as techniques to train the belt. We also discuss uses of belt-training hardware and installation of devices for belt training. Finally, equations used to calculate power consumption for belt-trainers are provided.

In an ideal world, a belt would be in good condition and center loaded and the conveyor structure would be well-engineered and -maintained; under those conditions, the belt would not wander but stay in perfect alignment (**Figure 16.1**).

However, belts that wander from the desired path are an everyday fact of life in many bulk-materials handling operations. A conveyor belt that wanders can cause material spillage, component failure, and costly damage to the belt and structures (**Figure 16.2**). A belt that runs to one side of the structure can greatly reduce its service-life, because it abrades one or both edges, becomes stretched, or folds over on itself (**Figure 16.3**). A wandering belt can run against steel chutes and structural members until the belt, other components, and steel structures are damaged, often beyond repair (**Figure 16.4**). Worse yet, belt-wander problems have contributed to fatalities.

In many ways, proper belt alignment is a precursor to, and a fundamental requirement for, resolving many of the fugitive-material problems discussed in this book. In this chapter, we will discuss many of the problems that cause a belt to wander and suggest solutions.

KEEPING THE BELT IN ALIGNMENT

Many terms are used when discussing the topic of belt wander. The terms tracking and training are often used interchangeably, as are their counterparts wander and misalignment. Here, training is defined as a procedure to make the conveyor belt track (or travel) on the centerline of the conveyor structure, both empty and fully loaded. Wander and mistracking can be defined as the tendency of the belt centerline to move from the conveyor structure's centerline, and misalignment is the amount that the belt wanders.

Belt tracking must be controlled before spillage can be eliminated; if the belt wanders to one side, or back and forth, as it passes through the loading zone, material is more readily released under the skirtboard seal on either (or both) sides (**Figure 16.5**). Belt mistracking is managed by "training

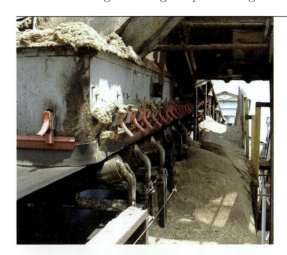

Figure 16.2

A conveyor belt that wanders out from underneath the sealing system can cause material spillage, component failure, and costly damage to its belt and structures.

Figure 16.3

A belt that runs to one side of the structure can greatly reduce its service-life, because it abrades one or both edges, becomes stretched, or folds over on itself.

Figure 16.4

A wandering belt can run against steel chutes and structural members until the belt, other components, and steel structures are damaged, often beyond repair.

16

the belt" and by installing components designed to limit or correct wander.

When a belt mistracks, it can cause large amounts of material spillage. These spillage piles can cause trip hazards. If an overhead conveyor mistracks, it can rain material of any size down on workers. The potential for worker injury and all of the associated costs would indicate that it is in an operation's best interest to solve the problem of mistracking belts (**Figure 16.6**).

BASIC BELT BEHAVIOR

Despite all of its various causes, mistracking is still unnecessary. It is a problem that can be controlled or, better yet, corrected. Understanding the basic patterns of belt behavior and undertaking a set of procedures to carefully align the conveyor's structure and components to correct fluctuations in the belt's path can, in most cases, prevent belt wander.

Belt behavior is based on simple principles. These serve as the guidelines for belt training, which is the process of adjusting conveyor structure, rolling components, and load conditions to correct any tendency for the belt to run off-center.

The fundamental rule of conveyor belt tracking is this: The belt will move toward the side that has more friction, or the side that reaches the friction first (**Figure 16.7**). When a side of the belt encounters that friction, that side of the belt moves slower. The belt's other side moves faster; a force imbalance occurs, which pivots the belt toward the slower moving side.

For example, if an idler set is installed at an angle across the stringers, the belt will move toward the side it reaches first. If one end of the idler set is higher than the other, the belt will climb to the higher side (because, as the belt is laid down on top of the idlers, it touches the higher side first).

16

Figure 16.5

Belt tracking must be controlled before spillage can be eliminated.

Figure 16.6

When a belt mistracks, it can cause large amounts of material spillage.

Figure 16.7

The fundamental rule of conveyor belt tracking is that the belt moves toward the high-friction side of the belt.

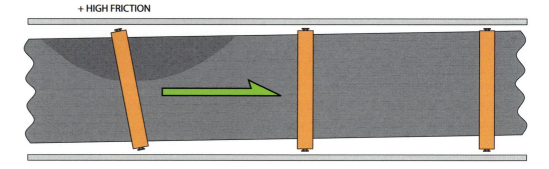

This can be demonstrated very simply by laying a round pencil on a flat surface, such as a table. If a book is laid across the pencil and gently pushed away from the experimenter, the book will shift to the left or right depending upon which end of the pencil is closer to the person doing the pushing: that is, the end the book is contacting first (**Figure 16.8**). This basic rule is true for both flat idlers and troughed idler sets.

In addition, troughed idlers exert a powerful tracking force. With their troughed configuration, a portion of each belt edge is held aloft. A gravitational force is exerted on that raised portion. If the belt is not centered in the set of rollers, the force on the higher edge will be greater than the force on the other side, steering the belt toward the center of the troughed idler set. This gravitational tracking force is so pronounced that bulk conveyors usually depend upon it as their major tracking influence.

Another constant rule of belt tracking is that the tracking of the belt at any given point is more affected by the idlers and other components upstream (the places the belt has already passed) than the components downstream (which the belt has not yet reached). This means at any point where mistracking is visible, the cause is at a point the belt has already passed. Consequently, corrective measures should be applied some distance before the point where the belt shows visible mistracking (**Figure 16.9**).

With these basic rules in mind, operators and maintenance personnel can make the adjustments to the conveyor that will bring the belt path into alignment.

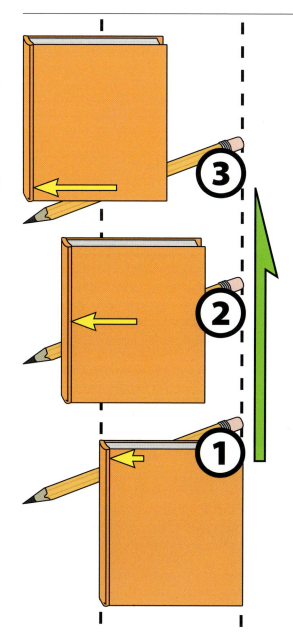

Figure 16.8

The basic rule of belt training can be demonstrated by laying a book on top of a round pencil. When pushed away, the book will shift to the left or right depending upon which end of the pencil is closer to the person doing the pushing: that is, the end the book is contacting first.

16

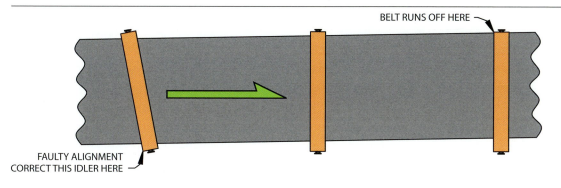

BELT RUNS OFF HERE

FAULTY ALIGNMENT
CORRECT THIS IDLER HERE

Figure 16.9

Because mistracking takes place after the point of cause, corrective measures should be applied some distance before the point where the mistracking is visible.

CAUSES OF MISTRACKING

The Avoidable Problem of Belt Wander

To properly train a conveyor, the first step is to survey the existing system to understand the state of the structure and components and to determine the causes of mistracking.

As Clar Cukor noted in the undated Georgia Duck (now Fenner Dunlop) monograph *Tracking (Reference 16.1)*:

> The problem of tracking should be approached from a systems point of view. The belt may well be at fault—however, it is more likely merely reacting to a structural defect or maladjustment in the system....[A conveyor belt] is flexible and if designed, manufactured, slit and cut properly, will "go where directed" by the conveyor system as designed and built. The conveyor belt serves as an indicator and should be so regarded.

Belt wander can be caused by a number of problems. Factors contributing to belt wander include misalignment of conveyor components, off-center loading of cargo, accumulation of fugitive material on rolling components, poor belt splices, structural damage caused by inattentive heavy-equipment operators, ground subsidence, and many others. And these problems may occur in any combination, greatly complicating the process of correction.

In spite of the complexity of these problems, they are solvable. Misaligned components can be straightened, chutes can be redesigned to load the cargo in the center of the belt, material accumulations can be prevented or removed, belt splices can be improved, and operators can be trained. The challenge comes in identifying which of the long list of possibilities is the specific cause of a given belt's problems. Once the cause of mistracking is identified, it can be corrected.

Causes of Mistracking

In many cases, the cause of mistracking can be determined from the form the mistracking takes. When all portions of a belt run off-center at one certain part of the conveyor length, the cause is probably in the alignment or leveling of the conveyor structure, idlers, or pulleys in that area. If one or more sections of the belt mistrack at all points along the conveyor, the cause is more likely in the belt construction, in the splice(s), or in the loading of the belt. If the belt mistracks when full and then tracks in the center when empty or vice versa, the cause is usually off-center loading or buildup in the chute that creates varying loading situations.

The most common causes of mistracking can be split into three groups: faults with the belt or its splices; faults with the conveyor structure, components, or the environment; and faults with material loading.

Faults with the Belt or its Splices

A. Belting
   a. The belt is bowed, cambered, or cupped.
   b. There are defects or damage in the carcass (plies or cords) of the belt.
   c. The belt edge or cover is damaged.
   d. There is belt degradation from exposure to the elements or to chemicals.

B. Manufacturing and application
   a. The belt is poorly matched to the structure or application.
   b. The belt has a "bow" or "camber" from its manufacturing process.
   c. The belt was not stored properly.

C. Splices
   a. There was poor installation of a vulcanized or mechanical splice, resulting in a splice that is not square to the belt.
   b. A belt was formed from several pieces joined at the wrong ends, resulting in a camber or crooked section.

16

c. Different types, thicknesses, or widths of belt have been spliced together.

d. The belt has splices that are damaged or coming apart.

Faults with the Conveyor Structure, Components, or the Environment

A. Structure

   a. The structure was not accurately aligned during its construction.

   b. The structure has settled on one side through ground subsidence.

   c. The structure has been damaged from plugged chutes, fires, or collisions with mobile equipment.

B. Components

   a. Rolling components (idlers and pulleys) are not aligned in all three axes.

   b. The gravity take-up is misaligned.

   c. Idler rolls have seized or been removed.

   d. Material buildup or wear has altered the profile of idlers or pulleys.

C. Environment

   a. The conveyor is subjected to high winds.

   b. Rain, frost, or ice and snow buildup altered the friction on one side of the belt.

   c. The sun shines on one side of the conveyor.

Faults with Material Loading

A. The load is not centered on the receiving belt.

B. The load is segregated, with larger lumps on one side of the belt.

C. There is intermittent loading on a belt that is tracked for a constant load.

Sometimes, a combination of these problems will produce belt wander, and the root cause will not be evident. However, if a sufficient number of belt revolutions are observed, the belt's running pattern generally will become clear, and the cause of mistracking will be disclosed. When a pattern does not emerge, the usual causes for belt mistracking are an unloaded belt that does not trough well or a belt that is unevenly loaded.

Wander Due To Faults with Splices or the Belt

Improper belt splicing is a significant cause of mistracking. If the belt is not spliced squarely, the belt will wander back and forth on the conveyor structure. This can usually be seen at the tail pulley. The belt will wander the same amount each time the splice reaches the tail pulley, only to return to its original position after passage of the splice. If the splice is bad enough, it can negate all alignment efforts. The solution is to resplice the belt squarely. *(See Chapter 5: Conveyors 101—Splicing the Belt.)*

A second significant cause of belt mistracking is a cupped belt. A cupped belt will track poorly because of differences in friction as it lays in the troughed idlers. Belt cupping is almost always a result of unequal shrinkage between the top and bottom covers of fabric belts. Heat, chemicals, trough angles, and over-tensioning can also cause belt cupping. This problem can usually be avoided by keeping the proper aspect ratio between top and bottom cover thicknesses: usually 3:1 or less. In some cases, the belt will cup as the top cover rubber properties change as the result of aging or exposure to chemicals. A cupped belt is hard to make track consistently, because tracking depends upon the friction between the belt and the rolling components. If the belt is so badly damaged that the contact area is reduced, the ability of the components to keep the belt in line is also reduced.

While manufacturing defects in the belt or failures of components are often blamed for many belt-alignment problems, most of these problems can be traced to improper application of the belt. A belt that is poorly matched to the application will usually track poorly on the structure.

16

16

Wander Due to Structural and Component Problems

To be able to keep the belt running straight, the structure must be properly erected and corrected if damaged. Most structural damage occurs when the conveyor structure is struck by mobile equipment. Structural damage can also occur as the result of corrosion or a settling of the foundations.

It is equally important that the components be properly installed and maintained in relation to the belt for reliable belt travel. One major source of belt wander is gravity take-up systems that are out of alignment or that have too much side-to-side movement, or "slop." The take-up pulley, like all other main pulleys, must remain in alignment with the belt throughout the take-up's travel, or the belt will mistrack.

Rotating components can have a significant mistracking effect on the belt. Rotating components that have become frozen or inoperative due to material buildup or those with material accumulations that alter their circumference can be major contributors to erratic belt tracking. Consequently, transfer points should be engineered, constructed, and maintained to prevent material spillage. An effective multiple-cleaner belt-cleaning system should be installed to prevent material carryback. If necessary, cleaners can be installed to clean snub, take-up, and other pulleys. *(See Chapter 14: Belt Cleaning.)*

Wander Due To Environmental Conditions

Strong winds on one side of the conveyor can provide enough force to move the belt off its center line or even blow the belt off the idlers. The solution is to install retaining rings known as "wind hoops" over the conveyor to keep the belt in place, provide a windbreak on the windward side, or enclose the entire conveyor.

Should rain, ice, or snow be blown onto one side of the conveyor, the result would be a difference in friction on the idlers. This difference may be enough to push lightly-loaded belts off the proper path. Even the difference created when the sun warms one side of a belt in the morning is enough to cause a belt to wander. Here again, the solution would be some form of conveyor cover.

In some cases, the conveyor's design was not sufficiently strong to withstand lateral winds, and the entire conveyor will sway back and forth in high winds. The path of a belt can also be greatly influenced by a slight shift of the take-up pulley due to crosswind.

Wander from Loading Faults

Mistracking that arises from loading problems is generally easy to spot, because the belt will run in one position when loaded and another position when unloaded (**Figure 16.10**). This observation may be confused on older conveyors where years of adjustments performed to "fix" the belt's path have altered the natural track of the belt.

The load's center of gravity will seek the lowest point of the troughing idlers (**Figure 16.11**). When the belt is not center-loaded, the weight of the cargo pushes the belt off-center toward the conveyor's more lightly-loaded side. This can be corrected by proper loading-chute arrangements, or through the use of deflectors, grids, or chute bottoms that can be adjusted to correct the placement of the load on the belt. *(See Chapter 8: Conventional Transfer Chutes.)*

Figure 16.10

A belt that is not loaded in the center will mistrack, running the risk of damage to belt and structure.

Wander on Reversing Belts

Reversing conveyors can be a special source of frustration. When the belt direction is reversed, the tension areas in the belting change location in relation to the drive pulley and loading area(s). Imagine having a conveyor that has a head drive, and at the flip of a switch, it becomes the tail drive. When the top side of the belt is running toward the drive pulley, the tight side of the belt is on top. However, when the belt is reversed and the top side is running away from the drive pulley, the tight side is now on the bottom. The carrying side of the conveyor actually changes from being pulled to being pushed. A belt being pushed is inherently more unstable than a belt being pulled; thus, it is more difficult to train.

This poses especially difficult problems, because all of the components now contribute differently to the tracking problems. The belt may run fine in one direction and wander all over when reversed, because different sets of rollers and pulleys control the steering of the belt. In order to overcome this type of problem, the system should be surveyed to determine which components are out of alignment. Corrections should be made as required to get all rotating components in alignment.

Other problems encountered and aggravated by reversing belts relate to off-center loading, multiple load points, and loading different materials on the same belt. Off-center loading can greatly aggravate tracking problems on reversing belts, especially if the load is applied closer to one end of the conveyor than to the other. This can be corrected by proper loading-chute design and the use of adjustable deflectors, grids, and chute bottoms that can be adjusted to correct the placement of the load on the belt.

Different materials on the same reversing belt can also cause problems. Suppose the belt has been "set" to track with a material with a specific bulk density. Now, reverse the direction of travel and introduce a material with a different bulk density, and all of the previously-applied training adjustments may be wrong. In order to overcome this type of problem, a survey of the structure should be conducted to determine if the take-up pulley counterweight is insufficient or if the components are out of alignment, and corrections should be made as required.

Problems with Traveling Conveyors

Conveyor systems that move (such as bucket-wheel reclaimers, traveling stackers, or tripper belts) are greatly influenced by the rail structure on which they ride. For instance, if one rail is higher or lower at a given point than the parallel point on the other rail, the traveling conveyor can tip or rock (sometimes several millimeters (in.) on tall structures), leading to belt-mistracking problems.

Many times this problem is overlooked when trying to find the cause of belt mistracking and the resultant damage. The "traveler" part of the system might be parked in an area where the rails are level when a survey is performed. The survey results then would show everything to be in alignment; however, when the traveling system is moved to a different location, the belt mistracks, because the supporting structure is not level.

The rail systems must also be checked for parallel alignment. Improper alignment may cause the carrying wheels to "ride up" on the inside or outside of the rail, causing the same effect as one rail point being higher than its opposite counterpart.

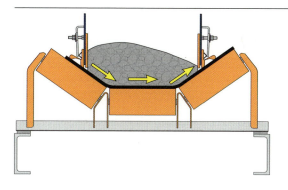

Figure 16.11

When the belt is not center-loaded, the weight of the cargo pushes the belt off-center toward the conveyor's more lightly-loaded side.

INVESTIGATING THE PROBLEM: THE SURVEY

The first, and most important, step in training a conveyor is to check and align the structure. The best way to begin this process is to make a detailed survey of existing conditions and the original design criteria. This allows measured corrections to be made returning the system to original specifications, rather than adopting an unplanned "let's 'tweak' the idlers a little more today" approach.

The traditional method of checking alignment has been to stretch a piano wire from one end of the conveyor to the other and use this wire as a baseline to take the measurements to evaluate alignment. However, this method has a number of potential problems. For example, the wire is vulnerable to shifts in its line. Changes in ambient temperature from the warmth of the sun, or even the actual weight of the wire itself, can stretch the wire, changing the line. Another problem is that there is no accurate way to measure a 90-degree angle from the wire. If the wire moves when touched when laying a ruler or square against it, the accuracy of subsequent measurements is destroyed.

Now, high technology, in the form of beams of light from a laser transit set in parallel to the conveyor structure, provides an unobstructed and repeatable reference for the alignment of the conveyor-structure components (**Figure 16.12**).

This laser-surveying technology avoids the problems encountered with the old "piano wire" technique. The laser generates a perfectly straight beam with an effective range of 150 meters (500 ft), with multiple set-ups allowing unlimited distance. To check objects set at angles to the baseline, prisms can be used to bend the beam. With a laser transit, the survey crew is no longer trying to measure a perpendicular line; they have created one. Since a laser beam cannot be touched, it cannot be moved accidentally when taking readings from it.

Most operations do not have the equipment and expertise to properly conduct a laser survey. Therefore, it is in the best interest of the operation to hire a specialty contractor or service with the hardware and experience to conduct this survey. A specialty contractor will laser survey the belt, inscribe a permanent series of benchmarks or alignment points, create a detailed report, and offer recommendations as to how to correct the major tracking problems.

The report should tell which components are out of alignment and by how much, so the plant maintenance crew or the specialty contractor can adjust these components to improve the belt's tracking (**Figure 16.13**). By doing repeat surveys of the same conveyor at regular intervals—annually, for example—plant management can provide a regular check of the condition of the conveyor structure. The survey will tell if the structure is deteriorating or if other circumstances—such as subsidence of the ground under the conveyor or change in

Figure 16.12

The beam of light from a laser transit provides an unobstructed and repeatable reference for the alignment of the conveyor-structure components.

Figure 16.13

The laser-survey report will tell which components are out of alignment and by how much, so the plant maintenance crew or the specialty contractor can adjust these components to improve the belt's tracking.

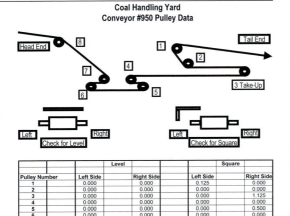

Coal Handling Yard
Conveyor #950 Pulley Data

	Level			Square	
Pulley Number	Left Side	Right Side		Left Side	Right Side
1	0.000	0.000		0.125	0.000
2	0.000	0.000		0.000	0.000
3	0.000	0.000		0.000	1.125
4	0.000	0.000		0.000	0.000
5	0.000	0.000		0.000	0.500
6	0.000	0.000		0.000	0.000
7	0.000	0.000		0.000	0.000
8	0.000	0.000		0.000	0.000

For "Level" a positive number indicates that side is higher.
For "Square" a positive number indicates that side is farther forward.

16

the counterweight mass—are occurring. This information can be used to prevent unexpected shutdowns and subsequent loss of production by alerting the plant's engineering and maintenance staff to problems as they develop.

TRAINING THE BELT

Getting the belt to track in the center of the conveyor's structure and components is a process of adjusting idlers and loading conditions to correct any tendency of the belt to run outside the desired path. The first step is to get the structure into alignment with the belt's theoretical centerline, as identified in the system survey. Once the structure is aligned, all the pulleys and idlers must be aligned so they are level and square to the center line. Then attention can be given to getting the belt to run true.

When training a belt, only one person should be in charge of the procedure. When more than one person adjusts the conveyor at the same time, it can lead to conflicting "corrections" that make the belt's path more difficult to correct. It is important that records be kept, noting the conveyor's problem areas and detailing the corrective steps taken. This will prevent, or at least identify, the problems arising from correction, re-correction, over-correction, and counter-correction when problems return to a specific area.

Procedure for Training

The following is a step-by-step process for training the belt to correct for component alignment and loading problems.

Determine Areas of Belt Tension

Adjustments to components in the low-tension areas have the highest impact on correcting the path of the belt. By identifying and starting in the low-tension areas, the training process can have the greatest impact with the least amount of changes. In high-tension areas, there is too much tension on the belt for relatively minor adjustments to have significant impact on the belt path. Belt tension is usually highest at the drive pulley (**Figure 16.14**). The area of lowest tension will vary on the location of the snub and take-up pulleys. The low-tension areas are completely dependent on the individual conveyor and must be identified for each application. Conveyor Equipment Manufacturers Association's (CEMA) *BELT CONVEYORS for BULK MATERIALS, Sixth Edition*, or an experienced conveyor engineer can be consulted for additional information.

It is important to make sure the take-up weight is applying the correct tension required by the current belt and capacity ratings. If the belt is inadequately or improperly tensioned by the take-up pulley, it is likely to have severe variations in its path.

Determine Locations of Mistracking

It is best if inspection for mistracking begins with the first rolling component directly after the highest-tension area (typically where the belt leaves the drive pulley), as the tension will usually be lower in that area, and continues along the path of the belt until a point where the belt is visibly off track.

It is important to remember that the track of the belt at any given point is affected more by the idlers and other components upstream (the points the belt has already passed) than the components downstream (the points the belt has not yet reached). This means where mistracking is visible, the cause of mistracking is at a point the belt has already passed.

Therefore, corrective measures should be applied at points the belt passes before the area where it shows visible mistracking. The movement of one idler generally has its greatest training effect in an area within 5 to 8 meters (15 to 25 ft) downstream.

Train the Belt

The conveyor must be locked out / tagged out / blocked out / tested out before making any adjustment to components or belt tension to correct mistracking.

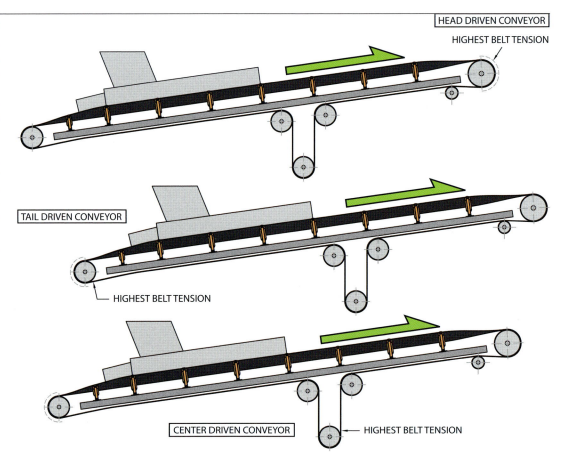

Figure 16.14

Belt tension is usually highest as the belt enters the drive pulley; areas of low tension will vary depending on the location of the snub and take-up pulleys. To train a belt, start inspection for mistracking directly behind the highest-tension area (where the belt leaves the drive pulley).

TAIL DRIVEN CONVEYOR

HIGHEST BELT TENSION

CENTER DRIVEN CONVEYOR — HIGHEST BELT TENSION

To correct the belt's running path, it is necessary to start in the areas of lower tension and move around the conveyor, making adjustments to idlers to move the belt back into the center. Then, following the route of the belt from the driven pulley toward the next rolling component in sequence, the belt path can be corrected by adjusting the idlers, one place at a time.

Starting with the first or second idler set before a point where the belt is visibly off track, the idler should be skewed in a direction opposite the misalignment. Then, the conveyor should be restarted to check for the belt's running alignment. The conveyor needs to be run to evaluate the effect of the correction, but it is important to wait two or three complete revolutions of the belt before further adjustments are made.

It is best to shift only one idler at a time, as pivoting additional idlers may cause over-correction or competing corrections. If the observation shows the belt path has been over-corrected, the path should be restored by moving the original idler back, rather than shifting additional idlers.

The belt should be tracked empty all the way around the conveyor, making especially sure the belt is centered as it enters the loading zone and discharge zone.

Figure 16.15

The most basic technique for training a belt is shifting the idler axis with respect to the path of the belt, commonly called "knocking the idlers."

Techniques for Training the Belt

The most basic training technique is to adjust idlers. Training a belt by using its return and carrying idlers is accomplished by shifting the idler axis with respect to the path of the belt. This is commonly called "knocking the idlers," because the idler base is shifted with a blow from a hammer (**Figure 16.15**).

Training a belt by shifting the position of one or more idlers is the same as steering a bicycle with its handlebars (**Figure 16.16**). When you pull one end of the handlebars (or the idler) toward you, the bicycle (or belt) turns in that direction. This is in keeping with the basic rule of belt training: The belt will steer to the side of the idler it touches first.

This handlebar principle of steering is sound, but only if the belt makes good contact with all three troughing rollers. So before training a belt, it is necessary to check to be sure the belt is troughing well at all points along the carrying side, even when unloaded. If the belt does not "sit down" in the trough, there may be a problem with its compatibility with the structure (**Figure 16.17**). A belt that is too thick and not suitable for a given conveyor might never track correctly.

Adjustments to the idlers should be small. Research at Australia's University of Newcastle has shown that once an idler is skewed past a certain point, it will not correct the belt path more, because the belt slides across the idler like a car skidding across a patch of ice *(Reference 16.2)*.

Obviously, such shifting of idlers is effective for only one direction of belt travel. A shifted idler that has a correcting influence when the belt runs in one direction will likely misdirect the belt when the conveyor is running in the other direction.

For unidirectional conveyors, shifting the position of idlers has benefits in belt training. However, there are drawbacks as well. It should be obvious that a belt

might be made to run straight with half the idlers knocked one way and the other half knocked in the opposite direction, but this would be at the expense of increasing rolling friction between belt and idlers. Idlers turned in all different directions in an effort to train the belt create extra friction, resulting unnecessarily in increased wear in the bottom belt cover and increased power consumption.

Adjustments should be made to idlers only—never to pulleys. Pulleys should be kept level with their axis 90 degrees to the intended path of the belt.

Other Techniques to Center the Belt

Another approach to centering the belt is to tilt the carrying idlers slightly, up to two degrees, in the direction of belt travel. The friction of the belt on the wing rollers creates a centering force that is directed to the centerline of the belt. This can be done by simply inserting flat metal washers beneath

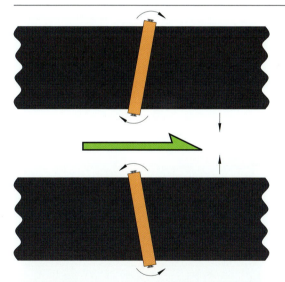

Figure 16.16

Training a belt by shifting the position of one or more idlers is the same as steering a bicycle. When you pull one end of the handlebars (or the idler) toward you, the bike (or belt) turns in that direction.

Figure 16.17

For a belt to track well, it must trough well along the conveyor's carrying side. If the belt does not "sit down" in the trough, it might never track correctly.

16

the back side of the idler frame. Many idler manufacturers build this tilt into their products. Just as with "knocking" the idlers, there is a limit to the effectiveness of this technique, and it does increase the power consumption of the conveyor and wear on the belt bottom cover and idlers (**Figure 16.18**).

An ill-advised method to center the belt as it approaches the tail pulley is to slightly skew in opposite directions (**Figure 16.19**) or raise opposite ends of the two return rolls nearest the tail pulley (**Figure 16.20**). The theory is that this deliberately-induced mistracking in opposite directions produces competing forces that work to center the belt. Though this may sound reasonable in the abstract, the practical application of it is problematic. This method incorporates instability into a system when the goal for optimum operation is stability. It could be argued that there is enough of a problem with getting the system square in order to

run true, without adding two more variables in the form of deliberately-misaligned idlers.

Training the Belt at Start-Up of New Conveyors

If a new conveyor system has been designed and built in accordance with sound engineering and installation practice, the belt will probably track at start-up on a path close to the desired one. There may be minor variations from the ideal structure that result in the belt not tracking perfectly; however, in these circumstances, the variations should be relatively minor, so the belt can be operated without damage long enough for a training procedure to take place.

The first movement of the belt through a new conveyor should be slow and intermittent, so any tendency of the belt to wander may be quickly recognized and the belt stopped before damage occurs. The first alterations must be made at points where the belt is in immediate danger of damage. Once the belt is clear of danger points, the conventional sequence for belt training, as noted previously, can be followed.

Insufficient attention at start-up can create problems, including serious runoff and edge damage, belt creasing or fold-over, spillage, and damage to other conveyor components. For conveyor start-up, observers should be positioned at locations where trouble might be expected or where the belt is at greatest risk—where it enters the discharge and loading chutes. These "spotters" should have a radio, telephone, or, at a minimum, a pull-rope emergency-stop switch within easy reach.

In severe cases, it may be necessary to shut the conveyor down, make any adjustments indicated, and reposition the belt before a new start-up is undertaken.

Training of Replacement Belts

A new belt—whether new belting on a new conveyor or a replacement belt on an established system—often has to be gradu-

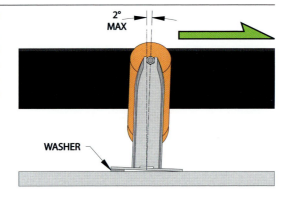

Figure 16.18

Tilting carrying idlers by inserting flat metal washers beneath the back side of the idler will increase the centering force.

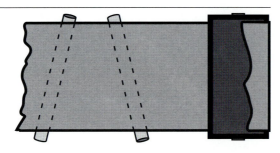

Figure 16.19

An ill-advised method to center the belt as it approaches the tail pulley is to slightly skew the two closest return rolls in opposite directions.

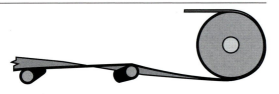

Figure 16.20

Raising the opposite ends of the two returns rollers nearest the tail pulley is an ill-advised method to help center the belt.

ally "worn in" like a new pair of shoes. It is relatively rare to pull a new belt onto an existing conveyor, splice it together, push the conveyor's start button, and have the belt track down the middle of the structure. All new systems must be run for several hours before the final training of the belt to run in the idlers and stretch the belt.

Some new belts will tend to run to one side in one or more portions of their length because of a permanent camber or a temporary unequal distribution of tension arising from the storage, handling, or stringing of the belt. In many cases, operation of the belt under tension for a break-in period will correct this condition. Loading the belt to 60 percent capacity will help the belt fit the conveyor.

The conveyor structure may not be neutral to the new belt, particularly in the case of a new belt going onto an existing conveyor. If numerous training adjustments have been made over time to correct the mistracking of the previous belt, these adjustments may have to be "undone" to allow the new belt to track correctly.

Training Feeder Belts

Feeder belts are normally short, high-tension, slow-moving belts that use flat rollers or picking-style idlers on the carrying side. A square splice is critical for tracking on these belts, and the head and tail pulleys must be perfectly aligned. Training can be done on only the return, or slack, side of a feeder belt because of the construction and high loads on the carrying side. If needed, a single training device can be placed in the center of the return where there is some slack in the belt to allow the training device to function.

Training Reversing Belts

None of the techniques such as knocking or tilting the idlers is effective on reversing belts. Any correction made to track a reversing belt in one direction will have the opposite mistracking effect when the belt reverses. This makes reversing belts one of the most difficult belt-training challenges.

Consequently, all idlers and pulleys must be in perfect alignment and the splice must be square to make the system as clean or neutral as possible. Only training devices designed for reversing belts should be installed.

HARDWARE FOR TRAINING THE BELT

Most conveyors need some tracking correction to account for unexpected or environment-induced belt wander. There are also occasions when the training procedure is not successful at providing a long-term solution to a mistracking problem. As a result, the operation is faced with repeating the training procedure on a frequent (sometimes daily) basis or installing some form of mechanical belt-training system to reduce this requirement. Engineered training solutions are devices that sense the position of a belt and, through a mechanism or geometry change, actively adjust the belt's path.

Belt Misalignment Switches

While not a corrective device, a belt misalignment switch is a hardware system that offers some control over belt tracking. These switches are electro-mechanical sensors that send a signal when activated by the mistracking belt. These switches are installed at intervals along the length of the conveyor on both sides of the belt near the outer limit of safe belt travel. When the belt moves too far in either direction, it pushes over the lever arm to activate a switch or send a signal interrupting the conveyor's power circuit, stopping the belt so the operator has the opportunity to make corrections (**Figure 16.21**). In many cases, plant personnel will need to walk the conveyor to manually reset the switch before operation can begin again. Some devices have the ability to send multiple signals: the first one an alarm indicating a pre-set amount of belt wander, and the second signal cutting drive power due to a more serious tracking problem.

16

Of course, the tripping of a belt misalignment switch is a signal indicating something is wrong with the conveyor system. It is like a light on a car instrument panel that shows red when the engine is too hot. It is possible to ignore this light, to reset the switch and resume conveyor operations, but both the car's warning light and the conveyor misalignment switch should serve as a warning that there may be more serious, more expensive, possibly catastrophic problems. Conveyor stoppages can be a nuisance and very costly; each outage creates downtime and lost production. Belt misalignment switches are not a solution to the problem of misaligned belts; they are an indicator of a severe problem.

When the belt moves too far in either direction, the misalignment switch will activate a switch or send a signal interrupting the conveyor's power circuit, stopping the belt.

16

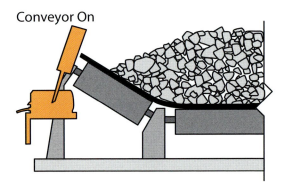

Conveyor On

Normal Operating Position

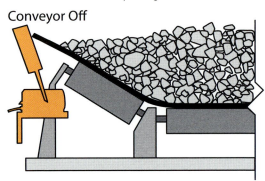

Conveyor Off

Mistracking Belt Has Pushed Switch to "Conveyor Off" Position

Passive Tracking Solutions

Vertical Edge Guides

The first impulse on seeing a wandering belt may be to install some sort of barrier to keep the belt straight, or at least keep it away from obstacles (**Figure 16.22**). One version of this simple approach to minor tracking problems is the vertical edge guide (**Figure 16.23**). These devices place a spool or roller on a simple frame close to the belt edge. The vertical edge guides are installed in a position approximately perpendicular to the belt's path to keep the belt edge away from the conveyor structure. These side guides do not train the belt. Rather than preventing belt wander, they perform a damage-control function, allowing the belt to strike a rolling surface rather than unyielding structural steel. Vertical edge guides are most effective on short, low-tension belt installations where the belt can be forced to stay in position through brute force on the edge of the belt. Vertical edge guides can allow severe belt or structure damage when the belt rides up over the guide into the structure or the guide causes the belt to roll over on itself. Vertical edge guides should not be used to compensate for persistent misalignment problems. They are not particularly effective on very thin belts.

"Vee" Idlers

Another hardware addition that can help remedy belt wander is the installation of "Vee" idlers on the belt return; they are becoming popular on longer, high-tension

Figure 16.22

The first impulse may be to install some sort of barrier to keep the belt away from obstacles.

Figure 16.23

Vertical edge guides place a spool or roller on a simple frame close to the belt edge to keep the belt away from the conveyor structure.

conveyors. They are available in two versions: traditional "Vee" rollers (**Figure 16.24**) and inverted "Vee" rollers (**Figure 16.25**). Both systems form the belt into a trough to assist in steering it into the center. They rely on a centering force to correct the belt path, so they place added stress on the belt, which can lead to damage. These systems are more expensive and require somewhat more maintenance than a conventional return idler.

"Crowned" Pulleys

Pulleys that have larger diameters at the center than at the edges are sometimes used to provide a centering effect (**Figure 16.26**). These "crowned" pulleys operate from the basic tracking principle, also. As the raised portion of the pulley (the crown) touches the belt first, it steers the belt into the center. The outer sections of the belt on both sides then produce a force driving it toward the center. If the belt is centered, these forces cancel each other out. If the belt misaligns and the belt wanders to one side of the pulley, the friction force will be greater on that side, acting to push the belt back toward the center.

Crowned pulleys are most effective on conveyors with short, low-tension belts. With higher-tension or steel-cable belts, little steering effect is obtained from the crown of the pulley. That is because the centering force created is smaller in magnitude than the forces of mistracking and most of the contact force between the belt and pulley is on the outer edges of the pulley due to the transition of the belt. Crowned pulleys are most effective where there is a long unsupported span—four times the belt width or greater—approaching the pulley. As this spacing is not often possible on the carrying side of the conveyor, the use of crowned head pulleys is relatively ineffective and may not be worth the stress it produces in the belt. They are somewhat more effective when used as a conveyor's tail pulley. *(See Chapter 6: Before the Loading Zone.)* Another problem with crowned pulleys is that they can create inef-

fective belt cleaning, because the cleaning blade(s) may not mate properly with the whole belt surface.

Dynamic Training Solutions

There are a number of dynamic belt-tracking systems: systems that when activated move a component to correct the belt path. These belt-training systems are designed to "self-align." That means the force of the mistracking belt causes an idler to reposition itself, creating a steering action that directs the belt back into the center.

As with adjusting fixed idlers, the correcting force of a skewed idler approaches a limit as the skew angle of the idler increases. All trainers will eventually reach this limit. It is more effective to stimulate quick, low-angle corrections of belt mistracking than to wait for one larger angle.

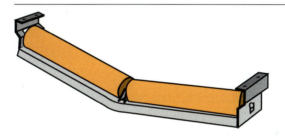

Figure 16.24

Traditional "Vee" idlers form a trough to keep the belt centered.

Figure 16.25

Although installed in an attempt to keep the return run of the belt in alignment, inverted "Vee" idlers risk damage to the belt.

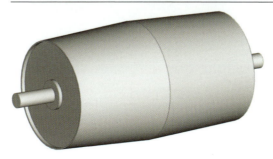

Figure 16.26

Pulleys with a "crown," or larger diameters at the center than at the edges, are sometimes used to provide a centralizing effect.

Many tracking solutions carry the seeds for their own destruction. Because they are designed to move to provide a correcting influence on the belt path, they are particularly vulnerable to the accumulations of fugitive material. Piles of spillage can block their range of motion or seize the pivot bearing (**Figure 16.27**). This can lock the belt-training idler into a position where it functions as a "misalignment" idler. It now pushes the belt out of the proper track, creating (or worsening) the problem it was installed to correct. To correct the now misaligned system, the maintenance crew may tie the training idler into (approximately)

the right position (**Figure 16.28**). In such case, when a tracking solution is not capable of functioning properly, it is better to remove it, rather than to just "tie it off."

All these systems work under the disadvantage of being "after the fact:" They correct mistracking after it has occurred. A certain amount of wander must happen before the required correction can take place. But these systems do function as a form of insurance against a problem becoming so severe that the belt suffers costly damage before the mistracking can be discovered and corrected.

In-Line Sensing-Roll Trainers

The simplest belt-trainer design, the in-line sensing-roll trainer, has a carrying roll in a framework mounted on a central pivot bearing (**Figure 16.29**). Vertical guide rolls that act as sensors to the belt's path are mounted on both sides of the belt in line with the roller, with their centerline running through the idler's pivot point. Movement of the belt against either of these sensing rolls causes that roll to move in the direction of the belt misalignment. This pivots the entire idler. In keeping with the basic rule of tracking that the belt always moves toward the side it contacts first, the pivoted idler then steers the out-of-track belt back to the proper path.

Yet these in-line sensing rollers have almost no leverage. They require considerable force from the edge of the moving belt to create a correction. With this design, the belt wanders from side to side; the correcting action is caused by the belt literally slamming into one side or the other. When the correcting action takes place, the idler may "kick over" with such force that the belt is then directed all the way over to the other side of the structure; the belt, in turn, contacts the roller on the other side of the tracking idler, which corrects the belt path back in the other direction. Because the tracking idler has a single, central pivot point, belt movement to one side brings the opposite guide roll into a hard, pinching contact against the belt, which can lead to

16

Figure 16.27

Piles of spillage can block a training idler's range of motion or seize its pivot bearing.

Figure 16.28

To correct a misaligned system, a training idler is sometimes "tied-off"—secured into what is thought to be the right position. But changes in conditions will likely render this position incorrect.

Figure 16.29

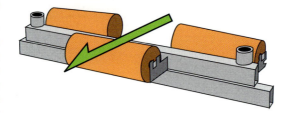

The in-line sensing-roll trainer has a carrying roll on a central pivot bearing with vertical guide rolls mounted on both sides of the belt in line with the roller.

edge damage. The belt can be kept constantly in motion, back and forth between the two sides, risking edge damage and overuse of the pivot bearing.

Leading Sensing-Roll Trainers

The most common belt-training design has a carrying roll (or troughing set) held in a framework that is mounted on a central pivot bearing (**Figure 16.30**). Guide rolls are mounted on short arms and positioned 25 to 75 millimeters (1 to 3 in.) from the belt on both sides. The rolls are positioned in advance of the pivoting roller; hence the designation leading sensing-roll trainers (**Figure 16.31**). Some designs tilt the pivot shaft slightly in the direction of belt travel to improve the sensitivity of the trainer. Leading sensing-roll trainers are available designed for use both on the upper (or carrying) side of the belt and on the lower (or return) side.

Movement of the belt against either guide roll causes the steering idler to pivot, correcting the belt path back toward the center. Again, as the belt always moves toward the side it contacts first, the pivoted roll steers the out-of-track belt back to the proper path.

Sensing rollers installed on short arms in advance of the steering idler have slightly more leverage than the in-line sensing idlers, but they still require considerable force from the belt edge to cause correction. Consequently, this trainer design suffers from all the delay, pinching, and fugitive-material problems of the in-line sensing idler.

The leading sensing-roll trainer is the most popular and most common tracking idler. It is supplied as original equipment on almost all new conveyors sold. It is typically installed at intervals of approximately 30 meters (100 ft) on both the carrying and return sides.

In the field, however, these trainers are commonly seen in two unsatisfactory conditions. The first condition is "frozen" from material accumulations or corrosion of the center pivot (**Figure 16.32**). This problem can be solved with better maintenance or a higher quality pivot point. The second condition is "tied off"— locked in place with a rope or wire—so the training device is the equivalent of a "knocked" idler (**Figure 16.33**). The reason these are "tied

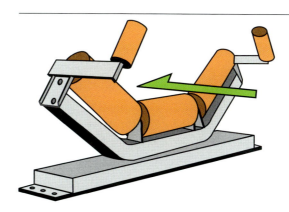

Figure 16.30

The most common belt-training design, the leading sensing-roll trainer, has a central idler mounted on a pivot bearing; guide rolls are positioned on short arms on both sides in advance of the positioning roller.

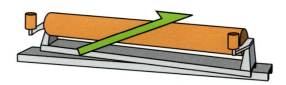

Figure 16.31

On a leading sensing-roll trainer, the guide rolls are mounted on short arms in advance of the pivoting roller.

Figure 16.32

Leading sensing-roll belt trainers are subject to material accumulation that can "freeze" the trainers in place.

Figure 16.33

To control erratic movement, belt trainers are often "tied off" or locked in place with a rope or wire.

off" originates in the design. The sensing rolls swing in an arc about the center pivot; therefore, the rolls must be spaced far enough apart to not pinch the belt when the rolls reach extreme positions. As the pivot becomes fixed in position from material accumulation, lack of maintenance, or corrosion, the idler will not react until the belt has mistracked a distance equal to this wide spacing. Consequently, the idler oversteers and, therefore, becomes an unstable control system. The idlers often overreact, providing unpredictable results, and, as a result, they are frequently "tied off."

Torsion-Spring Trainers

The torsion-spring trainer is an improved version of the leading sensing-roll trainer (**Figure 16.34**). This system removes one sensing roll and incorporates a spring into

Figure 16.34

The torsion-spring trainer uses only one sensing roll that is in continuous contact with the belt.

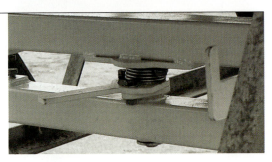

Figure 16.35

The torsion-spring trainer incorporates a spring into its pivot point.

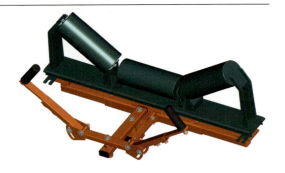

Figure 16.36

Multi-pivot belt trainers use a torque-multiplying system to supply a mechanical advantage to improve belt path correction.

the pivot (**Figure 16.35**). This spring keeps the one remaining sensing roll in contact with the belt edge at all times. As the belt mistracks in either direction, the idler will compensate by pivoting and steering the belt.

These spring-loaded leading-sensor trainers tend to have the sensing rolls installed on long arms in advance of the steering idler. This creates more leverage and a greater mechanical advantage in converting belt wander into steering torque. There is no delay in reaction of this trainer, due to the fact that the sensing roll is in constant contact with the belt. There is also no pinching, because there is only one sensing roll. Because of the constant "fine-tuning" action of the idler, it is harder for fugitive material to accumulate to the point it can impede the pivoting action of the tracking device.

One drawback of this trainer is the fact that it cannot function with a troughed idler set. In addition, because the single roller is in constant contact with the belt, this roller is subject to more frequent replacement than those on leading sensing-roll trainers.

Multi-Pivot Belt Trainers

There is another belt-tracking system that uses the force of the wandering belt to position a steering idler and so correct the path. This device uses a multiple-pivot, torque-multiplying system to supply a mechanical advantage to improve belt-path correction (**Figure 16.36**).

This style of training device transfers the motion of mistracking to the steering idler through a unique parallel linkage (**Figure 16.37**). This requires less force to initiate the correction, and as it steers, it needs less force to turn the belt. Belt training becomes a continuous, active, precise fine-tuning of the belt path. This design is available in models for the troughed (or carrying side) or the return side of the conveyor (**Figure 16.38**).

This multiple-pivot training device uses guide rolls that are set very close—6 millimeters (1/4 in.) from the belt (**Figure 16.39**). With the rollers set at the edge of the belt, the device can sense smaller movements of the belt and make corrections after very slight misalignments. Rather than waiting for a powerful mistracking force, the multi-pivot belt-training device adjusts constantly, reacting to smaller forces and providing continuous, precise corrections of the steering roller.

The sensing rollers of the multi-pivot trainer use longer arms to increase the distance from the guide rolls to the steering idler. This allows the unit's torque arm to act as a force multiplier, increasing the mechanical advantage of the steering action. As a result, this belt-training system can correct the belt line with one-half the force required for conventional tracking idlers.

Unlike the other training devices, the multiple-pivot device is installed so the belt crosses the steering roller before it reaches the guide rollers (**Figure 16.40**). This means the guide rolls adjust the "corrected" belt path rather than the mistracking belt path. The result is a roller that is continuously working to prevent the belt from moving very far from the proper path. The multi-pivot design allows the rollers to move perpendicular to the structure's centerline while directing the steering idler to the proper angle, instead of pivoting and pinching the belt edge.

Variations of Multi-Pivot Belt Trainers

Several manufacturers have created a slight modification to the multi-pivot belt-training device (**Figure 16.41**). These use the same force amplification geometry, but the idler slides laterally as well as pivoting. With the sliding-idler system, the sensing roll has to overcome the resistance to pivoting as well as the friction force of trying to move an idler from under a belt. This greatly decreases the overall steering force of this training system.

Free-Pivoting Trainers

Manufacturers have developed training idlers in which the steering roll also serves as the sensing roll. With this design, there is a bearing in the center of the roll, so the ends of the roll can pivot around the axis of the roller as well as rotate. The pivot shaft is usually tilted in the direction of

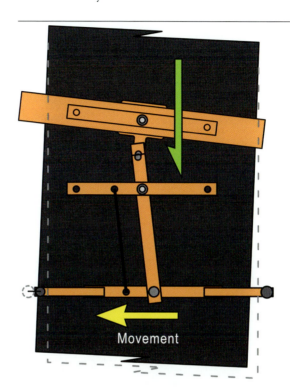

Movement

Figure 16.37

Because it transfers the motion of mistracking to the steering idler through a unique parallel linkage, the multi-pivot system requires less force to initiate the correction, and as it steers, it needs less force to turn the belt.

16

Figure 16.38

The multi-pivot trainer is available in models for the troughed (or carrying side) or the return side of the conveyor.

Figure 16.39

With the rollers set at the edge of the belt, the multiple-pivot training devices can sense smaller movements of the belt and make corrections after very slight misalignments.

belt travel to improve the sensitivity of this type of training idler. Some manufacturers have used a rubber-covered tapered roll to improve the performance of this tracking solution (**Figure 16.42** and **16.43**).

Figure 16.40

With the multiple-pivot training devices, the belt crosses the steering roller before it reaches the guide rollers. This way, the guide rolls adjust the "corrected" belt path rather than the mistracking belt path.

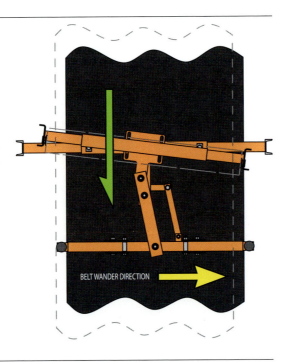

BELT WANDER DIRECTION

Figure 16.41

Some multi-pivot belt-training devices feature rollers that slide laterally as well as pivot.

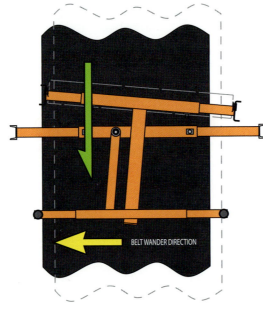

BELT WANDER DIRECTION

Figure 16.42

Some manufacturers use a rubber-covered tapered roll to improve the performance of the free-pivoting trainer.

When the belt wanders to one side of the roll, it creates a larger friction force on that side. The trainer's roll will pivot in reaction, moving in the direction that the larger force is pushing it. In accordance with the basic principle of belt steering, the pivoting roll will steer the belt back toward the center. When the belt is moving, the force on each side of the idler evens out, and the idler moves back into a position that is perpendicular to the path of the belt.

While this solution is effective and has very few moving parts, it still contains a complex bearing structure that is susceptible to airborne dust. As the forces that cause the unit to pivot are very small, the unit must be very free to pivot. Such freedom allows the unit to be influenced by many different environmental conditions, thus causing it to pivot when the belt is not wandering.

Trainers for Reversing Conveyors

Conveyors that run in two directions have always been the "last frontier" of belt tracking. With reversing conveyors, even experienced plant personnel are hesitant to adjust the idlers and perform the maintenance "tricks" typically used to train wandering belts. Conventional belt-training devices cannot be used, all for the same reason: because what works to centralize a belt's path when it runs in one direction may have the opposite effect when the belt direction is reversed. A pivoted idler that correctly steers the belt when the conveyor is operating in one way will work to mistrack a belt moving in the opposite direction.

Some manufacturers have developed trainers for reversing belts. The in-line sensing-roll trainers will correctly steer these belts, because the sensors are not direction dependent. The torsion-spring trainer can be modified to accommodate reversing belts. Adding a second arm and sensor in the opposite direction allows the torsion-spring trainer to switch sensing arms based on the direction of the belt movement (**Figure 16.44**).

These reversing trainers will have the benefits and shortcomings associated with their use on one-direction conveyors.

INSTALLATION OF BELT-TRAINING DEVICES

Training devices can be installed at any point the belt path needs adjustment. They should be installed approximately three to four times the width of the belt in advance of the point of the mistracking. The conveyor must be locked out / tagged out / blocked out / tested out before installing a belt trainer.

The typical places belt-training devices are installed include (**Figure 16.45**):

A. Just before the belt enters the tail pulley, to ensure it is centered on the pulley and into the loading zone

B. Shortly after the loading zone, to make sure the loaded belt is tracking in the center

C. Just before the discharge pulley, to make sure the belt is in the center before it enters the enclosure and discharges the cargo

Dynamic training devices can be installed over the entire length of conveyor, especially to cover any problems. Training

devices may need to be installed to correct the path at any place the belt enters an enclosure. They should not be positioned so close together that they will "compete," or contradict each other's steering action. There should be 21 to 50 meters (70 to 150 ft) between units, depending on the severity of the mistracking problem (**Figure 16.46**).

Figure 16.43

A rubber-covered roll improves the performance of this belt-training device.

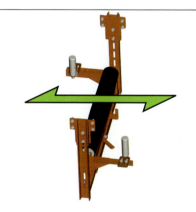

Figure 16.44

Adding a second arm and sensor in the opposite direction allows the torsion-spring trainer to be used on reversing belts.

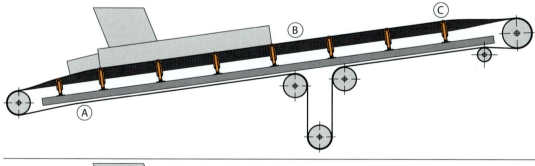

Figure 16.45

The typical places training devices are installed include:
A. Just before the belt enters the tail pulley
B. Shortly after the loading zone
C. Just before the discharge pulley

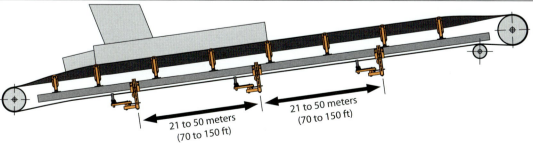

Figure 16.46

Belt trainers should be positioned from 21 to 50 meters (70 to 150 ft) apart to prevent them competing or contradicting each other's steering action.

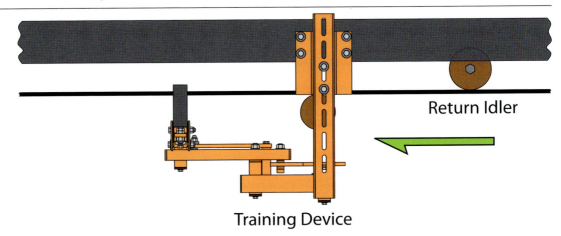

Figure 16.47

To improve the performance of return belt-tracking systems, a return idler can be installed before the device above the belt to increase the training idler's grip on the belt.

Return Idler

Training Device

16

When installing any form of dynamic training device, the center roll is typically elevated 12 to 19 millimeters (1/2 to 3/4 in.) higher than the rolls of the adjacent conventional idlers. This increases the belt's pressure on the tracking device and improves the corrective action. This is applicable to both troughed (carrying side) and flat (return side) self-aligning idlers. Some training-idler manufacturers build this feature into their various models.

Another technique to improve the performance of return belt-training systems is to reinstall a conventional return idler upstream of the tracking device above the belt to push the belt down, increasing the pressure on the training idler, allowing it to work more effectively (**Figure 16.47**).

Rubber-covered rollers are often useful on belt-tracking devices, particularly where the material is slippery or the belt wet from the climate or the process. These rollers may require replacement more often than "steel can" rollers, but may be necessary to achieve the friction needed to steer the belt.

SYSTEM MAINTENANCE

Tracking a conveyor is a maintenance function. Because a belt is moving and under load, belt training is subject to problems/corrections, requiring on-going inspection and probably maintenance.

Once a belt is properly trained, an operation should include checks on its alignment as part of the on-going maintenance program. The belt's alignment should be reviewed on a regular basis, and minor corrections should be made as needed and documented.

Mechanical training devices should also be inspected regularly. Most mechanical training devices contain moving parts that are vulnerable to contamination. The devices should be inspected to determine that all parts are free to move and the device is still tracking the belt properly. *(See Chapter 27: Conveyor System Survey.)*

Maintenance personnel should look for conditions out of the ordinary or things that have changed that might affect belt

⚠ SAFETY CONCERNS

In some facilities, it is common to make adjustments to correct belt tracking while the conveyor is in operation. However, a responsible safety program will always recommend that the conveyor be locked out / tagged out / blocked out / tested out before making adjustments to components or the belt tension in order to correct mistracking. While adhering to this practice may require several shutdowns and start-ups of the belt to observe the effect of corrections, it is the safe way to train the belt.

alignment. Changes could be anything such as the addition of a new piece of equipment, a large dent in the structure where a piece of heavy equipment collided with the conveyor, or changes in material condition that affect loading patterns. As seen above, misalignment has many causes, and small disturbances can manifest in a major mistracking incident.

TYPICAL SPECIFICATIONS

Belt-Training Device

A. Belt trainer(s)

To control the path of the belt and prevent belt mistracking, one or more belt-training devices will be installed on the conveyor.

B. Belt-path correction

The belt-training devices will sense any mistracking of the moving belt and use the force of that mistracking to articulate an idler. This idler will steer the belt back into the center of the structure.

C. Location

To keep the belt centered in the conveyor loading zone, one belt-training device will be installed on the conveyor return as the belt enters the tail pulley. To make certain the loaded belt is centered properly, a second belt-training device will be installed at the exit end of the loading zone. Additional training devices will be located along the conveyor as required to correct belt wander.

ADVANCED TOPICS

Power Consumption and Belt Trainers

Any alteration in a conveyor's rolling equipment, from skewing idlers to installing special mechanical training devices, has implications for the system's power requirements.

There are several styles of training idlers, all designed to exert a centering force on the belt perpendicular to its direction of travel. This centering force must be considered in calculation of the power consumption of the conveyor.

Analyzing the power consumption of a training idler requires knowledge of the load on the idler. This load is due to the weight of the belt and any component of the belt tension arising from the idler misalignment. In operation, the typical training idler can pivot from 2 to 5 degrees. It is common practice to install training idlers 12 to 19 millimeters (1/2 to 3/4 in.) above the standard idlers. This results in greater load on these idlers, which creates enough centering force to influence the travel of the loaded belt. This extra load is described by The Conveyor Equipment Manufacturers Association (CEMA) in *BELT CONVEYORS for BULK MATERIALS, Sixth Edition*, as idler misalignment load (IML).

When an idler is pivoted, it will exert a force on the belt in a direction perpendicular to the pivoted idler. This is called the

---

$Tr = PIW \cdot BW \cdot \tan \phi$

Given: A 450-millimeter (18-in.) belt with a tension of 17,5 newtons per millimeter (100 lb$_f$/in.) travels over an idler that is pivoted 3,5 degrees. **Find:** The misalignment force due to the idler.

	Variables	Metric Units	Imperial Units
Tr	Misalignment Force	newtons	pounds-force
PIW	Belt Tension per Unit of Belt Width	17,5 N/mm	100 lb$_f$/in.
BW	Belt Width	450 mm	18 in.
φ	Idler Misalignment	3,5°	3.5°
Metric: *Tr = 17,5 · 450 · tan 3,5 = 481*			
Imperial: *Tr = 100 · 18 · tan 3.5 = 110*			
Tr	Misalignment Force	481 N	110 lb$_f$

Equation 16.1

Calculating Misalignment Force

misalignment force and can be calculated (**Equation 16.1**).

The component of the misalignment force in the direction of belt travel is called the misalignment drag force and can also be calculated (**Equation 16.2**).

The misalignment drag force is used to find the power required to compensate for a tracking idler (**Equation 16.3**).

This additional power requirement should be multiplied by the number of tracking idlers installed.

It is interesting to note that an 1800-millimeter (72-in.) belt with 175 newtons per millimeter (500 PIW) slack-side tension would have a centering force of approximately 9640 newtons (2200 lb$_f$) and a centering force component in the direction of travel of approximately 589 newtons (134 lb$_f$). A tracking idler on this belt would require 1,177 kilowatts (1.6 hp) per tracking idler.

If the training idler becomes frozen and is neither rotating nor pivoting, it can add a substantial power requirement.

The power consumed by the tracking solutions should be considered when selecting a tracking solution. While some methods of training may be effective, the solution could draw more power than the drive of the conveyor can deliver. Most engineering companies include a healthy safety factor to account for unknowns such as this when designing a conveyor, but it is in an operation's best interest to verify its conveyor(s) have sufficient power to handle these increased loads.

16

Equation 16.2

Calculating Misalignment Drag Force

$Tm = Tr \cdot \sin \phi$

Given: A misalignment force of 481 newtons (110 lb$_f$) and an idler pivot of 3,5 degrees. **Find:** The misalignment drag force.

Variables		Metric Units	Imperial Units
Tm	Misalignment Drag Force	newtons	pounds-force
Tr	Misalignment Force (Calculated in Equation 16.1)	481 N	110 lb$_f$
φ	Idler Misalignment	3,5°	3.5°
Metric: $Tm = 481 \cdot \sin 3{,}5° = 29$			
Imperial: $Tm = 110 \cdot \sin 3.5° = 6.7$			
Tm	Misalignment Drag Force	29 N	6.7 lb$_f$

Equation 16.3

Calculating Power to Compensate for a Training Idler

$P = Tm \cdot V \cdot f \cdot k$

Given: A misaligned idler exerts 29 newtons (6.7 lb$_f$) on a conveyor system. The belt is traveling at 2 meters per second (400 ft/min). The interface friction between the belt and the idler is 1. **Find:** The power added to the drive due to the training idler.

Variables		Metric Units	Imperial Units
P	Power Added to Belt Drive	kilowatts	horsepower
Tm	Misalignment Drag Force (Calculated in Equation 16.2)	29 N	6.7 lb$_f$
V	Belt Speed	2,0 m/s	400 ft/min
f	Friction Coefficient	1,0	1.0
k	Conversion Factor	1/1000	1/33000
Metric: $P = \dfrac{29 \cdot 2 \cdot 1}{1000} = 0{,}058$			
Imperial: $P = \dfrac{6.7 \cdot 400 \cdot 1}{33000} = 0.081$			
P	Power Added to Belt Drive	0,058 kW	0.081 hp

BELTS IN THE REAL WORLD

In Closing...

In the real world, conveyor belts wander. But allowing a belt to chronically mistrack can lead to personal injury, release of fugitive materials, and belt and structural damage. However, training a belt without some knowledge of the effects of the training actions can result in increased energy usage, component failure, and belt damage.

There are a variety of self-aligning idlers that can help control belt tracking. But it is wise to note that conveyor operations should not depend on these training idlers to overcome gross misalignment of conveyor structure or significant and continuing loading problems. The continuous working of a training idler indicates more serious problems that should be identified and corrected. It is much better to discover what the real problem is and make the necessary corrections.

While belt wander is a complex problem, it can be controlled by systematically and proactively identifying the root causes of mistracking and eliminating them. Training a belt is a skill that takes time to learn and is best left to a qualified and experienced employee or specialty contractor.

Looking Ahead...

This chapter about Belt Alignment, the last chapter in the section Return Run of the Belt, explained how fugitive materials can cause belt mistracking and how, in turn, belt mistracking can cause increased fugitive material. The following chapter, Dust Management Overview, begins the next section about Dust Management.

REFERENCES

16.1 Cukor, Clar. (Undated). *Tracking: A Monograph*. Scottdale, Georgia: Georgia Duck and Cordage Mill (now Fenner Dunlop).

16.2 Barfoot, Greg J. (January/March 1995). "Quantifying the Effect of Idler Misalignment on Belt Conveyor Tracking," *Bulk Solids Handling*, Volume 15, #1, pp. 33–35. Clausthal Zellerfeld, Germany: Trans Tech Publications.

16

SECTION 4
DUST MANAGEMENT

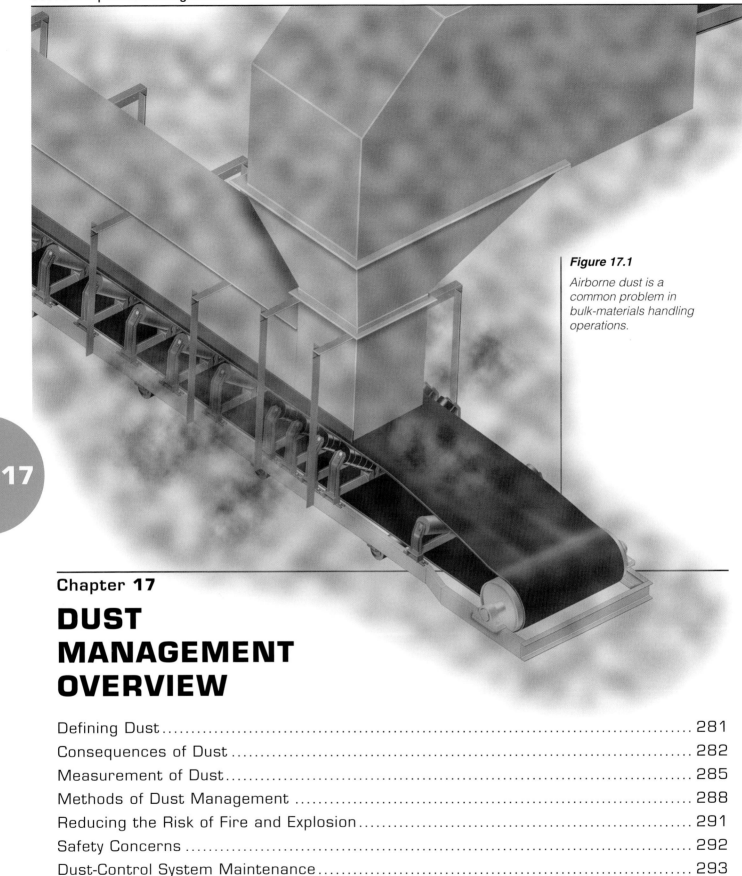

Figure 17.1

Airborne dust is a common problem in bulk-materials handling operations.

17

Chapter 17

DUST MANAGEMENT OVERVIEW

In this Chapter...

This chapter defines dust and provides an overview of the topic of dust management. It examines problems associated with dust, including fire and explosions; methods of measuring dust; and methods for minimizing and managing dust. The following three chapters will provide more detail of specific areas of dust control.

Fugitive dust, finely divided materials that have become airborne, is becoming one of the greatest concerns of bulk-materials handling operations. Dust creates problems in the process, in the plant, in employee performance, in employee and community health, and in relations with the neighbors surrounding the plant. This heightened level of concern may be because dust is more visible from outside the plant than spillage. Spillage is more localized; it affects affairs inside the plant. A cloud of airborne dust rising from an operation's conveyors or crushers is visible to outsiders and can create problems in health, safety, neighbor relations, and regulatory compliance (**Figure 17.1**).

To successfully combat the growing nuisance of airborne dust, one must understand how dust is generated, what the consequences of dust are, what agencies are monitoring dust, how dust is measured, and what methods are available to combat dust. When both material spillage and airborne dust are controlled, the operation will be cleaner, safer, and more productive.

DEFINING DUST

Confining this Discussion

This book discusses the control of dust from conveying material or loading and unloading belt conveyors. Other industrial and material handling operations will create dust, including crushing, milling, machining, and truck hauling. Some of the issues and technologies discussed in this and the following chapters may be useful in understanding and controlling dust from these sources.

Solving the problems of dust is complicated by the nature of dust. The creation of dust has a significant number of variables that are regularly altered by changes in environment and materials. The variety of process designs and plant layouts, of production techniques and technologies, of system options and equipment choices, and of differences in conveyors and the materials conveyed will affect conditions and results. These differences might even be detected on a daily basis within a single operation. Thus, the information presented here cannot be considered absolutely applicable in all circumstances. Any application of this information must be carefully reviewed in light of the specific circumstances before options are selected and investments are made. The first step is to understand the nature of the problem before considering the specific options for a given operation.

The Definition of Dust

Dust is defined by the Mine Safety and Health Administration (MSHA) in the United States as "finely divided solids that may become airborne from the original state without any chemical or physical change other than fracture."

This is a complicated way of saying dust is "material that, when disturbed, can take to the air and stay there." While it is nice to say that all disturbances can be eliminated, bulk-material industries are full of disturbances. Examples include, but are not limited to, a solid lump of material broken by impact, crushing, abrasion, or grinding; material transferred from one belt, vessel, process, or container to another; or material agitated by wind, workers, or machinery.

The size of a particle of dust is measured in microns (μm). Micron is a shortened form of micrometer, a unit of measurement that is one-millionth of a meter. The equivalent in Imperial measurements is 1/25400 (or 0.0000394) of an inch. A human hair is typically 80 to 100 microns in diameter.

Respirable dust is that dust which is small enough to enter the lungs when inhaled. The U.S. Department of Labor's Office of Occupational Safety & Health Administration (OSHA), in their handbook entitled Dust Control Handbook for Minerals Processing, defines respirable dust as follows:

> Respirable dust refers to those dust particles that are small enough to penetrate the nose and upper respiratory system and deep into the lungs. Particles that penetrate deep into the respiratory system are generally beyond the body's natural clearance mechanisms of cilia and mucous and are more likely to be retained.

Most regulating agencies define 10-microns or smaller as the size of respirable dust. According to OSHA, MSHA defines respirable dust as the fraction of airborne dust that passes through a 10-micron (3.28 x 10^{-5} ft) size-selecting device (sieve) (**Table 17.1**).

CONSEQUENCES OF DUST

Unlike material spillage, which generally stays close to the point on the conveyor where the material is released, airborne dust affects the entire operation. Once dust is released into the air, it will settle wherever the currents of air take it. There are many dangers, expenses, inconveniences, and inefficiencies associated with airborne dust.

It is in an operation's legal and financial best interest to deal properly with dust.

When an operation violates a safety regulation, there are legal ramifications for the parties accountable (including personal culpability and possible financial liability for executives of operations where safety violations occur); therefore, there is a personal incentive to eliminate dust.

Health Risks

The greatest danger of dust is in the exposure of workers, neighboring homes, and businesses to dust. If the material is toxic, carcinogenic, or otherwise hazardous, having it airborne can endanger large numbers of people. In addition to the toxic dangers of materials, there is a respiratory danger presented by airborne dust. Once respirable dust is taken into the lungs, it might not be expelled. Prolonged exposure will lead to buildup of material in the lungs. Most regulating agencies define 10 microns as the size of respirable dust. When airborne particles 10 microns or smaller are inhaled, they will stay in the lungs; therefore, dust particles of 10 microns or smaller have a much lower allowable concentration. With toxic materials, the allowable concentration is even lower. In the United States, silica is normally regulated to the point where maximum allowable concentrations are below 2 milligrams per cubic meter (2.0 x 10^{-6} oz/ft^3) per eight-hour day. Many government and private agencies have deemed that continued exposure to concentrations higher than these will cause silicosis.

OSHA has determined admissible dust levels for the United States (**Table 17.2**). The levels determined by OSHA are representative of levels of regulation seen, and increasingly enforced, around the world.

Explosion Risks

Another danger of dust is its potential for explosion. Materials that obviously have this potential are coal and other fuels. Even materials that are not flammable in their bulk state can combust when airborne as fine dust. For example, aluminum dust is flammable.

Table 17.1	Percentages of Particle Sizes Passing Through a 10-Micron Size-Selection Device					
	Particle Size (µ)	10,0	5,0	3,5	2,5	2,0
	% Passing Sieve	0	25	50	75	90

There are five contributing components necessary for a dust explosion to occur. The first three form the "triangle" of components of any fire:

A. Fuel (ignitable dust)

B. Ignition source (heat or electric spark)

C. Oxidizer (oxygen in the air)

The final two components are required to create a dust explosion:

D. Suspension of the dust into a cloud (in sufficient quantity and concentration)

E. Confinement of the dust cloud

If any one of these components is missing, there can be no explosion.

Many businesses offer products and solutions to counter the requirements for explosion, but the control of ignitable dust will decrease the chance for explosion as well as increase the effectiveness of these products.

It is the responsibility of the plant owners and management to be aware of the explosive properties of material in its various states and to actively eliminate the potential for explosion.

Safety Risks

The control of dust and other fugitive material is a key issue in preventing employee accidents. At any operation where dust reduces visibility and accessibility, there is an increased risk of problems in the operation of heavy equipment or the movement of personnel. The presence of dust requires cleaning and placing plant personnel in the vicinity of conveyors and other process equipment, resulting in a higher risk of injury.

Airborne dust generally creates an unpleasant working environment. Workers will have higher morale and increased productivity if conditions where they spend their working days are not seen as dirty, unpleasant, and possibly unhealthy.

In some plants, workers must wear a respirator to work in the vicinity of dust-generating material-handling systems. This increases the safety hazard due to impaired visibility, and it adversely affects morale. In addition to the dust problem, the company is viewed as not concerned with the employees' health and well-being, as they are forced to work in an environment that is possibly hazardous and certainly uncomfortable.

Table 17.2

Admissible Dust Exposure Levels per Eight-Hour Day per the Occupational Safety & Health Administration (OSHA) (USA)		
Substance	**Type**	**mg/m³**
Silca: Crystalline *Cristobalite: Use ½ the value calculated from the count or mass formula for quartz.* *Tridymite: Use ½ the value calculated from the formula for quartz.*	Quartz (respirable)	$\frac{10 \text{ mg/m}^3}{\%SiO_2+2}$
	Quartz (total dust)	$\frac{30 \text{ mg/m}^3}{\%SiO_2+2}$
Amorphous	Amorphous, including natural diatomaceous earth	$\frac{80 \text{ mg/m}^3}{\%SiO_2+2}$
Coal Dust	Respirable fraction < 5% SiO_2	2.4 mg/m³
	Respirable fraction > 5% SiO_2	$\frac{10 \text{ mg/m}^3}{\%SiO_2+2}$
Inert or Nuisance Dust	Respirable fraction	5 mg/m³
	Total dust	15 mg/m³

NIMBY and Neighbor Relations

In the "old days," the accepted excuse for visible signs of a plant's presence, like dust and odors, was that those were signals that the plant was "making money."

That is no longer true. Now, the "Not In My Backyard" syndrome—sometimes shortened to the acronym "NIMBY"—is stronger. Environmental groups are more organized. No one wants their property values impaired by the outputs of an industrial operation. Community groups are more likely to complain to, and seek support from, stockholders, environmental groups, and regulatory bodies. The need to expand an operation, such as an aggregate plant, often brings about lengthy and often contentious permitting hearings.

One solution for the increasingly-cumbersome permitting processes is for the operation to maintain good working relations with the communities in which they operate. Efforts like community donations, open houses, and plant tours work to showcase the operation and demonstrate the value the plant provides and the efforts it is making to be a good neighbor.

Those efforts can go for naught if dust clouds rise from operating equipment on a regular or even periodic basis.

Regulatory Agencies

In addition to the health and explosive dangers of dust, an operation has to be conscious of the visual nature of dust pollution. It is becoming increasingly common for industries that produce dust to be inspected and fined because of the visual pollution to surrounding homes and businesses. It is much easier to spot a dust plume from a distance than spillage. For this reason, a bulk-materials operation must be aware of the enviro-political climate in the area in which they are operating.

Regulatory agencies are empowered to protect the health of workers and others; accordingly, they will monitor and review dust-level results. Agencies will also act in the interest of other parties, including neighbors concerned about protection of their property and groups concerned with the overall environment. In addition, they are susceptible to pressure from interest groups and the media.

As a result, many companies and/or individual operations control the discussion of their regulatory limits and test results. Management does not want this information discussed outside the plant, because the results are "bad," the data is subject to misinterpretation by outsiders, or the discussion points out that dust is an issue for the facility.

Mention must also be made of other regulatory agencies, which, while not formally responsible for regulating airborne dust, are responsible for other aspects of the governmental control of industrial operations, such as land use permits and zoning. These regulatory bodies are subject to, and vulnerable to, outside influences including neighborhood residents, owners (and/or potential developers) of adjacent tracts of land, and environmental interest groups.

Problems in the Process

In addition to the environmental, safety, and health issues discussed above, there are a number of reasons to control fugitive dust in order to improve the process internally.

Dust affects the quality of an industrial operation and its output. It contaminates the plant and possibly even the finished product. Dust will settle on sensitive instruments and sensors, impairing the instruments' ability to monitor a process and confusing the data supplied to the operators. In some industrial operations, such as iron-ore sinter and pelletizing plants, material dust in the process is a contaminant that adversely affects results.

Another danger of airborne dust falls into the category of property damage. If a material is corrosive, so is the dust. As airborne dust settles on every surface within an operation, there is the potential

17

for massive damage due to operation-wide corrosion. Airborne dust will be pulled into the air intakes on motors and pumps, leading to premature failure of this critical and expensive equipment.

Dust represents a loss of valuable material, a material that has been paid for and, in many cases, has had some level of processing applied to it. Fugitive dust represents a lost opportunity for profit. In some plants, the airborne dust will have higher concentration of the operation's target mineral than the general body of material. Dust at large precious-metal mines was found to have more gold and copper than the raw ore, with concentrations increased from 25 to nearly 100 percent. The recovery of this valuable dust offers a significant payback on the investment in dust-control systems.

Dust also increases the amount of maintenance work required. It consumes labor man-hours of plant personnel, adding expense and distracting workers and managers from other responsibilities. It is important to take into account the extra man-hours needed to clean up areas where dust settles. Spillage falls below a conveyor; in contrast, dust will settle all over a plant, including elevations well above the dust's release point.

Fugitive dust can affect a plant's production capacity by reducing the availability of conveyors and equipment due to accidents, extra required maintenance, and downtime for cleanup.

MEASUREMENT OF DUST

Proper dust studies are needed to evaluate an operation's compliance with regulations as well as the effectiveness of its dust-control measures. The method of dust sampling is specific to the region and the agency doing the survey. More popular sampling methods include personal dust samplers, location-specific dust samplers, visual opacity readings, and handheld electronic dust-measuring devices. Those sampling methods are discussed below.

Personal Dust Sampling

The exposure of a worker to a concentration of dust can best be measured with a personal dust sampler (**Figure 17.2**). This is a small vacuum pump attached to a tube which is connected to the worker's collar or neck. The worker wears the sampler throughout the course of the workday. At the end of the shift, the amount of dust captured by the sampler is weighed. This weight is divided by total airflow the pump has pulled throughout the day and the time in operation to determine a concentration of dust in the air. This repeatable methodology is useful in determining the amount of dust in the air and the size of the dust particles, as well as measuring an individual's exposure to a dust hazard. Real-time particulate monitoring devices are also available.

Basic Location-Specific Dust Sampling

Basic location-specific dust sampling is usually completed by placing many pans or containers in a dusty area and leaving them for a specific period of time (**Figures 17.3** and **17.4**). The amount of dust that settles into the containers is weighed. This sampling is normally done before and after a dust-control solution is implemented.

Figure 17.2

Worn by a worker, a personal dust sampler will determine the individual's exposure to airborne dust.

17

These two values are compared to assess the relative effectiveness of the dust-control measure. While this is a basic and intuitive method to measure the effectiveness of a system, it does not provide any information about the concentration of dust in the air or the size of the dust particles. It should be noted that this type of sampling measures not the content of the dust in the air, but rather the amount of dust that settles in the place where the pans are located. The results can be affected by air currents above the pans.

Advanced Location-Specific Dust Sampling

Advanced location-specific dust sampling is a combination of the personal dust sampler and the basic location-specific dust sampler in that it uses a vacuum to capture air (**Figure 17.5**). The sampling device is placed in a fixed location for a given amount of time. The method used to analyze the concentrations is similar to basic location-specific sampling, but these systems can be much more accurate and controlled. The results can be output to a computer or some other monitoring device, allowing readings to be taken remotely.

Another version of the location-specific dust sampler is the microwave opacity tester. This device operates by releasing light or microwaves into an air stream. The light or radiation is deflected or absorbed by dust in the air. The energy is measured across the air stream, and the amount of dust can be calculated from this value by measuring the difference in the strength of the signal sent and the signal received (**Figure 17.6**). More elaborate equipment can be utilized to measure the size of the particles. While this equipment is very accurate, it tends to be expensive and not

Figure 17.3

The pans on the floor around this conveyor tail have been placed to collect samples of the dust in this location.

Figure 17.4

This drawing shows the plan for collecting the dust and spillage around the tail pulleys of two conveyors.

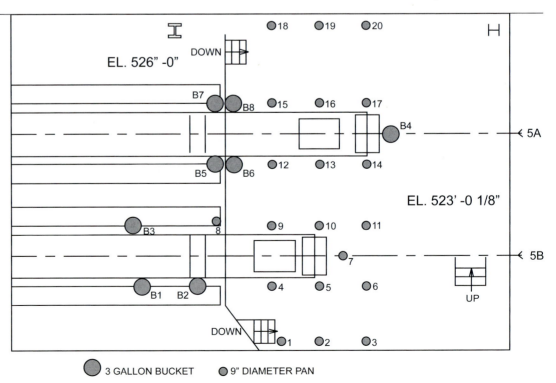

portable. This type of sensor would be employed by an operation to monitor a particular piece of equipment.

Visual Opacity Readings

A visual opacity reading is performed by a trained licensed inspector who observes the area for a set amount of time and documents the amount of visible dust in the air. While this method has been accused of being subjective, it is thoroughly regulated and generally accepted among environmental protection agencies in the United States.

Handheld Electronic Dust-Measuring Devices

Technology is always striving to make the "higher technology" (more scientific) methods of dust measurement more portable. Handheld devices can measure concentration, dust size, and many other properties of dust. As technology advances, these devices will become less expensive and more commonly seen in the field (**Figure 17.7**).

Standard Testing Methodologies

As the dust-measurement methodologies vary from region to region and application to application, it is in an operation's best interest to know who will be measuring their dust and what method they will be using. A number of national and international organizations have supplied standardized procedures for measuring dust. Some examples are listed below:

A. ASTM International (ASTM)

ASTM D4532-97 (2003) *Standard Test Method for Respirable Dust in the Workplace Atmospheres*

ASTM D6552-06 *Standard Practice for Controlling and Characterizing Errors in Weighing Collected Aerosols*

B. Deutsches Instit für Normnug (DIN-European Union)

DIN/EN 481 *Workplace Atmospheres: Size Fraction Definitions for Measurement of Airborne Particles*

C. International Organization for Standardization (ISO)

ISO 20988 *Air Quality—Guidelines for Estimating Measurement Uncertainty*

ISO 7708 *Air Quality—Particle Size Fraction Definitions for Health-Related Sampling*

ISO 12141 *Stationary Source Emissions—Determination of Mass Concentration of Particulate Matter (Dust) At Low Concentrations*

Consultation with standard organizations such as ISO, ASTM, and regulatory agencies for specific regions is advised to determine the current regulations and accepted testing methods.

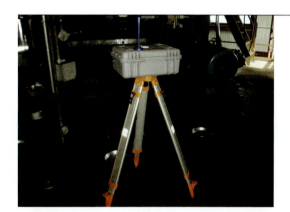

Figure 17.5

This advanced location-specific dust-sampling system uses a vacuum to pull in dust-laden air.

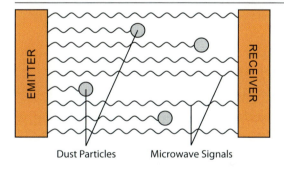

Figure 17.6

The microwave opacity tester uses the difference in the strength of the signal sent and the signal received to determine the amount of dust in the air.

Figure 17.7

Handheld devices can now measure the concentration and size of dust.

METHODS OF DUST MANAGEMENT

Minimizing the Creation of Dust

Airborne dust is created whenever a dry material is moved, manipulated, and subjected to air currents strong enough to raise or redirect the small particles within the body of material. One of the most common circumstances in which this occurs is conveyor transfer points where loading, unloading, or transit of the material creates air currents that carry the dust away from the material-handling system.

Dust emissions in conveying can be significantly reduced with an engineered transfer system, an effective sealing system, the addition of a dust-suppression system, and/or the use of an effective dust-collection system.

The first consideration in dust control should always be the minimization of the amount of dust actually created. While it is unlikely dust can be completely eliminated, any change in system design or production technique that will reduce the amount of dust produced should be considered. For example, if the energy released by the falling stream of material at the impact area can be reduced, then less energy will be imparted to the material, and fewer dust particles will be created or driven off. Consequently, it is best to design conveyor systems with minimal material-drop distances.

This type of improved engineering can be undertaken as a retrofit, or it can be considered at the initial plant-design stage. Methods to reduce the dust created through better engineering include, but are not limited to:

A. Shortening drops between conveyors

B. Loading material in the same direction as the receiving belt is moving

C. Avoiding drastic changes in material trajectory

D. Maintaining a cohesive material stream while controlling the flow of air into and out of the transfer point

These methods can be achieved through a combination of proper conveyor layout and creative design of conveyor transfers. For example, an engineered chute incorporating a "hood and spoon" can be of great assistance in achieving these methods. Other engineering improvements are available to make dramatic reductions in the creation and release of airborne dust. The success of these improvements can eliminate the need for, or greatly reduce the size and costs of, dust-collection and dust-suppression systems. *(See Chapter 7: Air Control and Chapter 22: Engineered Flow Chutes.)*

Three Ways to Control Dust

If an operation cannot prevent dust from becoming airborne, it must find ways to control it. Control can be accomplished by containing, suppressing, or collecting airborne particles. Before a dust-control system is selected, it is necessary to understand the contributing factors that create airborne dust.

The conditions that determine whether fine materials become airborne are air velocity, particle size, and cohesion of the bulk material. These characteristics contribute to the amount of dust generated by the following intuitive, relative relationship: The amount of dust generated is proportional to air velocity as divided by the factors of particle size and material cohesiveness (**Figure 17.8**).

This relationship emphasizes three important principles that can be utilized to control dust:

A. Dust creation can be minimized by lowering the air velocity around the bulk material.

Figure 17.8

Relationship in Creating Airborne Dust.

$$\text{Dust Generated} \; \alpha \; \frac{\text{Air Velocity}}{\text{Particle Size} \bullet \text{Cohesiveness}}$$

B. Dust creation can be minimized by increasing the particle size of the bulk material.

C. Dust creation can be minimized by increasing the cohesiveness of the bulk material.

Where one or more of these characteristics is a given, the ability to control dust depends on altering one or both of the other characteristics. For example, where the size of coal particles being transported cannot be changed, the air velocity or cohesive force of the particles must be altered to minimize dust emissions. Many dust-control systems combine several of these principles.

Minimizing Air Velocity

The easiest and most effective method to control dust is to minimize air velocity. Dust particles are heavier than air, and they will settle out if given still conditions and enough time. By reducing air velocity, particles have a chance to fall back into the material stream. Dust travels in the air stream, so it stands to reason that if air is controlled, dust can be managed.

Perhaps the earliest (and easiest) dust-control technology is simply to enclose the airborne dust (or the dust-generating location/operation) so the dust particles have the opportunity to settle before being carried outside the area. This is a method of minimizing air velocity, and thus preventing the pickup of fine particles from a body of material. As the enclosure volume increases, the velocity decreases, allowing the airborne particles to drop from the air.

An effectively-designed transfer chute reduces the air velocity by minimizing air drawn into the transfer point, sealing the leaks that allow dust-bearing air to escape, and allowing the dust time to settle out of the air. The traditional transfer-point enclosure is the most common method used to combat dust. The advantage of the traditional steel chutework is that it is rigid, permanent, and can completely enclose the transfer point. This makes it the ideal can-didate for an upgrade to any transfer point. Even a basic technique, such as the installation of dust curtains at the exit end of the chute, is one method to slow air movement.

An effective enclosure can theoretically be put on any transfer point. However, in some operations rigid, permanent, completely enclosed transfer points cannot be used, so the equipment cannot be enclosed. For example, many sand and gravel operations require equipment to be mobile, so a fixed transfer chute is undesirable. Other operations may need to visually monitor a transfer point, so a completely enclosed chute would not be suitable.

Increasing Particle Size

If enclosing the transfer point is not an option, then increasing the size of the particles of dust to make them heavier and more prone to dropping from the airflow may be the solution. Increasing the size of the particles of dust will make the dust particles heavier. A heavier particle will not be so easily picked up by air movement, and it will fall out of the air more readily when air velocity slows. A heavier particle also will have more momentum, so it will not be as affected by shifts in airflow.

Dust-suppression systems are generally based upon the principle of increasing dust particle weight to improve dust control and return the particles to the main material stream. These systems increase the weight of airborne dust particles by combining the particles with drops of water (or with a water-and-chemical solution). The wet, and now heavier, dust particles fall back into the material stream before they can escape into the atmosphere.

It is relatively difficult to capture dust particles once they have become airborne. Fog dust-suppression systems specifically target dust in this difficult state. *(See Chapter 19: Dust Suppression.)* A fog system needs time and relatively undisturbed space to bond with airborne dust particles. This necessitates an enclosed transfer point and relatively slow moving air. Fog systems are more successful when the dust particles and water droplets are of similar size. To achieve the small water-droplet size necessary to match the small particles of airborne dust, the water must be pumped at high pressure through atomizing nozzles or atomized with air. Both methods for making small water droplets to match small particles are expensive and complicated.

Dust collection is also used to increase material size. This method uses a vacuum to pull the air (and the dust it carries) out of the material-handling system. The dust agglomerates to itself or on the surface of the filter system and is then collected at a central location or deposited back onto the belt with the use of local collectors. *(See Chapter 20: Dust Collection.)*

Dust-collection systems require the enclosure of the transfer point(s) and a substantial amount of overhead space. Such systems do nothing to reduce the material's potential to create dust: When the material is agitated at the next transfer point, dust must be dealt with again.

Increasing Material Cohesiveness

The final common method used to minimize dust is to increase the cohesiveness of the material; that is, the material's "desire" (or ability) to stick together. The properties of the material must be altered in order to increase its ability to stick to itself. A real-life example of improved cohesiveness would be beach sand versus desert sand. Both types of sand have approximately the same size particles in solid form. Desert sand does not stick to itself; the particles can easily fracture off and become airborne as dust. The added moisture content of beach sand increases cohesion; the particles will stick together and not become airborne when the material is dropped.

A simple way to increase cohesiveness is to introduce water or another binding agent to the material. Care must be taken when applying moisture to a bulk material. If water is applied to the top of material lying on a pile or a conveyor belt, it will wet only the outside of the material. When this material is disturbed, by being reclaimed from a stockpile or by traveling through a transfer point, the particles are rearranged and dry surfaces are exposed to the air. Dust can then be released from these dry surfaces. The ideal application point for moisture addition is when the material is in free fall. This allows the water to penetrate the material and contact more of the material's surface.

The advantages of the application of water include the residual effect of the dust suppression. A wet material will retain its elevated cohesion level (and hence, the inability to generate dust) for as long as the material remains wet.

A disadvantage of the application of water is the large amount of water required to thoroughly wet most materials. The drawback is that, because wet material sticks to itself and system components, the moisture can create problems, including the blinding of screens, the plugging of chutes, and the carryback of material on the belt. Even the efficiency of a crusher is reduced with wet material. When designing a material-handling system or considering suppression as a solution for dust problems, the effect of added moisture must be considered.

Another concern of applying water is the added performance penalty that comes from wetting down a product that must be heated or burned. Individual operations must decide if the cost of fugitive dust is greater than the thermal penalty of suppression. An additional issue is that some materials, such as cement, cannot be exposed to water. A thorough understanding of an operation's material and process is required before a suppression system is selected.

One method to minimize the amount of water required for dust suppression is to improve the water's ability to wet the material with the addition of a surfactant to the water supply. The solution of water and chemical is then applied as a spray or foam. The addition of a surfactant will minimize

17

the amount of water necessary to do the job, but it increases operating costs. *(See Chapter 19: Dust Suppression.)*

REDUCING THE RISK OF FIRE AND EXPLOSION

Risks of Fire and Explosion

As evidenced by silo explosions in the grain-handling industry, dust explosions are very powerful and a very real risk. Consequently, extreme care must be taken to minimize this risk. For many dusts, a settled layer as thin as the thickness of a paper clip—only 1 millimeter (1/32 in.)—is enough to create an explosion hazard. A 6-millimeter (1/4-in.) layer is a bigger problem—big enough to destroy a plant.

For there to be a dust explosion, these factors need to be present: a confined combustible dust at the right concentration, a gas that supports ignition, and an ignition source. Many fine dusts, including chemicals, food products, fertilizers, plastics, carbon materials, and certain metals, are highly combustible, the first requirement for a dust explosion. By nature, any dust-collection device contains clouds of these fine particles suspended in air, which itself is a gas that supports ignition, the second requirement.

In any mechanical material-handling operation, there are a number of possible ignition sources, the third requirement for a dust explosion:

A. Mechanical failures that cause metal-to-metal sparks or friction

B. Fan blades that spark when they are struck by a foreign object

C. Overheating from a worn bearing or slipping belt

D. Open flames from direct-fired heaters, incinerators, furnaces, or other sources

E. Welding or cutting causing a point-source ignition or a hot-particle dropping (perhaps several floors) to a flammable atmosphere

F. Static electricity discharge

G. Migration of flammable dust into the hot region of a compressor or catalytic reactor

Categorizing Dust Explosions

There are several ways to look at dust-related conflagrations:

A. Flash fire

A flash dust fire is the sudden ignition of unconfined dust. A flash fire is usually localized and can cause significant damage or injury. A flash fire can create the conditions for a secondary explosion, which can cause catastrophic damage and fatal injuries.

B. Explosion

When dust is confined and ignited, an explosion is created. This rapid explosion of gases will generate significant and destructive over-pressures that can even demolish the building, leading to greater damage and injury.

C. Primary or secondary

An initial or primary explosion can cause secondary explosions by disrupting, dispersing, and igniting new sources of dust removed some distance from the original blast. Secondary explosions can be more destructive than the primary explosion, and every explosion can lead to additional secondary explosions.

D. Magnitude

The speed and force of an explosion are direct functions of a measurable characteristic called the deflagration index. Dust explosions can be more hazardous than explosions caused by flammable gases.

Control Mechanisms

Where the ingredients—confined combustible dust at the right concentration, ignition-supporting gas, and an ignition source—are present, precautions must be taken to avoid an explosion.

17

These precautions include:

A. Inerting

The addition of an inert gas (typically nitrogen or carbon dioxide, rather than air) into the collector

B. Suppressing

Adding a suppressant material as explosive pressure starts to rise

C. Venting

Adding an explosion relief-panel or bursting membrane, which releases the explosion energy out of the enclosure

Proper grounding of the dust-control systems will help reduce the risk by increasing conductivity through the system, allowing static charges to leak into the ground.

It is advisable to consult with equipment suppliers to design dust-collection systems for handling potentially explosive dusts.

Venting

The theory behind explosion vents is simple. The vent is a deliberately weakened wall that will release early in the pressure rise created by a rapidly rising temperature. Once this weakened area is opened, the burned and unburned dust and flame can escape the confined area, so the vessel itself does not experience the full rise in pressure. If the release is early and large enough, the pressure will remain low inside the vessel to protect it from damage. However, the fire or explosion can develop outside the vessel, and if dust is present, other equipment

SAFETY CONCERNS

As noted above, airborne dust is a safety issue in and of itself, but an operation must also be aware of the safety concerns associated with its dust-control equipment. In addition to the standard latent-energy risks associated with any piece of industrial equipment, a bulk-materials operation must be aware of the potentially explosive nature of its dust-control solutions.

If an operation is attempting to control dust to prevent an explosion, care must be taken to ensure the dust-control equipment has the proper hazardous-duty rating for the expected conditions. Any electrical enclosure or motor must be spark-resistant or rated for hazardous duty.

When air moves through a filter media, the media develops a static charge. If the media is close to a grounded structural member, the static charge could discharge as a spark, possibly igniting any flammable dust in the air. Filter manufacturers have created media that dissipate a static charge through stainless steel mesh embedded in the filter material or through conductive carbon fibers woven into the material (**Figure 17.9**). These additions allow any charge generated by the airflow to move to the ground before it can spark. If a filter is used in an explosive environment, care must be taken to select static-discharging filter media.

The established safety procedures for entry into any confined spaces, including transfer-point enclosures and dust collectors, should be closely followed. Proper lockout / tagout / blockout / testout procedures must be completed and water, chemical, and electrical sources must be de-energized prior to performing maintenance on dust-collection or dust-suppression systems.

Figure 17.9

Dust collector filters should include a conductive thread woven into the filter fabric to carry any static buildup safely to ground.

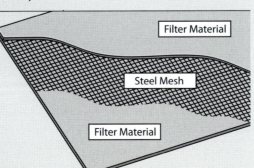

Filter Material

Steel Mesh

Filter Material

17

can be damaged, so venting systems do not eliminate the need for dust-control systems and high standards of housekeeping.

There are two types of explosion-venting devices. Rupture disks are thin panels that will open faster than other designs. They must be sized to withstand normal negative operating pressure—typically 2 to 3 kilopascals (0.29 to 0.44 lb_f /in.^{2})—yet rupture at a positive explosive pressure. A more widely used design is the spring-set door. This door—available in either hinged or unhinged designs—will vent (pop open) during a conflagration.

DUST-CONTROL SYSTEM MAINTENANCE

Adequate room for access and maintenance must be provided during installation of dust-control systems. The serviceability of a dust-control system must be considered when comparing systems prior to purchase. One dust collector may be less expensive than another, but the less expensive option may require a man lift and the removal of a wall to change a fuse, whereas the more expensive unit may have the fuses in an enclosure at ground level. It is in an operation's best interest to purchase a dust-management system that is maintenance-friendly. If a component causes a shutdown, every extra minute taken to repair that component affects the overall profitability of the facility.

Dust-control solutions are usually multiple-component systems that require multiple inputs. A dust collector typically requires electricity and compressed air; a foam-suppression system may need electricity, compressed air, water, and chemicals. With these elaborate systems, there are more parts that can wear out or break down. Particular attention should be paid

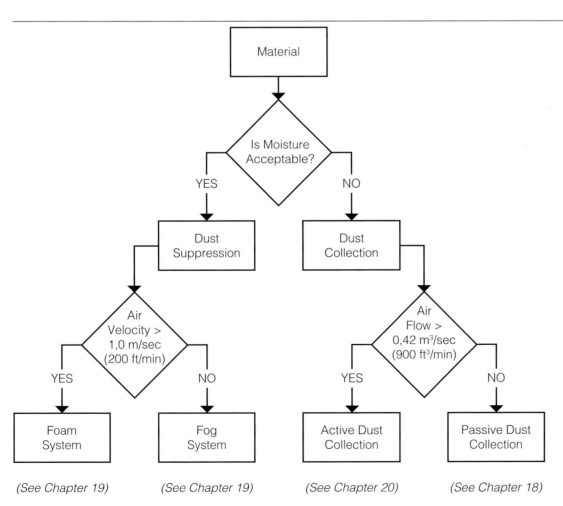

Figure 17.10

Dust-Management Selection Process Flow Chart

17

to dust-control system(s) in the operation's scheduled maintenance cycles. The operation has to take ownership of the service requirement or delegate it to a contract maintenance service company.

DUST-MANAGEMENT SELECTION PROCESS

The selection of the best dust-management technology to match the requirements of a given operation begins with an understanding of the material and the dimensions of the conveyor transfer point. There is a simplified approach to making the determination of what systems would be appropriate (**Figure 17.10**).

APPLICATION-SPECIFIC DUST MANAGEMENT

In Closing...

Every industry has preferred dust-control methods that are determined by application preference and regulations in the industry or geographic location.

The systems to control dust include containment, suppression, and collection. These systems can be used individually or in combination. There are a variety of techniques and technologies available to accomplish any one of these methods of dust control.

To successfully combat dust, any bulk-materials handling operation must understand all aspects of its problem. These aspects include the consequences, the sources, the methods of measurement, and the methods of control. An operation must select the most appropriate solution based upon the needs of the operation and the limitations of the application. Whatever solution is selected, the operation must be conscious of the safety and maintenance requirements to keep its dust-control system operating efficiently.

Looking Ahead...

This chapter introduced the section Dust Management and provided an overview of the topic while explaining the importance of controlling dust. The following three chapters continue the discussion of dust management, looking at various aspects more in-depth: Passive Dust Control, Dust Suppression, and Dust Collection. If all of the pieces of the dust-management system fit together correctly, the operation will become cleaner, safer, and more productive.

REFERENCES

17.1   Conveyor Equipment Manufacturers Association (CEMA). (2005). *BELT CONVEYORS for BULK MATERIALS, Sixth Edition*. Naples, Florida.

17.2   Any manufacturer and most distributors of conveyor products can provide a variety of materials on the construction and use of their specific products.

17

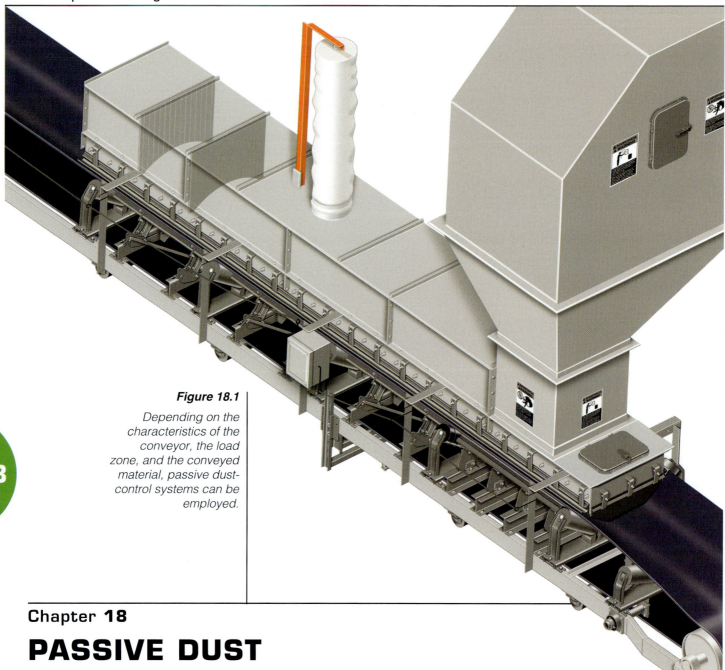

18

Figure 18.1

Depending on the characteristics of the conveyor, the load zone, and the conveyed material, passive dust-control systems can be employed.

Chapter **18**

PASSIVE DUST CONTROL

In this Chapter...

In this chapter, we discuss a variety of methods used for passive dust control that can be incorporated into initial conveyor design or added later as the need arises: methods to suppress dust and to capture it. Information about installation and ways in which those methods would be used in different applications is included.

Conveyor loading zones and discharge points are prime sources for the creation and release of airborne dust. There are a variety of systems to control airborne dust that can be installed at conveyor loading and unloading zones. Choice of the right system will depend on a number of factors, including the nature of the material carried, the height of drop onto the belt, and the speeds and angles of unloading and loading belts.

Depending on the characteristics of the conveyor, the load zone, and the conveyed material, passive dust-control systems—systems that do not require external supplies such as electricity or water—can be employed (**Figure 18.1**).

MINIMIZING DUST AT TRANSFER POINTS

While it is unlikely dust can be completely eliminated, the first consideration in dust control should always be the minimization of the amount of airborne dust created. Therefore, any change in system design or production technique that will reduce the amount of dust produced should be considered.

For example, minimizing the drop height reduces the amount of energy imparted to the fines and cuts the amount of dust driven off into the air. Consequently, it is best to design conveyor systems with minimum practical material-drop distances.

Since it is generally not possible to totally prevent the creation of dust, other systems to suppress and capture it must be employed. In their simplest form, these dust-control systems involve nothing more than attention during the engineering of the transfer point to the need to reduce airflow.

Airflow through the system can be managed by minimizing the amount of air entering the transfer point, building the enclosure large enough to slow or minimize airflow, and utilizing additional control measures to slow air movement. As air velocity is reduced, airborne particles are too heavy to be supported by the reduced air speed and begin dropping from the air stream.

CHUTES AND SETTLING ZONES

Enlarging the Settling Zone

As an example of Bernoulli's Principle, the Ventui effect is that a current of air speeds up as it passes through a constriction. This is due to the rise in pressure on the upwind side of the constriction and the pressure drop on the downwind side, as air leaves the constriction. In keeping with this basic physics principle, to slow airflow through the transfer point, the enclosed area should be made larger.

On conveyor transfer points, this enclosed area is called a settling zone (**Figure 18.2**). A settling zone is the area past the loading zone's impact area. The length of the settling zone is designed to slow the airflow and allow airborne dust to return to the main material cargo. *(See Chapter 11: Skirtboards, especially Equation 11.1.)* The

Figure 18.2

A settling zone is the area past the loading zone's impact area, where the airflow is slowed and airborne dust is allowed to return to the main material cargo.

height of the settling zone should be such that the calculated airflow through the transfer shall be slowed to less than 1,0 meters per second (200 ft/min). *(See Chapter 11: Skirtboards, especially Equation 11.2.)*

Modular Chutewall Systems

Skirtboard areas can be built or enlarged to serve as effective settling zones through the use of modular chutewall systems (**Figure 18.3**). These systems use formed wall panels with a bolt-together assembly method to combine the economy of prefabrication with the ease of on-site assembly. They come in standard sizes and can be combined to fit most settling-zone requirements.

Figure 18.3

Modular chutewall systems use formed wall panels with a bolt-together assembly method to combine the economy of prefabrication with the ease of on-site assembly.

Figure 18.4

The modular system can be used for design and construction of the settling zones on a new transfer point or the modification of existing transfer points.

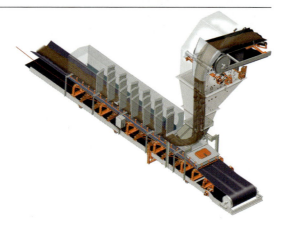

Figure 18.5

Engineered flow chutes typically incorporate a "hood and spoon" design that directs and confines the stream of moving material.

Existing settling zones can be readily enlarged using these modular chutewall or skirtboard systems (**Figure 18.4**). The modular system can also be used for the design and construction of the settling zones on new transfer-point construction to take advantage of its simplicity and off-site fabrication. Modular systems make it easier to update systems if transfer-point alterations are required by changes in material or conveyor specifications.

Engineered Flow Chutes

An advanced approach to passive dust control is the use of engineered flow chutes. These chutes usually incorporate a "hood and spoon" design that directs and confines the stream of moving material (**Figure 18.5**).

The hood minimizes the expansion of the material body, deflecting the stream downward. The spoon provides a curved loading chute that provides a smooth line of descent, so the material slides down to the receptacle, whether that is a vessel or the loading zone of another conveyor. The spoon "feeds" the material evenly and consistently, controlling the speed, direction, and impact level of the material in the load zone.

Basically, these "hood and spoon" chutes keep the material stream in a tight profile and minimize the disruption of the natural flow of material through the transfer. Keeping the material in a consolidated body reduces the amount of air that is induced into the transfer point; controlling the path of the material reduces impact and, therefore, dust generation.

By reducing the velocity and force of material impact in the load zone to approximate the belt speed and direction, this system mitigates splash when material hits the receiving conveyor. Therefore, there is less dust and high velocity air escaping. As the material is deposited more-or-less gently on the belt, there is minimal tumbling or turbulence of the material on the belt. There is less impact, which will reduce damage to the belt, and less side forces that push material out to the sides of the belt.

In some cases, either the hood or spoon is used, but not both. Sometimes there is not enough space to include both in the design. For very adhesive materials, the hood can be used to direct the stream downward for center loading. This variation is often seen on overland conveyors handling highly variable materials or when handling sticky materials, like nickel concentrate or bauxite. Gravity and the flow of material will tend to keep the hood from building up and plugging the chute. In other cases, with free-flowing materials, only the spoon is used to change the direction of the stream to minimize belt abrasion and skirt side pressure. Spoons are prone to backing up or flushing if the characteristics of the bulk materials are variable. Some leeway can be designed into the spoon for variability of materials. The main drawback to using the "hood and spoon" concept is initial cost for these specially-designed components. However, where they can be applied and maintained, they will offer significant benefits in reduced dust, spillage, and belt wear.

This "hood and spoon" system works best when the bulk-material flow rate is kept as uniform as possible. The success of this system may well eliminate the need for active dust-collection systems in many operations. *(See Chapter 22: Engineered Flow Chutes.)*

CONTROL OF AIR ENTERING THE TRANSFER

The material load begins to fan out as it leaves the head pulley, and as it spreads, it tries to pull in more air. Therefore, the chute entry area should be sealed off as much as possible to prevent additional air from being induced by the moving stream of material.

A technique employed to minimize induced air is to cover the inbound portion of the conveyor for several feet before it enters the head (discharge) chute.

This enclosure includes barriers between the carrying and return sides of the belt, between the return side of the belt and the chute, and between the chute and the top of the material load (**Figure 18.6**). Often these barriers are formed of rubber sheets or curtains. The idea is to close off as much area as practical to reduce the amount of air which will be pulled into the material flow as the material spreads out when it discharges from the head pulley.

Air-supported conveyors should be considered, because they have the advantage of an enclosed carrying side, which will restrict the flow of air into the discharge chute. Utilizing barriers to block the flow of air into the discharge chute will reduce the ability of the spreading material to entrain air into the material stream, air that would eventually be released when the material stream lands on another belt, storage vessel, or stockpile. It is equally important to control all other openings in the chute to minimize the open areas where air can be drawn into, or dust exhausted from, the chute. Access doors and openings around shafts and sensors need to be fitted with seals.

A technique to reduce air induction at the belt entry to a loading zone is the installation of a barrier—often formed from a dust curtain, used belting, or a sheet of rubber—between the return run and the carrying run. Placed from one side of the head chute to the other, the barrier partially encloses the head pulley, isolating it and reducing air flow. All openings in the head chute must be sealed to reduce induced air as much as possible. This could include shaft seals, inspection doors, belt-cleaner openings, and belt entry and exit areas.

Figure 18.6

Rubber curtains can be used to form barriers to prevent air intake around the conveyor belt as it enters the transfer system.

EXIT-AREA DUST CURTAINS

Another technique for passive dust control is the installation of dust curtains near the exit end of the transfer point's settling-zone area (**Figure 18.7**). Here, where the belt is leaving the transfer point, the rubber curtains provide a barrier, or baffle, that quiets air velocities, allowing airborne dust to fall back onto the belt. The curtains form a "settling zone" to reduce airflow and allow dust to settle out.

Most conveyors benefit from the installation of at least two curtains. Some installations, especially those where dust-collection and/or -suppression systems need to be isolated, can benefit from installing additional curtains.

These rubber curtains can be fabricated as individual curtains as wide as the skirtboard; or they can be fabricated as some fraction of the skirtboard width and then installed in alternating, or "staggered," fashion to slow airflow (**Figure 18.8**).

These curtains should be composed of 60 to 70 durometer elastomer and extend to approximately 25 millimeters (1 in.) from the top of the pile of conveyed product on the belt. The curtains are installed down through the top of the transfer-point enclosure. Curtains are often field trimmed to match trough angles and the profile of the material load.

Rather than place the curtains at the end of the covered chutework, it is better they be installed inside the covered skirtboard at a distance of 300 to 600 millimeters (12 to 24 in.) from the end of the chutework. The faster the belt speed, the further inside the chutework they should be installed. When the curtain is at the end of the steel enclosure, any material particles hit by the curtain can be displaced from the belt. By placing the curtains so the final curtain is inside the end of the enclosure, any material that contacts the curtains still has room to settle into a stable profile within the confines of the enclosed area. The curtains should be hung roughly 450 millimeters (18 in.) apart, forming an area where dust can settle or where dust-collection or -suppression systems can be applied. Use of dual dust curtains in combination with dust-suppression systems is a patented technology of The Raring Corporation (website: raringcorp.com). If curtains are used to isolate dust-suppression and/or dust-collection installations, it is better if they are located 900 millimeters (36 in.) apart.

In cases where two or more curtains are installed, the inner curtains can be solid (unslit) rubber to improve their air control capability. Only the final, or exit, curtain needs to be slit to reduce the risk of material being "kicked" off the belt.

The curtains should allow easy maintenance access to the chute and should be readily removable to allow replacement.

DUST BAGS

It is important that positive air pressure—the force of air moving through and away from the loading zone—is controlled in a manner that will minimize the outward pressure against the sealing system and reduce the release of dust.

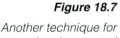

Figure 18.7

Another technique for passive dust control is the installation of dust curtains close to the end of the chutework, at the exit end of the transfer point's settling zone.

Figure 18.8

Dust curtains can be fabricated as some fraction of the skirtboard width and then installed in alternating, or "staggered," fashion to slow airflow.

One passive approach is the installation of at least one dust-collector filter bag (**Figure 18.9**). Dust bags can be protected from weather and still provide a method of collecting dust without the need for a baghouse central-collection system; they are often used when an enlarged settling zone is not possible or there is a large amount of generated air to control. These bags filter outgoing air to minimize the escape of dust into the plant environment. These systems consist of an open port in the roof of the skirted section with a filter bag, sock, or sleeve stretched over the top of the port (**Figure 18.10**). These bags can be attached with a simple circular clamp to the rim of the port. The positive air pressure is relieved through the dust bag, and the dust is captured on the inside of the bag. A transfer point may require installation of more than one of these dust bags, depending on the size and permeability of the bag and the airflow of the transfer point.

Dust bags usually feature a grommet at the top of the bag, which allows them to be hung from overhead supports (**Figure 18.11**). While the bag could extend without the support arm, the bag might be subject to wind or lay over on its side, vulnerable to damage. In installations where these pressure-relief bags are subject to environmental influences, such as snow or rain, the bags should be installed in a protective shelter.

Every bag has a certain airflow capacity based on the permeability of the filter material and the surface area of the bag. The size and number of bags required is directly related to the bag properties. *(See Advanced Topics: Calculating the Size of a Dust Bag.)*

A dust bag is typically installed at a point one-third the length of the transfer chute downsteam from the load zone. Installation of dust curtain(s) inside the skirted area, one on each side of the bag, is recommended, because that will slow the airflow, allowing more air to exit through the dust bag.

Care should be taken to make sure there is adequate clearance above the chutework to allow for full extension of the bag and for the installation of its support structure. The dust can be released from the filter bag mechanically, by shaking manually, or even by the partial collapse of the bag when the outflow of air stops during conveyor downtime.

Dust bags might generate a static charge when used. This charge could cause a spark, which could lead to an explosion if the conditions are predisposed to it. To combat this phenomenon, bag manufacturers are weaving a stainless steel grid into the material and grounding the grid to dissipate any charge that accumulates. This

Figure 18.9

Dust bags can be protected from the weather and still provide a method of collecting dust without the need for a baghouse central-collection system.

18

Figure 18.10

Dust bags consist of an open port in the roof of the skirted section with a filter bag, sock, or sleeve stretched over the top of the port.

Figure 18.11

Dust bags usually feature a grommet at the top of the bag, which allows them to be hung from overhead supports.

dissipation can also be accomplished by weaving conductive carbon fibers into the fabric. If there is a potential for explosive dust, static-dissipating dust bags must be used.

TYPICAL SPECIFICATIONS

A. Dust bags

The skirtboard cover will be fitted with one (or more) dust bags to relieve excess positive airflow and capture airborne dust. Each bag will be sized to relieve 0,5 cubic meters per second (1000 ft³/min) of airflow in the transfer point. The bag will fit over a port in the roof of the skirtboard and be hung from a support arm attached to the cover. The bag will incorporate static-dissipation technology to reduce the risk of dust explosion. The fabric used in the bag will be suitable for the bulk material being handled.

B. Dust curtains

The settling zone of the chutework will be fitted with at least two dust curtains to reduce airflow surges and increase the length of the airflow path. The curtains will be fabricated from elastomer rubber and will hang down from the skirtboard covering. The bottom edge of the curtains will be field-tailored to match the conveyor trough angle and the profile of material on the belt. The dust curtains will be spaced 450 millimeters (18 in.) apart, and the final curtain will be mounted no closer than 300 to 600 millimeters (12 to 24 in.) from the exit of the skirtboard.

C. Settling zone

The settling zone will be fabricated from materials suitable for the bulk material being handled. The length and height of the settling zone should be calculated to reduce the transfer point air velocity to less than 1 meter per second (200 ft/min). *(See Chapter 11: Skirtboards, especially Equations 11.1 and 11.2.)*

ADVANCED TOPICS

Calculating the Size of a Dust Bag

Dust bags can vent a finite amount of air per interval of time. This airflow rate is proportional to the permeability of the filter media and the area of the bag.

SAFETY CONCERNS

To reduce the risk of explosions, the fabric of the dust bag can be woven to incorporate a grounding wire inside the weave of the fabric (**Figure 18.12**). These wires will carry any static electrical charge to the ground. The wires provide low electrical resistance, in keeping with Deutsches Instit für Normung (DIN) Standard 54345 Parts 1 and 3. Depending on location and access, it may be necessary to use lockout / tagout / blockout / testout procedures on the conveyor before performing service work on dust bags. If the dust being collected is a health risk, appropriate personal protective equipment (PPE) must be used, and appropriate disposal methods must be followed.

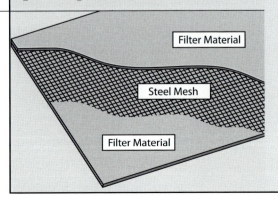

Figure 18.12

To reduce the risk of explosions, the fabric of the dust bag can be woven to incorporate a grounding wire inside the fabric.

Filter Material

Steel Mesh

Filter Material

The process of sizing a dust bag to an application is as follows:

A. Find airflow. This can be measured or calculated. *(See Chapter 7: Air Control for additional information on total airflow and specifically Equation 7.1 – Total Airflow Calculation.)*

B. Apply a reasonable safety factor to the airflow.

C. Select a filter media that will allow air through but will stop the dust present in an application.

D. Verify the need for static-dissipating fabric.

E. Find the area of the fabric by dividing the airflow required by the permeability of the fabric (**Equation 18.1**).

F. Design the bag to have the area required and to fit into the application geometry. This area can be achieved with a single or multiple bags.

WHEN PASSIVE CONTROLS ARE NOT ENOUGH

In Closing...

Because it is really not possible to totally prevent the creation of dust, passive dust-control methods can be successfully employed to suppress and capture dust. A variety of methods are available, and the use of any one system is dependent on characteristics of the conveyor, the load zone, and the conveyed material.

However, there are plenty of materials-handing systems in which material conditions and/or the design of the process will require additional dust-control systems. These systems will require active dust-management technologies, including dust-suppression and/or dust-collection systems. The choice of dust suppression and/or collection will be determined by other criteria, including the material, how it is being moved, and the next step in the process. *(See Chapter 19: Dust Suppression and Chapter 20: Dust Collection for more information.)*

Looking Ahead...

This chapter, Passive Dust Control, the second chapter in the section Dust Management, described methods of dust control that do not require external supplies such as electricity or water. The following two chapters continue this section and describe methods for active dust control: Dust Suppression and Dust Collection.

18

$$A = \frac{SF \cdot Q_{tot}}{P_f}$$

Equation 18.1

Area of Filter Bag Calculation

Given: A dust bag must dissipate 0,25 cubic meters per second (540 ft³/min). The material permeability is 0,127 meters per second (25 ft/min). Assume a 1.25 safety factor. **Find:** The area of filter media required.

	Variables	Metric Units	Imperial Units
A	Area of Filter Bag	square meters	square feet
SF	Safety Factor	1,25	1.25
Q$_{tot}$	Total Airflow	0,25 m³/s	540 ft³/min
P$_f$	Permeability	0,127 m/s	25 ft/min

Metric: $A = \dfrac{1{,}25 \cdot 0{,}25}{0{,}127} = 2{,}5$

Imperial: $A = \dfrac{1.25 \cdot 540}{25} = 27$

A	Area of Filter Bag	2,5 m²	27 ft²

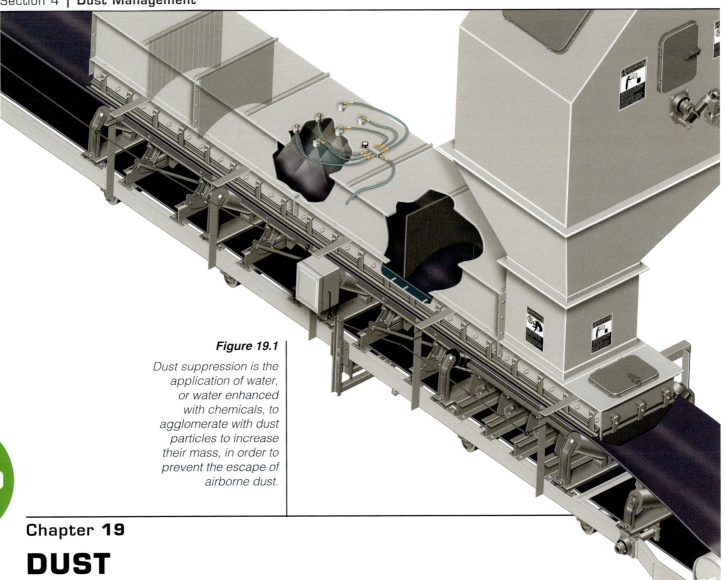

Figure 19.1

Dust suppression is the application of water, or water enhanced with chemicals, to agglomerate with dust particles to increase their mass, in order to prevent the escape of airborne dust.

19

Chapter **19**

DUST SUPPRESSION

In this Chapter...

This chapter examines various types of dust-suppression systems, including water or additive-enhanced water-spray systems, foam systems, and fog systems. Advantages and drawbacks of the systems, along with general guidance on applying the various methods, are addressed. This material is intended to be descriptive, rather than prescriptive. Any application of this information should be guided by experienced professionals with knowledge of the specific application.

Dust suppression is the application of water, or water enhanced with chemicals, to agglomerate with dust particles to increase their mass, in order to prevent the escape of airborne dust. The water, or water/chemical mix, can be applied to either a body of material—to prevent fine particles from being carried off into the air—or the air above a body of material—to create a curtain, or barrier, that returns the wetted airborne fines to the material.

There are a number of systems used for this purpose, ranging from "garden hose" water-spray systems through sophisticated, engineered, and automated systems that apply water—with or without added chemicals—as a spray, foam, or fog (**Figure 19.1**).

An advantage of dust-suppression systems is that the treated material does not have to be handled again in order to be reprocessed, as it would with a dust-collection system. The suppressed dust is returned to the main body of conveyed material and then proceeds on in the process, without requiring additional materials-handling equipment to reclaim the material.

A dust-suppression system cannot be recommended in any case where the material would react adversely to the addition of moisture or to the return of the dust to the process.

DUST SUPPRESSION

Evaluating the Options

The selection of the best dust-control solution for each application depends on a number of factors. The key is an understanding of the material, the application conditions, and the level of performance required.

Some general guidelines on the applicability of the various dust-suppression methods are available (**Table 19.1**).

Installation costs and continuous operating costs for power, chemicals, and maintenance should be reviewed. Another consideration in system selection is the availability of resources like water, compressed air, and electric power.

A plain water spray may present the lowest operating costs, but it may also be the least effective solution.

Size Matters

The basic principle of dust-suppression systems is that dust particles (whether airborne or contained in the body of the bulk materials being handled) are more likely to interact with water particles of the same relative size.

When water droplets mix and agglomerate with dust particles, the resulting heavier combined particles fall back to the body of material. For maximum efficiency, a dust-suppression system's water droplets must be kept within the specific size range of the airborne dust. If the water droplets are too large, the smaller dust particles will typically just "slipstream" around them, pushed aside by the air around the droplets (**Figure 19.2**). If the water droplets are properly sized and are provided in sufficient quantity for the given area, the droplets will bond with the material particles and drop from the air.

19

Table 19.1

Dust-Suppression Application Matrix

Type of Dust-Suppression System	Applications That Have:						
	Transfer Point	Crushers & Mills	Stock Piles	Rail Car Dump Station	Trippers	Ship Loading	Ship Unloading
Water Spray	X			X			
Water Fog	X						
Water + Air Fog	X			X		X	
Water + Surfactant Spray	X		X	X	X	X	X
Foam	X	X	X		X	X	
Hybrid System Dust Suppression + Passive Dust Collection	X	X					
Hybrid System Dust Suppression + Active Dust Collection	X	X		X			X

Notes: *Water + Surfactant Spray and Foam are best when a residual effect is needed (multiple application points, crushers, long distances between application points, stackers, etc). Water Spray, Water Fog, and Water + Air Fog are best when a residual effect is not required. Some kinds of materials and/or processes do not permit the addition of any chemical.*

19

The key is to provide dust-suppression system droplets at the same size as the particles of dust and in the same vicinity to provide the best opportunity for maximum interaction between the two. With the simplest water-spray systems, additional small droplets of water are created by spraying more water. The more water that is sprayed, the better will be the opportunity that properly-sized droplets of water are created. Water-and-surfactant-spray systems improve capture efficiency by controlling the wetting ability of the water through the addition of surfactants. These surface-acting agents make the water "wetter,"

increasing the efficiency of capturing dust, thus allowing a reduction in the amount of water supplied. Fog and foam suppression systems rely on other methods—atomization and chemicals, respectively—to create the small droplets necessary to effectively capture the dust.

A critical ingredient in the selection of a dust-suppression system is an understanding of the characteristics of the materials being handled. Some materials, such as cement, are incompatible with water; therefore, suppression should be avoided. Dust particles from some materials bond readily with water, whereas particles from other materials do not. The addition of a chemical surfactant can improve the ability of water to bond with these normally hydrophobic materials. A suppressant chemical supplier should be consulted to determine the effectiveness of a given chemical with any specific material. A thorough knowledge of the ramifications of adding moisture to a material and to a process must be obtained before implementation of any method of dust suppression.

Figure 19.2

Dust particles may "slipstream" around larger water droplets but join readily with droplets of similar size.

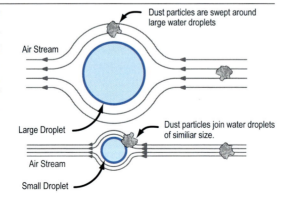

Dust particles are swept around large water droplets

Air Stream

Large Droplet

Dust particles join water droplets of similiar size.

Air Stream

Small Droplet

WATER SUPPRESSION

Suppression with Water

Perhaps the oldest method for controlling fugitive dust is the application of water sprayed over the body of material. By wetting fine particles as they lay in the bulk materials, or as they are being carried by the air, the weight of each particle is increased, making it less likely for the particle to become, or stay, airborne. Moisture increases the cohesive force between dust particles, making them more likely to agglomerate—creating larger, heavier groupings of particles—and making it more difficult for air movement to carry off fines. This is most effective when applying the water through a series of properly-sized spray nozzles at a point where the material expands and takes in air, such as during discharge from the head pulley in a transfer chute.

Water can also be applied to create a "curtain" around a transfer point. Dust fines that become airborne will come into contact with this water "barrier," increasing their mass—removing them from the air stream.

The most effective water sprays are low-velocity systems. High-velocity sprays can add velocity to the air and the dust particles in the air. This energy is counterproductive to the task of returning the dust to the material body.

The Pluses and Minuses of Water Sprays

Water-based suppression systems become more sophisticated as the engineering moves beyond "water hose" technology in the effort to improve results. The effectiveness of water-spray systems is dependent on the velocity of the applied water, the droplet size, the size of the nozzle opening, and the location and number of spray nozzles. Techniques to improve water-spray dust suppression include a reduction in droplet size, an increase in droplet frequency, or a decrease in the droplet's surface tension, making it easier for droplets to merge with dust particles.

Water-spray systems offer some advantages. The application systems are relatively simple to design and operate. Water is generally inexpensive, relatively easy to obtain, and generally safe for the environment and for workers who are exposed to it. Dust-suppression systems utilizing water are relatively simple systems and do not require the use of costly, elaborate enclosures or hoods. Changes can be made after start-up with minimum expense and downtime. Water-based suppression systems are simple to install, less subject to problems from wind or air velocity, and, due to the large orifices in their spray nozzles, do not normally require filtered water. The systems are typically cheaper to install and use far less space than "dry" dust-collection systems.

Unfortunately, the application of water has several liabilities as well. Restrictions on fresh water consumption are common in mine operating permits, as well as in many other industrial operations. Most water suppression systems must use recycled process water, rather than more expensive potable (drinking-quality) water. This process water may have contaminants or chemicals in it that can clog or corrode the spray components. The use of water may promote accelerated corrosion of conveyor structures and components.

Another drawback is that water has only a minimal residual effect—once the water evaporates, the dust-suppression effect is gone. In addition, large droplets of water are not good at attaching to small dust particles: To increase the result, more water is often applied, which can create disposal and cleanup problems.

Various levels of moisture are added in typical dust-suppression systems (**Table 19.2**).

With Water, Less is More

A water spray may appear to be the most inexpensive form of dust control available, as process water is available almost free in many operations, and it can be applied through low-technology systems. This cost justification can be a false assumption, as the addition of water may adversely affect materials-handling operations. Many bulk materials are hydrophobic—they have a high-surface tension and are averse to combining with water. In an effort to achieve effective suppression, the amount of water is increased. Because the material does not mix well with water, some particles will remain dry, and others will become very wet. The dry material will continue to create dust, possibly leading to the addition of even more water, worsening the problem. The overly-wetted material will lead to handling problems, including accumulations on chutewalls, plugged screens, reduced efficiency and shortened wear-life on crushers, and carryback on conveyor belts. Excess water may promote belt slippage and belt mistracking, and it may increase the possibility of wet (hence, sticky) fines accumulating within chutes and around the transfer point. When applying water to materials on conveyor systems, a good axiom is "less is more."

Another problem occurring in "process water" dust-suppression systems is the possibility of excessive moisture in the materials, which can downgrade performance in power-generation or other thermal-processing systems. Excessive water added to coal and coke used for boiler fuel results in a thermal penalty that can have a detrimental effect on utility heat rates. The more water added, the greater is this penalty.

The Thermal Penalty for Added Moisture

There is a substantial performance penalty added to combustion and other thermal processes when the water content of the fuel is significantly increased. In applications like coal-fired power plants and cement plants, water added to the material going into the process must be "burned off" by the process. This can dramatically reduce operating efficiency and increase fuel costs.

Some bulk materials are susceptible to naturally varying moisture contents from their exposure to weather in storage or during transport. Many bulk materials, such as coal, are hygroscopic, meaning they

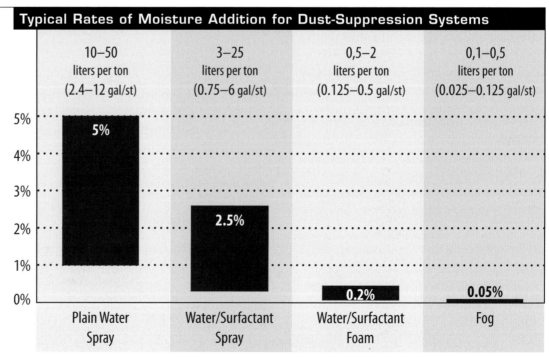

Table 19.2

Typical Rates of Moisture Addition for Dust-Suppression Systems			
10–50 liters per ton (2.4–12 gal/st)	3–25 liters per ton (0.75–6 gal/st)	0,5–2 liters per ton (0.125–0.5 gal/st)	0,1–0,5 liters per ton (0.025–0.125 gal/st)
5%	2.5%	0.2%	0.05%
Plain Water Spray	Water/Surfactant Spray	Water/Surfactant Foam	Fog

Dust-Suppression System Types

can absorb moisture from the air. Coals have the capacity to absorb free moisture at levels ranging from 2 to 45 percent of their weight. This absorption occurs rapidly, with from 1.5 to 5.5 percent weight gain in the first 15 minutes of exposure. A steady state occurs within 3 to 5 days of exposure. Often these natural changes are much more significant than the amount of water added by a well-designed and -maintained dust-control system. Any added water can present an added cost to the system and affect heat rate and plant efficiency; therefore, efforts to minimize moisture addition should be carefully considered.

With the thermal output of coal ranging from 16300 kilojoules per kilogram (7000 BTU/lb$_m$) for lignite to 27900 kilojoules per kilogram (12000 BTU/lb$_m$) for bituminous coal, a power plant loses roughly the heat from 1 to 1,5 kilograms per ton (1.9 to 3.3 lb$_m$/st) for every one percent of moisture added in the coal-handing operation. The plant must purchase, handle, and burn additional fuel to compensate for the added moisture. *(See Advanced Topics: Thermal Penalty in a Coal-Fired Power Plant.)*

Improving Water-Based Suppression

Because a water-only-spray system requires a high volume of moisture addition for effective dust suppression, it applies a high thermal penalty. Significant quantities of water can also create problems in materials handling.

Other methods to improve water-based dust suppression while limiting the addition of moisture should be considered. These solutions include generating a fine mist or "fog" spray, or using chemical additives to modify the water.

FOG SUPPRESSION

Fog Suppression Systems

The use of a water fog for dust suppression is one method to optimize the application of water to dusty materials. These systems use special nozzles to produce extremely small water droplets in a "cloud," or dispersed mist (**Figure 19.3**). These droplets mix and agglomerate with dust particles of similar size, with the resulting combined and heavier particles falling back to the material body. Fog systems are based on the knowledge that a wet suppression system's water droplets must be kept within a specific size to effectively control dust. If water droplets are too large, smaller dust particles typically just "slipstream" around them, pushed aside by the air around the droplets.

Fog systems supply ultra-fine droplets that maximize the capture potential of the water while minimizing the amount of water added to the product. Atomization reduces the surface tension of the water droplets, while increasing the number of droplets in a given area.

Fog systems generally add low levels of moisture to the material, typically in the range of 0.1 to 0.05 percent (1/10th to 1/20th of 1 percent) by weight of the material. These amounts, typically less than 0,5 liter per ton (1 pt/st), will minimize any degradation of the material.

There are two methods of producing a water fog:

A. Two-fluid atomization

One method produces fog from water and compressed air by passing them together through a two-fluid nozzle. Here the external air supply is the vehicle that fractures the water into the droplet mist used to capture the dust.

Figure 19.3

Fog dust-suppression systems use special nozzles to create a mist of fine droplets.

The supply of compressed air provides an additional expense for the installation and operation of this system. The cost of producing the compressed air must also be considered in the economics of the system. An additional concern is the consequence of injecting additional moving air into a transfer point's dust-control equation, which can further stimulate the movement of dust. However, this method allows the use of process water that has been simply filtered to remove any materials that might plug the nozzles.

B. Single-fluid atomization

The second system uses an ultra-fine stream of water pumped through single-fluid atomizing nozzles. It does not require compressed air or any additional power supply other than the electricity used to run its pump. It does require the use of clean, fresh water—or that the process water is filtered and treated—to reduce problems with nozzle clogging. The single-fluid nozzles use hydraulic atomization to generate the fog. In this method, a small stream of water is forced under high pressure—up to 14 megapascals (2000 lb_f/in.^{2}), although more typically 34 to 69 megapascals (5000 to 10000 lb_f/in.^{2})—through a small orifice that shatters the water droplets into microscopic particles. The energy created by the high-pressure pump is used to atomize the water droplets, rather than increase the water's velocity, thereby minimizing displaced air. By eliminating compressed air requirements, the single-fluid nozzles simplify installation and reduce operating costs. To keep the small orifices clear, suspended materials must be removed from the water, and the pH of the water must be controlled. The low volume of water applied makes this relatively easy to accomplish with filtration and ionization.

Location of Fog Systems

The installation of fog systems is a little unusual in that fog systems are designed to treat the air around the material, rather than the material itself. Therefore, the application point for the fog mist is generally near the end of the transfer point (**Figure 19.4**). This placement allows the material to settle and any pick-ups for active or passive dust-collection systems to see dust-laden air without risk of blinding the filtration media with the moistened particles.

Fog-generation nozzles are installed to cover the full width of the conveyor's skirted area (**Figure 19.5**). It is recommended that the height of transfer point skirtboard be at least 600 millimeters (24 in.) to allow the output cone of the nozzles to reach optimum coverage and fill the enclosure. The nozzle spray pattern should be designed so that airborne materials pass through the curtain of fog without putting spray directly onto the main body of material. The spray is directed above the materials, rather than at the materials.

The spray pattern from fog nozzles should not be directed onto any surface, and the nozzles should be shielded from being struck by the bulk material.

Figure 19.4

Fog suppression systems are applied near the end of the transfer-point enclosure.

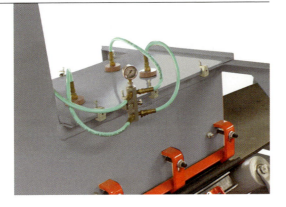

Figure 19.5

Fog-generation nozzles are installed to cover the full width of the conveyor's skirted area.

19

Pluses and Minuses of Fog Systems

Fog systems provide effective dust control combined with economical installation and operating costs. System operating costs are low when compared to conventional dust-collection systems.

A well-designed fogging system provides control of dust at the point of application without the need for chemical additives. This is especially important for processes such as the conveying of wood chips destined for papermaking. Many mills are concerned about the application of any chemical that might negatively affect the pulp or degrade the quality of finished paper. As fog systems add water without any additives, they protect the integrity of the process.

With fog systems, total moisture addition to the bulk material can be as low as 0.1 percent to .05 percent. This makes fog suppression systems attractive in industries such as cement and lime production that cannot tolerate excess moisture.

Because of the small orifice size of the nozzles, potable (drinking-quality) water is typically required for fog suppression systems. Filtration to remove suspended materials from the water supply is normally necessary. Nozzles can plug if the water supply is contaminated or if the water-treatment system is not serviced at required intervals. Preparations such as drains and heat-traced plumbing should be provided for plants in cold-weather environments.

Another consideration prior to choosing a fogging system is the air volume and air velocity at the open area surrounding the transfer point or chute. Fog systems using single-fluid nozzles (those that do not require compressed air) tend to be more compatible with engineered systems that control the air movement through a transfer point. These systems should not be used in "open area" applications. For truly effective performance, fog dust-suppression systems require a tight enclosure around the transfer point to minimize turbulent,

high-velocity air movement through the system. Since droplets are small, both fog and dust could be carried out of the treatment area onto surrounding equipment by high-velocity air leaving the chute.

Another potential drawback of fogging application is that this form of dust treatment is application-point specific. Dust control is achieved only at the point of application; there is little or no residual, or carry-over, benefit. Although one system can often control more than one transfer point, several fogging devices may be required for a complex conveyor system with multiple transfer points. The capital expenditure may preclude fogging if the conveyor system is too extensive.

ADDING CHEMICALS

Adding Chemicals to Water

It is a common practice to "enhance" the dust-suppression performance of water by adding surfactants—surface-acting agents. The addition of these chemicals will improve the wetting characteristics of water, reducing overall water usage and minimizing the drawbacks associated with excessive moisture addition.

If dust from coal, petroleum coke, or a similar material falls onto a puddle of water, the dust particles can, if undisturbed, lie on top of the pool for hours. This phenomenon takes place because these materials are hydrophobic—they do not mix well with water. Since it is not practical to alter the nature of the dust particles to give them greater affinity for water, chemicals are added to alter the water particles, so they attract, or at least join with, the dust particles more readily.

By adding surfactants, the surface tension of water is reduced, allowing dust particles to become wet. Surfactants are substances that, when added to water, improve the water's ability to wet surfaces and form fine droplets. Surfactants lower the water's surface tension and overcome the internal attraction between the molecules of water, ultimately resulting in improved droplet formation.

To understand surface tension, imagine a drop of water lying on a smooth, flat surface. It will usually form a liquid bubble with well-defined sides. It is the surface tension of the water that prevents the droplet walls from collapsing. A drop of water that has been mixed with a surfactant—such as dishwashing soap—will not form a liquid bubble, because its surface tension has been drastically reduced. The "walls" of the droplet cannot support the weight of the droplet, because the forces holding the walls together have been altered. This is the reason surfactant technology is applied to dust control. If the water droplets no longer have a surface that is a barrier to contact with the dust fines, then random collisions between droplets of water and dust will result in the wetting and enlargement of the fines to the point they will drop out of suspension in the air.

Choosing a Surfactant

The number of surfactants and surfactant blends currently in use is quite extensive. A number of specialty chemical companies have products formulated to address specific dust-control needs. Choosing the correct product and addition rate for a given application requires material testing as well as an understanding of the process and the method of application.

Objections to chemical-additive-enhanced water suppression systems include the ongoing costs of purchasing chemical additive. Costs can be higher, particularly when considering amortization and depreciation of the equipment. In addition,

these systems require regular maintenance, which adds labor expense to the continuing operating costs.

As contamination of the materials, or the process, can be a concern in some industries, the additive chemical must be reviewed in this light. It is important that chemical additives are compatible with the process, with the bulk materials, and with system equipment, including the conveyor belting. Although the use of a surfactant reduces the amount of water added to the dusting material, water/surfactant sprays may still add more water than is acceptable. It is common practice for a chemical supplier to provide samples to the customer for testing the effects on the end product.

Application by Spray or Foam

Once an efficient wetting agent has been selected, the decision must be made whether to apply the material as a wet spray, as discussed above, or as foam. Both systems offer advantages. Generally speaking, the moisture-addition rate of a wet-spray system is higher than that of a foam-generating system. Although the dilution rate is lower for the foam suppression system, the expansion of the foam allows it to provide effective suppression with less moisture added to the materials (**Table 19.3**). Recent developments have improved surfactant technology to the point that some mixtures can be applied as a spray at the lower moisture levels of a foam system while providing good dust suppression. This provides the benefit of limited moisture addition with minimal chemical cost, due to the higher dilution rates with the spray-applied surfactants.

Table 19.3	Maximum Typical Moisture-Addition Levels			
	Water Spray	Water with Surfactant	Foam	Fog
Nominal Rate of Moisture Addition	5%	2.5%	0.20%	0.05%
Water Addition	5455 l/h (1200 gal/h)	2725 l/h (600 gal/h)	218 l/h (48 gal/h)	54,5 l/h (12 gal/h)
Chemical-to-Water Ratio	N/A	1:5000	1:100	N/A
Chemical Usage Rate	N/A	0,44 l/h (0.096 g/h)	2,2 l/h (0.48 g/h)	N/A

FOAM SUPPRESSION

Foam Dust Suppression

The use of surfactants with water will improve the likelihood that fines will collide with droplets and that these collisions will result in suppression of the dust. It stands to reason that the objective is to maximize the surface area of available water droplets to make as much contact with dust fines as possible, thus limiting the amount of water needed. To do this, some suppliers offer dust-suppression systems that create chemical foam (**Figure 19.6**). As the moisture is in the form of foam, its surface area is greatly increased, improving the chance for contact between dust and water. Some foam bubbles attract and hold dust particles together through agglomeration. Other bubbles burst on contact with dust particles, releasing fine droplets that attach to smaller, more difficult to catch, and more hazardous to human health, dust particles. With moisture addition of 0.2 percent to 0.4 percent, foam systems add only 2 liters per ton (2 qt/st) of material. At these levels, foam suppression systems typically add less than 10 percent of the moisture that straight water-only spray systems apply.

Consequently, foam systems are welcomed where water supplies are limited or where excessive water can downgrade material performance, as in coal-fired power plants. In addition, the reduced water means fewer problems with screen clogging and materials adhering to mechanical components and enclosures.

Adding air to the surfactant and water blend and passing this compound through a mixing device creates the foam. Adjustment of the air/water/chemical ratio and other controllable factors allows the application engineer to generate foam ranging from very wet to "shaving cream" dry, in order to create the most efficient foam for each application. Well-established foam can expand the surface area of a quantity of water by 60 to 80 times. This allows for effective dust control with lower rates of moisture addition.

The system for the application of foam for dust suppression begins with mixing water with the foam-generating chemical. The water and additive are metered together through a proportioning pump, and the resulting mixture is pumped through a flow regulator to feed the system (**Figure 19.7**). A second flow regulator controls a supply of compressed air. The water/chemical solution and air arrive via separate hoses at a foaming canister, where they mix to create foam. The foam then travels through hoses to the application nozzles installed in the wall or ceiling of the equipment or transfer point (**Figure 19.8**).

Limitations of Foam Suppression

While many applications benefit from foam technology, there are some liabilities to the process. Surfactants that produce the most desirable foaming are not always the best wetting agents for the materials being treated. Some suppliers focus on chemicals to produce stable foam, without considering whether the resultant foam is

19

Figure 19.6

Foam dust suppression generates "dry" foam that expands the surface area of water 60 to 80 times its previous surface.

Figure 19.7

In the proportioning system, water and surfactant are mixed, and the resulting solution and compressed air are sent independently to foam canisters.

of any value in overcoming the hydrophobic nature of the material. It is critical the chemicals provide effective wetting of the material handled before foam generation is considered.

Foam generation requires compressed air. If a supply of compressed air is not readily available at the application site, a compressor must be installed and maintained.

Overall, foam-application equipment is slightly more expensive than conventional water-spray equipment and normally requires additional maintenance.

Finally, the amount of surfactant required to generate foam is somewhat greater than the amount of chemical typically added in a wet-spray system. The volume of surfactant in a given body of water is higher; however, due to the foam's expansion, the amount of moisture applied to the material is lower. The additional cost for this increased concentration of the additive chemical may be offset by a reduction in thermal penalty on fuel performance resulting from a substantial decrease in additional moisture (**Table 19.4**).

RESIDUAL CHEMICALS

Residual Chemical Suppression Agents

Surfactants wet the dust fines, so the particles agglomerate, thereby preventing

Figure 19.8

The water/surfactant solution and air are combined in the foaming canister and supplied to the application nozzles.

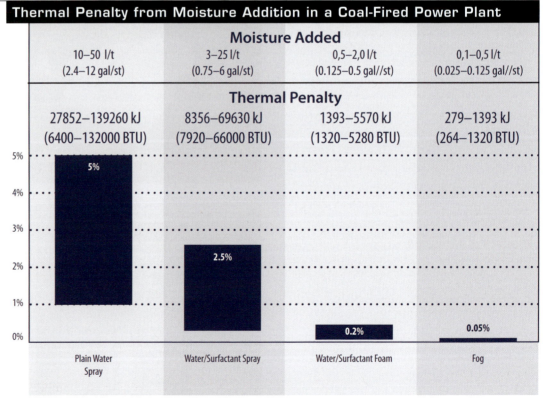

Table 19.4

Thermal Penalty from Moisture Addition in a Coal-Fired Power Plant			
Moisture Added			
10–50 l/t (2.4–12 gal/st)	3–25 l/t (0.75–6 gal/st)	0,5–2,0 l/t (0.125–0.5 gal/st)	0,1–0,5 l/t (0.025–0.125 gal//st)
Thermal Penalty			
27852–139260 kJ (6400–132000 BTU)	8356–69630 kJ (7920–66000 BTU)	1393–5570 kJ (1320–5280 BTU)	279–1393 kJ (264–1320 BTU)
5%	2.5%	0.2%	0.05%
Plain Water Spray	Water/Surfactant Spray	Water/Surfactant Foam	Fog

Dust-Suppression System Types

Note: Moisture Added—liters per ton (gal/st); Thermal Penalty—kilojoules (BTU)

them from becoming airborne. Once the solution evaporates, the suppression effect of normal surfactants is gone. In many cases, however, dust suppression is required not only as the materials move through multiple transfer points, but also after the materials reach the storage bins, railcars, barges, or stock piles. In these cases, it is wise to consider using a water/surfactant spray or foam system with a longer-lasting residual effect. Residual dust suppression is valuable when considering dust suppression for:

A. Large areas with multiple application points

B. Long distances between application points

C. Stackers or trippers

D. Crushers or mills

E. Elevated transfer points where it would be difficult to apply dust suppression

A well-designed residual suppression system makes it possible to control fugitive dust over a wide area by applying the solution at a few strategic points. In contrast, using water and/or fog systems for large areas will require multiple application points, including several pump stations; longer water, chemical, and air lines; higher pumping capacity; and more application nozzles—all of which can make the system considerably more expensive and, in some cases, not as effective.

Coal conveyed from unloaders to open storage piles might remain there for extended periods of time. Material stored in open stockpiles is subject to variations in climate, including wind, sun, and precipitation. The heat of the sun can evaporate moisture out of stored material, making it more likely to become wind-blown. Wind erosion creates large amounts of dust that can settle on nearby houses and yards. When stored coal is reclaimed, it may be dry and present greater dusting problems than it did during initial handling. Dusty materials, such as calcined coke or iron-ore pellets, may require dust control from the point of production to the point of end use. This could amount to several weeks and several thousand kilometers (miles) apart. In such cases, it may be more economical to apply a residual surfactant/binder to the materials than to apply surfactants and water at multiple sites throughout the materials-handling system. There are a variety of residual binders available.

Longer-Lasting Effects

The objective of a residual, or binder, suppressant is to agglomerate fines to each other, or to larger particles, and then hold the structure together, even after the moisture evaporates. In some cases, a hygroscopic material, such as calcium chloride, is used, which retards the ability of moisture to leave the treated material. The advantage to this approach can be a low treatment cost. More conventional binders include lignin, tannin, pitch, polymers, and resins. When combined with surfactants to aid wetting, these compounds coat larger particles and then act as a glue to attract and hold dust fines.

Application of residual binders tends to be more expensive than surfactant applications, because they must be applied with higher concentrations. Although binders are less expensive per kilogram (lb_m), they are typically applied at dilution rates ranging from 50-to-1 to 200-to-1 (2.0 percent to 0.5 percent).

It is important to mention that, with the use of a residual chemical, a plant can reduce the number of application points required, reducing, in turn, the amount of maintenance required.

When choosing a binder, it is especially important to know the effects the binder will have on transfer equipment and conveyor belts. If the binder adheres well to the material, it may do the same to the handling equipment. Proper application of the product becomes critical, because overspray of the binder onto process equipment or empty belts can result in considerable production and maintenance problems.

An important consideration in selecting a binder is the effect the chemical will have on both the material being treated and the environment. If the binder is applied to material going into a stockpile, and that stockpile is exposed to rain, portions of a water-soluble binder may end up in the runoff and provide an environmental concern. Most chemical manufacturers provide only binders that are compatible with the environment; however, this is an issue that should be raised with the chemical supplier.

SYSTEMS AND PLACEMENT

Hybrid Systems: Adding Suppression to Containment or Collection

The selection of a dust-control system should be based on the material, the causes of dust generation, and the specifics of the point(s) of application. A complete dust-generation analysis is important in order to detect not only the most problematic fugitive-dust-generation points, but to identify the true causes of dust generation and escape, in order to control the situation.

In some cases, a hybrid system combining a dust-suppression system with other dust-management systems—passive or active dust collection or containment—should be considered. This may yield the best possible performance with minimal installation, operation, and maintenance costs. It is important to consult with a specialist in the application of dust-suppression and -control systems to develop a solution for any specific application.

Location, Location, Location

In any dust-suppression system, it is important to select the best application point(s), not only to increase effectiveness, but also to reduce the costs of installation, operation, and maintenance. The sites chosen for nozzle placement and the pattern of delivery are as important, if not more important, than the selection of the chemical to be applied (**Figure 19.9**). Even the best-designed program will fail if the suppressant chemical is not delivered in the correct location to allow mixing of the suppressant and the dust fines.

Success of the suppression effort at the transfer point relies on properly mixing the materials and the suppressant. Whether the suppressant is simply water or a surfactant/water mixture as a spray or foam, it is best to locate the suppression system at the point the materials leave the head pulley. As the materials leave the head pulley, they spread out, and air is entrained into the stream of conveyed materials. The suppressant will be drawn into the materials by this negative air pressure. As the suppressant and the conveyed materials tumble through the chute, they will continue to mix, providing effective dispersion.

Foam suppression is normally most effective when applied at the discharge point of a crusher or conveyor, where the body of material is in turmoil and expanding (**Fig-**

Figure 19.9

The location of the application point is critical to the success of any dust-suppression system.

Figure 19.10

Foam suppression is most effective when applied where the material is in turmoil, as at the discharge of a crusher or conveyor.

19

ure 19.10). Here, the forces of the material movement will fold the suppressant into the material stream as it moves through the transfer point and down the conveyor belt. The application of suppressant at this point allows the foam to penetrate into the material stream and capture individual particles, rather than remaining on the external layer of material.

The Importance of Water Quality

Water quality plays an important role in the effectiveness of any dust-suppression program. The ability to generate acceptable foam is largely dependent on the quality of the water used. Depending on the dust-suppression system used, it is important to filter the water, removing particles between 5 and 40 microns, and to have the water as close to neutral pH as possible.

If the characteristics of the water available in the plant are known, the proper systems to filter the water can be applied. This knowledge will also make it easier to prevent possible failures—including the plugging of nozzles and premature failure of pumps—and to maintain the required flow rate.

SYSTEM MAINTENANCE

Without a doubt, one of the most common causes of failure for dust-suppression systems is a lack of preventive maintenance. Nozzles must be checked, filters cleaned, pumps oiled, chemical levels checked, application settings verified, and flow rates for water and air adjusted on a routine basis, or even the best system is doomed to fail. It is important to rely on

SAFETY CONCERNS

Central to any consideration of safety with dust-suppression systems is a proper regard for the relationship between water and electrical systems that power the suppression system and, indeed, the entire conveyor. Systems should be properly grounded, and water should not be sprayed directly onto them.

As many dust-suppression systems move water or air under some level of pressure, it is important to be wary of the plumbing system, whether it is pipe, hoses, or some combination. Pump or line pressure must be maintained at proper levels, and proper relief mechanisms must be available. Prior to work being performed on any piping system, care must be taken to make sure that pressure in the lines is relieved and electricity to the pump(s) is properly locked out.

Dust-suppression systems in cold-weather climates should incorporate measures to keep the system operating in freezing conditions, or the system should be designed to operate only when the temperature is above freezing. Systems should be designed to ensure they do not create safety risks, such as ice patches—on roads, walkways, or stairs—or frozen bodies of material inside a vessel that would require employee entry into confined spaces to remove material blockages.

Suppliers of chemical additives must provide all applicable Material Safety Data Sheets (MSDS) that spell out all safety concerns, health risks, and environmental issues.

It is important to follow established lockout / tagout / blockout / testout procedures when installing and maintaining dust-suppression systems.

the manufacturer's instructions for guidance on the proper service intervals and procedures for system components.

Some suppliers of dust-suppression equipment and chemicals now offer routine service as a part of their system package. It is wise to consider this solution, as it will free in-house maintenance and operating personnel for other duties, while guaranteeing the operation of the dust-suppression system.

TYPICAL SPECIFICATIONS

The following typical specifications pertain to foam dust-suppression systems only.

A. Foamed mixture

The conveyor loading zone will be equipped with a dust-suppression system that applies a foamed mixture of suppressant chemical and water to minimize the escape of airborne dust.

B. Additive

The dust-suppression system will work by the metering of a dust-suppression additive into a supply of water, generating a foam mixture of water and supplement, and applying this mixture over the body of material. This mixture will encourage the agglomeration of fine particles and inhibit the driving off of airborne dust.

C. Pump module

The dust-suppression system will include a pump module, containing a proportioning pump [0 to 76 liters per minute (0 to 20 gal/min)] with the addition of

0.2 to 1.5 percent additive, a regulator [170 to 520 kilopascals (25-75 lb_f/in.^{2})], a gate valve, and a flow meter [0 to 76 liters per minute (0 to 20 gal/min)].

D. Foaming chamber

The air and water/additive mixture will be combined in a foaming chamber. The inlets for the air and water/additive lines will be equipped with check valves to prevent backflow. An air gauge located on the foaming chamber will allow control of air pressure to create fully-developed foam.

E. Nozzles

The produced foam will be applied to the material on the conveyor through up to eight "duckbill" nozzles connected to uniform hose lengths. The nozzles will be held in position in the chutewall to allow a simple of removal for maintenance.

TYPICAL DUST-SUPPRESSION APPLICATIONS

Dust-Suppression Application 1

A belt is transporting mine refuse. This belt is properly supported in the load zone, and the transfer chute is effectively sealed. An anemometer reading at the outlet of the conveyor loading zone shows that the exit velocity of air is 0,25 meters per second (50 ft/min).

Material	Mine Refuse
Transfer Point	Effectively Sealed
Air Speed	0,25 m/s (50 ft/min)
Suppression Method	Fog

This is a good application for a fog system, because the material is not sensitive to water, containment is good, and air velocity is below 1,0 meters per second (200 ft/min). *(See Chapter 17: Dust Management Overview, Figure 17.10.)*

The nozzles should be placed on the top side of the transfer chute's settling zone. Dust curtains should be placed on each side

Figure 19.11

The nozzles should be placed on the top side of the transfer chute's stilling zone. Dust curtains should be placed on each side of the nozzles to slow the air stream, allowing the fog to remove dust from the air.

19

of the nozzles to slow the air stream, allowing the fog to remove dust from the air (**Figure 19.11**).

Dust-Suppression Application 2

A belt at an aggregate plant is transporting limestone. The transfer point has no enclosure.

Material........................Crushed Limestone
Transfer Point........... Open (No Enclosure)
Air Speed..................................... Unknown
Suppression Method .... Foam

This is a good application for a foam system, because the material is not sensitive to moisture, but there is no chute to control the air movement.

The foam could be applied to the limestone as it comes off the head pulley, while the material is in turmoil. This will allow the moisture to cover all surfaces of the material. Covering all surfaces with moisture will prevent the generation of dust when the material lands on the receiving belt (**Figure 19.12**).

Dust-Suppression Application 3

The conveyor is transporting coal. The transfer chute is properly sealed and supported. An anemometer reading at the end of the settling zone shows that the exit velocity of the air stream is 1,5 meters per second (300 ft/min).

Material..Coal
Transfer Point...........Sealed and Supported
Air Speed.................1,5 m/sec (300 ft/min)
Suppression Method .... Foam

This is a good application for a foam system, applied in combination with a reconstruction of the conveyor's transfer point(s). The high air velocity indicates that the transfer-point enclosure is not large enough to slow the air. High velocity usually means large amounts of dust will be generated. The transfer point should be lengthened and the height increased, to slow the air and allow the dust to settle.

The material is sensitive to moisture, so the amount of water should be minimized. The foam could be applied to the material while it is in turmoil. This will allow the moisture to cover all surfaces of the material. Covering all surfaces with moisture will prevent the generation of dust when the material lands on the receiving belt.

The moisture will also have a residual effect and may keep the coal moist all the way to the stackout conveyor (**Figure 19.13**).

Dust-Suppression Application 4

A bucket elevator is offloading coal from a barge. There is no "transfer chute," so the unloader is exposed to air currents.

Figure 19.12

Applying foam while the cargo is in turmoil allows the moisture to cover all of the surfaces of the material to prevent dust generation.

Before

Figure 19.13

The moisture will have a residual effect and may keep the coal moist all the way to the stackout conveyor.

After

Table 19.5

Thermal Penalty in a Coal-Fired Power Plant		
	Metric	**Imperial**
Unit to measure heat/energy	Kilojoule (kJ)	British Thermal Unit (BTU)
Weight of water	1 kg/l	8.33 lb_m/gal
Energy to vaporize water	2675 kJ to vaporize 1 kg (about 0,5 l) of water from Standard Temperature and Pressure (STP)	1150 BTU to vaporize 1 lb_m (about 1 pt) of water from STP
Coal unit	ton (1000 kg)	short ton (2000 lb)
Water required to raise moisture content of unit of coal by 1%	10 kg (10 l)	20 lb_m (2.4 gal)
Heat required to burn this 1% of additional water off the unit (ton/st) of coal	26750 kJ (2675 kJ/kg x 10 kg)	23000 BTU (1150 BTU/lb_m x 20 lb_m)
Heat content of coal Source: *Coal Data: A Reference* published by U.S. Department of Energy, Energy Information Administration. *Metric conversion by Martin Engineering.*	Bituminous = 27900 kJ/kg Subbituminous = 20900 kJ/kg Lignite = 16300 kJ/kg	Bituminous = 12000 BTU/lb_m Subbituminous = 9000 BTU/lb_m Lignite = 7000 BTU/lb_m
Amount of coal required to provide the heat required to burn off 1% water from 1 ton (1 st) of coal	Heat Required (kJ) divided by Heat Content (kJ/kg) = kg	Heat Required (BTU) divided by Heat Content (BTU/lb_m) = lb_m
	26750 / kJ/kg = kg	23000 / BTU/lb_m = lb_m
	Bituminous 0,96 kg Sub-bituminous 1,3 kg Lignite 1,6 kg	Bituminous 1.9 lb_m Sub-bituminous 2.55 lb_m Lignite 3.3 lb_m
Summary	It takes from 0,96 kg to 1,6 kg to burn off 1% of water added to a ton of coal.	It takes 1.9 lb_m to 3.3 lb_m of coal to burn off 1% of water added to a st of coal.
In percent	This is 0,0096 to 0,016 of the coal (1/10 to 1/6 of 1 percent)	This is 0.0095 to 0.0165 of the coal (1/10 to 1/6 of 1 percent)
Railcar contents	91 tons (91000 kg)	100 st (200000 lb_m)
Loss from every carload	~87 to 146 kg	~190 to 330 lb_m
Loss from every 120-car unit train	~10440 to 17500 kg/trainload or between 1/10 and 1/5 of a carload/train	~22800 to 39600 lb_m/trainload or between 1/10 to 1/5 of a carload/train
If this 270-megawatt (362000-hp) plant receives 60 unit trains per year:		
Annual loss	~625000 to 1,1 million kg or ~625 to 1100 tons or 6 to 12 carloads/year	~1.35 to 2.4 million lb_m or ~684 to 1188 st or 6 to 12 carloads/year

19

Material...Coal

Transfer Point..................................... None

Air Speed.......................................Ambient

Suppression Method ....Water with Surfactant

This is a good application for water-with-surfactant suppression, because the material is not sensitive to water, and containment around the material is poor. Water with a surfactant additive allows larger water drops than water alone and will not be as affected by air currents. The nozzles should be placed around the excavator to allow the water/surfactant mix to "rain down" on the barge as it is unloading (**Figure 19.14**).

ADVANCED TOPICS

Thermal Penalty in a Coal-Fired Power Plant

A 270-megawatt (362000-hp) power plant might burn approximately 82 tons per hour (90 st/h), 24 hours per day, seven days a week. This turns into 13776 tons per week (15120 st/wk). Even allowing a two-week maintenance outage, the annual coal consumption of this plant would be more than 688000 tons per 50-week-year (759000 st/50-wk-yr).

The plant receives its coal in unit trains composed of 120 cars, each with a capacity of 91 tons (100 st). With the total train capacity of 10920 tons (12000 st), the plant will need to receive roughly 1.25 trains per week, or 5 trains a month. That is approximately 60 trains per year. These figures will vary depending on the type (heat output) of the specific coal used.

Thermal penalty is the amount of coal that must be burned just to remove the moisture added to the coal by the dust-suppression system. It is equal to between 1,0 and 1,6 kilogram per ton (1.9 to 3.3 lb_m/st) for each one percent of water added to the coal.

At the rate of 0,1 of 1 percent of coal used to eliminate this one percent additional moisture, the plant will lose the heat

Figure 19.14

An unenclosed bucketwheel reclaimer is a good application for a water-surfactant dust-suppression system.

from 10440 to 17500 kilograms (22800 to 39600 lb_m) of coal per trainload, or roughly 0,1 to 0,2 of a carload. That amounts to 6 to 12 railcars per year—perhaps 1 railcar per month—burned just to drive off the added moisture (**Table 19.5**).

DUST SUPPRESSION: ONE PIECE OF THE PUZZLE

In Closing...

Dust suppression is best suited to enclosed spaces of reasonable size. It becomes difficult to apply and control any of the various forms of dust suppression in open areas or inside large structures such as railcar or haul-truck dumps. Acceptable results in these applications may require a combination of confinement, suppression, and collection.

Dust suppression cannot stand alone as the complete answer to controlling fugitive materials. Properly chosen, engineered, and maintained, a dust-suppression system can provide a critical portion of the total material-management program.

Looking Ahead...

This chapter about Dust Suppression is the third chapter in the section Dust Management, following Dust Management Overview and Passive Dust Control. The following chapter, the final chapter related to dust management, continues the topic of active dust control by focusing on Dust Collection.

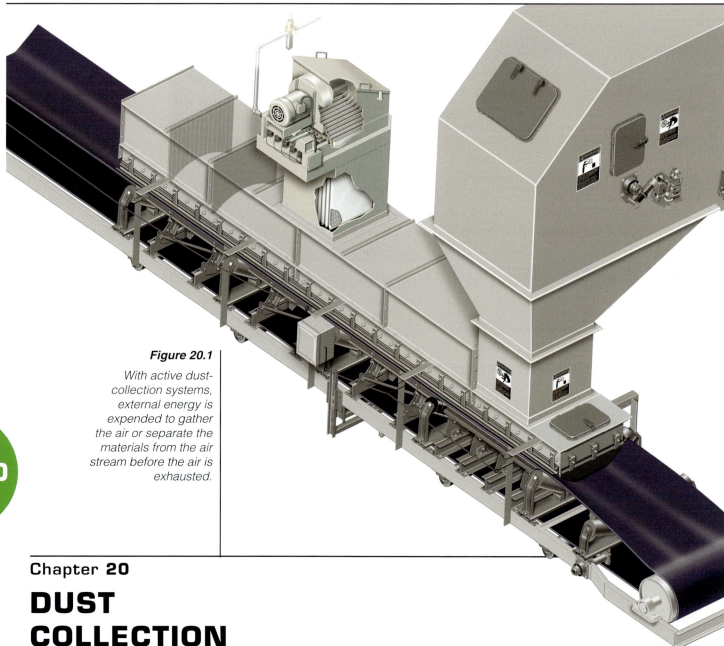

Figure 20.1

With active dust-collection systems, external energy is expended to gather the air or separate the materials from the air stream before the air is exhausted.

20

Chapter 20

DUST COLLECTION

In this Chapter...

In this chapter, we discuss the five main types of active dust-collection systems, looking at advantages and disadvantages of each. We also compare central, unit, and insertable dust collectors. Sizing and placement of dust collectors, along with some of the downfalls of dust-collection systems, are included. Advanced topics include selection and application of an insertable dust-collection system as well as three types of air velocity and their relationship to dust management.

Dust collection—the passing of dust-carrying air through some form of filtration or separation system—is the final piece in the dust-control system.

There are both active and passive dust-collection systems. A passive system merely allows air to move through the filtration system, whereas active systems work like a vacuum cleaner to pull or push air through a filtration method to remove the materials. *(See Chapter 18: Passive Dust Control for information on passive collection methods.)* This chapter discusses active dust-collection systems (**Figure 20.1**), in which external energy is expended to gather the air or separate the materials from the air stream before the air is exhausted.

DUST COLLECTION

Dust-Collection Systems

Mechanical dust-collection systems are installed to pull dust-bearing air away from a dust source, such as a conveyor loading zone; separate the dust from the air; and exhaust the cleaned air. A typical dust-collection system consists of four major components (**Figure 20.2**):

A. Exhaust hood(s) or pickup(s) to capture airborne dust at the source(s)

B. Ductwork to transport the captured air/dust mixture to a collector

C. Collector, filter, or separation device to remove dust from the air

D. Fan and motor to provide the necessary suction volume and energy

Considerations for Dust-Collection Systems

Dust-collection systems vary widely in design, operation, effectiveness, space requirements, construction, and costs for operation and maintenance. The selection of a system should include a review of the following factors:

A. Dust concentration

In bulk-materials handling operations, the dust concentrations typically range from 230 to 23000 milligrams per cubic meter (0.1 to 10.0 lb_m/ft^3) of dust, and the particle size can vary from 0,2 to 100 microns (μm is one millionth of a meter). The selection of a collector should be based on the level of cleanliness, or efficiency, required.

B. Characteristics of the air stream

The characteristics of the polluted (or dirty) air can have a significant impact on collector selection. Factors include temperature, moisture content, and relative humidity.

C. Characteristics of the dust

The properties of the dust itself are important to the choice of a dust-collection system. Moderate to heavy concentrations of many dusts, such as silica sand or metal ores, can be abrasive, hygroscopic, or sticky in nature. The size and shape of particles will determine the applicability of fabric collectors; the combustible nature of many fine materials rules out the use of electrostatic precipitators.

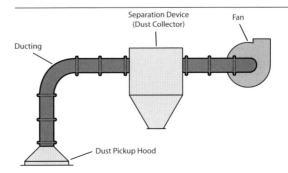

Figure 20.2

The basic components of a dust-collection system include pickups, ductwork, filter device, and fan and motor.

Ducting

Separation Device (Dust Collector)

Fan

Dust Pickup Hood

D. Method of disposal

The method to dispose of collected dust will vary with the nature and amount of the material, the overall plant process, and type of collector used. Collectors can unload continuously or in batches. Dry materials can create secondary dust problems during unloading and disposal. The disposal of wet slurry, or sludge, can be an additional materials-handling problem for a wet collector. Sewer or water pollution problems can result if wastewater is not properly treated.

COLLECTION TECHNOLOGIES

Dust-Separation Technologies

There are a number of specific "hardware" approaches used to remove dust from the air, each with its own benefits and drawbacks. The five main types of active dust-collector systems used in the industry include:

A. Inertial separators (usually called cyclones)

B. Wet scrubbers

C. Electrostatic precipitators

D. Cartridge filter collectors

E. Fabric dust collectors (often called baghouses)

Inertial Separators

Inertial separators separate dust from the air stream using a combination of centrifugal, gravitational, and inertial forces. These forces move the dust to an area where the forces exerted by the air stream are minimal.

The three primary types of inertial separators are:

A. Active settling chambers

B. Baffle chambers

C. Cyclones or centrifugal collectors

Active settling chambers operate under the assumption that confining the airflow to slow it will allow the airborne particles to fall out of the air. Baffle chambers, similar to active settling chambers but with baffles to slow and redirect the air, also allow the airborne particles to fall out of the air. Active settling and baffle chambers are not widely used as a sole collection method for a plant or process, due to their size requirements and poor efficiencies.

Cyclones are the most commonly used of these inertial-separator systems. They create a vortex—an internal tornado—that "flings" the dust out of the air stream (**Figure 20.3**). This whirling airflow created inside the structure creates a centrifugal force, throwing the dust particles outward toward the unit's walls. After striking the walls, particles agglomerate into larger particles and fall out of the air stream into a collection point, or discharge outlet, at the bottom of the unit.

There are single-cyclone separators that create two vortexes in one enclosure. The main vortex spirals coarse dust downward, while a smaller inner vortex spirals up from the bottom, carrying fine particles upward toward a filter. Multiple-cyclone units contain a number of small-diameter cyclones operating in parallel, with a common inlet and outlet. These units each create the two vortexes seen in the single-cyclone separator. These multiple-cyclone units tend to be more efficient, because they are taller—

Figure 20.3

Cyclone dust-collection systems create a vortex—an internal tornado—that "flings" the dust out of the air stream.

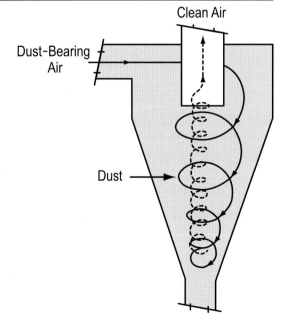

Clean Air

Dust-Bearing Air

Dust

providing greater time for air to remain inside—and are smaller in diameter—providing greater centrifugal force. Cyclones must maintain a high rate of airflow in order to maintain the separation process.

To improve the efficiency of some cyclones, particularly those handling fine particles, the collecting surface of these units may be wetted with water.

Inertial-separator systems are often used as pre-cleaners to reduce the workload on more efficient dust-collection systems, because they do not provide an adequately efficient collection of fine, or respirable, particles. Performance suffers in high-humidity conditions. In the absence of plugging problems, these systems can operate with low maintenance costs, because they have no moving parts.

Wet Scrubbers

In wet-scrubber systems, a liquid (most commonly water) is sprayed down into the stream of dust-bearing air (**Figure 20.4**). The dust particles are captured by water droplets and fall out of suspension in the air. The dust and water mixture is released out the bottom of the collector as slurry and passes through a settling, or clarification, system to remove the materials.

An advantage of wet scrubbers is that they can be used in high-temperature applications. There is little chance for the dust to escape and become airborne again, and there are minimal fire and explosion hazards associated with scrubbers. Scrubbers also provide the opportunity to collect both particulate matter (dust) and gases, so they provide a dual benefit for some operations.

Wet scrubbers have some disadvantages, also. One disadvantage is that these systems have high operating and maintenance costs and may require freeze protection for cold-weather operations. For heavy dust conditions, these systems often need a pre-cleaner, such as a cyclone. These systems will have high power requirements. There may be corrosion problems from the han-

dling of the water and the material slurry. Water treatment is usually required for the contaminated water from the system. Recovery of the materials from the scrubber waste is typically difficult.

Electrostatic Precipitators

Electrostatic precipitators are often used to handle large volumes of dust-laden air at wide ranges of temperature and pressure. These systems apply a negative electrical charge, ionizing the particles as they pass into the collection area (**Figure 20.5**). The charged particles are then attracted and adhered to positively-charged electrode plates positioned inside the collection zone. "Rapping," or vibrating, of these electrodes then discharges the agglomerated dust by allowing it to move downward on the plates by gravity.

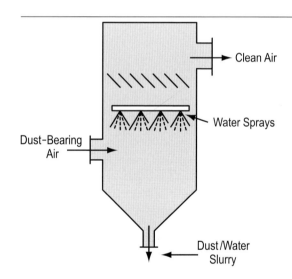

Figure 20.4

In wet-scrubber systems, a liquid (most commonly water) is sprayed down into the stream of dust-bearing air. The dust particles are captured by water droplets and fall out of suspension in the air.

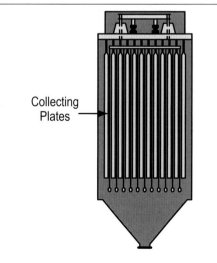

Figure 20.5

Electrostatic precipitators apply a negative electrical charge, ionizing the particles as they pass into the collection area. The charged particles are then attracted and adhered to positively-charged electrode plates positioned inside the collection zone.

The four main components of all electro-static precipitators are:

A. A power supply, to provide high-voltage, unidirectional current

B. An ionizing section, to impart a charge to particulates in the air stream

C. A means of removing the collected particulates

D. A housing to enclose the precipitator zone

There are two main types of precipitators:

A. High-voltage, single-stage

Single-stage precipitators combine ionization and collection into a single step. They are commonly referred to as Cottrell precipitators.

B. Low-voltage, two-stage

Two-stage precipitators use a principle similar to single-stage precipitators; however, the ionizing section is followed by a collection section.

Precipitators can be 99 percent effective on dust, including sub-micron particles, but they do not work well on fly ash from combustion of low sulfur coal, due to its high electrical resistivity. They work on other materials, including sticky and corrosive materials, in high-temperature or in high air flow environments, with minimal energy consumption. These systems require a large capital investment. Safety measures are required to prevent the exposure of personnel to the system's high voltage. Precipitators can cause an explosion threat when combustible gases are collected around the electric system.

Cartridge Filter Collectors

Cartridge filter collectors place perforated metal or plastic cartridges containing a pleated, non-woven filter media inside the dust-collector structure. The filter media used in these systems provides a larger collection surface area in a smaller unit size than other dust-collection systems. As a result, the size of the overall system can be reduced.

These systems are available in single-use systems—changing the filter while off line—and pulse-jet cleaning systems—allowing continuous-duty cleaning.

The drawbacks of these systems include the relatively high cost of replacement cartridges. High moisture content in the collected materials may cause the filter media to blind (become plugged), and the system itself requires higher levels of maintenance than other collection methods. Cartridge filters are generally not recommended for abrasive materials or applications where high temperatures are seen.

Fabric Dust Collectors

Perhaps the most common dust-separation technology is the use of fabric collectors, which are placed in structures commonly called baghouses (**Figure 20.6**). Fabric collectors use filtration to separate dust particulates from airflow. They are one of the most efficient and cost-effective types of dust collectors available and can achieve a collection efficiency of more than 99 percent for particulates 1 µm or less.

Fabric collectors utilize the dust itself to help perform filtration. A "cake" of the collected dust forms on the surfaces of the filter bags and captures dust particles as they try to pass through the bags.

Figure 20.6

The most common dust-separation technology is the use of fabric collectors, which are placed in structures commonly called baghouses. Fabric collectors use filtration to separate dust particulates from airflow.

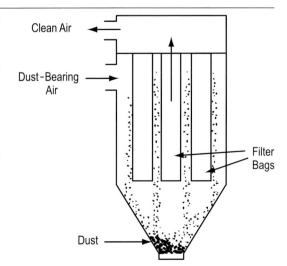

Clean Air

Dust-Bearing Air

Filter Bags

Dust

The bags—constructed of woven or felted cotton, synthetic, or glass-fiber material in a tube or envelope shape—need to be cleaned periodically, to reduce the level of dust cake and allow air to be drawn through the bag without overworking the exhaust fan.

Basic Principles

There are three basic principles of fabric dust collector, or baghouse, operations:

A. Cleaning efficiency depends on the dust-cake buildup on the filter surface: Performance is better from a filter with some cake buildup than from a new filter.

B. The quantity of airflow depends on the filter medium's permeability, the amount of dust in the airflow, the amount of buildup before filter cleaning, and the power of the reverse-cleaning blow.

C. The more permeable the filter cloth, the less efficient its collection, with or without a dust cake.

Collectors can be designed for "upflow" —in which dirty air passes up through the collector with clean air exiting the filters at the top of the collector—or "downflow"— with dirty air entering at the top and passing down through the collector with clean air exiting at the bottom. The downflow design operates in favor of the cleaning action and is generally more efficient.

Filter Cleaning

The cleaning of filters in a baghouse can be performed on-demand—when the filter is fully loaded, as determined by a specified pressure drop across the filter media. Automated cleaning can be performed off-line—when the collector is shut down— or on-line—which allows for uninterrupted collector operation.

Three common methods of cleaning are:

A. Mechanical shaking

With mechanical shaking, the air passes from the inside to the outside of the bag, with the dust captured on the inside of the bag. The bags are cleaned by shaking the top mounting bar from which the bag is suspended. This is performed off-line: The system needs to be stopped for cleaning.

B. Reverse airflow

With reverse-air systems, the bags are fastened at the bottom. Air moves up through the bag from inside, with the material collecting on the inside. Bags are cleaned by injecting clean air into the dust collector in a reverse direction, so the bag partially collapses, causing the dust cake to crack off the bag wall and fall into the hopper bottom. The system needs to be stopped for cleaning.

C. Reverse jet

Reverse-jet systems provide for on-line cleaning. With this reverse-jet system, the filter bags are fastened from the top of the baghouse and supported by metal cages. Dirty air flows from the outside to the inside of the bags, leaving the dust on the outside of the bag. The cleaned air moves up through the bag and exits at the top of the cage (**Figure 20.7**). Bags are cleaned by discharging a burst of compressed air into the bags at the top. A venturi nozzle at the top of the bag accelerates the compressed air. Since the duration of the compressed-air burst is short (typically one-tenth of a second), it acts as a rapidly moving bubble, which flexes the bag wall and breaks the dust cake off, so it falls into the collection hopper (**Figure 20.8**).

Reverse-jet systems provide more complete cleaning than shaker or reverse-air cleaning designs. The continuous-cleaning feature allows them to operate at a higher air-to-media ratio, so cleaning efficiency is higher, and the space requirements are lower than for other designs.

A baghouse collection system with fabric filters can be up to 99 percent effective in removing respirable dust emissions. The filtration bags are relatively inexpensive

Figure 20.7

With this reverse-jet system, the filter bags are fastened from the top of the baghouse and supported by metal cages. Dirty air flows from the outside to the inside of the bags, leaving the dust on the outside of the bag. The cleaned air moves up through the bag and exits at the top of the cage.

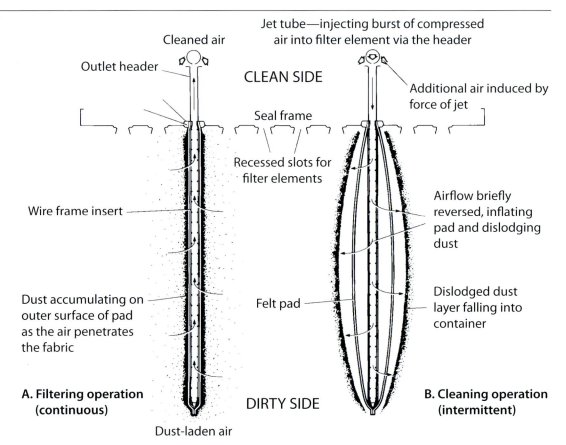

Jet tube—injecting burst of compressed air into filter element via the header

Cleaned air

Outlet header

CLEAN SIDE

Additional air induced by force of jet

Seal frame

Recessed slots for filter elements

Wire frame insert

Airflow briefly reversed, inflating pad and dislodging dust

Dust accumulating on outer surface of pad as the air penetrates the fabric

Felt pad

Dislodged dust layer falling into container

A. Filtering operation (continuous)

DIRTY SIDE

B. Cleaning operation (intermittent)

Dust-laden air

20

compared to other methods, and the large number of manufacturers in the marketplace ensures competitive pricing. The disadvantages of these systems include problems in applications above 260 degrees Celsius (500°F) or in high-humidity conditions. Some systems require entry into the baghouse for replacement of filter bags, with employee exposure to a confined space—with high dust levels and the possibility of a spark igniting an explosion being major concerns.

Figure 20.8

Reverse-jet systems provide for on-line cleaning. With reverse-jet cleaning, the filters are cleaned by discharging a burst of compressed air into the bags at the top. This air flexes the bag wall and breaks the dust cake off so it falls into the collection hopper.

Fans and Motors

The fan and motor system supplies mechanical energy to move contaminated air from a dust-producing source through a dust collector. Centrifugal fans and axial-flow fans are the two main types of industrial fans used to move air through dust-collection systems.

These fans are driven by electric motors. Both open and totally enclosed motors are available in models that are dust-ignition-proof and rated for hazardous duty, to guard against fire hazards in potentially hazardous dust-laden environments.

Suppliers of dust-collection systems will have recommendations for suitable fan and motor size and type.

CENTRAL, UNIT, AND INSERTABLE

Central Systems

The central method of handling dust collection from the total air for a conveyor system would be to connect all the individual collection points by means of ducting to a single dust collector that is installed at a single, remote location (**Figure 20.9**). This collector contains fans, filters, and a collection hopper. The filtration system would handle all the dust extracted from the entire conveying system, collecting it for disposal, or feeding it back onto the conveyor or into the process at a convenient point.

Central systems are particularly suitable when the process has all dust-generating points operating at one time, and/or it is desirable to process all dust at one site. It is also useful when there is limited space near the conveyors for dust-collection and -processing equipment or where the explosion risk requires the dust collector to be positioned at a safe distance. In some processes, it is better to remove fine particles from the main material flow. Central dust collectors may be preferred when handling hot dust, because its temperature may be reduced as the dust travels to the central collector or by adding "fresh air" into the flow.

The drawbacks of the central dust-collection system are its requirements for more complex engineering and lengthy systems of ducting. As all dust-gathering points (pickups) must operate at once, the central method may present higher operating costs. The need to service any one component requires that the entire system be shut down. The fan motor needed for a central collector could be much larger, due to the increase in static pressure and the losses

from ductwork as the system grows. The collected dust will require an additional materials-handling system, which—if not properly sized and operated—can, in turn, create its own dust problem.

Unit Systems

Unit systems consist of small, self-contained dust collectors applied at individual, or conveniently grouped, dust-generation points (**Figure 20.10**). The collector units are located close to the process machinery they serve, reducing the need for ducting. Typically, these unit dust-collection systems employ fabric filters for fine dust, with cyclone collectors used for coarse dust applications.

The unit systems benefit from reduced ducting and the resulting reduction of engineering and installation expense. These systems offer reduced operating expense,

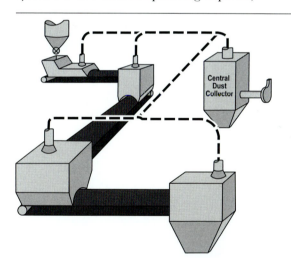

Figure 20.9

A central dust-collector system uses a single collector system (baghouse) to remove dust from a number of different points or operations in the plant.

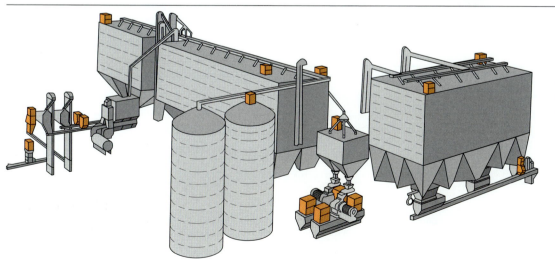

Figure 20.10

The unit system places smaller dust collectors near individual or closely grouped dust-generation points. (Note: Dust collectors are shown in orange.)

because some of the units may need to run only intermittently. Each unit can be serviced independently, without the need to shut down the entire dust-collection system.

The unit method requires space adjacent to each dust source. The disposal of dust from each of the unit collectors may require additional dust-handling mechanisms.

The advantages of unit collectors include low space requirements, the return of col-

lected dust to the main material flow, and low initial cost. However, their individual dust-holding capacities, servicing facilities, and maintenance periods are sacrificed to achieve this small size.

Insertable Collectors

An extension of the unit concept is the insertable system, in which the dust-collection system is incorporated within the dust-generation point itself (**Figure 20.11**). The filter is built into the enclosure around the dust-creation point, with the aim of controlling the dust at its source. The dust is not "extracted;" it is collected and periodically discharged back into the material stream within the enclosure.

Insertable dust collectors control contamination at its source. Installed above transfer points or other dust sources, they are small and self-contained, consisting of a fan and some form of dust filter (**Figure 20.12**). These collectors can use the positive pressure of the conveyor air, or they can be fan powered. The systems are designed to allow the filter bags or cartridges to be arranged vertically, horizontally, or at any angle. Insertable collectors eliminate ducting, reducing installation costs as well as energy costs during operation. They are suitable for individual, isolated, or portable dust-producing operations, such as bins, silos, transfer points, or mobile conveyors (**Figure 20.13**).

A principal advantage of this system is the elimination of ductwork. The insertable system is often more economical than centralized or unit systems, unless there are many points in close proximity that require dust control. As the static pressure is much lower, and there are no losses in pressure due to the ductwork, the fan motor is normally smaller than with other systems. The insertable system will operate only when needed—when the piece of equipment it is installed upon is running—reducing energy requirements. The dust is returned to the process at the point of generation, so there is no need for a separate dust-handling and disposal system.

Figure 20.11

Insertable systems place insertable dust collectors within the dust-generating equipment, such as conveyor transfer points.

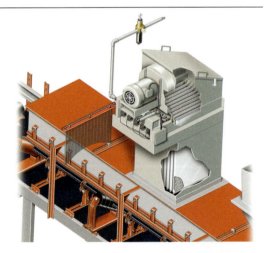

Figure 20.12

Insertable dust collectors control contamination at its source. Installed above transfer points or other dust sources, they are small and self-contained, consisting of a fan and some form of dust filter.

Figure 20.13

Insertable collectors are suitable for individual, isolated, or portable dust-producing operations, such as bins, silos, transfer points, or mobile conveyors.

20

Drawbacks to the insertable system include the use of compressed air to clean the filter. The compressed-air systems at many plants are already operating at a capacity, so addition of the system-standard reverse-jet cleaning system may overtax the plant air system. In addition, the use of compressed air to return dust to its source can cause a puff of airborne dust to escape from the system's entry and/or exit areas.

SIZING AND PLACEMENT

Filter Material and Size

After specifying a style of dust-collection system, the next step is selecting the material for the filter media. Selecting a filter media of the correct material and size is a critical function. Advancements in filter design allow the designer to pinpoint the proper style and material for an application based on the specifications of the dust to be collected. For example, if the collected dust is at a temperature that is near the limits of a standard filter, a high-temperature medium can be selected. With combustible materials, an antistatic filter media should be used.

Many filter manufacturers publish lists of air-to-cloth ratio for their various products. The air-to-cloth ratio is defined as the flow of air in cubic meters per second (ft³/min) divided by area of filtration media in square meters (ft²). The proper air-to-media ratio depends on the type and concentration of dust and the type of filter media. These lists should serve as guidelines, to be modified by variables such as dust particle size, process temperature, and moisture presence. A representative of a system supplier can provide detailed application information.

Most filter media require a dust cake on their surface to attain the desired collection efficiency. Too many people think that the cleaner the bags, the lower the emissions. This is not true: Over-cleaning the dust bags will lose the benefit of a dust cake on the filter and, therefore, reduce the operating efficiency.

Dust-collector manufacturers typically offer a valuable option in a Delta P (ΔP) controller, a device that automatically "pulses the baghouse" to clean the filters when the pressure differential—the difference between the clean and dirty sides of the filter—increases above the recommended value.

Dust Pickup Size and Location

An old saying goes, "The three most important things for a retail business are location, location, and location." The same is true in dust collection: The most critical element in the design of the dust-collection system is the location of the pickups.

It is important that the material fines be allowed a chance to settle, either of their own accord or with the addition of a dust-suppression system, before the dust-collection points are reached. Otherwise, energy will be wasted removing dust that would have quickly settled on its own, and the dust-collection system will be larger and more expensive than would otherwise be necessary. The location should be selected to minimize the capture of coarse particles, which settle quickly, and instead capture only fine dust.

For transfer points, multiple dust-collector pickup points are usually required (**Figure 20.14**). The main dust-collection pickup point is positioned approximately two times the belt width after the load point to collect three-fourths of the volume of moving air. Often, a secondary dust-collection pickup is located at the belt entry area of the transfer point (at the tail box and directly before the load zone). This pickup should take in approximately one-fourth of the total calculated air movement (**Figure 20.15**).

The Size of a Collection System

The American Conference of Governmental Industrial Hygienists' book *Industrial Ventilation* is a widely-used resource for information on dust-control systems. Originally published in 1951, this book offers standard calculations for many dust-control situations, including conveyor transfer points. However, while the specifications provided in *Industrial Ventilation* may prove useful in some circumstances, this volume should not be considered a reference for conveyor systems. Experience has shown that much of the data in *Industrial Ventilation* is no longer appropriate for transfer points. Reputable suppliers who have practical field-experience in designing, installing, and maintaining dust-collection systems for belt conveyors have developed new and more sophisticated methods of sizing and placing dust-collection systems.

DOWNFALLS

Downfalls of Dust-Collection Systems

Dry collection systems to clean dust-laden air work well in both warm and cold climates. These dust systems, regardless of the selection of a central, a unit, or an insertable collection system, may require a large amount of space for equipment and ductwork, making them expensive to install. Operating and maintenance costs are multiplied as the size of the system increases. Changes or alterations required after system start-up may be hard to implement without modification of the entire system. Even filter bag replacement can be costly and time-consuming. A leak in a filter bag can affect the efficiency of the entire collector; it can be difficult to identify and replace the leaking filter. If the collected dust must be returned to the material flow, care must be taken to prevent the dust from becoming re-entrained into the air, requiring collection at the next pickup point.

Perhaps the biggest problem in sizing a dust collector is the variation in the properties and quantities of the bulk materials being conveyed. The dust-collection system must be designed and operated for the worst conditions, even though those conditions are expected to occur on only rare occasions.

Handling of Collected Materials

The final requirement in any dust-control system is providing a mechanism to dispose of the dust after it has been collected. The steps that must be considered include removing the dust from the collector, moving the dust, storing the dust, and treating the dust for reuse or disposal.

The handling of collected material can be a problem, particularly if the material is to be returned to the process. Care must be taken to avoid affecting the process through the introduction of an overload of fine particles at any one point. In addition, collected dust must be returned into the main material body in a manner that avoids re-energizing the dust so it would need to be collected again at the next pickup point. Because the collected dust particles are small enough to readily become airborne again, they are often subjected to an extra combining process. The collected dust can be put through a mixer, pug mill, or pelletizer before re-introduction to the general material-handling system. In some industries, collected dust cannot be re-introduced to the process and, therefore, must be sent to a landfill or otherwise disposed of as waste material.

Figure 20.14

Many transfer points require more than one dust pickup location, with the main pickup positioned approximately two-thirds of the belt width after the load point.

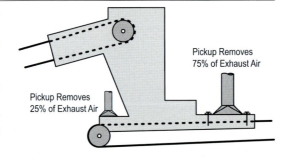

Pickup Removes 75% of Exhaust Air

Pickup Removes 25% of Exhaust Air

Figure 20.15

A secondary dust pickup is located at the belt entry area of the transfer point (at the tail sealing box and directly before the load zone).

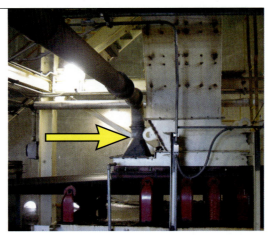

20

SYSTEM MAINTENANCE

Consultation should be made with dust-control system suppliers to determine the proper service intervals and procedures to maintain efficient operation.

It is important that dust-control systems be designed to allow efficient access to the enclosure and to allow inspection and service of filter bags and other components.

TYPICAL SPECIFICATIONS

A. Dust collector

The conveyor transfer point will be equipped with a dust-collector system to capture airborne particles and return them to the main body of material without the use of additional dust-handling equipment.

B. Transfer-point location

This dust collector will be installed inside the transfer-point enclosure, so it can operate without ducting or the high-powered fans that would be required to move dust-laden air to a central baghouse.

C. Integral fan

The dust collector's integral fan will pull dust-laden air through its filter elements on the inside of the enclosure.

D. Filter bags

The dust-collection system will incorporate a set of wire-frame mounted, envelope-shaped filter bags for optimum airflow and thorough cleaning. This filter system will capture 99 percent of all particles larger than one micron. The filters shall be serviceable from the clean-air side of the collector unit.

E. Reverse-jet cleaning

Periodic cleaning of the filters will be accomplished with an automatic reverse jet of compressed air into the filter bags. This will create a momentary reversal of air flow, inflating the filter element to dislodge the accumulated dust. The collected filter cake will return to the main material stream.

F. Access doors

A removable access door will allow access to the clean-air chamber.

G. Safety measures

To minimize the risk of explosion or fire, a spark-free fan with a motor rated for the appropriate hazardous duty, grounded dust filters, and a stainless steel lining inside the dust-collection hopper should all be utilized with any dust-collection system.

SAFETY CONCERNS

As noted elsewhere in this book, there are significant risks of fire and explosion in any enclosed area, such as inside a dust collector, where airborne particulates can become concentrated. *(See Chapter 17: Dust Management Overview.)*

It should be noted that many dust-collection systems constitute enclosed spaces requiring confined-space entry permits and procedures. The corresponding precautions should be taken when dispatching personnel to inspect or maintain the filters or other components.

TYPICAL DUST-COLLECTION APPLICATION

A belt at a concrete plant is suffering from an excessive amount of dust escaping from the transfer chute. The transfer chute is well constructed but cannot be expanded in any way.

Material. . . . . Cement

Transfer Point . . Good, but not expandable

Airflow. . . . . 0,75 m³/s (1600 ft³/min)

Collection Method . . . . . . Insertable Dust Collector

This is a good application for an insertable dust collector. There is a dust problem. The transfer point is already established, but it can not be expanded to utilize passive dust control. The airflow is too large to use a passive filter bag.

The insertable will place the collected dust back onto the conveyor as larger solids, eliminating the need for a secondary operation to handle the collected dust. An insertable will also keep the relatively valuable cargo on the conveyor belt.

The insertable would be installed on the conveyor belt transfer point near the outlet of the transfer chute. It would pull the excess air and dust through a set of filters. At regular intervals, air would be pulsed through the filters, and the now-agglomerated dust particles would fall back onto the conveyor.

ADVANCED TOPICS

Selection and Application of an Insertable Dust-Collection System

An insertable collector is a self-contained piece of dust-collection equipment. To properly size one, total airflow, combustibility of the material, and basic size restraints of the area on top of the transfer chute need to be known. The designer also needs to know the air-to-cloth ratio recommended for the material and the collector being considered. This value can be obtained from the collector manufacturer and is usually based on the material conveyed.

Below are two examples demonstrating the process needed to select an insertable collector and the effect the air-to-cloth ratio has on that selection:

A. Anthracite coal chute

Given: An anthracite coal chute generates 50 cubic meters per minute (1750 ft³/min) of air; the collector requires an air-to-cloth ratio of 2,75 meters per minute (9 ft/min) for this coal

Find: The basic requirements of an insertable dust collector

Solution: To find the total area of filter media required, divide the airflow by the air-to-cloth ratio; consider the combustibility of the material

Given that the material is coal, this application would require an insertable dust collector that is rated for the appropriate hazardous duty.

Answer: This application would need an insertable dust collector that could pull 50 cubic meters per minute (1750 ft³/min), have 18 square meters (194 ft²) of filter media, and be rated for the appropriate hazardous duty.

B. Sub-bituminous coal chute

Given: A sub-bituminous coal chute generates 50 cubic meters per minute (1750 ft³/min) of air; the collector requires an air-to-cloth ratio of 2,1 meters per minute (7 ft/min) for this coal

Find: The basic requirements of an insertable dust collector

Solution: To find the total area of filter media required, divide the airflow by the air-to-cloth ratio; consider the combustibility of the material

Given that the material is coal, this application would require an insertable dust collector that is rated for the appropriate hazardous duty.

20

Answer: This application would need an insertable dust collector that could pull 50 cubic meters per minute (1750 ft³/min), have 24 square meters (250 ft²) of filter media, and be rated for the appropriate hazardous duty.

The 24 square meters (250 ft²) insertable dust collector unit will be substantially larger than the collector in the previous example, so physical room—area on top of the transfer chute—must be verified.

Air Velocity and Dust Management

Understanding and controlling the speed of the air—pick-up velocity, capture velocity, and transport velocity—will greatly influence the amount of dust that becomes airborne.

Pick-Up Velocity

A material's pick-up velocity is the speed of the surrounding air required to lift the dust particle from a resting position into the air stream. The pick-up velocity for a material is dependent on the size and moisture content of its fine particles. The pick-up velocity for most materials is in the range of 1 to 1,25 meters per second (200 to 250 ft/min), with smaller, drier dust particles closer to the low end of the speed range, and larger, wetter dust particles closer to the high end.

Capture Velocity

Once the dust particle is suspended in the air, the amount of air speed required to gather the moving dust particle into the dust-collection system is called capture velocity. The capture velocity is dependent on how far the dust particle is located from the capture device (pickup) and the size and moisture content of the dust particle. Most properly designed collection hoods require the capture velocity to be in the range of 1 to 3,5 meters per second (200 to 700 ft/min), with the higher capture velocities required for heavier, wetter dust particles and lower capture velocities for lighter dust particles with less moisture.

There is a simple formula to determine the capture velocity of dust particles based on their density and diameter (**Equation 20.1**).

It is possible to calculate the exit velocity of air from a given transfer point and then work back to calculate the size of particle that would fall out in the air stream before it leaves the transfer point.

Transport Velocity

Transport velocity is the speed of the air required to keep an airborne dust particle in suspension in the ducts transporting the dust to the dust collector. These transport velocities are based on the size of the dust particle (**Table 20.1**).

DUST COLLECTION: ONE PIECE OF THE PUZZLE

In Closing...

This chapter has presented only an overview of the capabilities and considerations of dust-collection and -control equipment. It would be wise to consult with suppliers specializing in this equipment to receive specific recommendations.

While valuable additions to materials-handling systems, dust-collection systems are only one piece of the dust-management puzzle. The more successful a transfer point is in minimizing the amount of induced air and in loading cargo in the direction and at the speed of the receiving belt, the less airborne dust will be created. The more successful the enclosure and sealing of a conveyor or transfer point, the less fugitive dust will be released. The more successful a dust-suppression system, the less dust will be present in the air to be collected. Successful application of the principles of enclosure and suppression work to minimize the required size of a dust-collection system—and reduce the wear and tear and the risk of overload on that system. A pyramid composed of the three systems—enclosure, suppression, and collection—allows a plant to be successful

Capture Velocity of
Dust Particles

$v_t = k \cdot \rho_s \cdot D^2$

Given: *A particle that is 0,006 meters (0.020 ft) in diameter and has a particle density of 800 kilograms per cubic meter (50 lb$_m$/ft^3).* **Find:** *The capture velocity of the particle.*

Variables		Metric Units	Imperial Units
v_t	Capture Velocity of a Falling Particle in Still Air	meters per second	feet per second
k	Conversion Factor	$3,187 \times 10^3$	15.6×10^3
ρ_s	Particle Density	800 kg/m^3	50 lb$_m$/ft^3
D	Diameter of the Particle	0,006 m	0.02 ft

Metric: $v_t = 3,187 \times 10^3 \cdot 800 \cdot 0.006^2 = 91,8$
Imperial: $v_t = 15.6 \times 10^3 \cdot 50 \cdot 0.02^2 = 312$

v_t	Capture Velocity of a Falling Particle in Still Air	91,8 m/s	312 ft/s

Dust Transport Velocities Based on Dust Particle Size		
Material	**Metric**	**Imperial**
Fine Light Dusts *(flour, PRB, coal)*	10 m/s	2,000 ft/min
Fine Dry Dusts and Powders *(foundry sand, cement)*	15 m/s	3,000 ft/min
Average Industrial Dust	18 m/s	3,500 ft/min
Coarse Dust *(quarry dust)*	20 to 23 m/s	4,000 to 4,500 ft/min
Heavy or Moist Dust *(underground coal)*	23 m/s and more	4,500 ft/min and more

20

in controlling the amount of dust released into the environment.

Many dust-control projects have produced less than expected results when equipment is misapplied or simple "rules of thumb" are used for sizing systems. Successful selection, installation, and maintenance of dust-control systems require specialized knowledge, which is available from equipment suppliers or their authorized representatives.

Looking Ahead...

This chapter on Dust Collection is the fourth and final chapter in the section on Dust Management. The following chapter begins the section on Leading-Edge Concepts with a discussion about Clean, Safe, and Productive Conveyors by Design, followed by Engineered Flow Chutes in Chapter 22 and Air-Supported Conveyors

in Chapter 23. Chapters 24 and 25 focus on Belt-Washing Systems and Material Science.

REFERENCES

20.1 *Dustcollectorexperts.com* offers a detailed and useful tutorial on various dust-collection systems. This noncommercial web site provides background information and links to a number of suppliers of dust-collection equipment.

20.2 Mody, Vinit and Jakhete, Raj. (1988). *Dust Control Handbook (Pollution Technology Review No. 161),* ISBN-10: 0815511825/ISBN-13: 978-0815511823. Park Ridge, New Jersey: Noyes Data Corporation.

20

SECTION 5
LEADING-EDGE CONCEPTS

Figure 21.1

A new hierarchy for design decisions is useful in devoloping conveyors that are productive, safe, service-friendly, fugitive material-free, and cost-effective.

21

Chapter **21**

CLEAN, SAFE, AND PRODUCTIVE CONVEYORS BY DESIGN

In this Chapter...

In this chapter, the traditional conveyor design hierarchy—1: capacity, 2: minimum code compliance including safety, and 3: lowest price—is questioned. A new design hierarchy, in step with designing conveyors that are not only of the proper capacity and code compliant but also 1: clean (control of fugitive materials), 2: safe (service-friendly), and 3: productive (cost-effective and upgradable), is proposed (**Figure 21.1**).

Since the invention of the belt conveyor, there have been substantial changes in safety rules, pollution regulations, construction standards, and required carrying capacity of these systems. Unfortunately, the details of design and fabrication of belt conveyors are still governed by "rules of thumb" and design methods that have been passed down from one generation of designers to the next. Other than advances in computers to predict performance, synthetic carcasses for belts, and improved control technologies, conveyor systems are designed much the same way they were 50 years ago.

Most engineering and construction project contracts are awarded on a low-bid basis. Current supplier practice is to base a bid on the price per kilogram (per pound) of fabrication, with minimal design time, in order to be competitive in this low-bid system. Due to these competitive pressures, it is common practice for suppliers to base a proposal on specifications, drawings, and designs that were completed previously for a similar system. Regrettably for the owners, operators, and maintainers of conveyors, this practice often produces a 50-year-old design at state-of-the-art prices. Since the system was designed with old thinking, it will likely fail to meet today's expectations.

This chapter demonstrates how designing components and critical sections of conveyors can lead to clean, safe, and productive bulk-material transportation systems.

THE DESIGN PROCESS

As expressed by George E. Dieter, "To design is to pull together something new or arrange existing things in a new way to satisfy a recognized need of society" *(Reference 21.1)*. Design is just as much art as it is science. Every company's design process will differ, but most include:

A. Problem definition

B. Information gathering

C. Concept generation and evaluation

D. Modeling and simulations

E. Material selection

F. Risk, reliability, and safety reviews

G. Cost evaluation

H. Detail design

I. Communicating the design

The intricacies of this process will not be discussed in detail here, but it is important to note that the process begins with the identification of a need and a definition of the problem. This first step, although critical, is often overlooked. Based on how the problem statement is defined, the final result can differ greatly.

The purpose of a belt conveyor system is to provide a means of moving one or more bulk materials from one point to another. The total belt conveyor system can be broken down into several sections or zones with the detail and design of those sections being examined from new and different points of view. A traditional problem definition would be to transport a specific type, size, and amount of material from point A to point B. If the requirements are expanded to include considerations for safety and the minimization of the escape and accumulation of fugitive materials, then the entire conveyor system takes on a different perspective. When additional factors—such as ease of installation, maintenance, and cleanup; standardization of components; and the creation of a cost effective, upgradeable design—are included, a conveyor belt system designed under these criteria

21

is quite different from the typical conveyor system provided today.

In order to initiate a change to new, more modern designs—designs where cleanliness, safety, and serviceability are also included in the initial design considerations—a new, more comprehensive view of how bulk materials are handled must be investigated.

SAFE DESIGN

Personnel are the single most important resource of any mine or industrial operation; therefore, engineers and designers should incorporate functionality into designs that will improve safety. While designs have changed little, the work-place environment has changed significantly. Restrictions regarding lifting, requirements for lockout / tagout / blockout / testout, regulations on confined-space entry, and a host of other safety procedures have been established. At the same time, there is increasing pressure for continuous and ever-increasing production.

Applying design principles to help ensure worker safety should include the use of barrier guards and the implementation of new designs that will improve the ease of cleaning around and changing out equipment. Employee training for enhanced awareness up to and including qualification requirements should be instigated as well.

Barrier Guards

In order to better protect personnel from coming in contact with moving conveyor components, the trend is to install barrier guards (also called area guarding) around the entire conveyor (**Figure 21.2**). These barrier guards should be installed around all pinch point locations and anywhere personnel could come into contact with moving parts. Barrier guards should be designed for easy installation and removal to allow for authorized service personnel to perform required functions safely and efficiently and to ensure the guards are returned to place when the work is completed. *(See Chapter 2: Safety.)*

Maintenance During Operation

With many conveyors operating around the clock, scheduled downtime is at a premium. When handling bulk materials, problems occur and equipment fails prematurely, resulting in lost productivity, emergency cleanup, and repair requirements.

Many safety standards around the world recognize that certain maintenance procedures must be performed while equipment is in operation. These standards allow for exceptions to the rules requiring equipment to be shut down for service. Exceptions are written so only personnel authorized and trained for awareness of the potential hazards can adjust equipment that is in operation. Indeed, the trend in safety standards (as specified in the International Organization for Standardization (ISO) document ISO/EN 14121) is away from task-specific restrictions toward risk-ranked restrictions based on a formal risk analysis. When a case can be made that the risk of personal injury of servicing equipment while the equipment is in operation is actually equal to or less than the risk of personnel injury while servicing equipment that is stopped and locked and tagged out, newer safety standards will recognize that the lowest-risk procedure is the preferred approach.

Some conveyor belt system components require frequent service to maintain optimal efficiency (for example, belt cleaners). In the control of fugitive materials and the ability to run a conveyor continuously, belt cleaners are critical. Due to safety concerns, most operations prohibit the servicing of belt cleaners while the conveyor is

Figure 21.2

Barrier guards (also called area guarding) are installed to keep plant personnel from dangerous contact with equipment.

in operation. The inability to service a belt cleaner can lead to carryback and spillage problems that create safety hazards. Belt cleaners and other conveyor components can be designed to be safely serviced while the belt is running. Specialized tools can be designed and service techniques can be taught to develop authorized maintenance employees or service contractors who can safely service certain components while the belt is running (**Figure 21.3**).

CLEAN DESIGN

Clean designs are critical to operating a safe and productive material-handling system. However, in today's normal industrial facility or mine, it is not possible to operate a conveyor system that is 100 percent free of fugitive material (**Figure 21.4**). Poor initial designs, lack of maintenance follow-up, variability of the properties of bulk materials, conveyor overloading, and constant wear on system components are strong contributors to unexpected releases of fugitive materials.

Many design details contribute to creating a conveyor system as free of fugitive materials as possible. Incorporating dust-resistant structures, proper skirtboard design, external wear liners, appropriate pulley sizing, and belt-tracking alternatives; ensuring the working area is clean and free of utility components; and allowing for future upgradability are issues that will be discussed to improve material-handling operations. There are a number of leading-edge technologies that can be incorporated into a conveyor system to improve its control of material. These options include engineered flow chutes *(see Chapter 22: Engineered Flow Chutes)*, air-supported conveyors *(see Chapter 23: Air-Supported Conveyors)*, and belt-washing systems *(see Chapter 24: Belt-Washing Systems)*.

Modern 3D drafting and fabrication techniques make it feasible to arrange components in non-traditional ways without greatly increasing the costs of these systems. One of the simplest details is to ensure components are oriented in a manner that provides as few flat surfaces as possible upon which fugitive material can accumulate (**Figure 21.5**).

Dust-Resistant Structures and Components

Cleaning around conveyors is a necessity. By eliminating places where fugitive materials accumulate, cleaning requirements are reduced and simplified. Horizontal structural members should be angled at 45 degrees whenever possible in order to shed material, thus making it unlikely that cleanup crew members will have to reach under the belt with tools to remove buildup.

Structural members that cannot be oriented to reduce dust buildup should be fitted with dust plates or caps to reduce material buildup in hard-to-clean areas (**Figure 21.6**).

Figure 21.3

Specialized tools and safe designs make belt-cleaner service easier.

Figure 21.4

Fugitive material accumulates on flat surfaces.

21

Figure 21.5

The minimization of flat surfaces, including stringers and skirt supports, can reduce the buildup of material.

Deck plates and drip pans should be designed to shed material toward the outside of the conveyor where fugitive material can be more easily collected (**Figure 21.7**). In order to assist in the reduction of buildup of dust and ensure any fugitive material will flow to the outside of the conveyor, these pans should be designed for the application of vibration.

Figure 21.6

Dust caps are installed to reduce the accumulation of fugitive material.

21

Figure 21.7

Angle deck plates under the conveyor load zone will direct fugitive materials to the outside of the structure.

Figure 21.8

A conventional skirtboard design places the wear liner on the inside of the skirtboard.

Skirtboard Height

The height of the skirtboard (chutewall) cited in the Conveyor Equipment Manufacturers Association's (CEMA) *BELT CONVEYORS for BULK MATERIALS, Sixth Edition*, and in other references and standards, is based on the largest size lump that will be carried on the conveyor without skirtboard covers. Today, many skirtboards are covered in order to contain dust. It is recommended that skirtboards be designed to accommodate the air flow above the bulk material. (*See Chapter 11: Skirtboards.*) This leads to a requirement that is at least two times the height CEMA recommends for open-top skirtboards. *(See Chapter 11: Skirtboards for more information on calculating the proper height for covered-top skirtboards.)* Skirtboard tops should be designed to include significant pitch in order to avoid material buildup.

External Wear Liner

The practice for years has been to attach the wear liner to the inside of the vertical metal skirtboards. The wear liner is then positioned between the bulk material and the metal skirtboards (**Figure 21.8**). The skirtboard serves as the structural member that supports both the wear liner and the skirtboard seal. If incorrectly mounted, wear liners will fail to protect the skirtboard seal from wear and sometimes trap material against the belt, thus grooving or otherwise damaging the belt. In this traditional setup, with the wear liners mounted to the inside of the skirtboard, inspection and replacement are difficult due to the placement of the liners behind the skirtboard. Replacing wear liners mounted on the inside of the skirtboard is a complicated job requiring manual manipulation of heavy sections in tight quarters and sometimes even involving confined-space entry.

Wear liner repositioned so it is placed on the outside of the skirtboard—where it can be easily inspected, accurately installed, and easily replaced—is a simple modification potentially saving thousands of maintenance hours (**Figure 21.9**). Skirtboard provides structural support; raising it above the normal flow pattern of the bulk material and implementing a small design change to the skirt-seal clamps enables the wear liner to be installed on the outside of the skirtboards. The wear liner can also be made adjustable for accurate installation.

Pulley Sizes

For decades, tail, bend, and discharge pulley sizes have been selected from tables published by belting manufacturers with minimum pulley diameters based on minimizing costs and providing safe stress levels for the belt. Determining the correct size of pulleys should include consideration of ease of access for service. A larger-diameter pulley—one with a minimum pulley diameter of 600 millimeters (24 in.)—would allow adequate space between the carrying and return runs for installation of a tail pulley-protection plow and, if necessary, a return belt plow (**Figure 21.10**). The additional space that would be supplied between the carrying and return runs of the belt allows for easier inspection of plows and provides adequate space for the plows to eject fugitive materials from the belt. A larger-head pulley at the belt discharge provides needed space for installing belt cleaners in the optimal working position. The added cost of adding larger pulleys is offset by the cost savings derived by effectively controlling fugitive materials and by requiring shorter down times and less maintenance.

Indexing Idler

Belt mistracking is a major cause of spillage; therefore, much attention is given to belt-training devices in order to keep the belt centered in the structure. In an effort to keep the overall cost of a new installation down, training idlers are often supplied in lieu of belt-training devices on many new installations. Training idlers often end up tied off to one side or the other in an attempt to either compensate for a situation beyond the capability of the device or protect the device from excessive wear due to the belt continuously running to one side.

Conveyor belts often run to one side or the other due to conditions such as off-center loading, conveyor structure alignment issues, conveyor component alignment problems, weather conditions, or a variety of other factors. Loose pieces of wire or rope, used to tie off training idlers, in the vicinity of the moving belt are safety hazards. This issue can be exacerbated by changing conditions or operator preferences that require an indexing idler tied off to hold the belt in one direction in the morning to be switched to tie off the belt in the opposite direction in the afternoon.

In the absence of properly aligning the conveyor structure, replacing and/or aligning conveyor components that are causing belt-alignment issues, ensuring the load is properly centered, or installing one or several belt-training devices to properly align and track the belt, these training idlers can be fitted with a mechanism that allows them to be "indexed," or locked into position, without resorting to unsafe wire or rope tie-downs (**Figure 21.11**).

Conduit and Piping

Conveyors provide convenient paths for running utilities and electrical components. For decades, utilities and electrical piping have been installed along the conveyor's

Figure 21.9

Wear liner installed on the outside of the skirtboards is a simple modification potentially saving thousands of maintenance hours.

Figure 21.10

The tail-pulley diameter should be selected to have plenty of clearance for installation and service of a plow.

21

structure with little regard to the effects of this location on the installation, maintenance, and operation of conveyor components. This issue is particularly noticeable in the discharge and loading zone areas of the conveyor belt system. For example, it is common to see plows buried behind a web of conduit that was installed after the plow was positioned (**Figure 21.12**). Plows need to be able to eject foreign objects from the conveyor in the location selected by the designer.

The utility conduits in the discharge and load zones in particular should be run in locations where they do not interfere with access to components that are essential to the control of fugitive material. The main conduit could be run overhead with flexible conduit dropped down where required to provide power to or communicate with the components. Along the carrying run of the conveyor, the structure can be used for supporting conduits as long as the conduit does not interfere with access for service or reduce the effectiveness of the individual components.

Figure 21.11

This training idler can be adjusted and locked in place safely and without tying it off with wire or rope.

Figure 21.12

Conduit placed alongside the conveyor makes access for maintenance impossible.

PRODUCTIVE DESIGN

Following design principles that establish safe, service-friendly, and easy-to-clean belt conveyor systems leads to better and more productive operating systems. A cleaner, safer operation is normally a more productive operation in the long run. Safety issues normally correspond to unsafe operating conditions, which are also detrimental to the equipment. Airborne dust can find its way into lungs and bearings; material can accumulate under and on walkways and conveyors, leading to trip, slip, and fall hazards. These unsafe operating conditions are not only hazards to health, but also to the condition of the conveyor equipment. When equipment is shut down for unscheduled repairs, it cannot be productive.

Cost Effective

The total cost of ownership, including the cost per kilogram (per pound) of dealing with fugitive material releases, should be considered in making design and purchasing decisions. Unfortunately, the lowest-bid process discussed earlier, that considers only initial purchase price, has slowed the evolution of clean, safe, and productive designs. While initial purchase price may be lower for a system with no adjustment capabilities and no consideration for future wear-component replacement, the higher costs required to properly install and maintain components, clean up fugitive materials, and cover additional equipment downtime will far exceed the costs of a system which takes these factors into consideration in the initial design.

Utilizing standard components where possible in the design may make economic sense, because some economy of purchase may be realized. With some forethought and some slight design changes, standard components (structure, cradles, skirting, etc.) can often be adapted to these new design principles. Use of standard components can provide for ease of installation and replacement due to standardization across the plant. Designing the system for ease of upgradability, by making compo-

21

nents track mounted (**Figure 21.13**) and service-friendly, can reduce down time and control fugitive materials.

Upgradeable

Designers routinely consider capacity upgrades, but they rarely include provisions for component upgrades. A track-mount system provides flexibility for quickly installing different problem-solving components. The use of a pre-engineered mounting hole pattern in the structure around the conveyor's transfer point allows for the installation of a new or improved system quickly and easily (**Figure 21.14**). A uniform-hole pattern for accessory mounting will encourage component suppliers to adapt modular, bolt-on, or clamp-on designs for easy retrofits. Utilizing structural platform designs—which incorporate tracks, modularity, and easy retrofitability—will encourage designers to continue to modernize the way bulk materials are handled today and in the future.

Figure 21.13

This universal track allows for slide-in/slide-out maintenance.

Figure 21.14

A clamp-on bracket allows for the simple installation of a track system for belt-support components.

A NEW HIERARCHY

In Closing...

Modern design techniques—such as 3D modeling for fabrication, Finite Element Analysis (FEA) for structure, and Discrete Element Modeling (DEM) for chute design—can be used to improve conveyor reliability, productivity, and safety while reducing the total cost of ownership. To achieve clean, safe, and productive designs, designers should consider a new hierarchy for design decisions:

A. Capacity

B. Safety and code compliance

C. Control of fugitive materials

D. Service friendliness

E. Cost effectiveness

F. Upgradability

Decisions related to the design of the conveyor system or the selection of individual components should follow a hierarchy to ensure the best design possible is created.

In the future, all bulk-material handling systems should incorporate designs to safely move the required amount of material from point A to point B in a service-friendly, cost-effective manner that controls dust and fugitive materials for now and ever more.

Looking Ahead...

This chapter, Clean, Safe, and Productive Conveyors by Design, the first chapter in the section Leading-Edge Concepts, discussed the wisdom of designing bulk-materials handling systems that may cost more initially but save money in the long run. The next chapter, Engineered Flow Chutes, is the first of three chapters that present designs for cleaner, safer, more productive conveyor systems.

REFERENCES

21.1 Dieter, George E. (1999). *Engineering Design: A Materials and Processing Approach*, Third Edition. McGraw-Hill.

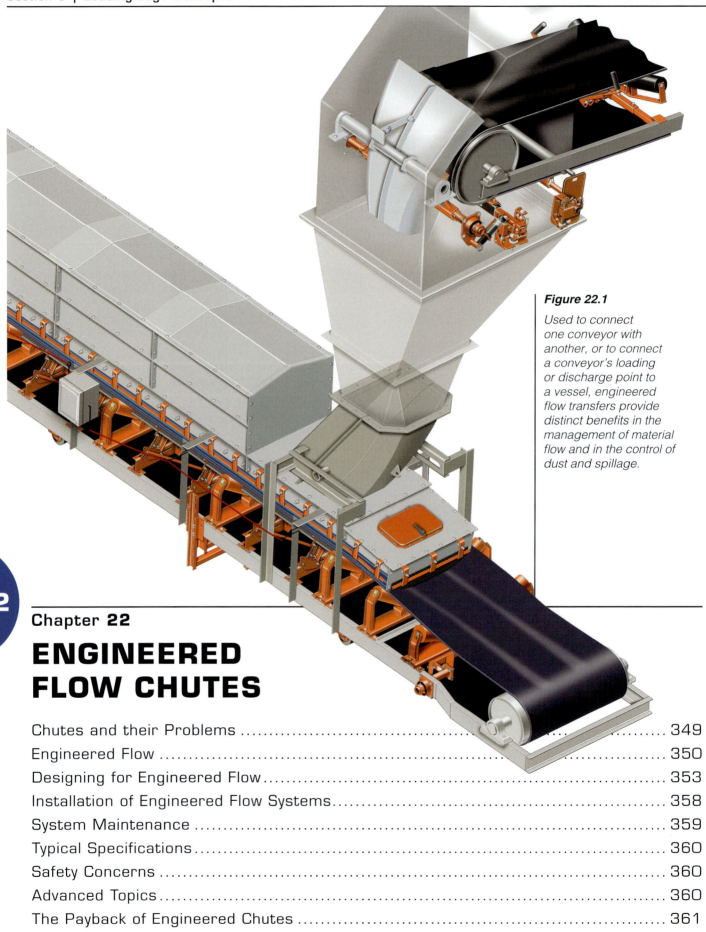

Figure 22.1

Used to connect one conveyor with another, or to connect a conveyor's loading or discharge point to a vessel, engineered flow transfers provide distinct benefits in the management of material flow and in the control of dust and spillage.

22

Chapter 22

ENGINEERED FLOW CHUTES

In this Chapter...

In this chapter, we discuss the benefits of engineered flow chutes and the ways they resolve problems common with transfer chutes. The components of engineered chutes—the hood, spoon, and settling zone—are defined. We also describe the process used to design them, along with information required by designers to do so.

One leading-edge development that improves the conveying of bulk materials is the advent of engineered flow chutes (**Figure 22.1**). Used to connect one conveyor with another, or to connect a conveyor's loading or discharge point to a storage vessel or other process step, engineered flow transfers provide distinct benefits in the management of material flow and in the control of dust and spillage.

Custom designed for each individual application, engineered flow chutes control the material stream from the discharge conveyor to the receiving conveyor. *(See Chapter 8: Conventional Transfer Chutes.)* A well-designed engineered flow chute maintains a consolidated material profile that minimizes dust generation and wear, by accomplishing all of the functions of a transfer chute:

A. Feeding the receiving conveyor in the direction of travel

B. Centering the material load

C. Minimizing impact on the receiving belt

D. Supplying the material at the speed of the receiving conveyor

E. Returning belt scrapings to the main material flow

F. Minimizing the generation and release of dust

Although the initial investment in an engineered flow chute may be greater than the cost of a traditional transfer chute, the return on investment to the plant will be prompt, through reduced operating and maintenance expenses. Problems such as belt damage, premature wear of belts and chutes, chute plugs, spillage, dust, sponta-

neous combustion, and material degradation are greatly reduced, if not eliminated, with the controlled material stream that travels through an engineered flow transfer chute.

CHUTES AND THEIR PROBLEMS

The engineering of bulk-materials handling systems has previously been largely based on experience, "rules of thumb," and educated guesses. But now sophisticated computers and software packages provide the design and modeling technologies that allow better understanding and management of material flow. These software and hardware systems allow the designer to work through a range of iterations that determine how a system will work with a specific material—in a range of conditions from best to worst case. A computer provides the kind of calculation power required for developing the models and generating the iterations—making small, step-by-step design adjustments that allow for the comparison of alternative solutions to improve bulk-materials handling.

Traditionally, there has been little thought given to the flow of materials through the chute beyond making sure the chute was big enough to accommodate the material stream and minimizing wear. It was a common practice for chutes to be generous in size to reduce plugging and control dust, but this actually represented a shortcoming in design methodology. Chutes were kept box like to avoid running up the expense for fabrication. Because these chute angles were designed based on the angles of repose, they were prone to buildups and blockages. With changes in flow direction from conveyor to conveyor and from the downward energy of the material movement, the chutes would suffer wear in their metal walls and on the surface of the receiving belt or vessel.

Traditionally-designed chutes generate dust by throwing a stream of uncontrolled material off the end of the conveyor and allowing it to spread. The movement of

22

material displaces air as the body of material is diffused. The air passes through the material stream, thus dispersing and entraining the small particles of dust. The traditional chute essentially can create a "chimney effect" by adding the dust to the displaced and moving air.

In addition, the receiving areas were typically small and unsupported, and they released dust. When the stream of material "crash lands" on the receiving conveyor, the profile of the material is compressed, and the induced air is driven off. This air takes with it the smaller particles of material as airborne dust. A loosely-confined stream will carry larger amounts of induced air, so more dust is driven off. If the material has been allowed to move through the chute in a turbulent stream—with what might be called "billiard flow," where the lumps bounce off each other and the chutewalls—the material lumps will degrade, creating more dust that can be carried out of the enclosure.

ENGINEERED FLOW

What is Engineered Flow?

Chutes with "engineered flow" are based on the application of the principles of fluid mechanics and an understanding of particulate movement. Engineered material flow is based on controlling the material's movement as it exits a discharging conveyor or a silo, bin, or hopper. The direction and speed of flow can be steered through subtle changes by guiding it down surfaces with known friction values. The gradual course modifications will minimize dust generation and center load the belt. This allows the energy lost through friction to be calculable and accountable.

What is an Engineered Flow Chute?

Developed from sophisticated material tests and computer flow simulations, engineered flow chutes are designed to satisfy a plant's operating requirements, so the material stays in continuous motion though the transfer chute, with the material moving as a tight, coherent stream.

This will minimize the amount of induced air carried along with the stream of material. As a result, there is less air released and less airborne dust created (**Figure 22.2**). In addition, the stream is directed or channeled, so the material is placed gently onto the receiving belt, minimizing impact and belt abrasion.

The material moves smoothly—like water through a faucet. The material slides in unison in a "fluid-like flow," rather than allowing the lumps to bounce off each other in the traditional "billiard-flow" fashion.

Benefits of Engineered Flow

There are a number of benefits to accrue from the installation of an engineered flow chute in a facility. These include:

A. Passive dust control

They reduce dust escape while minimizing, or eliminating, the need for active collection methods.

B. Increased material flow rate

They eliminate chutes as a production bottleneck.

C. Reduced material buildups and blockages

They reduce or prevent chute plugging.

D. Reduced loading impact

They extend belt-life by reducing damage and abrasion.

22

Figure 22.2

In an engineered flow transfer, the material is kept as a tight, coherent stream, minimizing the amount of induced air. Therefore, there is less air released and less airborne dust created.

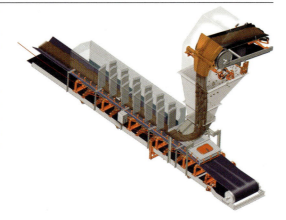

E. Reduced degradation of material

They minimize creation of dust.

F. Controlled load placement

They prevent mistracking, spillage, and belt-edge damage.

It should be noted, however, that engineered flow chutes are designed to accommodate a narrow range of parameters. Changes in the performance of these chutes (and in the wear life of the linings inside them) will occur when conditions vary, including:

A. Inconsistent flow rates

Variations of more than 20 percent from the stated flow, other than at start up and shut down

B. Inconsistent material characteristics

Variations of more than 20 percent in any attribute from the material samples tested prior to system design

C. Inconsistent environmental conditions

Variations that create alterations in the material, such as precipitation that changes the moisture content by more than 10 percent from the stated characteristics

Components of Engineered Flow Transfers

An engineered flow chute incorporates geometry that captures and concentrates the material stream as it travels through the chute, which has the dual benefit of minimizing aeration and preventing accumulation of materials inside the chute. Preventing accumulation of materials within a chute is particularly important when dealing with combustible materials, such as coal.

Engineered chutes typically employ a design called "hood and spoon" transfer. This design is composed of a "hood" discharge chute, at the top of the system, and a "spoon" receiving chute, which places the material onto the belt being loaded. The hood and spoon are typically installed as a pair, although a particular material-handling situation might require only one or the other. These components are custom-designed using the characteristics of the conveyed material and of the materials used for chute construction. The goal of hood and spoon is to confine the moving material stream, reducing the entrainment of air and minimizing the impact forces, while placing the material in the proper direction on the receiving belt with minimal impact—or "splash"—to reduce spillage, abrasion, dust, and damage. This controlled loading also prevents side loading of material, which causes belt mistracking.

In addition, many engineered flow chutes incorporate an additional area for dust confinement—called a settling zone or stilling zone. Here the air current above the material stream is slowed so that the residual dust can settle back onto the conveyor.

Hood

Installed at the discharge, a hood captures and confines the moving material stream at a low impact angle (**Figure 22.3**). This minimizes impact force, buildup, and wear. The hood redirects the material stream vertically, so it flows smoothly toward the conveyor system below (**Figure 22.4**). Once flow is vertical, then the direction of the material stream is gently modified to align the flow with the receiving conveyor.

Spoon

A spoon is installed at the bottom of the transfer chute, where it receives the mate-

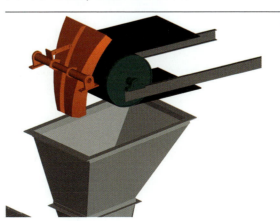

Figure 22.3

Installed at the discharge, a hood captures and confines the moving material stream at a low impact angle.

22

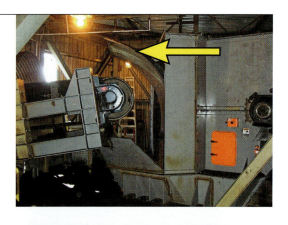

Figure 22.4

A hood is installed to redirect the material stream vertically, so it flows smoothly toward the conveyor system below.

Figure 22.5

A spoon is installed at the bottom of the transfer chute, where it receives the material stream and places it on the receiving belt.

Figure 22.6

By directing the concentrated stream of material onto the center of the receiving belt with the proper speed and angle, the spoon reduces impact on the belt, belt abrasion, dust creation, off-center loading, wear on wear liners, and other problems.

Figure 22.7

The settling zone, typically installed after the spoon on the receiving conveyor, corresponds to the conventional skirted and covered portion of the receiving conveyor.

rial stream and places it on the receiving belt (**Figure 22.5**). The spoon is designed to gently load the material onto the receiving conveyor, so the cargo is moving in the same direction as, and near the velocity of, the belt. By directing the concentrated stream of material onto the center of the receiving belt with the proper speed and angle, the spoon reduces impact on the belt, belt abrasion, dust creation, off-center loading, wear on wear liners, and other problems (**Figure 22.6**).

Another benefit of loading via an engineered spoon is that the belt may require less belt support in the load zone. Loading the material onto the belt at a similar speed and in the same direction as the belt is traveling provides less impact onto the belt and, consequently, less need for impact cradles and belt-support cradles.

In some complex chutes or transfers with large drop distances, more than one "hood and spoon" pair might be used to control flow.

Settling Zone

The settling zone, typically installed after the spoon on the receiving conveyor, corresponds to the conventional skirted and covered portion of the receiving conveyor (**Figure 22.7**). This area is carefully engineered to provide for optimum settling of dust-laden air and settlement of any airborne dust, by holding the air long enough to slow its velocity. The settling zone typically uses a higher, covered skirtboard to allow any airborne dust to settle out of the air, returning most of the dust to the main material bed without being released to the outside (**Figure 22.8**). The air currents are slowed by the larger area of the settling zone and the use of dust curtains within the area.

Some system designers omit a settling zone from their designs, using only conventional covered skirtboard designs. However, it is almost impossible to design a chute that will handle every possible material condition. Therefore, it is safer to include the

22

settling zone to accommodate unforeseen circumstances or to handle future changes in material characteristics.

DESIGNING FOR ENGINEERED FLOW

Even if two conveyors run at the same speed, gravity can cause the velocity of the material to increase during a transfer from one conveyor to the other if the flow is left unrestrained. Both the hood and the spoon must be designed to intercept the material trajectory at a low angle of incidence. This uses the natural forces of the material movement to steer the flow into the spoon for proper placement on the receiving belt with reduced impact and wear. Because the hood and spoon are designed with both the material specifications and the flow requirements as criteria, the chute can operate at the required flow with reduced risk of plugs or chute blockages that will choke operations.

To achieve the proper design of hood, spoon, and settling area, engineered flow chutes are created using three-dimensional (3D) computer-based modeling to define the geometry of the chute (**Figure 22.9**). The angle and force of impact should be minimized to maintain as much momentum as possible. Ideally, the impact angle should be no more than 15 to 20 degrees. This design must be based on rigorous processes and procedures to provide a precise, accurate, and complete design. Dimensional data can be determined from a site survey or—particularly for new facilities—from a review of the site plans and conveyor specifications.

It is essential for the designer of an engineered flow chute to have detailed information about the material that will be flowing through the chute and the parameters of the conveyor system itself. This information includes:

A. Feed system

   a. Type of feed system (e.g., crusher, vibratory feeder, stockpile, reclaim)

   b. Number of feed systems

   c. Angle of incline or decline (**Figure 22.10**)

   d. Belt speed

   e. Belt thickness

   f. Belt width

   g. Trough angle

   h. Transfer capacity

   i. Type of conveyor structure (channel, truss, cable)

   j. Method by which material is delivered to plant (e.g., barge, railcar, truck)

B. Transfer

   a. Interface angle (**Figure 22.11**)

   b. Horizontal distance to loading point (**Figure 22.10**)

   c. Drop height (**Figure 22.10**)

   d. Transfer capacity

   e. Number of transfers

Figure 22.8

The settling zone is carefully engineered to provide for optimum stilling of dust-laden air and settlement of any airborne dust, by holding the air long enough to slow its velocity.

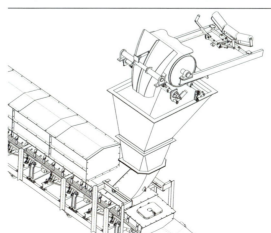

Figure 22.9

To achieve the proper design of hood, spoon, and settling area, engineered flow chutes are created using 3D computer-based modeling to define the geometry of the chute.

22

Figure 22.10

The designer of engineered flow chutes needs detailed information about the conveyor system and the material it carries.

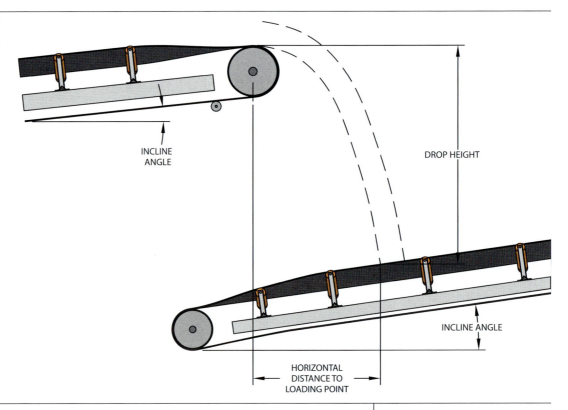

INCLINE ANGLE

DROP HEIGHT

INCLINE ANGLE

HORIZONTAL DISTANCE TO LOADING POINT

Figure 22.11

The interface angle of a transfer point is a key element in the design of engineered chutes.

22

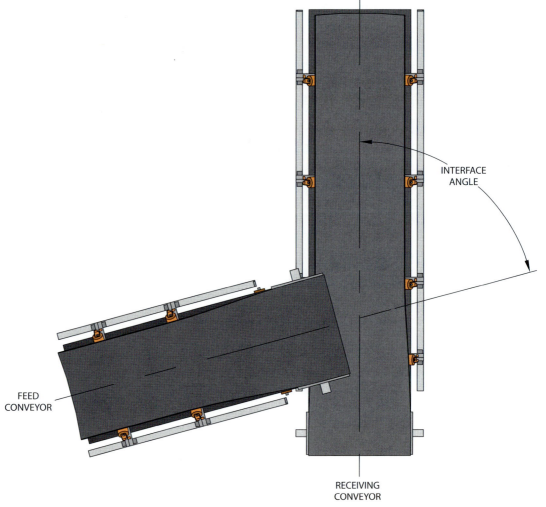

INTERFACE ANGLE

FEED CONVEYOR

RECEIVING CONVEYOR

f. Number of gates and purpose (e.g., splitting the flow or changing direction of the flow)

g. Interference due to surrounding structure

C. Receiving system

   a. Type of receiving system

   b. Number of receiving systems

   c. Belt speed

   d. Belt thickness

   e. Incline/decline angle of conveyor (**Figure 22.10**)

   f. Belt width

   g. Type of conveyor structure (channel, truss, cable)

   h. Trough angle

   i. Transfer capacity

   j. Belt/load support system

   k. Distance of conveyor to curve or interference for settling zone

D. Material conveyed

   a. Material type

   b. Temperature ranges (high and low)

   c. Moisture content

   d. Environmental conditions that affect material condition (including distance from source/supplier and location where sample was collected)

   e. Material size

   f. Bulk density

   g. Interface friction

   h. Cohesion/adhesion properties

   i. Particle size and percentage distribution

   j. Average lump size and maximum lump size

   k. Surcharge angle

   l. Angle of repose

E. Construction materials

   a. Chute construction materials

   b. Chute liner materials

   c. Tolerances for fabrication and installation

   d. Interface friction values for construction materials in contact with the bulk material

Design of Engineered Flow Transfers

Engineered flow transfer chutes are developed in a three-step engineering process. Phase one is testing of the conveyed material properties and the interface friction values in relation to the belt and construction materials, to establish the material characteristics and its performance in materials-handling systems. After the various conveyor and material parameters are defined, the material discharge trajectory can be determined using conventional methods such as the Conveyor Equipment Manufacturers Association (CEMA) method.

The second phase of the process includes verification of current field dimensions and development of preliminary engineering. A set of two-dimension conceptual drawings and a three-dimension pictorial representation of the chutework using 3D software are created, and the flow characteristics are verified using Discrete Element Modeling (DEM) method.

The third and final phase is the creation of the final design, followed by the detailed engineering and then, in turn, by the fabrication and installation of the system.

Phase 1: Material Analysis

The first step in the design of an engineered chute is testing of the actual conveyed material that will be passing through it. Information obtained includes material composition and physical properties, moisture content, lump size range, and fines size. Testing usually includes analysis of the bulk-material strength at several moisture contents—from "as-received" to "saturation" level—to allow for changing material conditions. There are typically at least three different types of tests, including direct shear, interface friction, and bulk density, at each of these moisture content levels. Direct linear or rotational shear testers are often used to measure the material flow and interface properties. The fine components of the material are usually used in testing, because the fines define the worst-case flow properties.

Testing samples of the actual material to be conveyed in relation to the actual belting and construction materials to be used must be performed to provide this important data. *(See Chapter 25: Material Science for additional information on material testing and analysis.)*

Material testing concludes with a recommendation for the chute angles, based on boundary friction required to find a balance between reliable flow through a transfer chute and acceptable levels of chute and belt wear. Recommendations for the material(s) to be used as liners inside the chute may also be included.

The various conveyor and material parameters and the material discharge trajectory are used to develop the transfer chute design.

Phase 2: Discrete Element Modeling (DEM) Method

The parameters developed in Phase 1 are used in developing a computer-generated 3D discrete element model of the chute system (**Figure 22.12**).

DEM is a design verification tool. The basic operating equation is Newton's Second Law: Force = mass times acceleration (F = ma), solved for every interaction between particle and particle, and particle and chutewall, as modified with the properties of the particles and of the interacting

elements. The forces, which act on each particle, are computed from the initial data and the relevant physical laws. Some of the forces that affect the particle motion include:

A. Friction

When two particles touch each other or move against the wall

B. Impact

When two particles collide

C. Frictional, or viscous, damping

When energy is lost during the compression and recoil of particles in a collision

D. Cohesion and/or adhesion

When two particles collide and stick to each other

E. Gravity

Solutions based on a DEM approach are more insightful than those based on basic design equations and "rules of thumb," because they enable the designer to more accurately evaluate important issues such as center loading of a receiving conveyor. The chute designer is also able to predict areas in the chute that may be prone to low material velocity—therefore plugging—and take corrective action to prevent them. When coupled with basic equations, DEM enables a designer to quickly determine the optimum chute design through a series of iterations. A minor downside of DEM is that only relatively few particles, compared to the total number of particles in the material stream, can be simulated in a reasonable length of time with computers that are commonly available, although advancements in computer technology may rapidly eliminate this problem.

An additional advantage of this computer-based system is that changes can be quickly developed to compensate for changes in the system characteristics.

Of course, the "garbage in, garbage out" principle still applies. If the data going into the software is not accurate, the

Figure 22.12

The parameters developed in Phase 1 are used in developing a computer-generated 3D discrete element model (DEM) of the chute system.

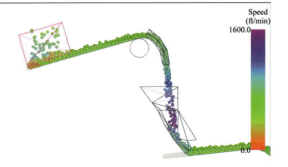

Speed
(ft/min)
1600.0

0.0

design coming out will not be accurate. That is why testing of the actual material to be conveyed, in the various conditions in which it will be handled—including "worst-case"—is critical.

Phase 3: Final Design

The use of computer-based modeling techniques allows the quick and efficient turnaround of a chute design to meet the requirements of a specific belt-to-belt transfer. The 3D model is used to produce the fabrication and installation drawings.

The completed engineered chute project includes hood(s), drop chute, spoon(s), wear liner, belt-support cradles, belt-tracking system, belt-cleaning systems, dribble chute, access doors, skirtboard seal, tailgate sealing box, and settling zone.

Other Items

Other items to be considered during chute design are the requirements for heaters, insulation, access to the interior of the chute, lighting, access platforms, plugged-chute switches, appropriate guards, and adequate space for replacement of belt cleaners, flow aids, or other components.

Other Design Considerations

In its simplest sense, a transfer chute should have internal surfaces that are sufficiently steep and smooth, with rounded corners, to prevent flow problems—such as material buildups and choking—even when transporting material with worst-case flow properties. Ideally, this geometry would be governed by the effects of gravity only. The reality is that there are a number of other considerations that should be included and calculated when planning for the installation of engineered flow transfers. These factors include:

A. Material trajectory

Calculation of the trajectory of the material stream as it leaves the discharge conveyor involves consideration of the center of mass of the material, velocities, the point on the discharge pulley where the trajectory begins, and the shape of the load. (A detailed discussion of discharge trajectory can be found in Chapter 12 of CEMA's *BELT CONVEYORS for BULK MATERIALS, Sixth Edition*.)

B. Wear

Impact, corrosion, and abrasion are primary contributors to chute wear, which takes place where the material stream hits the chute surface. Sliding abrasion is the passing of the material stream along the surface of the chutewall. The amount of abrasion that takes place is dependent on the difference in hardness between the material stream and the wear liner and on the amount, velocity, and force of the load on the wear liner surface. Because the design of engineered flow chutes links the material behavior with the interface at the chutewalls, analysis of impact and sliding abrasion is important in controlling the shape and speed of the material stream.

C. Tolerances

Even small differences in the installation of the components can affect the smooth flow of material and air through the transfer point. Manufacturers' recommendations for installation of components and materials must be strictly followed.

D. Two-phase flow analysis

Two-phase flow analysis takes into consideration the movement of both the material stream through a transfer chute and the induced air that travels with it into the settling zone of the receiving conveyor. If the material stream remains in contact with the chute surface—rather than bouncing off from it—there is less aeration and reduced impact force in the loading zone. During the chute's design phase, the analysis of the movement of both material particles and air through the transfer chute enables the chute designer to minimize induced air, which, in turn, reduces dust generation.

A variety of computer-based techniques, including DEM, Computational Fluid Dynamics (CFD), and Finite Element Analysis (FEA) are used to model two-phase flow. This analysis should include the displaced air, induced air, and generated air. *(See Chapter 7: Air Control.)*

Depending on the calculated airflow and the properties of the material, including particle size distribution and cohesion level, various systems—from rubber curtains to dust suppression and filter bags—can be utilized to minimize the effects of air currents in the transfer chute.

E. Structural concerns

Design of the support structure for a transfer chute generally requires analysis of four factors:

a. Dead load

Weight of chute (and structure) itself

b. Live loads

Wind, snow, and ice accumulations and fugitive material accumulating on flat surfaces

c. Dynamic load

The forces resulting from the movement and impact of material in the chute and other process equipment

d. Loaded capacity

Weight of the material in the chute—calculated using the highest value of material bulk density in the worst-case scenario of chute plugging

The objective of this analysis is to efficiently and effectively support the transfer chute without spending excessive amounts on the support structure. Developing a support structure that complies with local building codes is another important consideration.

INSTALLATION OF ENGINEERED FLOW SYSTEMS

Project Installation

Engineered chutes can easily be designed into new conveyor systems. They can be pre-assembled and aligned into manageable assemblies that can easily be rigged, hoisted, and bolted-in-place to reduce construction cost.

Engineered flow chutes can also be retrofit into an existing operation as a way of controlling dust to improve operations and achieve regulatory limits on dust, usually without installation of expensive "baghouse" systems. Regardless of whether it is a new or retrofit installation, the design and installation of engineered chutes should be left to companies experienced with the technology.

Chutes for Retrofit Applications

One of the earliest applications of engineered flow chutes was in the improvement of the transfer points in existing conveyor systems. The incorporation of these engineered systems into existing plants can pose some problems with fitting within existing structures.

To ensure accurate designs as well as to ensure that the engineered system will fit properly into place without requiring field adjustments, a site survey using laser measurement techniques is recommended (**Figure 22.13**). This precise survey uses a pulsed-laser technology to scan target areas and return a 3D "point cloud," which looks like a detailed rendering of a scene (**Figure 22.14**). Because this point cloud is three-dimensional, it can be viewed from any perspective, and every point has accurate

Figure 22.13

To ensure accurate designs as well as to ensure that the engineered system will fit properly into place without requiring field adjustments, a site survey using laser measurement technique is recommended.

x-, y-, and z-axis coordinates. The geometry of the points can then be exported to 3D modeling software packages as a starting point for the development of chute geometry. This will ensure the engineering of systems that will fit within the existing clearances.

In a retrofit application, before and after release of fugitive materials testing and analysis can also be performed, allowing the opportunity for performance to be compared and for improvements to confirm the justification for the project.

Flow Aids and Engineered Chutes

Even a well-engineered chute should make provision for the future installation of flow-aid devices by incorporating mounting brackets in the original design. Changes in material flow properties, or less-than-optimum design constraints, may lead a designer to require flow-promotion devices, such as vibration or air cannons, in a given design. It is difficult, especially in retrofit applications, to have the luxury of an optimum design. Compromises are often inevitable, because the locations of the feed and receiving conveyors are set, and moving them would be economically unfeasible. Potential flow problems, caused by variations in material characteristics in the future, can then be accommodated with the installation of vibrators or air cannons. Including the brackets during the initial installation of the chute will save money and time over retrofitting a bracket (**Figure 22.15**).

Flow aids enhance material flow in those situations where compromises are made to what would have been an optimum design. *(See Chapter 9: Flow Aids.)*

SYSTEM MAINTENANCE

An operation should keep accurate records of chute and liner design and positioning to simplify the fabrication and installation of replacement liners as they become needed.

In order to simplify the replacement of liners, the chute should be designed with an easy-opening flange system that allows one wall—in most cases, the back wall and liner-bearing wall—of the chute to slide away from its position (**Figure 22.16**). This will allow more efficient access for inspection and replacement of liners inside the chute structures (**Figure 22.17**).

Figure 22.14

Pulsed-laser technology is used to scan target areas and return a 3D "point cloud," which looks like a detailed rendering of a scene.

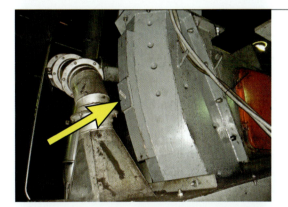

Figure 22.15

Including brackets for the installation of flow aids during the initial construction of the chute will save money and time over retrofitting a bracket.

22

Figure 22.16

To simplify the replacement of liners, the chute should be designed with an easy-opening flange system that allows one wall—in most cases, the back wall and liner-bearing wall—of the chute to slide away from its position.

Figure 22.17

The flanged back of the chute will allow more efficient access for inspection and replacement of liners inside the chute structures.

TYPICAL SPECIFICATIONS

A. Material specifications

The material-transfer system will incorporate belt-to-belt transfer chutes custom engineered to match material specifications and flow requirements. Through testing of material properties, the chute system will be designed to provide the required flow rate without plugging and to eliminate the creation of additional dust from the degradation of material and the entrainment of air.

B. "Hood" and "spoon"

Included in the chute system will be a "hood" discharge chute and a "spoon" receiving chute. The "hood" will take the flow of material from the discharging belt, confining it to limit air entrainment and creating a consistent inertial flow through its trajectory onto the receiving "spoon." The "spoon" receiving chute will receive the material stream and place the material on the receiving belt with the proper direction and speed to minimize material turbulence, impact, belt abrasion, and belt mistracking.

C. Volume

The volumetric design of the head chute and skirted area will be calculated to reduce air speed and turbulence. Fugitive and respirable dust levels will be greatly reduced through the settling features of the design.

D. Access

The chute will be fitted with an easy-opening flange closure system to enable simplified inspection and replacement of liners inside the chute structures.

E. Settling zone

The exit of the receiving conveyor will be fitted with an extended covered skirtboard system to form a settling zone. The settling zone will incorporate multiple dust curtains to form a serpentine plenum that reduces the air velocity and provides time for airborne particles to return to the main material cargo by gravity.

⚠ SAFETY CONCERNS

Engineered chutes should be designed with an access opening on the non-flowing side of the enclosure. These doors should be fitted with restricted-access screens to reduce the hazard from materials flying out of an opening, and warning labels should be applied.

Personnel entry to any chute should be governed by confined-space entry regulations.

ADVANCED TOPICS

Engineering Calculation: Continuity

The continuity calculation determines the cross section of the material stream within a transfer chute and is important in determining the ideal chute size (**Equation 22.1**). This helps to keep the cost of chute fabrication under control. The industry and CEMA's standard indicates the chute should be at least four times the material cross-sectional area at any position.

More important than the calculation of the area is the acknowledgment of the relationship between velocity and cross-sectional area. A designer must keep this continuity relationship in mind when the velocity of the material needs to match the speed and direction of the receiving belt (**Equation 22.1**). Material velocity is influenced by many things, such as fall height, change in direction of flow, surface friction, internal friction, and instantaneous density to name a few. These factors will alter the stream velocity in a predictable way, but it is important to note that this change in velocity will influence the cross-sectional area of the stream. Conversely, the area can be altered to influence the velocity. The cross-sectional area of the stream is vitally important when designing to prevent problems with chute blockage.

THE PAYBACK OF ENGINEERED CHUTES

In Closing...

An engineered transfer chute can be applied in virtually any transfer chute application, so facility management often will use a cost justification procedure to evaluate its payback for the operation. Applications in which there is a significant drop height from the discharge conveyor to the receiving conveyor will usually warrant the investment. Facilities that are attempting to meet regulatory requirements or satisfy environmental and safety concerns may find the investment in an engineered flow chute has a short-term payback. The additional investment required for an engineered flow chute over the cost of a traditional transfer chute is promptly repaid through increase in productivity, accident reduction, and meeting environmental regulations rather than cleaning up fugitive materials, coping with plugged chutes, or tracking an improperly loaded belt.

Looking Ahead...

This chapter, Engineered Flow Chutes, the second chapter in the section Leading-Edge Concepts, provided information about another method of reducing fugitive materials. The next chapters continue this section, focusing on Air-Supported Conveyors and Belt-Washing Systems.

$$A = \frac{Q \cdot k}{Y \cdot v}$$

Equation 22.1

Continuity Calculation for Cross-Sectional Area of Material Stream

22

Given: *A coal stream carrying 1800 tons per hour (2000 st/h) with a density of 800 kilograms per cubic meter (50 lb$_m$/ft³) is traveling at 4,0 meters per second (800 ft/min).* **Find:** *The cross-sectional area of the coal stream.*

Variables		Metric Units	Imperial Units
A	Cross-Sectional Area	square meters	square feet
Q	Flow Rate	1800 t/h	2000 st/h
Y	Material Bulk Density	800 kg/m³	50 lb$_m$/ft³
v	Average Materials Velocity at Cross Section in Question	4,0 m/s	800 ft/min
k	Conversion Factor	0,278	33.3

Metric: $A = \dfrac{1800 \cdot 0{,}278}{800 \cdot 4{,}0} = 0{,}16$

Imperial: $A = \dfrac{2000 \cdot 33.3}{50 \cdot 800} = 1.67$

A	Cross-Sectional Area	0,16 m²	1.67 ft²

Note: The stream cross-sectional area will be different from the cross-sectional area when the material is on the belt due to the differences between conveyed density and loose bulk density. (See Chapter 25: Material Science for additional information.)

REFERENCES

22.1 Stuart, Dick D. and Royal, T. A. (Sept. 1992). "Design Principles for Chutes to Handle Bulk Solids," *Bulk Solids Handling*, Vol. 12, No. 3., pp. 447–450. Available as PDF: www.jenike.com/pages/education/papers/design-principles-chutes.pdf

22.2 Roberts, A.W. and Scott, O.J. (1981). "Flow of bulk solids through transfer chutes of variable geometry and profile," *Bulk Solids Handling*, Vol. 1, No. 4., pp. 715–727.

22.3 Roberts, A.W. (August 1999). "Design guide for chutes in bulk solids handling operations," *Centre for Bulk Solids & Particulate Technologies*, Version 1, 2nd Draft.

22

22

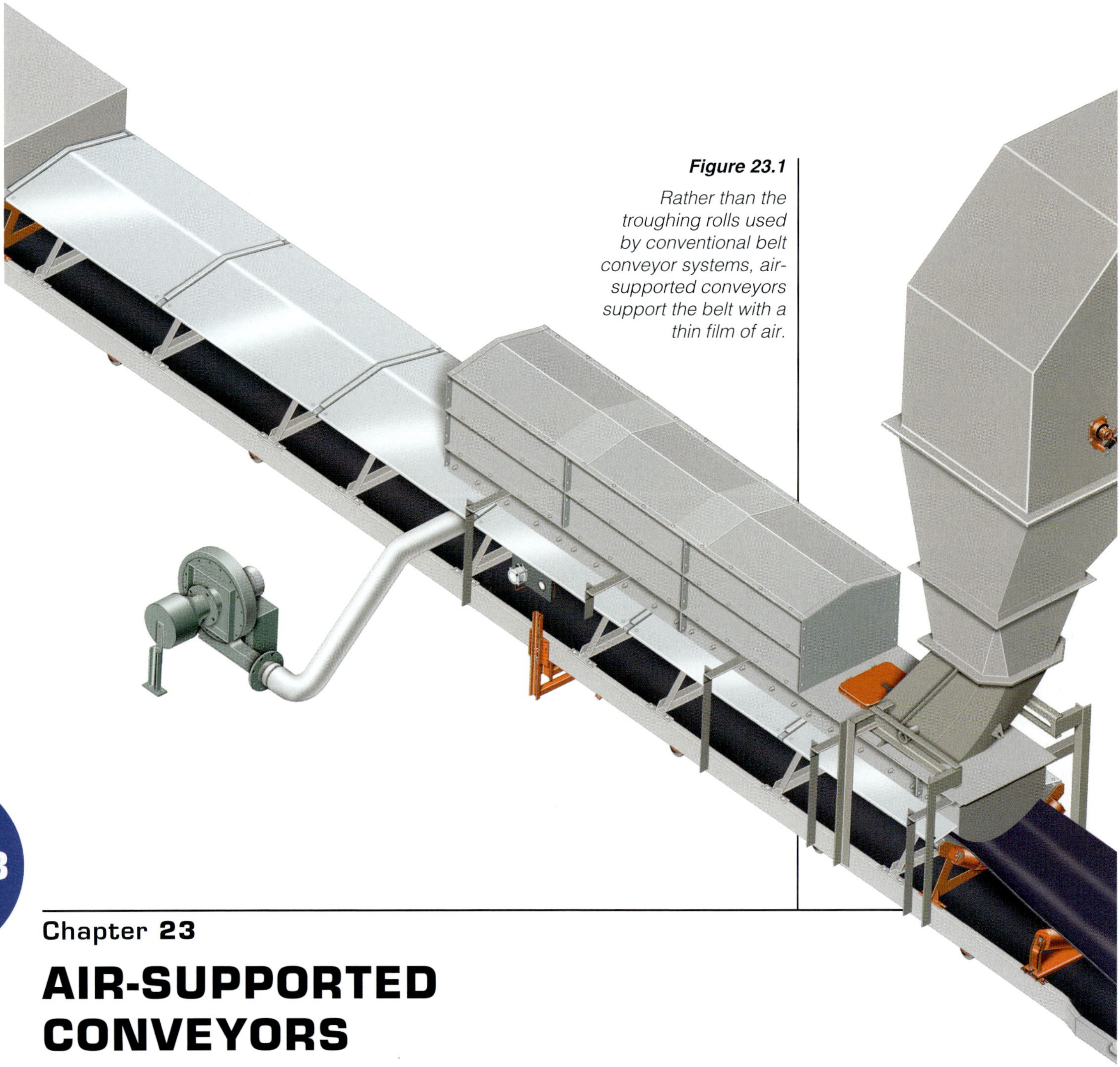

Figure 23.1 |

Rather than the troughing rolls used by conventional belt conveyor systems, air-supported conveyors support the belt with a thin film of air.

23

Chapter 23

AIR-SUPPORTED CONVEYORS

In this Chapter...

This chapter focuses on the basic concepts of air-supported conveyors and applications for which they would be appropriate. We also present both the benefits and drawbacks of their use, along with information about the size of fan needed for various conveyor lengths and widths.

One example of "leading-edge" conveyor technologies is air-supported conveyor systems (**Figure 23.1**). Rather than the troughing rolls used by conventional belt conveyor systems, air-supported conveyors support the belt with a thin film of air. This method of conveying bulk materials limits the areas of mechanical friction, which results in a dramatic reduction in maintenance and operating costs. A fully-enclosed, weather-resistant, air-supported conveyor requires less structural support than a traditional conveyor, and it minimizes material segregation, spillage, and dust (**Figure 23.2**). While not suitable for all applications, air-supported belt conveyors offer a number of advantages, including a smooth ride for the bulk materials and containment of dust. Air-supported conveyors, like conventional conveyors, must be designed by an experienced conveyor engineer.

BASICS OF AIR-SUPPORTED CONVEYORS

An air-supported conveyor uses low pressure air to raise and support the belt and cargo. The air is supplied by a low-pressure centrifugal fan and released through a trough-shaped pan below the conveyor belt (**Figure 23.3**). A series of holes drilled in the center of the pan along the length of the conveyor—between the air-carrying chamber (plenum) and belt—enables the air, supplied by the blower through the holes in the pan, to lift and support the loaded belt (**Figure 23.4**). The edges of a troughed belt act as a pressure regulator, automatically balancing the pressure required to lift the load. The air film eliminates the need for most idlers on the carrying side of the conveyor; conventional

return idlers may be used for the return run of the belt. With no troughing idlers, budgets typically designated for replacement of rolling components and maintenance labor needed to accomplish that replacement are reduced.

The plenum runs below the pan. As the top of the plenum, the pan provides the form for the trough of the belt. The most common and economical trough angles are 30 and 35 degrees. The plenum can be a box or a V shape that sits on conventional conveyor structural stringers (**Figure 23.5**). These plenums can be modular to simplify installation (**Figure 23.6**).

Figure 23.2

A fully-enclosed, weather-resistant, air-supported conveyor will minimize material segregation, spillage, and dust.

Figure 23.3

An air-supported conveyor utilizes a stream of air to raise and support the belt and cargo.

23

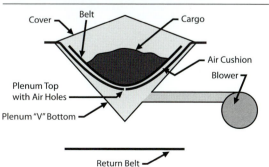

Figure 23.4

Holes drilled in the center of the pan along the length of the conveyor between the plenum and belt enable the air supplied by the blower to lift and support the loaded belt.

Figure 23.5

The plenum can be a box or a V shape that sits on conventional conveyor structural stringers.

Figure 23.6

The plenums of an air-supported conveyor can be modular in construction to simplify installation.

Figure 23.7

The typical air-supported conveyor, of less than 180 meters (600 feet) in length, requires a single centrifugal fan.

Figure 23.8

A plant may specify the installation of a redundant, or back-up, fan to assure conveyor operation in the event of a fan failure.

23

Because an air-supported conveyor uses a thin film of low-pressure air—approximately 1 to 2 millimeters (0.04 to 0.08 in.) thick—to support the conveyor belt, air consumption is low. Consumption is typically 180 to 270 liters per minute per meter (2 to 3 ft³/min/ft) of belt length. The film of air is created by a blower supplying 5 to 7 kilopascal (0.7 to 1.0 lb_f/in.², or 20 to 30 in. of water gage) of air pressure.

The speed and pressure of the air film are sufficient to help keep material from accumulating between the belt and pan but low enough so that additional dust is not created.

SYSTEM COMPONENTS

Plenum

The plenum, through which the air from the fan flows, can be formed of plastic or galvanized (or stainless) steel troughs, sized to match the belting required for the conveyor application.

The plenum must be smooth, without irregularities in profile or surface. The plenum sections should be flush and sealed at each connection of the modular units. The structure must be designed to minimize deflection under various loads and climate conditions, to protect the integrity of the seal between plenums.

Air Supply

The air to support the belt is provided by one or more centrifugal fans (**Figure 23.7**). The typical conveyor of less than 180 meters (600 ft) requires a single fan, although a plant may specify the installation of a redundant, or back-up, unit to assure conveyor operation in the event of a fan failure (**Figure 23.8**).

It is important the air supply to support the belt is sufficient to handle the entire range of loading conditions for that particular conveyor. The number of blowers required depends on both the length of the conveyor and the width of the belt. For long conveyors, more than one air supply may be required to prevent loss of volume and static pressure. The volume of air is minimal, because the thickness of the air film required to raise the belt is only 1 to 2 millimeters (0.04 to 0.08 in.).

The size of the centrifugal blower required depends on the width of the belt and the length of the conveyor, with sizes ranging from 2,5 to 12 kilowatt (3 to 15 hp) common (**Table 23.1**). Direct-drive fans as specified improve efficiency and reduce the maintenance problems that can arise with mechanical couplings. In hazardous-duty situations, such as handling grain or coal, no-spark blades and hazardous-duty motors must be specified.

The spaced holes in the center of the pan allow the air to raise the belt (**Figure 23.9**). The size and spacing of the holes in the pan are critical to proper operation, because they directly affect the static pressure and volume at the interface between the belt and the plenum/pan.

For best results, the air source(s) should be located in the middle of the system—equal distance from the head and the tail of the conveyor; if there are two or more

fans, they should be located equidistant from each other and from the head and tail of the conveyor.

The fan is controlled by a pressure switch, typically located at the conveyor's head section, close to the electric supply, to save conduit and labor costs. The fan is interlocked with the conveyor, so the fan must be running before the drive can start. The conveyor's normal start up procedure is to start the fan first and allow it to come up to pressure before engaging the drive motor. If the fan fails to start or come up to pressure, the pressure switch will sense low air pressure, and the conveyor will not run.

The intake air for the fan should be from

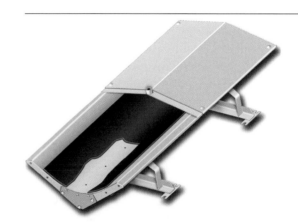

Figure 23.9

The spaced holes in the center of the pan allow the air to raise the belt. The size and spacing of the holes in the pan are critical to proper operation.

Belt Width	Conveyor Length				*Table 23.1*
mm (in.)	**Up to 45 m (150 ft)**	**45-90 m (150-300 ft)**	**90-140 m (300-450 ft)**	**140-185 m (450-600 ft)**	
500-650 (24)	Fan Size A	Fan Size B	Fan Size C	Fan Size D	
650-800 (30)	Fan Size A	Fan Size B	Fan Size C	Fan Size D	
800-1000 (36)	Fan Size A	Fan Size B	Fan Size C	Fan Size D	
1000-1200 (42)	Fan Size A	Fan Size B	Fan Size C	Fan Size D	
1200-1400 (48)	Fan Size A	Fan Size B	Fan Size C	Fan Size D	
1400-1600 (54)	Fan Size A	Fan Size B	Fan Size C	Fan Size D	
1600-1800 (60)	Fan Size B	Fan Size C	Fan Size D	Fan Size D	
1800-2000 (72)	Fan Size B	Fan Size C	Fan Size D	Fan Size D	

Typical Sizes for Centrifugal Fans Used with Air-Supported Belt Conveyors

Fan Size	Power Output
Fan Size A	2,5 kW (3 hp)
Fan Size B	6 kW (7.5 hp)
Fan Size C	7,5 kW (10 hp)
Fan Size D	12 kW (15 hp)

Metric measurements and fan size ratings are conversions of Imperial specifications.

Fan size represents the size of the centrifugal fan only (which supplies air to raise belt and reduce friction). It does NOT include conveyor drive power.

23

a fresh air source and filtered to reduce buildup of dust in the fan and pan. In some cases, the air must be heated to avoid condensation, which can cause the belt to stick to the pan or allow fines to choke the holes in the pan.

Conventional or Air-Supported Return

The return run of an air-supported conveyor may also be air supported (**Figure 23.10**), or it may have traditional return idlers (**Figure 23.11**).

Without the idlers on the return side, a completely air-supported conveyor has reduced maintenance costs. In fact, this system may allow the elimination of walkways along the conveyor, due to its minimal maintenance requirements. Because an air-supported return run is totally enclosed and the belt is visible only at the head and tail of the conveyor, it can provide a cleaner system.

Figure 23.10

An air-supported belt conveyor may incorporate an air-supported return side.

Figure 23.11

The return run of an air-supported conveyor may have traditional return idlers.

Roller return systems may be preferred in applications where optimal belt-cleaner performance cannot be maintained, because fugitive material can interfere with the operations of the air-supported return. Return rollers can be hung from brackets below the conveyor or enclosed in the structure below the air-supported plenum. A typical return run has idlers installed every 3 meters (10 ft).

Enclosing an air-supported conveyor's return run is recommended only when contamination is a critical problem. An enclosed return run can use as much energy as the carrying side, and the cost of the enclosure often outweighs any benefit. In addition, there is the problem of accumulation of dust and fines in the return-run chamber. It is usually more economical to install and maintain a good belt-cleaning system. On an air-supported return run, the belt tends to want to lift in the center, and the edges touch the pan if the belt is not of the proper stiffness. It is sometimes difficult to balance the airflow and pressure required for the return run and the carrying run with one fan. Air support of the conveyor return also increases the cost of fabrication. The cost of an effective belt-cleaning system and related maintenance is usually much less than the added cost of enclosing the return run.

Support Structure for Air-Supported Conveyors

Compared to conventional stringer or truss conveyors, air-supported conveyors can span longer distances with less structure because of the structural strength of the air-supported system plenum/pans (**Figure 23.12**). This provides the benefit of reducing the capital investment in the conveyor system.

In a traditional conveyor, for example, a support pier is required approximately every 15 meters (50 ft). Because of the strength of its plenums, an air-supported conveyor may require fewer support piers, thus reducing the investment in concrete pillars and structural steel. In one example,

23

an air-supported belt conveyor system was installed at a wood waste-fired power plant close to the North Sea near Emden, Germany. Designed with a triangular-truss system, this air-supported conveyor spans distances of approximately 50 meters (160 ft) and covers the conveyor's 167-meter (550-ft) length with only two intermediate supports. Each application must be reviewed by qualified engineers to determine the requirements for foundation and structure.

Conventional Components

Air-supported conveyors can use standard take-up conveyor drives, loading and discharge chutes, and support structures. This allows the conversions of, or the connections to, many existing standard belt conveyors to air-support systems.

Although an air-supported conveyor will use conventional conveyor belting, the belt should be vulcanized rather than joined with mechanical splices. This will prevent damage to the pan and the splice from metal-on-metal contact associated with mechanical splices passing over the system. Mechanical belt fasteners can be used as long as the splices are properly recessed and then dressed with belt patching rubber.

Loading an Air-Supported Conveyor

Because of the low friction against the belt, misalignment from forces such as off-center loading is particularly troublesome for air-supported conveyor systems. Consequently, proper placement of the cargo is critical to the successful operation of an air-supported conveyor. The load must be properly centered and placed with as little impact as possible. This may require loading through a spoon to place the material gently on the belt, with the proper speed and direction. In many ways, air-supported conveyors are ideal for use with "hood and spoon" engineered flow transfers (**Figure 23.13**). *(See Chapter 22: Engineered Flow Chutes.)*

To regulate the delivery of cargo to an air-supported conveyor, feeders or flow-controlling gates are sometimes used in conjunction with a load-centering spoon. These gates help to deliver a consistent load to the air-supported conveyor and prevent material from piling up in one area. A regulated delivery of material to the belt eliminates the "starve and flood" conditions that impede smooth operation of the system.

Operating the air-supported conveyor when not loaded is not recommended. When there is no load on the belt, the air gap under the belt increases, which increases the volume of air used. The pressure goes down; however, the volume contributes more to the power consumed than does the pressure.

Air-supported conveyors should not be subjected to loading impacts above the light-duty impact ratings as found in the Conveyor Equipment Manufacturers Association's (CEMA) publication CEMA STANDARD 575-2000 *Impact Cradle/Bed Standard*. One solution to high-impact loading conditions is to use conventional trans-

Figure 23.12

Using a triangular-truss system, this air-supported conveyor needs only two intermediate supports to cover the system's 167-meter (550-ft) length, much fewer than a conventional conveyor would need.

Figure 23.13

Proper placement of the cargo is critical to the successful operation of an air-supported conveyor, making "hood and spoon" engineered flow transfers ideal for air-supported systems.

23

fer-point components (e.g., impact cradles and impact idlers) to cushion impact in the loading zone, and then switch to the air-supported system outside the loading zone. Sections of conventional conveyor can be easily inserted in air-supported-conveyor systems to allow the use of accessories such as scales. It is still important to have the load properly centered in the air-supported portion of these hybrid systems.

ADVANTAGES OF AIR-SUPPORTED SYSTEMS

The Benefit of a Smooth Ride

Traditional belt-support systems in load zones consist of standard or impact idlers (rollers) that are placed as close together as possible. Even in the best of installations, however, the troughing idlers provide a less-than-perfect belt line. The material follows a path similar to a roller coaster (**Figure 23.14**). The belt moves up and down as it crosses over the idlers. This up and down motion agitates the material, allowing some particles to become airborne, causing the material to segregate by size, or pushing some material to the outsides of the belt where it can be spilled from the belt.

If the rolls are spaced just 225 millimeters (9 in.) apart, the belt can still sag between rollers, allowing dust and spillage to escape from the belt. In addition, this sag creates entrapment points between the belt and vertical steel in skirtboards or wear liners. These pinch points can catch material that can then abrade the belt surface. In many cases, the sealing system is blamed for belt damage, when it is material entrapment that has actually caused this abrasion.

As air-supported conveyors use a pan rather than rollers to create the belt line, they present a smooth surface and level belt line that when combined with center loading may allow the elimination of skirtboard and sealing systems. Stable belt support and the elimination of skirtboards prevent entrapment points that allow material to become wedged or jammed.

On conveyors with a steep incline, the movement of the belt over the idlers may disturb the material sufficiently that it causes lumps of material to roll back down the conveyor as the belt progresses up the incline. With its stable path, the air-supported conveyor eliminates the disturbance of the cargo as it goes over the rollers in a conventional conveyor. This smooth path will allow the air-supported conveyors to operate at a steeper angle than roller conveyors. This benefit is of interest to operations handling bulk materials that tend to roll back on the conveyor. A typical gain in slope is three degrees. This increase in angle acts to reduce the overall length of the conveyor, reducing the installed cost when compared to a roller conveyor.

23

Figure 23.14

The idlers of a conventional conveyor provide a less-than-perfect belt line, so the material follows a "roller coaster" path. An air-supported conveyor uses a troughed pan to provide a smooth, stable ride for the belt and cargo.

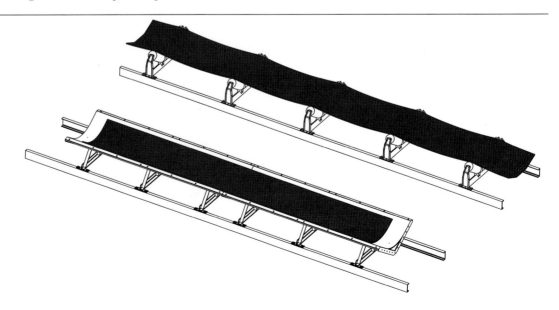

Containment of Dust

Dust is generated when the material stream encounters air movement, which can result from the velocity of the material drop, from mechanical equipment, or from other outside influences. Higher velocities of air moving across the material stream may entrain greater quantities of dust. A well-designed and properly-installed air-supported conveyor has a totally-enclosed

---

Benefits and Drawbacks of Air-Supported Belt Conveyors

Benefits

A. Effective Dust Control: When the air-supported system is utilized from the loading area to the head chute, total dust control can be achieved.

B. Improved Belt Tracking: Air-supported conveyors have self-centering action.

C. Stable Belt Path: Troughing idlers spaced along the conveyor create hills and valleys in the belt line where the cargo is agitated and begins to segregate; the fines end up on the bottom and larger pieces on top. The air-supported conveyor offers a smooth ride for the cargo, with less spillage, segregation, and degradation of material.

D. Lower Operating Cost: On horizontal conveyors, the air-supported conveyors can use up to 30 percent less energy; on inclined conveyors, the energy saving is up to 5 percent.

E. Reduced Maintenance Expense: There are no carrying-side idlers, so there are no rollers to replace and no idler lubrication required.

F. No Skirtboard Seal: No skirting is required in the loading area, because the chutewall/wear liner forms a barrier to contain the material being loaded.

G. Retrofit Availability: New designs allow air-supported conveyors to be installed on existing (conventional conveyor) stringer and support systems. Air-supported and conventional roller sections can be mixed in a single conveyor, to allow for loading zones, tracking idlers, belt scales, or other requirements.

H. Improved Product Condition: An air-supported belt is gentle to the cargo. There is no bumpy "roller coaster" ride over the idlers, so there is no material segregation, no product degradation, and no breakage. Because the conveyor is fully enclosed, there is no contamination of conveyed material.

I. Greater Inclines Allowed: By eliminating load agitation, air-supported conveyors can allow for steeper inclinations, depending on the bulk-material properties.

J. Savings on Walkways: By eliminating troughing idlers and so reducing routine lubrication and conveyor maintenance, air-supported conveyors may allow omission of walkways.

K. Improved Safety: The system has fewer moving parts that pose risk to workers.

Drawbacks

A. Required Engineered Belt-Cleaning Systems: Air-supported conveyors require aggressive belt-cleaning systems to ensure carryback is controlled. Carryback may also blind the air supply holes when allowed into the plenum area.

B. Tracking Affected by Material Accumulation: Belt tracking can be affected by fugitive material building up on system components.

C. Necessary Center Loading: The air-supported conveyor must be center-loaded, or belt mistracking will occur. No belt tracking devices can be installed within the air-supported system.

D. Required Stable Flow: Surges of material must be avoided, because the system is totally enclosed and blockage and system shut down could occur.

E. Limitations in Impact Loading: Impact must be minimized in the loading zone, or plenum damage will occur.

F. Higher Initial Investment: The initial cost is higher than for a conventional conveyor system.

G. Reduced Access for Observation: The conveyor is totally enclosed, so it is difficult to inspect cargo or interior of system.

H. Unsuitability for Heavy-Duty Applications: The system may not be suitable for heavy-duty applications.

I. Reduced Margin for Error in Design or Installation: Success of installation may depend on belt path and joints between plenum/pans.

23

conveying system that may prevent generated dust from being expelled into the environment (**Figure 23.15**). Air-supported conveyors generally need a smaller dust-collection system, such as an insertable collector, than comparable conventional conveyors and transfer points (**Figure 23.16**).

"Hood style" conveyor covers installed over the trough side of conventional conveyor belts will not prevent the wind from blowing material off the belt, but in many cases, the air velocity will be increased due to wind rushing up from the return

Figure 23.15

A well-designed and properly-installed air-supported belt conveyor is totally enclosed on its carrying side to limit the escape of dust.

Figure 23.16

An insertable dust collector can be used with an air-supported conveyor to prevent the escape of airborne dust.

Figure 23.17

This cement plant air-supported belt conveyor incorporates a totally enclosed gravity take-up for complete containment of fugitive material.

side of the conveyor. A properly designed air-supported system is totally enclosed on its carrying side; consequently, there are no outside influences to "fluff" the material or blow it off the belt.

As the length of the enclosure created by the air-supported conveyor system is increased, the airborne dust gains more time to "settle out" and return to the bed of material on the belt. As a result, air-supported conveyors are well suited for carrying materials that present fire or explosion hazards, including pulverized coal or grain.

To improve dust control, some operations select air-supported conveyors that are fully enclosed on both the top and bottom strands of the belt and also on the take-up tower (**Figure 23.17**). Total enclosure of the conveyor's load-carrying side will improve the performance of dust-collector systems, because it will reduce to a minimum the open area and prevent outside air from entering the collector's intake.

APPLICATIONS AND INSTALLATION

Ideal Applications for Air-Supported Conveyors

An application where air-supported conveying may provide the most advantageous return on investment is one in which the cargo is a lightweight material that is easily entrained in the air. These materials would include ground cement, pulverized coal, wood chips, bark fuel, and grain.

The air-supported system is even more advantageous when there are safety concerns about exposure to the material itself, or where any spillage or dust presents an environmental hazard. Because of their fully-enclosed nature, air-supported conveyors are well suited for carrying dusty materials that present fire or explosion hazards, including pulverized coal or grain.

Applications Not Suitable for Air-Supported Conveyors

Merely changing from a standard conveyor to an air-supported conveyor system will not eliminate pre-existing problems. Although air-supported conveyors have been successfully installed and operated in a wide variety of industrial settings, there are certain applications where this equipment is not recommended:

A. High degree of impact

Situations where there is a high degree of impact in the loading zone are not conducive to air-supported conveying.

B. Prone to plugging

Applications where the material or chute design is prone to plugging are not good applications for air-supported conveyors.

C. Power circuit tripping

If a conveyor power circuit was tripping because the operators are overloading a conventional system, it will probably also "trip" as a result of the operators overloading the air-supported system.

D. Significant head load pressure

Applications where there is significant head load pressure, as might be found under a feeder hopper or a fully-loaded chute, are not conducive to air-supported conveying.

E. Heavy load at loading point

Air-supported conveyors are capable of lifting 975 kilograms per square meter (200 lb_m/ft^2). If the load on the belt at its loading point exceeds that amount, a conventional conveyor with idlers may be more appropriate for the application.

F. Large lumps

Material containing occasional lumps larger than 125 millimeters (5 in.) should include a significant portion of fines to be suitable for air-supported conveying.

G. Lack of maintenance or sticky materials

Plugging of the plenum and the holes in the pan can occur when there is a lack of belt cleaner and fan filter maintenance or sticky materials.

H. Tight curves

Installations with tight horizontal or convex vertical curves are generally not good applications for air-supported conveyors. Convex curves are possible with the use of conventional idlers in the curved section.

Installation of an Air-Supported Conveyor

Regardless of whether the air-supported conveyor is new construction or a retrofit, the installation will require some special details and a high level of workmanship to assure efficient operation. Placement of the plenum may require heavy equipment or cranes to lift the sections into position. The plenums will need to be carefully aligned, and the air passage through the base of the sections should be tightly sealed (with caulk or gasket materials) to prevent air leakage. The edges of the pan need to line up precisely to prevent any raised edge from shaving off the belt cover.

Retrofit vs. New Construction

The modular construction of air-supported conveyors makes them suitable for retrofit applications. Because their design matches the CEMA or International Organization for Standardization (ISO) profiles of the existing idlers, the air-supported sections may easily be incorporated into existing conveyor systems (**Figure 23.18**). The plenums may be installed on top of existing stringers. This allows the air-supported conveyors to be used for a retrofit upgrade of an existing system, and the air-supported conveyors' compatibility with CEMA or ISO standards will allow the systems to

Figure 23.18

Because their design matches the CEMA or ISO profiles of the existing idlers, the air-supported sections may easily be incorporated into existing conveyor systems.

upgrade portions of existing belt conveyor systems. It is possible to convert an existing conveyor to the air-supported system without taking the whole conveyor off-line by installing one section at a time. The fan is sized for the completed installation and the airflow adjusted with a damper to match the number of sections installed.

For greenfield projects (new construction), the air-supported conveyor plenums may be integrated into the conveyor support structure.

SYSTEM MAINTENANCE

By eliminating (or nearly eliminating) the idlers on an air-supported conveyor, the expense of both the replacement rolling components and the manhours of labor required to maintain the system is significantly reduced.

Another opportunity for reduced expenses for maintenance and replacement components is the elimination of a skirt-board-sealing system. With their stable belt path, air-supported conveyors will allow the placement of wear liners very close to the belt. This might eliminate the need for a skirtboard-sealing system, or at least reduce the length of the system required.

It is essential that the belt-cleaning system on an air-supported conveyor function at an optimum level to eliminate fugitive material. Effective belt cleaning is even more important on a system with an air-supported return to prevent material residue from building up on the return plenum or choking the air holes in the pan.

If the air holes become plugged, they can be cleaned by blowing them out with compressed air, or, in a worst case, by re-drilling. In extreme cases, new plenum holes can be drilled with the belt in place by drilling through the belt and plenum and then covering the holes in the belt with an elastomer patch.

Regular maintenance of the intake air filter is required to maintain fan output.

TYPICAL SPECIFICATIONS

A. Design

The bulk-materials handling system will incorporate an air-supported conveyor system. This air-supported belt conveyor will be designed by an experienced conveyor engineer and constructed to CEMA standards.

23

⚠ SAFETY CONCERNS

Because every rolling component on a traditional conveyor system is not only a maintenance concern but also a safety issue, air-supported conveyors are inherently safer to operate and maintain, because they have fewer moving parts. The enclosed conveyor also poses less risk to plant personnel, because there is less danger of becoming entangled in the moving conveyor belt or entrapped in rolling components.

However, there are still pinch points that will need to be guarded. Proper lockout / tagout / blockout / testout procedures must be followed with air-supported conveyors.

Air-supported conveyors can be less noisy than traditional conveyors, because they have fewer rolling components (idlers and bearings) that generate noise when the belt passes over them. The fan is the noisiest part of the system, typically operating at 75 to 85 decibels; the air-supported conveyor operates at a very quiet 60 decibels.

B. Air support

This conveyor will use a film or stream of air released through a trough-shaped pan below the conveyor's belt to support the belt and the cargo without need for idlers on the carrying side. The air will be supplied by a low-pressure centrifugal fan.

C. Idlers

Conventional idlers will be used for the belt's transitions and return run.

D. Plenum

The air-supported conveyor will use a "V"-shaped plenum to allow air movement along the conveyor length. The pan will trough the belt at a 30- or 35-degree angle without distortion of the belt line.

E. Retrofit applications

The structural integrity of the plenum shall allow its use in retrofit applications without requiring modification/re-engineering/reinforcement of the existing conveyor structure.

F. Enclosed carrying side

Constructed of galvanized mild (or stainless) steel, this air-supported conveyor assembly will be totally enclosed on the belt's carrying side to prevent the release of fugitive material. The structure will be modular in construction to simplify installation.

G. Loading zone

The loading zone of the air-supported belt conveyor will incorporate an engineered chute system to load the material onto the conveyor with centralized placement and minimal impact levels. Proper placement of the material will allow material loading without requiring rubber skirting.

H. Belt cleaning

The air-supported conveyor will incorporate a suitable multiple-element belt-cleaning system. This system will be composed of a minimum of a urethane primary cleaner installed on the head pulley below the material's discharge trajectory and one or more secondary cleaners incorporating tungsten carbide cleaning elements. The cleaning system will also include a rubber-bladed V-plow to protect the tail pulley. Additional and/or specialty cleaners shall be incorporated to maintain effective cleaning as determined by material characteristics and operating conditions.

I. Manufacture/installation

To achieve uniform belt support, the air-supported conveyor plenums should be manufactured to strict tolerances, and the sections must be carefully aligned during installation.

THE RIGHT CONVEYOR FOR THE RIGHT CIRCUMSTANCES

In Closing...

While not suitable for all circumstances, air-supported belt conveyors offer significant improvements over conventional conveyors, including improved control of dust and spillage. The key to a successful air-supported conveyor system is a commitment to provide suitable belt loading conditions. By addressing concerns such as high-impact or off-center loading with the installation of load-centering spoons, a plant may reap the benefits of clean, efficient, low-maintenance air-supported conveying. Air-supported belt conveyors can be particularly beneficial when installed in combination with engineered flow loading chutes.

Looking Ahead...

This chapter about Air-Supported Conveyors, the third chapter in the section Leading-Edge Concepts, explained how they can improve control of dust and spillage. The following chapter continues this section, focusing on Belt-Washing Systems.

23

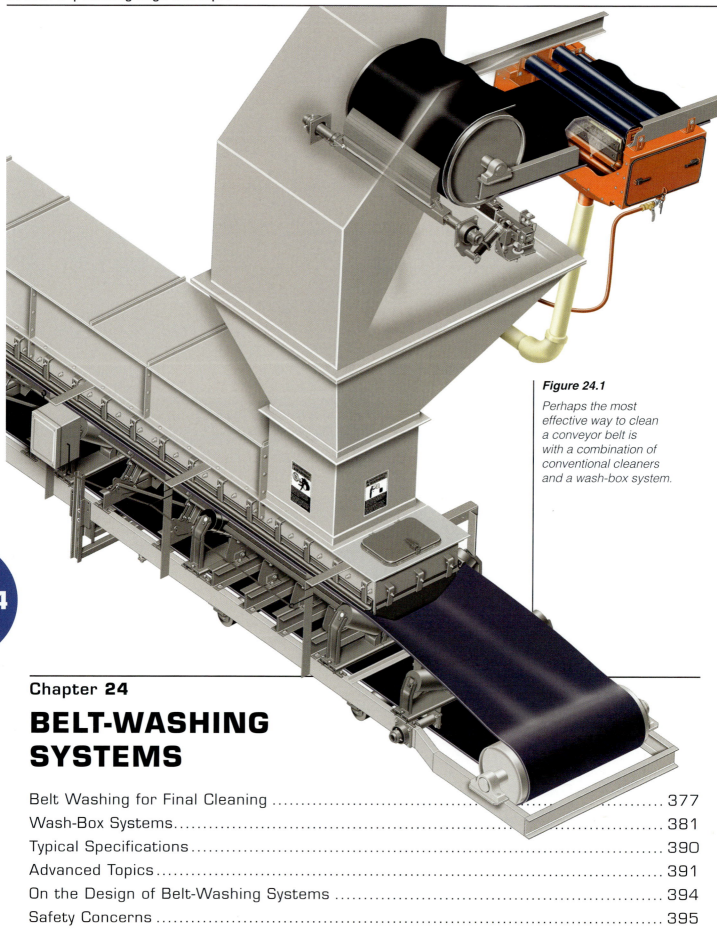

Figure 24.1

Perhaps the most effective way to clean a conveyor belt is with a combination of conventional cleaners and a wash-box system.

24

Chapter **24**

BELT-WASHING SYSTEMS

In this Chapter...

This chapter will cover the principles of wash-box systems, discuss how washing systems are specified and designed, and review the options for water handling, belt drying, and recycling of water and solids.

Perhaps the most effective way to clean a conveyor belt is with a combination of conventional cleaners and a wash-box system (**Figure 24.1**).

Belt-washing systems are a proven method to remove residual material from conveyor belts in applications where environmental issues, regulatory concerns, or other issues mandate high-efficiency cleaning. The typical belt-washing system, or wash box, will contain some configuration of water-spray bars or nozzles covering the load-carrying width of the belt, followed by any of a variety of belt-cleaning devices, from scrapers to rotating brushes. Some variation of a belt-drying system, from pressure rollers to squeegee blades to forced-air nozzles, may follow. In addition, the system must include arrangements for handling the discharge of the effluent (the slurry of water and removed solids) and for the separation, recycling, and/or disposal of the water and removed material. The system will also include an enclosure, sealing components to reduce overspray, controls, and access to allow inspection and maintenance (**Figure 24.2**).

BELT WASHING FOR FINAL CLEANING

Water in the Belt-Cleaning Process

Water assists the belt-cleaning process in a number of ways. *(See Chapter 14: Belt Cleaning.)* The addition of water to the belt-cleaning process has its own drawbacks, but ones that can be overcome with other belt-washing system components and features. With the proper design, a belt-washing system can dramatically reduce the amount of material that is carried back through the conveyor system.

The main concern is the use of water, which is frequently limited in industrial operations. Many plants have severe restrictions on how much water can be consumed in the plant or added to the material.

Other operations have stringent requirements about what must happen to the removed effluent (solids/water mix). Water recycling is a viable option in these cases. Some plants will use a settling pond or a settling basin to separate the fine materials from the water so that the water may be reused. Others will collect the water/solids for disposal. The effluent material is then run through a water-recycling system (or material-separation system) to remove the solids and return the "clean" water back to the system for reuse. The solids can then be returned to the material-handling system.

A second drawback of adding water to the cleaning process is that water itself can cause problems "downstream" on the conveyor. Water will prematurely "age" bearings, rollers, and other equipment vital to the conveyor's operation. Even small amounts of residual water remaining on the belt can cause problems. Methods for drying the belt have been developed that can help reduce these problems, keeping the water local to the washing system and not allowing it to be carried back into the conveyor system or plant.

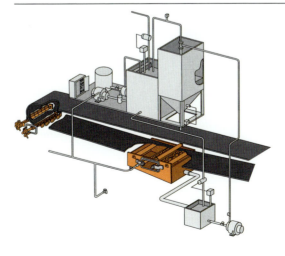

Figure 24.2

The typical belt-washing system contains water-spray bars or nozzles, belt-cleaning devices, and possibly a belt-drying system. It also includes arrangements for handling discharge of the effluent and for separation, recycling, and/or disposal of the water and removed material, along with an enclosure, sealing components, controls, and access.

24

Many plants or operations will be reluctant to add water into the material-handling system at any point, mainly because of a somewhat limited understanding of the effect water has on the flow of bulk materials. While it is true that an increase in the moisture content of the bulk material can have a dramatic effect on the behavior of the material that is detrimental to other processes and handling equipment, the amount of water added to the system for carryback removal is very small in proportion to the total conveyed cargo. Carryback causes far more problems than the addition of these small amounts of water to the system.

Most instances in which conveyor belt(s) are washed by some version of the technology discussed in this chapter are applications where high levels of belt-cleaning performance are required. These would include ship loading and unloading systems, where the escape of carryback might pollute the environment and lead to issues with regulatory agencies, neighbors, and environmental activists. Another application in which the same concerns lead to use of belt-washing systems are overland conveyors, where the cross-country nature of the conveyor's path might allow material to escape into the outside environment. Belt-washing systems are also seen on conveyor systems used to carry several different cargos; the belt is washed to eliminate the potential for cross contamination.

The Principles of Belt Washing

The principles of belt-washing systems are not significantly different from the principles of belt-cleaning systems in general. However, washing systems are technically more sophisticated and are far more effective than traditional mechanical methods of belt cleaning. Water improves the effectiveness of a cleaning system in a number of ways:

A. Water "softens" the bulk material, making it easier to remove.

B. Water keeps the belt-cleaner blades free from buildup, maximizing their cleaning efficiency.

C. Water reduces friction between the belt and cleaning blades, decreasing the forces that generate blade and belt wear, which improves the life expectancy of the blades and so extends the maintenance interval.

Softening the Bulk Material

During belt conveying, the motion of the belt across the idlers will cause the fines and moisture present in the cargo to sift downward and become compacted on the belt surface. The mission of the water in the belt-washing system is to soften the bulk material and reduce its internal strength (cohesion) and its ability to stick to the belt (adhesion). This allows the cleaning elements to remove material more effectively from the belt.

The addition of water will typically increase the cohesion and adhesion of a bulk material up to a maximum level, at which point these properties decrease in a dramatic manner (**Figure 24.3**). This critical point is the saturation moisture of the bulk material. The strength of "buildup" properties of the bulk material depend on its cohesion and adhesion properties. Consequently, the strength of a bulk material will decrease dramatically once the material is beyond its saturation point. At this point, the material becomes more of a slurry. If the material can be "wetted" enough, it is far less likely to build up or stick to any surfaces, including the belt and the belt-cleaning blades. Wetting the material makes the belt-cleaning process far more efficient than using mechanical scraping alone.

24

Figure 24.3

The addition of water increases the cohesion and adhesion of a bulk material up to a maximum level, at which point these properties then decrease in a dramatic manner.

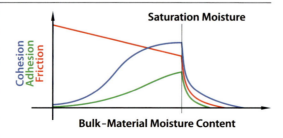

Keeping the Blades Free from Buildup

A second benefit of water in belt cleaning is keeping the leading edge of the cleaning blades free from buildup (**Figure 24.4**). On conventional ("dry") belt-cleaner installations, this region of stagnant material will almost certainly form a buildup of material on the tip of the blade (**Figure 24.5**). Unless it is "cleared," this material will either eventually pass through the cleaner blades and be carried back through the conveyor system or continue to grow larger, increasing the surface area in contact with the belt and reducing the cleaning pressure, allowing more carryback to be carried through the system. Water sprays are used to keep the material from forming this stagnant layer on the surface of the belt-cleaner blade (**Figure 24.6**).

Reducing Blade-to-Belt Friction

Water also improves the performance of a belt-cleaning system or belt-washing station by acting as a lubricant between blade tip and belt surface (**Figure 24.7**). This has a number of advantages. The presence of water reduces the drag, or frictional forces on the belt-cleaner blades and on the belt itself. The reduction of these forces increases the wear-life of the cleaner blades: Less friction means less blade wear.

Another advantage is that the reduction in these frictional forces will reduce heat buildup at the tip of the belt-cleaner blades, minimizing the thermal breakdown of the blades and so extending their life.

In addition to improving the wear-life of the blades, the presence of water will also minimize wear on the conveyor belt.

Field trials have shown that a single, low-volume water spray on the pre-cleaner of a dual-cleaning system increases the system's cleaning efficiency by seven to ten percentage points and can double the interval between required maintenance procedures. In a paper presented to the 1990 International Coal Engineering Conference in Australia, J.H. Planner reported that adding a water

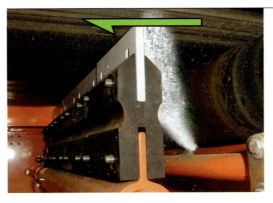

Figure 24.4

One benefit of water in belt cleaning is keeping the front edge of the cleaning blades free from buildup.

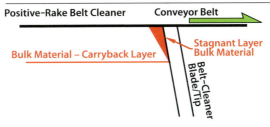

Figure 24.5

On conventional ("dry") belt-cleaner installations, a region of stagnant material will form a buildup of material on the tip of the blade.

Figure 24.6

*Water sprays are used to prevent the formation of the stagnant layer on the surface of the belt-cleaner blade.
Left: Functioning water spray. Right: Non-functioning water spray.*

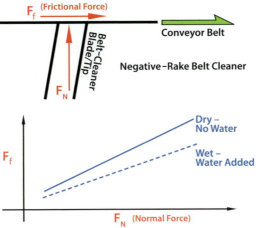

Figure 24.7

Water also improves the performance of a belt-cleaning system or belt-washing station by acting as a lubricant between blade tip and belt surface.

24

spray to various conventional cleaning systems raised cleaning efficiency from the 85 percent range to the 95 percent range *(Reference 24.1)*.

Methods for Washing the Belt

Several methods have been used to wash conveyor belts. As described by Dick Stahura in a 1987 paper *Conveyor Belt Washing: Is this the Ultimate Solution?*, the methods are flood, bath, and wash box *(Reference 24.2)*.

Flood Method

The flood method utilizes jets of water that literally blast the particles off the belt (**Figure 24.8**). Pressures of 400 to 700 kilopascals (60–100 lb_f/in.^{2}) are used, and compressed air can be added to increase the effect. High-pressure sprays can be difficult to use in a belt-washing system,

because they require specialty nozzles and clean water for operation. Behind the water blast, a squeegee-type blade is used to remove the water.

Belt speed (that is, the time the belt is exposed to the spray) and the adhesiveness of the carryback are factors that generally limit the application of this approach to conveyors that operate at less than 5 meters per second (1000 ft/min) *(Reference 24.3)*. Water consumption can be quite high with this method.

Bath Method

The bath method consists of pulling the belt through an enclosure filled with water (**Figure 24.9**). This enclosure could be located along the belt return or even at the gravity take-up, where the weight of the "bath tub" of water can become part of the conveyor's counterweight tensioning system. There are no spray jets or nozzles, only a method of maintaining the water level. The water is exchanged as necessary to keep sediment from building up in the bath. The length of the "bath tub" has to be considerable to achieve any significant "dwell" (belt in the water) time and resultant cleaning effect.

This system poses some difficulties, including carcass damage and problems in maintenance and in drying the belt as it leaves the bath.

Wash-Box Method

The state-of-the-art in belt washing is the wash-box method. In this system, a water-spray method is combined with one or more conventional belt cleaners in an enclosure installed as a tertiary belt-cleaning system (**Figure 24.10**). The design and specification of a wash-box system will depend on application specifics (such as belt speed, material conveyed, belt width, and belt composition); the desired level of cleaning (and drying); and the presence of any site constraints (limits on the use of water or compressed air and/or environmental requirements) (**Figure 24.11**).

Figure 24.8

The flood method utilizes jets of water that literally blast the particles off the belt.

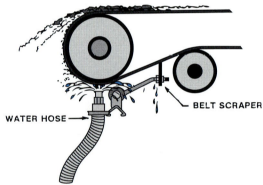

Figure 24.9

The bath method consists of pulling the belt through an enclosure filled with water.

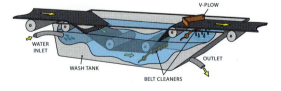

Figure 24.10

In the wash-box method, a water-spray method is combined with one or more conventional belt cleaners in an enclosure installed as a tertiary belt-cleaning system.

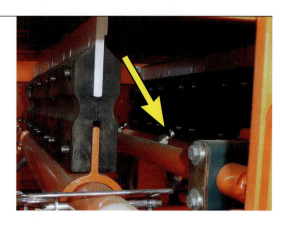

24

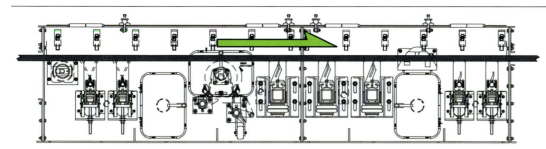

Figure 24.11

The design of a wash-box system will depend on application specifics, the desired level of cleaning, and the presence of any site constraints.

WASH-BOX SYSTEMS

The typical "wash-box" configuration is one or two spray bars for applying water followed by two or three secondary belt cleaners of a more-or-less conventional design (**Figure 24.12**). The wash-box system is engineered so that the adjustment of the cleaner's angle of attack and cleaning pressure can be performed from outside the enclosure, with the operator looking in through an access door (**Figure 24.13**).

These cleaning elements might be conventional secondary belt cleaners or brush cleaners. Brush cleaners may be more effective in cases where the belt is significantly grooved or damaged, making cleaning by flat-edged blades difficult, if not impossible (**Figure 24.14**). Depending on the application, brush cleaners can also require a significant volume of water to keep the brush clean and free from the material buildup that would render it useless.

Historically, the vast majority of wash box applications have been custom-designed out of necessity, due to each application's unique blend of conveyor specifications, material characteristics, and space limitations. A more recent development has been the concept of a "modular" wash-box system. Unlike the custom wash boxes that are designed on an application-by-application basis, the modular wash boxes use a number of "standard" components and configurable modules to combine increased flexibility and ease of use with economy in engineering and construction. The concept includes the "basic elements" of a wash-box system in a modular container (**Figure 24.15**). These modular units can then be "joined" to form more elaborate systems and customized solutions.

The modular approach allows for a number of features to be incorporated with minimal increase in the cost of the system. The modular approach includes options that provide for improved accessibility, simplified installation, easier maintenance, and the ability to easily swap components as application requirements change over

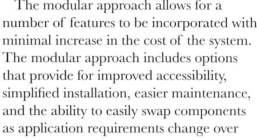

Figure 24.12

The typical "wash-box" configuration is one or two spray bars for applying water, followed by two or three secondary belt cleaners of a more-or-less conventional design.

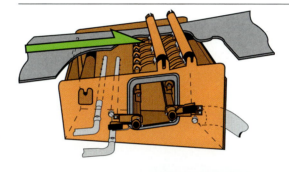

Figure 24.13

The wash-box system should allow the inspection and adjustment of belt cleaners from outside the enclosure by the operator looking in through an access door.

24

Figure 24.14

Brush cleaners may be more effective in a wash box in cases where the belt is significantly grooved or damaged.

Figure 24.15

A modular wash box uses a number of "standard" components and configurable modules to provide flexibility with economy in engineering and construction.

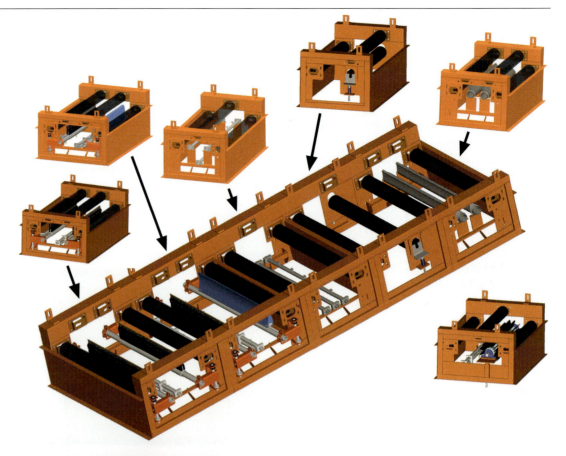

Figure 24.16

The modular approach allows the separation of components to place greater distance between the components or for installation around conveyor structural members and other obstructions.

Figure 24.17

A wash-box system can require a distance of more than 2 meters (7 ft) of belt length and at least 0,6 meters (2 ft) of headroom for the installation at a point where the belt is free of the head and bend pulleys.

time. In fact, the modular approach allows the separation of components—putting the drying mechanisms in a different enclosure from the scraping components, for example—to allow greater distance between the components or to allow installation around conveyor structural members and other obstructions (**Figure 24.16**). An additional benefit is the modular wash-box approach allows for system expansion with different or additional modules added later, as material characteristics, cleaning requirements, or budgetary limitations change.

The drawbacks of wash-box systems include the problems that the belt-washing system can require a distance of more than 2 meters (7 ft) of belt length and at least 0,6 meters (2 ft) of headroom for the installation at a point where the belt is free of the head and bend pulleys (**Figure 24.17**). The drain for the effluent must be as vertical as possible with minimal bends to prevent it from becoming plugged (**Figure 24.18**).

24

Belt Washing for Final Cleaning

Belt-cleaning systems installed so the belt passes through them before it reaches the belt-washing system have an effect on the amount and pressure of water required and on the effectiveness of the wash box. It is strongly recommended that at least one primary cleaner and one or two secondary cleaners be used on any conveyor where a washing station is being considered. These cleaners—installed upstream (closer to the material discharge) of the point where the wash box will be installed—will greatly reduce the amount of carryback to be removed in a washing station, with resulting savings on water usage and operational costs (**Figure 24.19**). Without these cleaners, there will be more material to be removed from the belt by the wash box and more solids in the effluent. Belt-washing stations are intended as the ultimate in cleaning the belt; they are designed to deal with only the final removal of any residual amount of material that passes the upstream cleaning equipment.

Applying the Water

The challenge for any belt-washing application is to get the water to the correct place(s) in the cleaning system in an effective and efficient manner. There are a number of ways to apply water to the belt and material. They range from a simple hose pointed at the belt, to a pipe with drilled holes (**Figure 24.20**), to a more elaborate system of nozzles and spray bars. Engineered nozzles accomplish the application of water in a far more effective manner than a hose or pipe with holes. While the latter are effective methods of water delivery, the water usage requirements of a hose or pipe are far higher than for a system utilizing engineered nozzles. The question then becomes: What is the most effective combination of water pressure, spray pattern, contact angle, and the other variables?

The most effective and efficient way of spraying the cleaning water in these systems is a series of engineered nozzles

placed along a pipe (**Figure 24.21**). Selection of a specific nozzle typically depends on a number of factors, including the type and amount of carryback material, the speed of the belt, the cleanliness of the water supply, the spray pattern needed to achieve uniform spray across the belt's width, the impact pressure of water needed to saturate the material, and the water pressure and flow rate required to keep the blades clean. As with many other aspects of

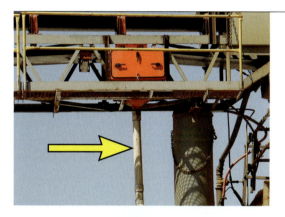

Figure 24.18

The drain for the effluent must be as vertical as possible with minimal bends to prevent it from becoming plugged with solids.

Figure 24.19

Conventional belt cleaners installed in advance of the wash box will greatly reduce the amount of carryback to reach the washing station.

Figure 24.20

There are a number of ways to apply water to the belt and material: e.g., a simple hose pointed at the belt or a water pipe with drilled holes.

conveyor system design, the washing system must be designed to function when carry-back conditions are at their worst.

The two most critical factors in the choice of spray nozzles are the amount of carryback present and the speed of the belt. The higher speed belts require more water to thoroughly cover the belt and soften the carryback during the shorter time the belt is exposed to the spray. High levels of carryback will require more water: The thicker carryback layer will require more water to "soften" the material, because there is more material to be softened. High levels of carryback also require water to be delivered to the system at higher pressures so the water will penetrate the material mass to reach the belt surface. The pressure does not need to be high enough to remove the material, but it must be sufficient to allow the water to reach the belt surface.

Nozzles are available offering a wide variety of spray pattern, flow rates, and pressures. Factors such as distance from the belt to the spray bar, spray pattern, rate, and pressure must be considered in determining the configuration of the spray bars.

Typically, wide spray angles are used to maximize the coverage area while minimizing water consumption (**Figure 24.22**). The nozzle selected, with its spray pattern and spray angle, will control the spacing and mounting distance of the spray bars used in the washing system. In some cases, specialty nozzles are required. Nozzles that are resistant to corrosion, abrasion, or chemicals encountered in the process can be specified.

A typical wash box, operating with sprays at moderate pressure—138 kilopascals (20 lb_f/in.^{2})—will require approximately 63 liters per meter (5.1 gal/ft) of belt width per minute of operation (**Table 24.1**). As noted above, appropriate water pressure and volume should be selected after consideration of both belt speed and carryback (material adhesion) levels.

The need for and use of additional water-spray nozzles to maintain material movement by flushing the wash box and drain system will typically double the required water volume.

The engineering of a belt-washing system can be a complicated process with a number of options compounded by wide-ranging variables in conveyors and materials. A comprehensive understanding of system, material, and process requirements is required. Trained and experienced personnel need to be involved to assure the system will meet customer expectations and applications requirements.

Water Quality

The quality of the water is perhaps the most critical part of designing a high-performing system and if neglected, can render the system non-functional or prone to maintenance intervals and clean-out requirements that are not acceptable.

Belt-washing systems are best when designed for the water-flow rate and pressure required, as determined by material testing and application specifics. Some plants have severe limitations on water usage and flow

24

Figure 24.21

The most effective and efficient way of spraying the cleaning water in these systems is a series of engineered nozzles placed along a pipe.

Figure 24.22

Wide spray angles are typically used to maximize the coverage area while minimizing water consumption.

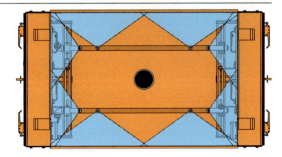

rate/pressure available. These constraints can limit the wash box effectiveness to well below what was designed or specified.

Since engineered nozzles are typically "optimized" to provide a wide spray area, minimized flow rates, and optimal pressure for a given application, the orifice size of the nozzles is typically small and of a unique shape. If the water to be used in a washing system is not "clean enough," the water quality must be evaluated to ensure that there are no particulates large enough to plug the spray nozzles. This is often far easier to say than it is to accomplish, because plant water quality can change dramatically in a matter of minutes. Consequently, a water-filtration system is a valuable addition to the belt-washing system.

Drying the Belt

Following the addition of water to the cleaning process, many applications will require that the belt be dried before it leaves the washing system. In some cases, this is simply to prevent carryback suspended in the water on the belt from being flung from the return rollers. In other cases, the material-handling process requires a dry belt. In still other applications, the belt is used for transporting several bulk materials, and cross-contamination cannot be allowed, so the belt must be clean and dry before the cargo is changed.

Water Consumption for Typical Belt-Washing Systems									*Table 24.1*
Belt Width mm (in.)	Nozzles Used	Approximate Liters per Minute (gal/min)						
		34 kPa (5 lb$_f$/in.²)	69 kPa (10 lb$_f$/in.²)	103 kPa (15 lb$_f$/in.²)	138 kPa (20 lb$_f$/in.²)	207 kPa (30 lb$_f$/in.²)	276 kPa (40 lb$_f$/in.²)	414 kPa (60 lb$_f$/in.²)
400-500 (18)	6	16 (4.3)	23 (6.0)	27 (7.2)	32 (8.4)	39 (10.2)	45 (12.0)	57 (15.0)
500-650 (24)	8	22 (5.7)	30 (8.0)	36 (9.6)	42 (11.2)	51 (13.6)	61 (16.0)	76 (20.0)
650-800 (30)	9	24 (6.4)	34 (9.0)	41 (10.8)	48 (12.6)	58 (15.3)	68 (18.0)	85 (22.5)
800-1000 (36)	11	30 (7.8)	42 (11.0)	50 (13.2)	58 (15.4)	71 (18.7)	83 (22.0)	104 (27.5)
1000-1200 (42)	13	35 (9.3)	49 (13.0)	59 (15.6)	69 (18.2)	84 (22.1)	98 (26.0)	123 (32.5)
1200-1400 (48)	15	40 (10.6)	57 (15.0)	68 (18.0)	79 (21.0)	97 (25.5)	114 (30.0)	142 (37.5)
1400-1600 (54)	16	43 (11.4)	61 (16.0)	73 (19.2)	85 (22.4)	103 (27.2)	121 (32.0)	151 (40.0)
1600-1800 (60)	18	48 (12.8)	68 (18.0)	82 (21.6)	95 (25.2)	116 (30.6)	136 (36.0)	170 (45.0)
1800-2000 (72)	22	59 (15.6)	83 (22.0)	100 (26.4)	117 (30.8)	142 (37.4)	166 (44.0)	208 (55.0)
2000-2200 (84)	26	70 (18.4)	98 (26.0)	118 (31.2)	138 (36.4)	167 (44.2)	197 (52.0)	246 (65.0)

24

There are three basic methods for drying a moving conveyor belt that can be applied to the conveyor as it exits the washing station: evaporation, mechanical water removal, and forced-air drying.

Evaporation

Evaporation is a natural process that will dry the belt (**Figure 24.23**). Evaporation can be accelerated by forcing heated air over the moving belt. However, evaporation of the water film by forced air alone is not a feasible means of complete water removal for typical bulk-materials handling conveyor belt applications, because there is a limit to how fast water can be evaporated.

Mechanical Drying

There are a number of mechanical methods to remove water from the belt. The first is mechanically wiping the belt, using what is commonly called a "squeegee" blade. This is similar to a car's windshield wipers.

A squeegee blade placed as the final cleaning device in the wash-box system will remove a significant amount of excess water. The result will vary depending on the type of squeegee used, its material of construction, and its location, as well as application specifics such as belt speed and the amount of water present on the belt. In general, the squeegee blade is an effective and economical means of water removal (**Figure 24.24**).

The use of squeegee rollers, either as single or dual rolls, is also an effective way of removing excess water from the moving conveyor belt (**Figure 24.25**). A study from the University of Newcastle Research Associates (TUNRA) explored the effectiveness of a single-roll squeegee system and examined the effect of using different diameter rolls on various belt speeds *(Reference 24.4)*. The results of this study clearly showed the smaller the roller, the better the squeegee action, regardless of belt speed (**Figure 24.26**). Squeegee rollers are generally effective in reducing the thickness of the film of water on the belt to approximately 50 microns, with an effective lower limit of 20 microns (**Figure 24.27**).

Figure 24.23

The natural process of evaporation can be accelerated by forcing heated air over the moving belt.

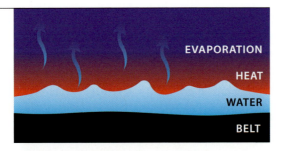

Figure 24.24

A squeegee blade is an effective and economical means of removing water from a conveyor belt.

24

Figure 24.25

The use of squeegee roller systems, with either single or dual rolls, is an effective way of removing excess water from the moving conveyor belt.

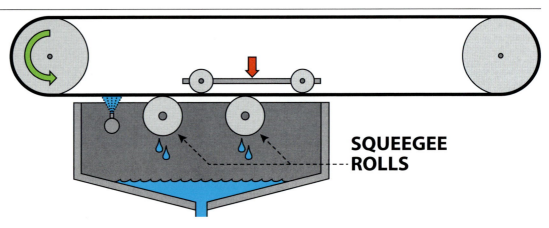

SQUEEGEE ROLLS

Forced-Air Drying

The third technique for drying the belt is using high-velocity air to separate the water film from the belt. Two mechanisms can be dominant: hydrodynamic instability and disjoining. Hydrodynamic instability occurs when the water film is exposed to moving air. The film will form a wave, which grows rapidly, causing the formation of droplets that then leave the surface. Disjoining occurs when the water film is exposed to high-velocity air, and the water is "peeled" from the belt surface (**Figure 24.28**).

High-velocity air can be highly effective in the removal of thicker films of water. There are a number of systems commercially available, including air "knives" that use blowers to generate the air velocity and pressure required; other systems operate from compressed air lines (**Figure 24.29**).

To remove the largest quantity of water, the velocity of air must be maximized. However, the achievable velocity of air is limited by several factors, including the power consumed to generate high velocities using a blower or compressed air line as

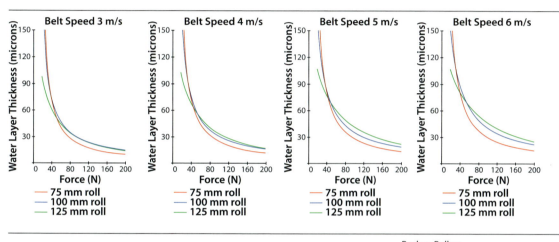

Figure 24.26

The results of this study clearly showed the smaller the roller, the better the squeegee action, regardless of belt speed.

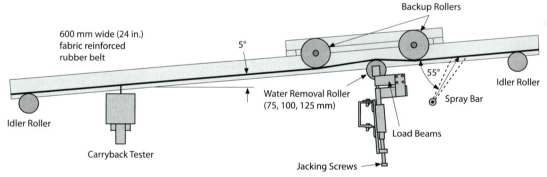

Figure 24.27

Squeegee rollers are generally effective in reducing the film of water on the belt to approximately 50 microns, with an effective lower limit of 20 microns.

24

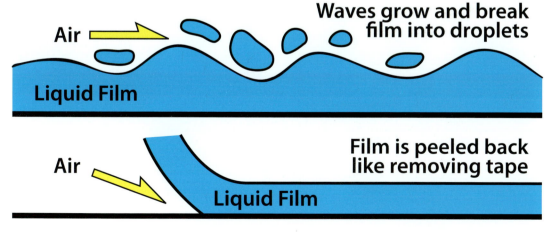

Figure 24.28

Hydrodynamic instability occurs when moving air forms the water film into a wave, from which droplets then leave the surface. Disjoining occurs when the water film is exposed to high-velocity air so the water is "peeled" from the belt surface.

well as the noise associated with extremely high velocities of air.

Research has indicated the dominant factor in water removal was the relative speed of the air; the angle of contact was not critical in terms of water removal. A feasible range of air velocity at the belt is 80 to 100 meters per second (15000 to 20000 ft/min). Within this air speed range, experimental results show that water can be removed from a moving belt down to a film thickness of 7 to 11 microns *(Reference 24.5)*. These velocities can be reached with specially designed nozzles and regenerative blowers for about 7,5 kilowatts per meter (3 hp/ft) of belt width dried. Compressed air can also be used—with other air nozzle types—with similar power requirements. As typical belt speeds are from 1 to 5 meters per second (200 to 1000 ft/min), belt speed is not a major parameter compared to the speed of the air.

Performance of Water-Removal Systems

The relative performance of the various water-removal systems can be assessed and compared (**Table 24.2**). These three water-removal methodologies can be used individually, but the best approach may be to use a combination of the different possibilities.

Reclaiming the Water

Once the basic components of a belt-washing system are established, it is possible to examine the systems for dealing with the effluent—the dirty water—removed from the belt. In many industrial environments, the amount of water used and the quality of water released are strictly controlled. In other cases, the material has a high value and, therefore, it is cost effective to recover the solids. In both cases, a system to separate the solids from the water is often required.

In choosing a mechanical water-separation system, several factors need to be considered. Principal among them is the quantity of water and its solids content, as well as the location in which the water-recycling system can be installed. Depending on the method of treatment, the rate of settlement of the solid in water can be the main criteria, but due to the size of the devices, relying solely on settling is often impractical.

In the most simple water-treatment system, the effluent is channeled to a settling pond and, by the process of sedimentation, the water is clarified and filtered for reuse as plant water (**Figure 24.30**). This has several potential problems, including keep-

Figure 24.29

There are a number of systems commercially available, including air "knives" that use blowers to generate the air velocity and pressure required to dry the belt.

24

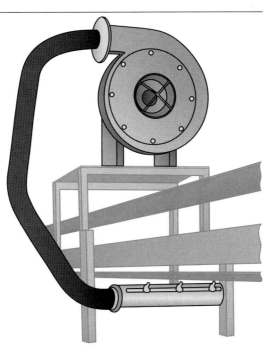

Table 24.2	Comparison of Various Water-Removal Methods			
Method	**Water-Removal Efficiency**	**Energy Use**	**Purchase Cost**	**Operating Cost**
Squeegee Blade	1	2	1	1
Squeegee Roller	2	1	3	2
Compressed Air Knife	3	3	2	3
High Pressure Blower	4	4	4	4

1 = Lowest, 4 = Highest

ing the drainage system from plugging with solids, the periodic dredging of the solids from the settling pond, and the subsequent disposal of this recovered material.

Concrete settling basins are sometimes used close to the point of effluent generation. These can be designed so a front-end loader can drive into the basin and collect the settled solids. On a smaller scale, dumpsters can be used as the location for settling, with the advantage the solids can often be returned to the material-handling system simply by emptying the container (**Figure 24.31**).

Engineered water-separation and -reclamation systems are available (**Figure 24.32**). Modular water-recycling systems can provide up to 1250 liters per minute (300 gal/min) of continuous recycling; the modules can be combined to provide higher volumes of clean water.

In some cases, a chemical additive can be used to expedite solids settling, but this will require periodic inspection and service of the equipment to assure that the chemical is available to the system at all times.

Mechanical-filtration or chemical-additives systems are occasionally necessary for bulk materials that do not wet easily or that have a specific gravity close to or less than water. There is a variety of mechanical means available including filter presses, dewatering screens, hydrocyclones, and clarifiers. However, most bulk materials that are handled in large quantities are heavier than water and so can be separated using a simple and effective inclined screw separator system.

When designing a complete wash-box system, the washing portion should be designed first, to define the system's operational requirements. Following that, the water-recycling system can be developed to provide the water-handling capacity to meet the washing requirement. One detail that is often overlooked is that the discharge from a wash box is prone to plugging. For this reason, the discharge should be either

Figure 24.30

In the most simple water-treatment system, the effluent is channeled to a settling pond and, by the process of sedimentation, the water is clarified and filtered for reuse as plant water.

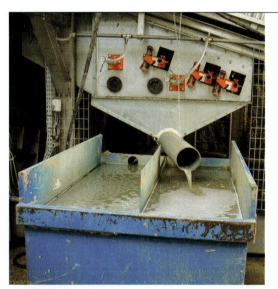

Figure 24.31

Dumpsters can be used as the settling tank, with the solids returned to the material-handling system simply by emptying the container.

Figure 24.32

Engineered water-separation and -reclamation systems can provide up to 1250 liters per minute (300 gal/min) of continuous recycling.

24

an open channel or a large-diameter pipe with minimal bends. It should also have many removable fittings or plugs to allow clean-out and use plenty of flushing water (**Figure 24.33**).

Recovering the Solids

The materials in the wash-box effluent can be recovered. This is important in those operations in which the cargo is especially valuable and/or the cargo has already been subjected to some processing or treatment.

If the addition of water to the process is not a concern, the slurry can be returned to the conveyor cargo or plant process directly from the wash box by means of a pump. If the plant needs to minimize the water added to its process, the water can be recycled and the recovered solids can then be placed on the belt or process through a mechanical means such as a screw conveyor. A simple settling test, in which the material is placed in a container of water and the rate at which it settles is observed, will give a good indication of the dwell time needed for settling and whether or not chemical additives might be needed to promote settling.

24

Figure 24.33

The discharge from a wash box should be either an open channel or large diameter pipe that has minimal bends, has several removable fittings to allow clean-out, and uses plenty of flushing water.

TYPICAL SPECIFICATIONS

A. Spray-wash system

The conveyor system will be equipped with a spray belt-washing system installed directly after the head chute to provide final removal of any residual cargo from the belt. This spray-wash system will be contained in a watertight metal enclosure fitted with water supply and an oversized drain.

B. Size

The belt-washing system will be sized based on the amount and properties of expected carryback per square meter (/ ft^2) of belt.

C. Water-spray bar

The enclosure shall be fitted with at least one water-spray bar with engineered nozzles that are positioned to wet the entire cargo-carrying portion of the belt and to flush removed material out of the box through the drain.

D. Secondary cleaners

The wash box will be fitted with a minimum of two secondary cleaners to remove fines and water from the belt's load-carrying surface.

E. Access

The wash box shall be fitted with watertight access door(s) to allow easy inspection and service.

F. Hold-down rollers

The spray-wash system shall include a minimum of three hold-down rollers above the belt that hold the belt in position against the spray-applied water and cleaning edges.

G. Drain system

The volume and flow rate of flushing water and design of the drain system shall be sufficient to prevent settling of bulk solids in the drain system.

ADVANCED TOPICS

The Process for Developing a Wash-Box System

When developing a belt-washing station, it is desirable to provide a complete system analysis that takes into account a number of factors including the physical layout of the conveyor and water-recycling system, the amount of energy required for drying and water recycling, and the ability of the solids to be separated from the water.

When considering the installation of a wash box, there are a number of questions that must be considered. These include:

A. How much water will the wash box use?

B. How clean will the belt be as it enters the wash box?

C. How clean must the belt be as it leaves the wash box?

D. How dry will the belt be?

E. What will be done with the effluent (the mix of solids and water)?

These questions can be answered with reasonable accuracy if there is detailed information available about the properties of the bulk material, the belt and bulk material interface conditions, the amount of carryback present, and the general choice of equipment in the wash box.

Sample Problem

The approach to develop a preliminary design of a Wash-Box System is as follows:

A. Determine the amount of carryback entering the wash box per day (Cb_{day-in}).

B. Determine the desired amount of carryback leaving the wash box per day ($Cb_{day-out}$).

C. Determine the amount of effluent to be handled per minute.

D. Consider options and other questions.

These four stages can be answered by following the four steps below.

Step 1. Calculate the carryback on the belt entering the wash box per day (Cb_{day-in}) (**Equation 24.1**)

$$Cb_{day-in} = BW \cdot CW \cdot S \cdot T \cdot Cb_{in} \cdot k$$

Given: A 1,2-meter (48-in.) belt with a cleaned width of 67 percent traveling 3,5 meters per second (700 ft/min) has a measured carryback of 100 grams per square meter (0.33 oz/ft²) in a 24-hour operation. **Find:** The carryback entering the wash box per day.

Variables		Metric Units	Imperial Units
Cb_{day-in}	Carryback Entering the Wash Box per Day	tons	short tons
BW	Belt Width	1,2 m	4 ft
CW	Cleaned Width of Belt	0,67 (67%)	0.67 (67%)
S	Belt Speed	3,5 m/s	700 ft/min
T	Time in a Day	86400 s	1440 min
Cb_{in}	Amount of Carryback Reaching Wash Box	100 g/m²	0.33 oz/ft²
k	Conversion Factor	1 x 10⁻⁶	3.12 x 10⁻⁵

Metric: $Cb_{day-in} = 1,2 \cdot 0,67 \cdot 3,5 \cdot 86400 \cdot 100 \cdot 1 \cdot 10^6 = 24,3$

Imperial: $Cb_{day-in} = 4 \cdot 0.67 \cdot 700 \cdot 1440 \cdot 0.33 \cdot 3.12 \cdot 10^5 = 27.8$

| Cb_{day-in} | Carryback Entering the Wash Box per Day | 24,3 t | 27.8 st |

Step 1 Answer: There are 24,3 tons (27.8 st) of carryback entering the wash box per day.

Equation 24.1

Calculating the Amount of Carryback Entering the Wash Box per Day

24

Step 2. Calculate the desired carryback on the belt as it leaves the wash box per day ($Cb_{day\text{-}out}$) (**Equation 24.2**)

Step 3. Determine the amount of effluent to be handled per minute (**Equation 24.3**)

Step 4. Consider options and additional questions

A more detailed study and theoretical analysis combined with field testing at the actual site would produce additional factors and variables that could be used to further investigate options. Additional questions that can now be considered include:

A. Is 10 grams per square meter (0.033 oz/ft²) too much carryback material left on the belt as it leaves the wash box? (Carryback is measured as the dry weight of the material.)

B. How wet is the belt as it leaves the wash box?

C. How can the overall water usage be reduced?

For the purpose of keeping this example short, some assumptions regarding moisture content of the carryback (50 percent) and of the effluent (15 percent) must be made to answer these questions. These assumptions are based on experiences in wash-box design.

Is 10 Grams per Square Meter Too Much Carryback Material to Leave on the Belt Leaving the Wash Box?

As noted in the discussion of carryback levels, 10 grams per square meter (0.033 oz/ft²) is considered a clean belt. *(See Chapter 31: Performance Measurements.)* Testing has shown that on average, only about 50 percent of the carryback left on the belt at this level of cleanliness will fall from the belt on the return run.

Belt cleaning is a process with results in a bell-shaped curve. A 10-grams-per-square-meter carryback level could range from 20 grams per square meter (0.066 oz/ft²) to sometimes 0 grams per square meter. To achieve a belt cleaner than 10 grams per

---

Equation 24.2

Calculating Desired Carryback Exiting the Wash Box per Day

24

$$Cb_{day\text{-}out} = BW \cdot CW \cdot S \cdot T \cdot Cb_{out} \cdot k$$

Given: A 1,2-meter (48-in.) belt with a cleaned width of 67 percent traveling 3,5 meters per second (700 ft/min) has a desired carryback of 10 grams per square meter (0.033 oz/ft²) in a 24-hour operation. **Find:** The carryback exiting the wash box per day.

Variables		Metric Units	Imperial Units
$Cb_{day\text{-}out}$	Desired Carryback Exiting the Wash Box per Day	tons	short tons
BW	Belt Width	1,2 m	4 ft
CW	Cleaned Width of Belt	0,67 (67%)	0.67 (67%)
S	Belt Speed	3,5 m/s	700 ft/min
T	Time in a Day	86400 s	1440 min
Cb_{out}	Amount of Desired Carryback Exiting the Wash Box	10 g/m²	0.033 oz/ft²
k	Conversion Factor	1 x 10⁻⁶	3.12 x 10⁻⁵

Metric: $Cb_{day\text{-}out} = 1{,}2 \cdot 0{,}67 \cdot 3{,}5 \cdot 86400 \cdot 10 \cdot 1 \cdot 10^{-6} = 2{,}4$

Imperial: $Cb_{day\text{-}out} = 4 \cdot 0.67 \cdot 700 \cdot 1440 \cdot 0.033 \cdot 3.12 \cdot 10^{-5} = 2.8$

$Cb_{day\text{-}out}$	Desired Carryback Exiting the Wash Box per Day	2,4 t	2.8 st

Step 2 Answer: The desired carryback exiting the wash box is 2,4 tons (2.8 st) per day.

square meter (0.033 oz/ft²) of carryback, so much cleaning pressure would need to be applied that it would endanger the belt's top cover. Therefore, 10 grams per square meter is an acceptable and practical lower limit for carryback material remaining on the belt.

How Wet is the Belt as it Leaves the Wash Box?

The amount of water left on the belt can be estimated based on the type of water-removal system used. The most effective method is a high-velocity air-knife system.

Testing has confirmed that for moving conveyor belts, the theoretical value of 6,0 grams per square meter (0.020 oz/ft²) of water left on the belt is about as low as is practical to obtain.

In the example, assuming 10 grams per square meter (0.033 oz/ft²) (dry weight) of carryback is left on the belt at 50 percent moisture content means there will be an equal amount, or 10 grams per square meter, of water left on the belt. *(Note: A film of carryback or water 1,0 micron thick, with a specific gravity of 1.0, is equal to 1,0 grams per square meter.)*

How Can Overall Water Usage Be Reduced?

Overall water usage can be reduced by recycling all the water from the effluent and adding only the required makeup water to the wash-box system. Theoretically, the amount of makeup water needed would equal the amount of water left on the belt as it leaves the wash box and in the effluent. However, there will be other system losses

$$E_m = \frac{(Cb_{day-in} - Cb_{day-out}) \cdot k}{\rho \cdot T} + (W_{SB} + W_F)$$

Equation 24.3

Calculating the Amount of Effluent Handled per Minute

Given: A wash box where 24,3 tons (27.8 st) of carryback enter and 2,4 tons (2.8 st) of carryback exit, and the density of the material is 1 kilogram per liter (62 lb$_m$/ft³). The spay bar and the flush system in the wash box each consume 100 liters per minute (25 gal/min) in a 24-hour operation. **Find:** The amount of effluent per minute of operation.

Variables		Metric	Imperial
E_m	Effluent Handled per Minute	liters per minute	gallons per minute
Cb_{day-in}	Carryback Entering the Wash Box per Day	24,3 t	27.8 st
$Cb_{day-out}$	Desired Carryback Exiting the Wash Box per Day	2,4 t	2.8 st
k	Conversion Factor	1000	14960
ρ	Bulk Material Density	1 kg/l	62 lb$_m$/ft³
T	Time in a Day	1440 min	1440 min
W_{SB}	Water Consumption for Spray Bar	100 l/min	25 gal/min
W_F	Water Consumption to Flush Wash Box	100 l/min	25 gal/min

Metric: $E_m = \dfrac{(24,3 - 2,4) \cdot 1000}{1 \cdot 1440} + (100 + 100) = 215$

Imperial: $E_m = \dfrac{(27.8 - 2.8) \cdot 14960}{62 \cdot 1440} + (25 + 25) = 54$

E_m	Effluent Handled per Minute	215 l/min	54 gal/min

Step 3 Answer: The system handles 215 liters per minute (54 gal/min) of effluent.

in a wash box such as leaks and splashing as well as evaporation. The amount of water that leaves the wash box, contained in the carryback and recycled solids, is usually at least half of the required makeup water. Because the addition of makeup water is usually controlled by some type of level indicator in the settling tank, the demand is not constant. Therefore, the makeup water system should be oversized to keep the tank at the proper level without having to run continuously.

The makeup water required can be calculated (**Equation 24.4**).

The wash box requires 200 liters per minute (50 gal/min) of water. By using only the required makeup water and recycling the effluent water, the operation will consume only 8,6 liters per minute (2.5 gal/min). This produces a water savings of 191,4 liters per minute (47.5 gal/min) of water. The recycled water can be used

to flush the wash box and/or in spray bars with large orifices. Most of the makeup water can be added as a low-volume clean water spray on the last belt cleaner or squeegee inside the wash box. This example, although simplified, is typical of a conveyor belt-washing system for this belt width and speed using a mechanical means of recycling the water and solids.

ON THE DESIGN OF BELT-WASHING SYSTEMS

In Closing...

Combining effective belt cleaners, spray-washing technology, effluent treatment, and belt-drying systems produces a state-of-the-art belt-washing station. Such a station can meet the need for keeping the belt reasonably clean and dry, provide for the recovery and recycling of the solids at a reasonable cost, and use a minimum amount of clean

---

Equation 24.4

Calculating Required Makeup Water per Minute

24

$$M_W = \left[\left(\frac{Cb_{day-out}}{\left(\frac{1 - M_{Cb}}{M_{Cb}} \right)} \right) + \left(\frac{Cb_{day-in} - Cb_{day-out}}{\left(\frac{1 - M_E}{M_E} \right)} \right) \right] \cdot SF \cdot k$$

Given: A wash box where 24,3 tons (27.8 st) of carryback enter and 2,4 tons (2.8 st) of carryback exit, with a carryback moisture content of 50 percent and an effluent moisture content of 15 percent. **Find:** The amount of makeup water needed per minute.

Variables		Metric Units	Imperial Units
M_W	Makeup Water per Minute	liters per minute	gallons per minute
Cb_{day-in}	Carryback Entering Wash Box per Day	24,3 t	27.8 st
$Cb_{day-out}$	Desired Carryback Exiting Wash Box per Day	2,4 t	2.8 st
k	Conversion Factor	0,69	0.17
M_{CB}	Moisture Content of Carryback	0,5 (50%)	0.5 (50%)
M_E	Moisture Content of Effluent	0,15 (15%)	0.15 (15%)
SF	Safety Factor to Account for Other Losses	2	2

$$\text{Metric: } M_W = \left[\left(\frac{2,4}{\left(\frac{1 - 0,5}{0,5} \right)} \right) + \left(\frac{24,3 - 2,4}{\left(\frac{1 - 0,15}{0,15} \right)} \right) \right] \cdot 2 \cdot 0,69 = 8,6$$

$$\text{Imperial: } M_W = \left[\left(\frac{2.8}{\left(\frac{1 - 0.5}{0.5} \right)} \right) + \left(\frac{27.8 - 2.8}{\left(\frac{1 - 0.15}{0.15} \right)} \right) \right] \cdot 2 \cdot 0.17 = 2.5$$

M_W	Makeup Water per Minute	8,6 l/min	2.5 gal/min

water. Where space is limited, the elements can be designed to fit into confined spaces, but the difficulties may result in corresponding reductions in cleaning effectiveness and increased difficulty in operation and maintenance.

Belt-washing systems combine all of the desired features of a complete belt-cleaning system into a single operating system. By selecting the equipment appropriate for the application, the cost can be minimized and a return on investment can be calculated based on meeting environmental regulations, recovering the carryback material, reducing cleanup expense, and increasing component-life.

Looking Ahead...

This chapter about Belt-Washing Systems, the third chapter in the section Leading-Edge Concepts, discussed using water with belt-cleaning systems to reduce both carryback and the damage it can cause to the conveyor. The following chapter, Material Science, is the final chapter in this section.

REFERENCES

24.1 Planner, J.H. (1990). "Water as a means of spillage control in coal handling facilities." In *Proceedings of the Coal Handling and Utilization Conference: Sydney, Australia*, pp. 264–270. Barton, Australian Capital Territory, Australia: Institution of Engineers, Australia.

24.2 Stahura, Richard.P, Martin Engineering. (1987). "Conveyor belt washing: Is this the ultimate solution?" *TIZ-Fachberichte*, Volume 111, No. 11, pp. 768–771. ISSN 0170-0146.

24.3 University of Illinois. (1997). *High Pressure Conveyor Belt Cleaning System*. Unpublished Study done for Martin Engineering.

24.4 University of Newcastle Research Associates (TUNRA). Untitled, unpublished study done for Engineering Services and Supplies P/L (ESS).

24

24.5 University of Illinois. (2005). *Design of Conveyer Belt Drying Station.* Unpublished Study done for Martin Engineering.

24.6 Swinderman, R. Todd, Martin Engineering. (2004). "Standard for the specification of belt cleaning systems based on performance." *Bulk Material Handling by Conveyor Belt 5*, pp. 3–8. Edited by Reicks, A. and Myers, M., Littleton, Colorado: Society for Mining, Metallurgy, and Exploration (SME).

24.7 Roberts, A.W.; Ooms, M.; and Bennett, D. *Conveyor Belt Cleaning – A Bulk Solid/Belt Surface Interaction Problem.* University of Newcastle, Australia: Department of Mechanical Engineering.

24.8 Spraying Systems Company (http://www.spray.com) contains a variety of useful material on the basics and options available in spray nozzles.

24

24

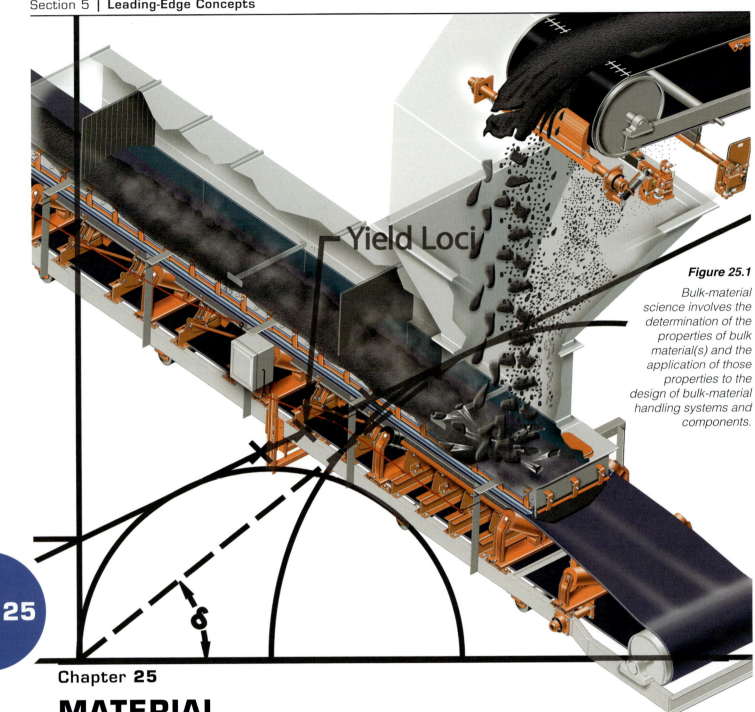

Yield Loci

Figure 25.1

Bulk-material science involves the determination of the properties of bulk material(s) and the application of those properties to the design of bulk-material handling systems and components.

25

Chapter **25**

MATERIAL SCIENCE

In this Chapter...

In this chapter, we discuss the importance of testing the actual bulk materials to be conveyed for proper conveyor design. We describe both basic and advanced properties of bulk materials and the test methods used for measuring those properties, along with typical applications for which these tests are performed.

Bulk-material science is an interdisciplinary field involving the determination of the properties of bulk material(s) and the application of those properties to the design of bulk-material handling systems and components. This science investigates the interaction between bulk material(s)—both as a body and as individual particles—and the surfaces over which the material(s) will flow.

Since the design of the first conveyors, the basic properties of bulk materials, such as bulk density and angle of repose, have been used to size equipment and to calculate the power requirements of bulk-material handling systems. Modern bulk-material science traces its roots to Andrew W. Jenike's work at the University of Utah, in which the critical dimensions required for bins to operate in a mass flow condition were determined based on the strength of the bulk material under various conditions. The methods developed by Jenike are used to determine the internal strength of bulk materials and the friction between the bulk material and the surfaces it will contact (e.g., the belt or chute). These properties are used with increasing success to predict the behavior and flow of bulk materials as they travel on conveyors and through chutes, thus allowing the design of cleaner, safer, and more productive systems.

A number of references are published with typical properties for many bulk materials *(Reference 25.1)*. This reference data is normally a general description, and while useful for preliminary equipment design, it does not represent a specific bulk material under the actual conditions of use. Serious errors can be made by designing a materi-al-handling system without determining the appropriate basic and advanced properties of the specific bulk material.

There are many applications for bulk-material science, such as the design of bins, screw conveyors, and stockpiles. This chapter will discuss the importance of the application of the properties of bulk materials to the design of belt conveyor systems and components for handling bulk materials (**Figure 25.1**).

BASIC PROPERTIES OF BULK MATERIALS

Many of the basic properties and tests for bulk materials are described in the Conveyor Equipment Manufacturers Association (CEMA) publication CEMA STANDARD 550-2003. The properties most often used (or misused) in the design of belt conveyor systems are described below.

Bulk Density

Bulk density (ρ) of a bulk material is the weight per unit of volume—kilograms per cubic meter (lb_m/ft^3). Differences in bulk density will occur at different moisture contents and as the bulk material travels on the conveyor belt and is compacted due to vibration.

Loose Bulk Density

Loose bulk density (ρ_1) of a bulk material is the weight per unit of volume that has been measured when the sample is in a loose or non-compacted condition (**Figure 25.2**). The loose bulk density must always be used when designing the load-zone chutes and the height and width of the skirtboards, or the design capacity may not flow through the transfer point, due to the increased volume in the loose state.

Consolidated Bulk Density

Consolidated bulk density (ρ_2)—sometimes called vibrated bulk density—is normally the heaviest density that can be found in the conveying of bulk materials (**Figure 25.3**). This is achieved by apply-

Figure 25.2

Loose bulk density (ρ₁) of a bulk material is the weight per unit of volume that has been measured when the sample is in a loose or non-compacted condition.

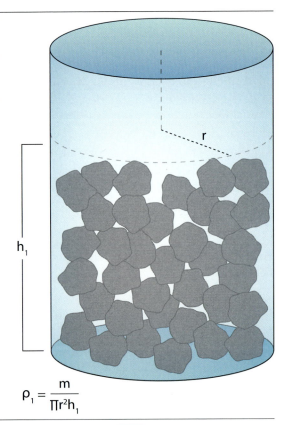

$$\rho_1 = \frac{m}{\Pi r^2 h_1}$$

Figure 25.3

Consolidated bulk density (ρ₂)—sometimes called vibrated bulk density—is normally the heaviest density that can be found in the conveying of bulk materials, achieved by applying a compressive force (F) or vibratory energy to the body of material.

25

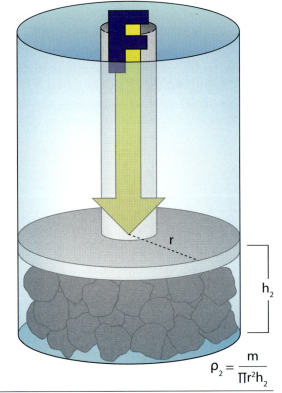

$$\rho_2 = \frac{m}{\Pi r^2 h_2}$$

Figure 25.4

Angle of Repose

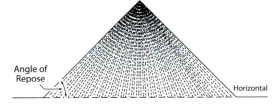

ing a compressive force (F) or vibratory energy to the body of material. The consolidated bulk density is used for determining the weight of material conveyed on the belt based on surcharge angle. A compressibility percentage may be found by taking vibrated bulk density minus loose bulk density divided by vibrated bulk density times 100. The above ratio is rarely above 40 percent and may be as low as 3 percent, indicating that a caution must be used when making density-related calculations.

There are a number of standards published for determining bulk density, such as American Society for Testing and Materials International (ASTM) ASTM D6683-01 *(Reference 25.2)*, but it is recommended that the test methods described in CEMA STANDARD 550-2003 be used when the density will be applied to the conveying of bulk materials.

Angle of Repose

The loose angle of repose for bulk materials is that angle between a horizontal line and the sloping line from the top of a freely formed pile of bulk material to the base of the pile (**Figure 25.4**). The angle of repose for a given material may vary, however, depending upon how the pile is created and the density, particle shape, moisture content, and size consistency of the material. Because the angle of repose is relatively easy to measure, it is often used as a convenient design parameter. However, this can lead to serious errors due to large variations in the angle for a given category of bulk material. For example, the angle of repose range for various types of coal as listed in CEMA STANDARD 550-2003 runs from 27 to 45 degrees. The application of the angle of repose should be limited to the shape of freely-formed stockpiles.

Surcharge Angle

The surcharge angle is the angle of the load cross section measured by the inclination in degrees to the horizontal (**Figure 25.5**). The symbol Θ_s is frequently used to represent the surcharge angle. The angle of surcharge of a bulk material on a moving conveyor depends upon the kind of conveyor involved. With a troughed belt conveyor, the top surface of the load cross section of the bulk material is assumed to be part of a circular arc, the ends of the arc meeting the inclined sides of the belt at the free-edge distance. *(See Chapter 11: Skirtboards for a discussion about edge distance.)* On conveyors with vertical pan sides, the top of the load cross section may be assumed as a portion of a circular arc, the ends of which meet the vertical sides (**Figure 25.6**). The surcharge angle is measured by the inclination of the line tangent to the circular arc. On a flat belt or apron conveyor, the top surface of the bulk material is presumed to be triangular in cross section (**Figure 25.7**).

The surcharge angle is useful in conveyor design for determining the profile of the load on the belt for various belt widths and trough angles, which therefore provides the theoretical carrying capacity of the belt. Standard test methods for determining surcharge angles for bulk materials are described in CEMA STANDARD 550-2003.

Material Size

The size of the bulk material is often described using either the material's maximum lump size or as the percent of particles that pass a series of defined sieves through a process typically called screening. Both measurements are important to conveyor design.

The material size is often described as the width and height of the largest lump. For example, a material with a maximum lump width and height of 50 millimeters x 50 millimeters (2 in. x 2 in.) would be described as 50 millimeter (2 in.) minus material. However, it is common practice to assume the length of the lump can be as much as three times the larger of the width or height; the above example yields a length up to 150 millimeters (6 in.) long. This information is useful in determining the size for various components, including the width of chutes and skirtboards. A common "rule of thumb" is that chute or skirtboard width should be at least two

times the largest lump dimension in order to prevent plugging.

A screening analysis provides the most complete representation of the size of the bulk material (**Figure 25.8**). ASTM D6393-99(2006) *(Reference 25.3)* provides

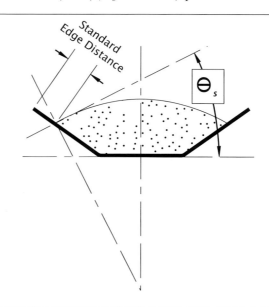

Figure 25.5

Surcharge Angle (Θ_s) for a Troughed Belt

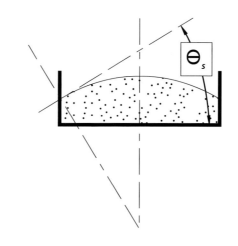

Figure 25.6

Surcharge Angle (Θ_s) for a Pan or Apron Feeder

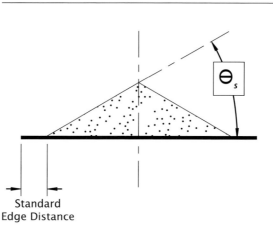

Figure 25.7

Surcharge Angle (Θ_s) for a Flat Belt

25

one test method for screen analysis of bulk materials. Particle size distribution is a tabulation of the percent of the material represented in each size range as a part of the total sample, as demonstrated by passing through a given screen size and being retained on the next smaller screen. A particle-size distribution curve is usually a semi-log plot using the particle size as the abscissa on a logarithmic scale and the cumulative percentage by weight passing a given screen size as the ordinate (**Figure 25.9**). The shape of the curve and the slope of any straight portion indicate the relative uniformity of size distribution of the sample. This information is useful for determining the particle size needed for calculating airflow. *(See Chapter 7: Air Control for calculating induced air that includes particle diameter (D) in the denominator.)* The amount of induced air is inversely proportional to particle diameter; the smaller the diameter the more induced air. Knowing the average particle diameter presents a simplified way to calculate induced air, based on the percentage of particles at each sieve size.

ADVANCED BULK-MATERIAL PROPERTIES

Moisture Content

Moisture content is the total amount of water present in a bulk material. A bulk material can have surface (or free) moisture content and inherent moisture content. Surface moisture is the mass of water that is between particles, on the surface, and in open pores. The surface moisture content can have a major effect on the material's values for adhesion, cohesion, and wall friction angle. Inherent moisture content is

Figure 25.8

Screening analysis provides the most complete representation of the size of the bulk material.

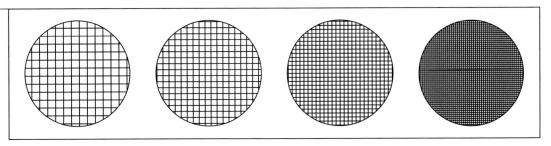

Figure 25.9

A particle-sized distribution curve is usually a semi-log plot.

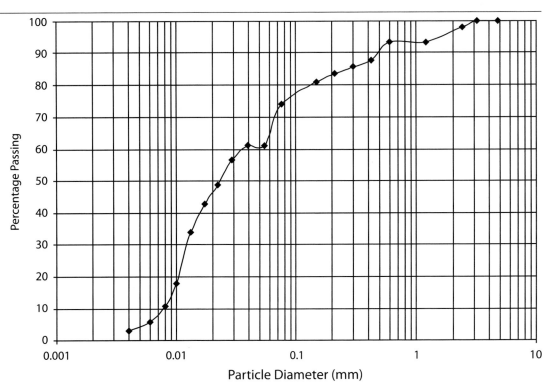

the mass of water contained within closed pores but does not include moisture that is chemically bound within the particles. Moisture content is defined on a wet basis in the bulk-materials handling industry; the moisture content is expressed as a percentage of the total wet weight. The most common method of determining surface moisture is to dry a sample in an oven until equilibrium is reached and then measure the weight loss.

Wetting and Settling Rate

The wetting ability of a solution is a measure of its ability to "wet" (spread across) and penetrate a bulk material. This is important, because it affects and reflects the performance of dust-suppression systems and chemicals with the particular material.

Sedimentation measurement methods are based on the application of Stokes' Law, which describes the terminal velocity for an isolated sphere settling in a viscous fluid under the influence of a gravitational field (i.e., free falling). For materials with low Reynolds numbers (i.e., laminar flow conditions) the terminal velocity depends on the density contrast between the particle and medium, the viscosity, and the particle size.

Sedimentation or settling rate is similarly important to help assess the performance of belt-washing and dust-suppression systems with specific materials. A simple and common test for the settling rate of bulk materials in water is commonly called a "Jar Test." In its simplest form, a sample of the bulk material is put into a beaker of water, and the time for the material to settle to the bottom is recorded. Detailed procedures are described in ASTM D2035-08 *(Reference 25.4)*.

Internal Stresses

Internal stress in a material cannot be measured directly. It must be deduced from the force acting across a unit area of a bulk material, as it resists the separation, compacting, or sliding induced by external forces. Normal stress refers to the stress caused by forces that are perpendicular to a cross-sectional area of the material. Shear stress arises from forces that are parallel to the plane of the cross section (**Figure 25.10**). Stress is expressed as a force divided by an area.

The original work by Jenike centered on the properties of bulk materials as derived from the shear stress capacity of the material. Jenike's work focused on determining the outlet dimensions of a storage bin for reliable gravity flow and the stresses on the hopper wall for safe bin design. Many of the same methods from his original work are successfully applied to the transport of bulk materials on conveyor belts.

Flow properties of a bulk material can be derived from measuring the force to shear the bulk material, using a shear cell (**Figure 25.11**). There are several shear cell manufacturers and test methods used to determine bulk material properties such as ASTM D6128-06 *(Reference 25.5)* or ASTM D6773-02 *(Reference 25.6)*. Normally, only the fines from the bulk material are tested, because they typically yield the greatest adhesion and cohesion values. Shear cell tests are time-consuming, due to the large number of tests required to determine the properties at different moisture levels and consolidating pressures. Repeatability of the test results requires careful sample preparation and testing procedures by a skilled technician.

When conducting these tests, the values that are known are the principal or consolidating force (V) and the sample area.

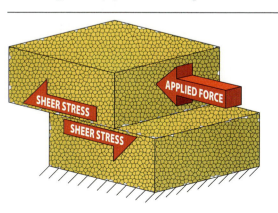

Figure 25.10

Shear stress arises from forces that are parallel to the plane of the cross section.

Figure 25.11

Using a shear cell, flow properties of a bulk material can be derived from measuring the force to shear the bulk material.

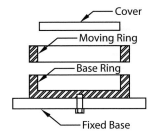

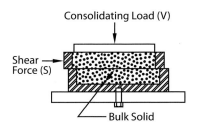

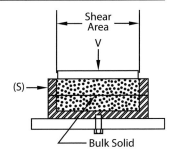

Figure 25.12

Mohr's Circle Concept

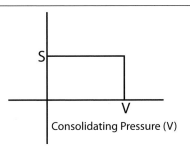

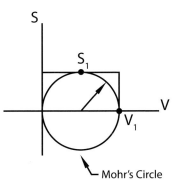

Figure 25.13

A Series of Shear Tests of Various Consolidating Pressures

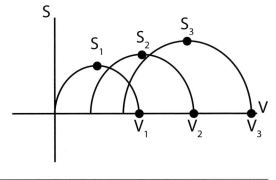

Figure 25.14

Internal Friction (Φ)

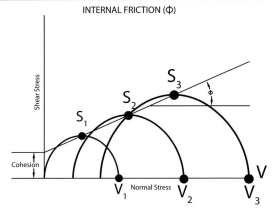

What is measured is the shear force (S) required to shear the bulk material. The shear force and shear area are used to determine the shear and normal stresses at different consolidating pressures and moisture contents using a Mohr stress circle. The Mohr stress circle represents the stresses in cutting planes that are inclined through all possible angles. The position of the Mohr stress circle is defined by the two principal stresses (**Figure 25.12**). It is important to note that a bulk material can transmit shear stresses even if it is at rest. Also, in bulk materials, compressive stresses are defined as positive stresses.

Internal Friction Angle

By running a series of shear tests at various consolidating pressures, values of internal friction can be determined. Internal friction angle is the angle at which the particles within a bulk material slide over one another within a pile, or, in other words, failure due to shearing. This angle is between the horizontal and the tangent of the line defining the change in shear stress as the consolidating stress is increased (**Figure 25.13** and **Figure 25.14**).

The Jenike Method exclusively uses an effective angle of internal friction, which is the angle from the horizontal of a line passing through the origin while remaining tangent to the Mohr's circle representing the consolidation condition (**Figure 25.15**).

Bulk materials do not conform to the stress strain relationships that metal, glass, and rigid plastic materials do. Bulk materials do not have a unique yield stress like steel or other materials, but rather have a yield surface. This surface is built up of

25

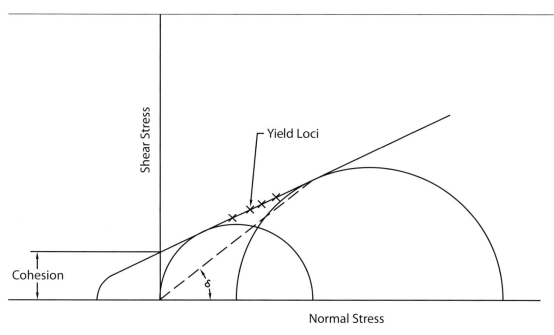

Figure 25.15

Effective Internal Friction Angle

δ = EFFECTIVE INTERNAL FRICTION ANGLE

yield points or loci. This yield locus increases in length as the consolidating stress increases at a fixed particle size and moisture content. Increasing the consolidating stress increases the bulk density, giving the graph a three-dimensional failure surface.

Interface Friction

Interface friction (Θ) for chutes handling bulk materials can be determined with a shear cell and a sample of the actual interface material—that is, the material that will be in contact with the bulk material (**Figure 25.16**).

The interface friction (sometimes referred to as the wall friction) is usually high at low consolidating pressures and reduces rapidly as the pressure increases (**Figure 25.17**). The significance of this in chute design relates to the depth of the bed of the material flowing in the chute. This property is especially critical when determining the slope needed for a chute to be self-cleaning when the flow stops and the depth of the bed of material approaches zero. In this situation, the resistance to the flow of the material off the chute is at a maximum.

Two values of friction (μ) are important in chute design: the coefficient of friction

between the bulk material and the chute-wall and the coefficient of friction between the bulk material and the belt. The coefficient of friction is equal to the tangent of the interface friction angle (determined the same way as the effective angle of internal friction) (**Figure 25.18**). Bulk materials, particularly the fines, have the ability to

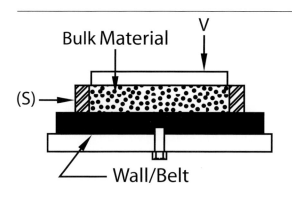

Figure 25.17

Effect of Consolidating Pressure on Wall Friction

25

cling upside down on horizontal surfaces and thus exhibit strength even under negative consolidating forces greater than that of gravity. The shear force at negative consolidating forces is of particular interest in chute design in determining adhesion and cohesion values. An adaptation of the shear cell to apply negative consolidating forces can be used, or the wall yield locus can be extrapolated to estimate these values (**Equation 25.1**).

Cohesion

Cohesion (τ) is the resistance of the bulk material to shear at zero compressive normal stress. Cohesion can be thought of as the ability of the particles to stick to each other. Moisture content (surface tension), electrostatic attraction, and agglomera-

tion are the three principle conditions that affect the level of cohesive stress in a bulk material. Cohesive stress increases as moisture is added to the bulk material until a maximum value is reached (**Figure 25.19**). As more moisture is added, the ability of the bulk material to withstand shear—its cohesion—begins to decrease. Cohesive stress can be determined from shear cell tests. It is given by the relationship that cohesive stress is equal to the consolidating stress times the tangent of the internal friction angle plus constant [$\tau = \sigma_c \tan \Phi + k$].

Adhesion

Adhesion (σ) is the resistance of the bulk material to movement at zero shear stress. Adhesion can be thought of as the stickiness of the material to surfaces, such as chutes and belts. Surface condition, moisture, and impurities such as clay are the three principle conditions that affect the level of adhesive stress in a bulk material. Adhesive stress can be determined from shear cell tests and is very useful in determining the likelihood that a material will stick or cling to surfaces.

Figure 25.18

Coefficient of friction is the slope or tangent of the wall friction angle.

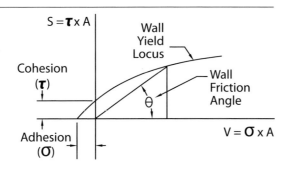

Equation 25.1

Interface Friction Relationship

$$\tan \Theta = \frac{S}{V} = \frac{\tau}{\sigma} = \mu$$

Figure 25.19

Testing has shown that cohesion and adhesion increase as moisture increases until enough moisture has been applied to begin fluidizing the material and reducing the cohesion. The exact variation in adhesion and cohesion with moisture content will vary from material to material and from site to site. Note: Moisture content is the % of weight loss between the wet material and the material after it has dried.

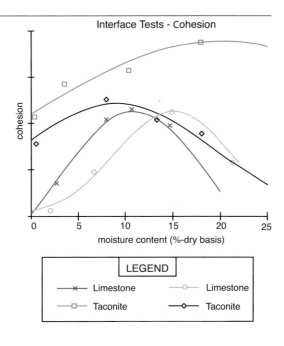

TYPICAL APPLICATIONS OF BULK-MATERIAL PROPERTIES

Conveyor Capacity

Conveyor capacity, usually expressed as tons per hour (st/h), is one of the basic design parameters directly calculated from the density—kilograms per cubic meter (lb_m/ft^3)—of the bulk material.

Density is a familiar term used when referring to material like steel or concrete; this is called the particle density. In conveyor design, both loose and vibrated densities must be considered. A material's bulk density changes from its stockpile state, both through transfer points and while it is conveyed. If a transfer point is designed using the vibrated bulk-density value, it is likely the chute will plug at less than rated capacity. As the bulk material falls, air is induced, increasing the material's volume;

there is just too little space for the loose material to move at the full flow rate. The loose bulk density can be as little as half of the vibrated bulk density.

If the designer looks in a general engineering handbook for a material's density, the value listed will most likely be particle density, which can be compared to vibrated density. If this value is used, it will lead to an undersized design by a factor of 2 to 4 times. The designer needs to be aware of changes in densities and design accordingly.

Chute Design

Chute design is more than a matter of having the correct cross-sectional area based on the loose bulk density. The reliable flow of bulk materials through a chute depends, among other factors, upon the friction between the bulk material and the chutewalls and wear liners. The design of curved chutes for reliable flow depends upon knowing the properties of the bulk material in relation to the flow surfaces. When a typical value of chute angles is used, the result is often material buildups, leading to chute blockage. For example, lignite has a significantly higher coefficient of friction on stainless steel than bituminous coal, but the coefficients of friction are similar when Ultra-High Molecular Weight (UHMW) polyethylene is the liner. Serious flow problems can result from not testing the actual bulk material and liners being considered for use in the design.

One significant property of bulk materials that is not normally considered in chute design is the effect of time and consolidating pressure on the strength of the material. Bulk materials generally gain strength in storage. However, in chute design, the consolidating pressures are usually low, and the time spent in the chute should be minimal. The effect of changes in moisture content is significant, however, especially in regard to the accumulations of material in chutes. One result of this phenomenon is that carryback usually gains strength as it dries on belts or dribble chutes.

Belt Cleaning

Adhesion and cohesion are important properties used to predict the nature of the challenges in belt cleaning. Knowing how the strength of material is affected by changes in moisture content gives guidance in the use of water to weaken the bulk material so carryback can be efficiently cleaned from the belt. Knowing the critical moisture content allows a designer to calculate the volume of water needed. Absent this knowledge, the point of view that "adding water to the process is a bad thing" will remain, even though there may be significant advantages to using it.

Designing a functional belt-washing system requires knowledge of how the bulk material behaves in water. The transport and treatment of the effluent is directly related to the rate at which the material settles in water. The size of the separation tank or settling pond is directly related to the material's sedimentation rate. Heavy materials such as iron ore need a lot of flowing water to keep wash boxes and piping from plugging. Other materials that will not settle, such as some coals, may not be suitable for a belt-washing station.

Dust Suppression

Selection of a dust-suppression method requires knowledge of how the material will react with water and with the various chemicals used to improve the wetting and agglomeration of the particles. Some bulk materials do not react—or react too slowly—to be good candidates for water alone as the suppression agent. Testing must be done to determine if chemical additives need to be used to provide effective dust suppression.

TYPICAL SPECIFICATIONS

When testing bulk materials for design of material-handling systems, the bulk material shall be tested for the range of conditions that are anticipated to occur during normal and extreme operating conditions and for all variations expected in material source, quality, and properties. These tests would include:

A. Particle size

A sieve analysis shall be performed for all expected qualities and variations of the bulk material in accordance with test methods described in ASTM D6393-99(2006) *(Reference 25.3)* or CEMA STANDARD 550-2003.

B. Density

The bulk density of the material shall be determined at three different consolidating pressures—representing the loose, average, and maximum expected bulk densities—in accordance with test methods described in ASTM D6393-99(2006) *(Reference 25.3)* or CEMA STANDARD 550-2003.

C. Angles of repose and surcharge

The angles of repose and surcharge shall be determined for the bulk material in accordance with test methods described in ASTM D6393-99(2006) *(Reference 25.3)* or CEMA STANDARD550-2003.

D. Strength of the material

The adhesion and cohesion values shall be determined at a minimum of three different moisture contents by testing at minimum, average, and saturation moisture levels and at each of three different consolidating pressures—zero, average, and maximum pressure—in accordance with test method ASTM D6128-06 *(Reference 25.5)* or ASTM D6773-02 *(Reference 25.6)*.

E. Interface friction values

The interface friction values shall be determined for the bulk material and the chutewall and wear-liner material(s) at a minimum of three different moisture levels and three different consolidating pressures, in accordance with test method ASTM D6128-06 *(Reference 25.5)* or ASTM D6773-02 *(Reference 25.6)*. The interface friction values for the bulk material and the belt shall be determined at a minimum of three different moisture levels and three different consolidating pressures in accordance with test method ASTM D6128-06 *(Reference 25.5)*.

SAFETY CONCERNS

Testing a bulk material's properties improves the ability of a designer to create safe methods for storage and conveyance. For example, it is well known that flowing bulk materials can create unequal wall pressures on silos. Without testing the specific materials under the expected conditions for storage, a designer can only guess at the forces involved. Many examples of failure in storage vessels demonstrate the wisdom of testing the material and of using the structure for only the specified materials. Less catastrophic, but just as damaging to productivity, are systems that fail to deliver on design capacity due to the use of typical or "handbook-published" values for a material's bulk density.

Most bulk materials are inert. Generally, the testing of the properties of bulk materials is a relatively safe process if the procedures in the test standards are followed. Some materials will pose chemical, explosive, or health hazards. Material Data Safety Sheets are a good source of information about the safe handling of a particular material.

25

ADVANCED TOPICS

Belt Capacity with Different Coal Properties Example

The sixth edition of CEMA's *BELT CON-VEYORS for BULK MATERIALS* gives detailed equations for calculating the capacity of a conveyor based on the trough angle and the surcharge angle. The same formulas can be used with the angle of repose to determine the capacity of the bulk material in a loose state such as when the material is first transferred from one belt to another.

CEMA STANDARD 550-2003 lists nine different classifications for coal. The loose bulk densities listed for these different classifications run from 720 to 960 kilograms per cubic meter (45 to 60 lb_m/ft^3); the angles of repose vary from 27 to 40 degrees.

The angles of surcharge are typically 10 to 15 degrees less than the angles of repose. *(Note: CEMA offers only Imperial measurements; the metric measurements are conversions by Martin Engineering.)*

In this example, the design capacities are compared across the range of properties of nine different coals (**Equation 25.2**). This demonstrates how sensitive a conveyor or transfer-point design is to the properties of the bulk material. Example 1 analyzes the densest coal; Example 2 analyzes the coal with the least density.

For these examples, a comparison of the cross-sectional areas found using the values near the extremes of nine different coals demonstrates how sensitive a design will be to the properties of the bulk material.

$$Q = A \cdot \rho_{lb} \cdot S \cdot k$$

Equation 25.2

Calculating Belt Capacity with Different Coal Properties

Given #1: *A conveyor belt transporting coal with a density of 960 kilograms per cubic meter (60 lb_m/ft^3) is traveling 2,5 meters per second (500 ft/min). The coal has a surcharge angle of 30°.*
Find: *The capacity of the conveyor belt.*

	Variables	Metric Units	Imperial Units
Q	Belt Capacity	tons per hour	short tons per hour
A	Cross-Sectional Area of Load (per CEMA)	0,195 m²	2.1 ft²
ρ_{lb}	Loose Bulk Density	960 kg/m³	60 lb_m/ft^3
S	Conveyor Speed	2,5 m/s	500 ft/min
k	Conversion Factor	3,6	0.03

Metric: Q = 0,195 · 960 · 2,5 · 3,6 = 1685
Imperial: Q = 2.1 · 60 · 500 · 0.03 = 1890

Q	Belt Capacity	1685 t/h	1890 st/h

Given #2: *A conveyor belt transporting coal with a density of 720 kilograms per cubic meter (45 lb_m/ft^3) is traveling 2,5 meters per second (500 ft/min). The coal has a surcharge angle of 20°.*
Find: *The capacity of the conveyor belt.*

Q	Belt Capacity	tons per hour	short tons per hour
A	Cross-Sectional Area of Load (per CEMA)	0,168 m²	1.804 ft²
ρ_{lb}	Loose Bulk Density	720 kg/m³	45 lb_m/ft^3
S	Conveyor Speed	2,5 m/s	500 ft/min
k	Conversion Factor	3,6	0.03

Metric: Q = 0,168 · 720 · 2,5 · 3,6 = 1089
Imperial: Q = 1.804 · 45 · 500 · 0.03 = 1218

Q	Belt Capacity	1089 t/h	1218 st/h

Our examples assume:

- Loose Bulk Density: 720 to 960 kilograms per cubic meter (45 to 60 lb_m/ft^3)
- Angle of Repose: 27 to 45 degrees
- Angle of Surcharge: 20 to 30 degrees
- Belt Width: 1200 millimeters (48 in.)
- Trough Angle: 35 degrees
- Edge Distance: Standard CEMA edge distance
- Belt Speed: 2,5 meters per second (500 ft/min)

Analysis

If a conveyor were designed by using the published values for coal from a book rather than testing the actual coal, the design capacity could be off by more than 600 tons per hour. This discrepancy would have a major effect on the rest of the process and the desired outputs. This example shows that the actual material properties must be measured.

MATERIAL SCIENCE FOR IMPROVED DESIGN

In Closing...

No matter what the generic classification is, no two bulk materials are the same. Therefore, physical testing of the actual materials is of critical importance to the proper design of the systems that will handle the bulk materials. The typical costs to determine the flow properties required to properly design a chute are from $1,000 to $3,000 USD per sample per moisture level. The cost of this testing is a minor part of the overall cost of engineering and constructing a conveyor system. Having this basic data will be an important tool for future troubleshooting of the conveyor, as when processes or raw materials change.

Looking Ahead...

This chapter about Material Science, the fifth and final chapter in this section Leading-Edge Concepts, explained how to test properties of bulk materials to help design conveyor systems for total material control. The following chapter, Conveyor Accessibility, begins the new section Conveyor Maintenance.

REFERENCES

25.1 Density Standards: Aggregates– ASTM C29 / C29M-07, Crushed Bituminous Coal–ASTM D29-07, and Grains–U.S. Department of Agricultural Circular #921.

25.2 ASTM International. (2001). *Standard Test Method for Measuring Bulk Density Values of Powders and Other Bulk Solids*, ASTM D6683-01; Work Item: ASTM WK14951 – *Revision of D6683-01 Standard Test Method for Measuring Bulk Density Values of Powders and Other Bulk Solids*. West Conshohocken, Pennsylvania. Available online: http://www.astm.org

25.3 ASTM International. (2006). *Standard Test Method for Bulk Solids Characterization by Carr Indices*, ASTM D6393-99(2006). West Conshohocken, Pennsylvania. Available online: http://www.astm.org

25.4 ASTM International. (2001). *Standard Practice for Coagulation-Flocculation Jar Test of Water*, ASTM D2035-08. West Conshohocken, Pennsylvania. Available online: http://www.astm.org

25.5 ASTM International. (2006). *Standard Test Method for Shear Testing of Bulk Solids Using The Jenike Shear Cell*, ASTM D6128-06. West Conshohocken, Pennsylvania. Available online: http://www.astm.org

25.6 ASTM International. (2002). *Standard Shear Test Method for Bulk Solids Using Schulze Ring Shear Tester*. ASTM D6773-02; Work Item: ASTM WK19871 – *Revision of D6773-02 Standard Shear Test Method for Bulk Solids Using the Schulze Ring Shear Tester*. West Conshohocken, Pennsylvania. Available at: http://www.astm.org

25

25

SECTION 6
CONVEYOR MAINTENANCE

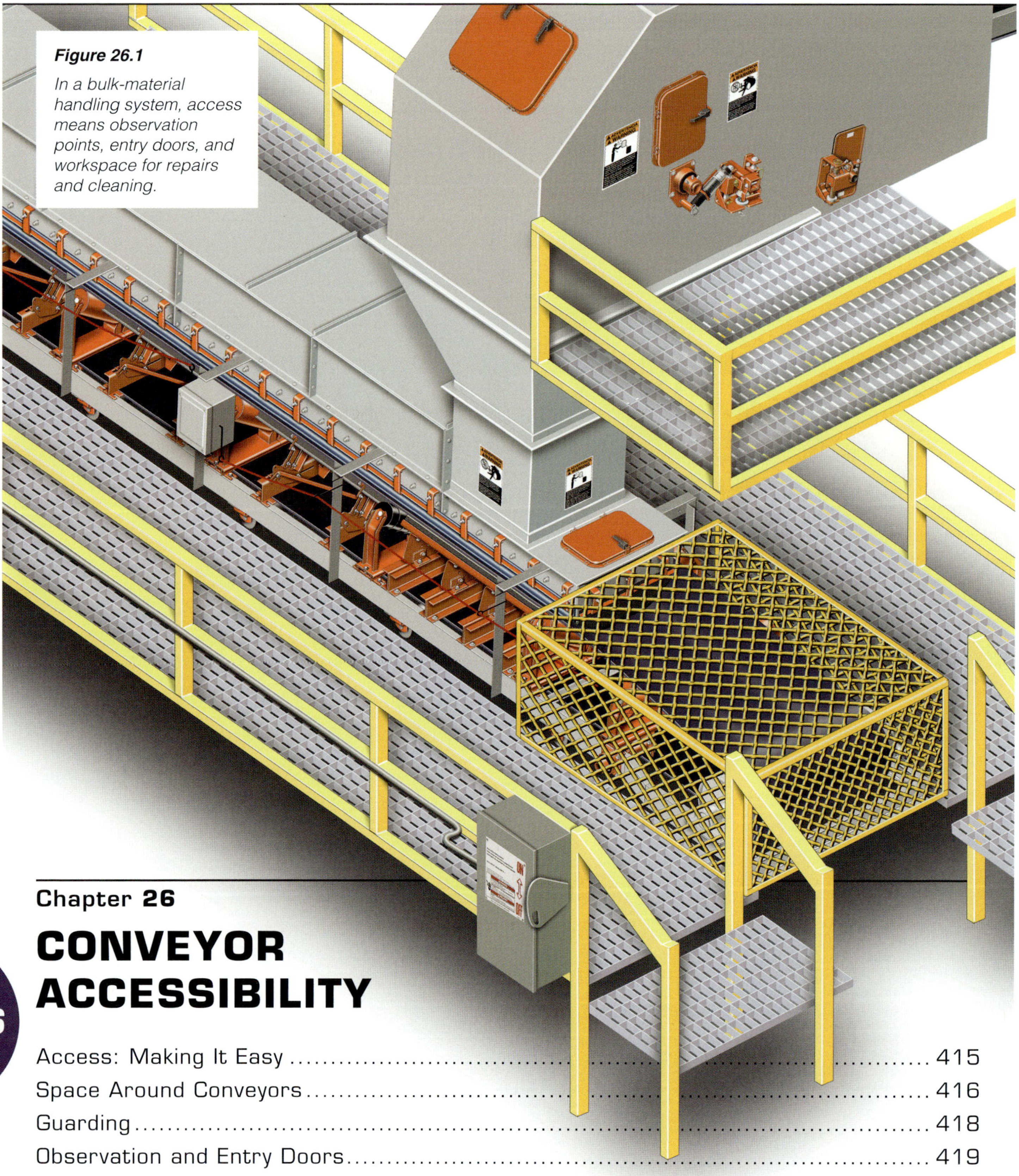

Figure 26.1

In a bulk-material handling system, access means observation points, entry doors, and workspace for repairs and cleaning.

Chapter 26

CONVEYOR ACCESSIBILITY

26

In this Chapter...

In this chapter, we discuss the importance of easy access to the conveyor system—to increase safety, ease maintenance, and save money. We focus on the amount of space needed around various parts of the conveyor system for different maintenance needs. We also look at requirements and different types of barrier guards and observation/ entry doors. Issues around confined space and safety concerns are also presented.

Access can be defined as the "right to enter or use." In a bulk-material handling system, access is used to mean observation points, entry doors, and workspace for repairs and cleaning (**Figure 26.1**).

To maintenance and operations personnel, proper access is critical to productivity. This means safe, quick, and easy access to a problem should outweigh other concerns, like cost. It has been estimated that providing proper access in the design of a bulk-materials handling system can account for as much as 15 percent of the capital cost of a project.

Yet, when a conveyor system is being designed, there is rarely enough money allocated to do more than provide the minimum access required by code. This practice results not only in lost production time and in increased time required for maintenance, but also in increased safety and health costs. From the perspectives of ownership and management, inadequate access contributes to continuing problems with lost productivity and unnecessarily high maintenance costs. Lack of proper access leads to poor maintenance practices; poor maintenance often leads to emergency outages, which in turn affect the operation's productivity and profitability.

Of course, adding proper access later— after the materials-handling system has been completed and the access mechanisms have been found wanting—will cost substantially more.

Insufficient access to equipment results in lost productivity and dirty systems, due to the difficulty of cleaning and making required repairs. It has been estimated that poor access could add as much as 65 percent to the maintenance and cleaning costs of a bulk-materials handling system over its lifetime.

ACCESS: MAKING IT EASY

It is frustrating to maintenance personnel when they cannot work on equipment— that would require minimal time to repair—because they cannot gain safe and proper access to it. Delays in access may be due to a requirement for a confined-space permit, for air testing, for scaffolding or man lifts, for cranes or hoists, or for special tools required to open access doors. In some cases, it is necessary to remove the entire system just to gain access to the component requiring service attention. These delays can be mitigated through the design of proper access and by staging tools and parts close to the required location.

Three "Easy's" should be included when designing proper access into a materials-handling system:

A. Easy to see

   If equipment develops a problem that cannot be seen by plant personnel, the problem tends to grow unseen into a catastrophic situation.

B. Easy to reach

   If a piece of equipment develops a problem, but the equipment is difficult for maintenance personnel to reach, repair is likely to be postponed, again risking a catastrophic situation.

C. Easy to replace

   If an equipment problem is known, but unnecessarily requires an outage to correct, the broken equipment is likely to remain out of order for an extended period.

When systems are too difficult to see, reach, and replace, plant operations or maintenance personnel may attempt short-

26

cuts during repairs. Such shortcuts often increase risks to safety, as well as add the potential for additional damage to equipment. Taking shortcuts—whether intentionally or because of the lack of proper access and, therefore, the inability to follow proper maintenance procedures—can easily result in reduced safety, shorter equipment life, reduced process efficiency, and an increase in the emission of fugitive materials.

SPACE AROUND CONVEYORS

In order to save costs, conveyor equipment is often placed in small galleries or enclosures. (**Figure 26.2**). One side of the conveyor is typically butted against a wall, an adjacent conveyor, or other equipment. It is extremely difficult to service this type of installation. If the conveyor is installed flush against a wall, vessel, or other structure, basic service requirements, such as bearing lubrication or idler replacement,

become major operations requiring extended production outages.

Generally, there must be sufficient room to allow access to all sections of the conveyor system and, in particular, both sides of the conveyor. Failure to provide access to both sides of the conveyor is a common deficiency in making conveyors maintainable.

The open space along the more critical side of a conveyor (the work area side where major tasks must be performed or where there is access to lifting equipment or other resources) should be at least the width of the belt (BW) plus 300 millimeters (12 in.), with a minimum of 1000 millimeters (36 in.). The walkway side of the conveyor should have space equal to at least one-half the width of the belt (BW/2), with a minimum space of 750 millimeters (30 in.) along the entire length. This dual access facilitates replacement of equipment such as idlers that cannot be easily handled by one worker. If a pulley or other large object needs replacement, then the open space should be expanded to at least the width of the belt (BW) plus 300 millimeters (12 in.).

Chapter 2: "Design Considerations" in the sixth edition of Conveyor Equipment Manufacturers Association's (CEMA) *BELT CONVEYORS for BULK MATERIALS* includes detailed recommendations for minimum clearances.

Walkways and Work Spaces

Proper access requires the provision of walkways and work platforms beside conveyors. These should provide a firm path adjacent to the conveyor and around head and tail pulleys, with easy access to all points where observation, lubrication, cleanup, or other maintenance chores are required.

As noted above, walkways should be a minimum of 750 millimeters (30 in.) wide for passage and 1000 millimeters (36 in.) wide in areas where major service work must be performed (**Figure 26.3**). Both areas should have ample headroom; anywhere a person must stand or kneel to

Figure 26.2

Conveyor equipment is often placed in small galleries or enclosures, butted against a wall, an adjacent conveyor, or other equipment.

Figure 26.3

Access space should be a minimum of 750 millimeters (30 in.) wide for the walking passage (B) and 1000 millimeters (36 in.) wide in areas where service work must be performed (D).

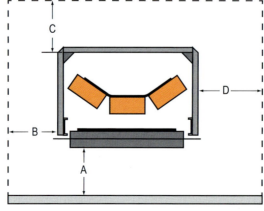

A. To Allow Cleaning	600 millimeters (24 in.)
B. Walkway	750 millimeters (30 in.)
C. Headroom	1200 millimeters (48 in.)
D. Work Area	1000 millimeters (36 in.)

26

perform service or inspection, the "overhead" or "head room" should be at least 1200 millimeters (48 in.) (**Figure 26.4**). Areas where frequent service or cleaning is required should have solid flooring, rather than an open grate.

When conveyors run parallel to each other, the space between them should be a minimum of 750 millimeters (30 in.) or the width of the belt, whichever is greater, to allow for belt repair and removal of idlers.

Another common deficiency in conveyor design is failure to allow adequate space for cleanup. A study of conveyor-related accidents in mining showed that one-third of all accidents occurred to workers trying to clean under, or around, the carrying run of the conveyor. Areas that require frequent cleanup should allow for mechanical cleaning, such as the use of a skid-steer loader or vacuum truck under the conveyor. If this is not practical, a minimum clearance of 600 millimeters (24 in.) between the bottom of the return rollers and the floor should be provided.

Access Requirements Around Equipment

The minimum access around equipment is the area necessary to accommodate the largest piece of equipment. This is determined by measuring the largest item and adding 450 to 600 millimeters (18 to 24 in.). There should also be access on both sides of the structure, with a minimum of 1000 millimeters (36 in.) on the second (non-critical) side, with a clear path to a lifting area, or overhead trolley system, for removal of the equipment. A convenient staging area for large replacement parts is a good idea.

Access for Belt Repair and Replacement

Access for belt maintenance requires an area convenient for lifting or lowering the vulcanizing equipment and exposing the conveyor belt itself. Removal of conveyor covers or load-zone skirtboards and chutewall will add significant time to the process. At least 1000 millimeters (36 in.) plus the width of the belt is required on each side of the conveyor. In addition, an area 3 meters (10 ft) long where the belt is exposed (that is, not enclosed) is required to facilitate repair operations.

Access at the Take-Up

Maintenance and repair of a gravity take-up system may be dangerous and time-consuming. Two chain fall hoists of sufficient capacity to raise and lower the counterweight are typically required. Because many of these take-up systems are close to the head pulley, the maintenance access area on inclined conveyors may be elevated. Access platforms that provide adequate space for maintenance of bearings and pulleys and for rigging chain fall hoists are essential. A lifting mechanism that can remove the pressure of the gravity take-up counterweight from the belt might save many man-hours during conveyor repair.

Crossovers

A surprising number of accidents happen to personnel who are crossing over or under the conveyor. To eliminate this concern, access platforms, ladders, crossovers, and other equipment necessary to reach the far side of the conveyor should be included in the system's design and specifications. Although written for unit conveyors, CEMA SPB-001 (2004) *Safety Best Practices Recommendation (Reference 26.3)* provides useful guidelines for the design of conveyor crossovers for bulk-materials handling conveyors. As always, any local regulations must take precedence.

26

Figure 26.4

Anywhere a person must stand or kneel to perform service or inspection, the "head room" should be at least 1200 millimeters (48 in.).

Encroachment of Other Systems

It is not uncommon to see a conveyor or transfer point captured in a web of electrical conduit, dust-suppression piping, control panels, or sprinkler systems (**Figure 26.5**). Any attempt to reach the components of the transfer point must first bypass the "thicket" surrounding the transfer point. The interruption of these other systems results in a variety of other complications to the operations of the plant.

To control the growth of these auxiliary systems around the conveyor and transfer point, the designer should specify the equipment to which access is necessary. By including specific areas in the conveyor plans for the installation of control panels, gate actuators, and other equipment, unnecessary obstacles can be avoided.

GUARDING

Barrier Guards

Simply walking up the walkway offers risks that could lead to fatality: Walkways close to the conveyor provide areas where employees can become entrapped in the conveyor. There have been instances in which employees slipped on a walkway and then became entangled in the conveyor, simply by trying to break the fall or catch their balance.

While it may not be a government requirement, the belts should be guarded along walkways to prevent unwary employees from becoming trapped in the moving conveyor.

A barrier guard is a fence, or other obstacle, to keep personnel away from a conveyor and its components (**Figure 26.6**). Barrier guards, used extensively in Australia and Europe, are designed to prevent injury by making it physically impossible to reach a potentially dangerous piece of equipment—such as the pinch points of pulleys, idlers, sprockets, chains, or belts—or the suspended-load hazard of a take-up counterweight.

It is important that guards be provided at every dangerous part of a conveyor normally accessible to personnel. Hazards are considered to be "guarded by location" if they are located a sufficient distance above the ground or walkway to prevent contact.

The challenge to designing and installing effective guarding is to protect operations and maintenance personnel without interfering with the operation of the plant. A well-designed barrier guard will prevent personnel from reaching moving machine parts, yet its size and shape will permit the safe handling of the parts during removal or replacement (when the guard has been removed or disabled and the conveyor locked out / tagged out / blocked out / tested out.)

Typically, barrier guards require a tool for removal and may include a sign indicating which equipment should be isolated (that is, locked out / tagged out / blocked out / tested out) before the guard can be removed. Visual inspection of conveyor components and access to lubrication points should be possible without removal of the barrier guards (**Figure 26.7**).

26

Figure 26.5

It is not uncommon to see a conveyor or transfer point captured in a web of electrical conduit, dust-suppression piping, control panels, or sprinkler systems.

Figure 26.6

A barrier guard is a fence, or other obstacle, to keep personnel away from a conveyor and its components.

Barrier guards may be hinged, allowing them to be moved out of position, or they may be bolted in place (**Figure 26.8**). They should be constructed of materials that are both wear- and corrosion-resistant, and they should be able to withstand the vibration of system operation.

New Technology in Guarding

In many facilities, conventional physical barriers are being replaced with newer technologies that monitor area perimeters and detect entry to hazardous areas.

However, these newer technologies, such as laser beams and infrared light curtains, are effective only when power is on, and they can be fooled by someone determined to gain entrance to a secure area. These systems require care and good maintenance in order to be reliable. For example, they might not function well where airborne dust can settle onto the electric eye receivers. It is wise to have mechanical access-control systems as redundant or back-up systems, as a failure of the "high-tech" devices may not be obvious.

OBSERVATION AND ENTRY DOORS

Observation Requirements

Access systems, including doors and work platforms, should be installed to make it easy to reach and observe equipment. Flow problems within chutes may be more easily solved if the material path can be observed. The actual path of material within a chute cannot always be predicted, so observation is necessary to allow adjustment of diverters, gates, and grizzly bars (**Figure 26.9**). Many transfer chutes have only one inspection door. This is usually installed near the head pulley, where it does not permit a view of the actual material path in the lower chute and skirted area, where problems often occur (**Figure 26.10**).

The chute should incorporate observation openings with easy-to-operate covers located away from the material path. These openings should allow safe observation

Figure 26.7

Visual inspection of conveyor components and access to lubrication points should be possible without removal of the barrier guards.

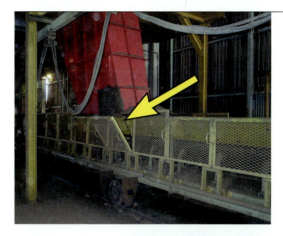

Figure 26.8

Barrier guards may be hinged, allowing them to be moved out of position, or they may be bolted in place.

Figure 26.9

Flow problems within chutes may be more easily solved if the material path can be observed through an inspection door.

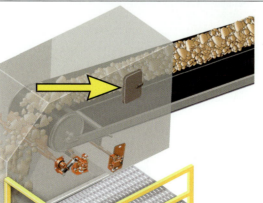

Figure 26.10

A poorly placed door will not allow observation of the path of material flow.

26

of both the material flow and component wear at critical areas of the installation. The openings should be limited in size and/or protected with fixed bars or screens to prevent personnel from reaching in or material from flying out.

Consideration must also be given to providing ample lighting at access points. In some cases, overhead lighting is sufficient, but in other applications, it may be necessary to provide high-power spotlights or strobes that can be aimed into the chute to observe material movement. For general maintenance and machine repair, light of 540 to 1080 lux (50 to 100 foot-candles) is recommended. As many bulk materials absorb light, and due to the distance to the problem being observed, it is common for an installation to need lighting rated at 10 million lux (900000 foot-candles) to achieve this level of illumination.

Doors

Inspection doors should be side opening and sized so personnel can easily and safely view the components inside the structure. Doors must be installed on the non-wearing side(s) of the chute (that is, the side(s) away from the flow of impacting, or abrasive, material).

Doors should be designed for easy operation in tight clearances, with corrosion-resistant hinges and latching systems. It is important that all ports be dust-tight through use of a securely sealing door. Hinged metal doors with easy-opening latches are now available to provide access

Figure 26.11

Hinged metal doors with easy-opening latches are available to provide access.

(**Figure 26.11**). Flexible rubber "snap-on" doors provide a dust-tight closure while allowing simple, no-tool opening and closing, even in locations with limited clearances.

Doors should be easy to close securely, once a service procedure is finished.

Door sizes should be large enough to provide the required access. If the observation and service requirements are limited to systems such as belt cleaners, a door that is 225 by 300 millimeters (9 by 12 in.) or 300 by 350 millimeters (12 by 14 in.) is usually sufficient. If service to major components, such as chute liners, will be necessary—or if personnel will need to use the door as an entry into the structure—then door sizes of 450 by 600 millimeters (18 by 24 in.), 600 by 600 millimeters (24 by 24 in.), or larger will be necessary.

Poorly-designed doors may have hinges and latches that are difficult to operate and, as a result, impede access. In addition, door seals unable to withstand abrasion and abuse from materials and implements used to reach through the doors become sources of dust. Some access doors also have small ledges or flat shelf areas, where combustibles such as coal might accumulate, creating risk of fire and explosion.

CONFINED SPACE

Any discussion of access to equipment, whether for routine maintenance or emergency repair, should include the topic of confined space. The U.S. Department of Labor Occupational Safety & Health Administration (OSHA) defines "confined space" as an area that:

A. Is large enough and configured so that an employee can enter and perform assigned work

B. Has limited or restricted means for entry or exit

C. Is not designed for continuous employee occupancy

26

"Permit-required confined space" (shortened to "permit space" in common use) means a confined space that has one or more of the following characteristics:

A. May contain a hazardous atmosphere

B. Contains a material that has the potential for engulfing an entrant

C. Has an internal configuration that could trap or asphyxiate an entrant, with inwardly-converging walls or a floor that slopes downward and tapers to a smaller cross-section

D. Contains any other recognized serious safety or health hazard

"Non-permit confined space" means a confined space that does not contain— or, with respect to atmospheric hazards, have the potential to contain—any hazard capable of causing death or serious physical harm.

"Permit-required confined spaces" require cumbersome and costly safety procedures, including personnel training, safety harness and rigging, and added personnel for a "buddy system." Consequently, designing systems to minimize permit-required confined spaces may provide a significant return on investment: When maintenance and repair work can be done without requiring permits or specially trained personnel, the labor expense associated with such tasks is minimized.

Designs that incorporate the following features of "non-permit confined space" are cost effective over time:

A. Easy access and sufficient access for entering and exiting the enclosure

B. Natural ventilation of the internal work areas

C. Materials that do not create a hazardous atmosphere

SAFETY CONCERNS

The obvious benefit of both access doors and barrier guards is their ability to prevent access to moving parts, regardless of whether or not the power is on or the controls are working properly. These guards are easy to see and can be made impossible to bypass without suitable tools.

It is important that plant personnel observe all appropriate shutdown and lockout procedures when opening access doors for observation or entry. It is also essential that access doors and covers be closed following use, to avoid the escape of material and the risk of injury to unsuspecting personnel.

In the sixth edition of *BELT CONVEYORS for BULK MATERIALS*, CEMA recommends that "warning labels are to be placed in a conspicuous location near or on the access door that warn of potential

hazards that may be encountered when the door is opened" (**Figure 26.12**).

It makes sense to locate the attachment points for safety harnesses in line with the access doors.

Enclosed areas that have flow-aid devices, such as air cannons, must be locked—with correct signage in place— to prevent injury to employees who would open the door for inspection.

Figure 26.12

CEMA recommends that "warning labels are to be placed in a conspicuous location near or on the access door that warn of potential hazards that may be encountered when the door is opened."

26

Figure 26.13

Another approach is to design "non-confined-space chute access" that allows whole sections of chutes to be easily opened.

Another approach is to design "non-confined-space chute access" that allows whole sections of chutes—especially those portions that see continuous abrasive wear—to be easily opened (**Figure 26.13**). This type of design allows repairs to be accomplished without requiring workers to be enclosed inside the chute. The same type of access can be designed for feeders, gates, silos, or bunkers.

TYPICAL SPECIFICATIONS

A. Walkways

To facilitate inspection and service, the conveyor will be fitted with walkways on both sides. These walkways will provide sufficient space for the required maintenance activities at points where service work will need to be performed.

B. Guards

Rolling components and pinch points will be fitted with effective guards to prevent the encroachment of personnel while the conveyor is running. Guards should be locked securely in place, but

they should be removable to permit service activities.

C. Access

Suitable access for viewing and service will be provided to the conveyor, the transfer point, and other related systems. Sufficient openings will be provided to allow inspection of all interior areas.

D. Viewing ports

Doors will be positioned out of the path of the material stream. These viewing ports will be fitted with dust-tight doors. The doors will be provided with screens to prevent material from being cast out of the opening if observation is to be made while the conveyor is operating.

THE BENEFIT OF ACCESS

In Closing...

Well-designed conveyor access need not be a trade-off between safety, accessibility, and cost. Access to equipment for maintenance and repair is essential for a clean, safe, and productive system. Safe access that is carefully located and adequately sized will increase dependability, reduce downtime and manhours required for maintenance, and minimize hazards such as dust and confined-space entry. Over time, well-designed access improves safety and saves money.

Looking Ahead...

This chapter regarding Conveyor Accessibility is the first chapter in the section Conveyor Maintenance. The following chapter, Conveyor System Survey, continues the discussion about the type of maintenance needed to extend the life of the conveyor system and reduce dust and spillage.

26

REFERENCES

26.1 Mine Safety and Health Administration (MSHA). (2004). *MSHA's Guide to Equipment Guarding*. Other Training Material OT 3, 40 pages. U.S. Department of Labor. Available as a free download: http://www.msha.gov/s&hinfo/equipguarding2004.pdf

26.2 Giraud, Laurent; Schreiber, Luc; Massé, Serge; Turcot, André; and Dubé, Julie. (2007). *A User's Guide to Conveyor Belt Safety: Protection from Danger Zones*. Guide RG-490, 75 pages. Montréal, Quebec, Canada: IRSST (Institut de recherche Robert-Sauvé en santé et en sécurité du travail), CSST. Available in English and French as a free downloadable PDF: http://www.irsst.qc.ca/files/documents/PubIRSST/RG-490.pdf or html: http://www.irsst.qc.ca/en/_publicationirsst_100257.html

26.3 Conveyor Equipment Manufacturers Association (CEMA). (2004). CEMA SPB-001 (2004) *Safety Best Practices Recommendation: Design and Safe Application of Conveyor Crossovers for Unit Handling Conveyors*. Naples, Florida.

26.4 Conveyor Equipment Manufacturers Association (CEMA). (2005). *BELT CONVEYORS for BULK MATERIALS, Sixth Edition*. Naples, Florida.

26

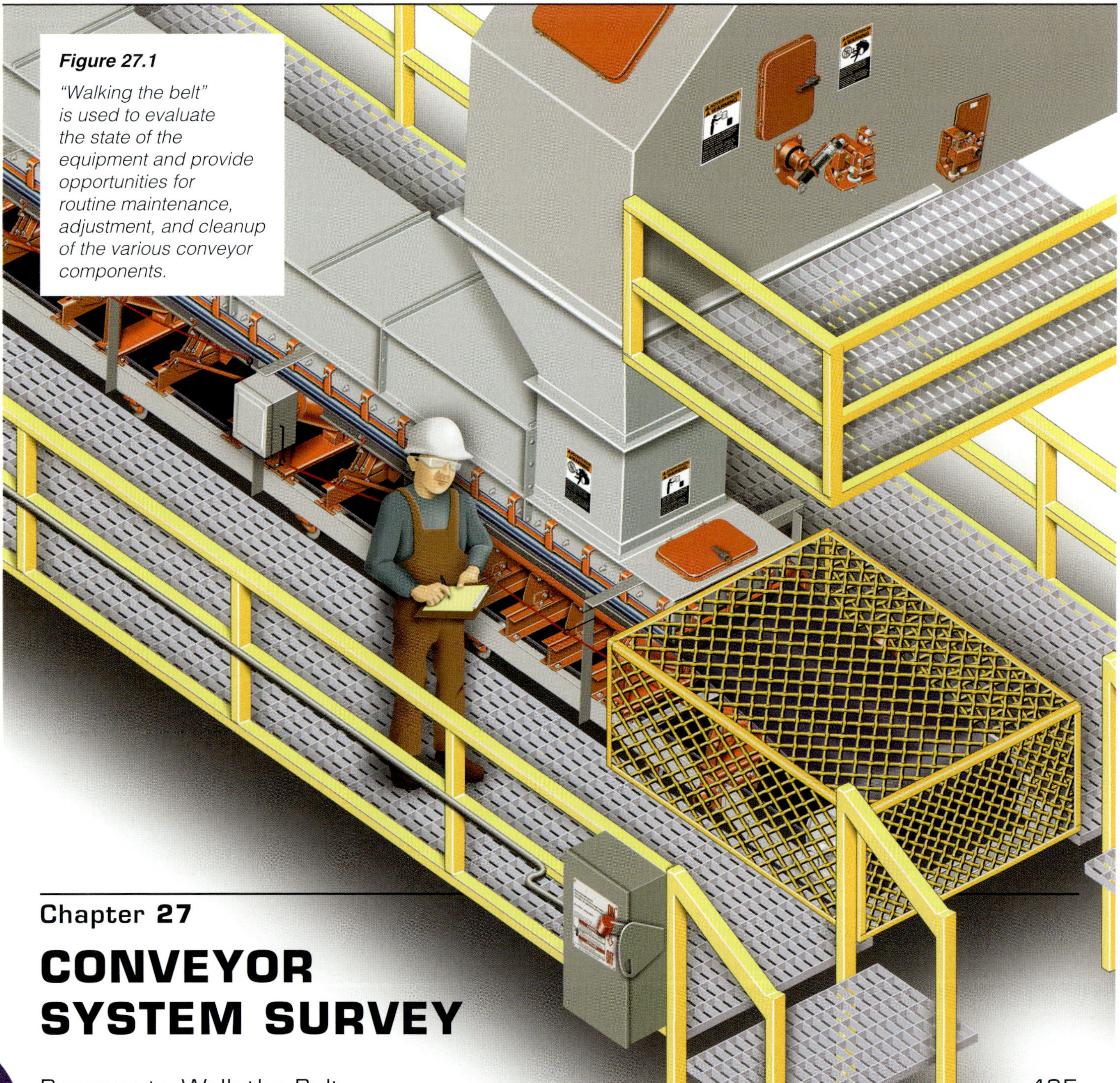

Figure 27.1

"Walking the belt" is used to evaluate the state of the equipment and provide opportunities for routine maintenance, adjustment, and cleanup of the various conveyor components.

Chapter **27**

CONVEYOR SYSTEM SURVEY

27

In this Chapter...

In this chapter, we discuss the importance of routine "walking" inspections of a plant's conveyor(s) and the different requirements of three types of "conveyor walk." They are system census, maintenance inspection, and site survey. Also included are considerations for safety as well as benefits of using outside contractors for these conveyor inspections.

"Walking the belt" is a phrase used in industry to describe a trip along the plant's conveyor system for the purpose of inspection and maintenance. The trip, starting at one end and continuing along the entire length of an individual conveyor, or from one end through the entire chain of conveyors, is used to evaluate the state of the equipment and provide opportunities for routine maintenance, adjustment, and cleanup of the various conveyor components (**Figure 27.1**).

Walk the Belt^{SM} is a term that Martin Engineering uses to describe an assessment of the condition of a belt conveyor system and its components.

REASONS TO WALK THE BELT

This chapter will discuss three distinct types of "conveyor walks." These different inspections can be combined, but the mission, frequency, and equipment required to conduct each type will be different. The three types are:

A. System census

The system census is often the initial inspection, in which the primary mission is fact-finding to capture the system's basic specifications and nameplate information. This provides an inventory of the equipment and components in the conveyor system.

B. Periodic maintenance inspection

On the periodic maintenance inspection, the daily operations of the conveyor are checked, and routine or simple maintenance, adjustment, and cleanup are performed.

C. Site survey

The site survey is a detailed inspection and performance analysis of some or all of the materials-handling system, often preparatory to projects for system improvement.

SYSTEM CENSUS

First Steps: Fact Finding

Too often, the details of a conveyor system—the specifications of components, even basic information about their make and model—are lost or forgotten over time. Plant ownership or management changes; operations, maintenance, and engineering personnel change responsibilities or leave the company. Even the components of a conveyor system will change from the original equipment, evolving as replacement parts or improvements are incorporated into the system. The "nameplate information" about various components will not have been kept up to date in the maintenance department, the engineering office, or the purchasing department. The file becomes outdated, or perhaps, worse, the information is retained only in the mind of one person.

Consequently, the capture of current information in a consistent format should be a principal requirement of a plant's first belt walk or initial conveyor survey. The initial belt walk becomes a fact-finding mission in addition to being a maintenance inspection. The collection of all pertinent information can make this walk a lengthy one, which often leads to a supplier, contractor, or consultant completing the survey. This "belt walker" is undertaking a census of the system.

Special equipment may be required to properly complete a site survey (**Table 27.1**). Having all the tools necessary to capture key data allows the survey to be completed without repeat trips to the site,

27

Table 27.1

Tools for the Site Survey	
Tool	**Function**
Personal Protective Equipment (PPE)	Provide for personal safety, as appropriate for locations to be surveyed. **ALWAYS USE**.
Anemometer	Capture air velocities used in dust suppression
Angle Finder	Document the incline of structures
Decibel Meter	Measure noise levels
Durometer Reader	Determine the surface hardness of conveyor belt and skirtboard seal
Infrared Thermometer	Find temperature of components
Level	Determine if components are installed level
Soap Stone	Temporarily mark steel components
Tachometer	Measure belt speed
Tape Measure	Measure distances
Flashlight	Allow inspection of dark or enclosed areas
Camera	Capture still and/or video images of conditions
Pencil & Paper	Record data
Tool Belt/Harness	Carry tools safely

Table 27.2

Conveyor System Census		
Performed By:		Date:
Basic Data	Conveyor ID *[name or number]*	
	Location	
	Length-Pulley Centers *[m (ft)]*	
	Conveyor Capacity *[tons/h (st/h)]*	
	Total Lift (Elevation Change) *[+/-m (ft)]*	
	Hours Conveyor Runs/Day *[hours]*	
	Days Conveyor Operates/Week *[days]*	
	Belt Width *[mm (in.)]*	
	Belt Speed *[m/sec (ft/min)]*	
	Trough Angle *[degrees]*	
Belting	Belt Manufacturer & Type	
	Belt Rating *[kN/mm (PIW)]*	
	Belt Installation *[date]*	
	Belt Thickness *[mm (in.)]*	
	Belt Splice Type *[mechanical or vulcanized]*	
	Number of Splices (Joints) in Belt	
Cargo	Material Conveyed	
	Material Size (Maximum Lump) *[mm (in.)]*	
	Material Temperature *[Celsius (F°)]*	
	Material Drop Height *[m (ft)]*	
	Moisture Content (Maximum) *[%]*	
	Moisture Content (Normal) *[%]*	
Conveyor Drive	Manufacturer & Model	
	Power Output *[kW (hp)]*	

27

looking for "just one more" dimension or serial number.

Recording the Information

During a conveyor system census, a variety of manufacturer, model, and system specification information is collected and recorded (**Table 27.2**). Once this information is collected, it should be archived in a single, central location in the plant maintenance file.

It makes sense to locate this information, complete with the various documents, in one central library of material related to the belt conveyor system. This would include maintenance manuals, service procedures, parts lists, operating instructions, supplier information, and drawings. This captured information can be transferred into a plant's computerized maintenance-management system (CMMS).

Some specialized computerized systems allow the incorporation of materials such as engineering drawings, schematics, parts lists, operator manuals, and other documents into a digital asset library. This system becomes a centralized library for personnel, both plant and contractor, to quickly access information and shorten the time required performing maintenance tasks.

Some maintenance contractors and suppliers offer data collection as a specialized service. This service can be performed as part of a regular assessment of a plant's materials-handling system, as part of a contract maintenance service, as a stand-alone service, or as a part of a CMMS selection and implementation process.

It is important to keep the data updated; it is also a good practice to retain original

Conveyor System Census (continued)		
Idlers	Make, Model, & CEMA Class	
	Carrying Idlers	
	Impact Idlers	
	Impact Cradles	
	Belt-Support Cradles	
	Return Idlers	
Airflow	Loading-Zone Exit Area [m² (ft²)]	
	Max Loading-Zone Exit Air Speed [m/s (ft/min)]	
Components	Make & Model	
	Sampler	
	Belt Scale	
	Magnet	
	Rip-Detection System	
	Pull-Cord Safety Switches	
	Misalignment Switches	
	Carrying-Side Training Devices	
	Return-Side Training Devices	
	Primary Belt-Cleaning Systems	
	Secondary Belt-Cleaning Systems	
	Tail-Protection Plows	
	Dust Collectors	
	Dust-Suppression Systems	
	Chute Flow-Aid Devices	
	Access Doors	

Table 27.2
Continued

27

design information and any change records. Best practice is to regularly update the data with changes as part of a regular inspection cycle.

MAINTENANCE INSPECTION

"Walking the Belt" for Routine Inspection

In its most typical usage, "walking the belt" is a routine inspection and service opportunity of the conveyor(s) in a given operation. The belt inspector—the plant's belt person, or "belt boss"—walks the system: inspecting its operation, performing minor adjustments or cleaning activities, and noting more significant conditions or problems for later attention.

The belt walker should take notes of the problems observed: "The center roller on idler number 127 on Conveyor B is not turning," for example, or "There is a lot of spillage at the tail of Conveyor 3." It is better to make notes as the problem is seen, rather than waiting until the inspector is finished with a particular conveyor, or even when back in the maintenance room.

Information can be recorded on a pad of paper or on a personal digital assistant (PDA) or "smart phone." A conventional cell phone can be used to call the conveyor inspector's own voice mail to record things noticed on the walk, especially those things that might be forgotten by the time the inspector returns to the office to be bombarded with other daily details. A digital camera—or even many cell phones— will allow the inspector to take photos of problems. These photos will let the belt inspector send the images of the problem to others for evaluation.

Standing in One Place

Standing still while walking the conveyor is not the oxymoron it might seem. It is important for the conveyor inspector to observe the running of the belt as it completes at least one revolution of the structure. This will let the inspector study the condi-

tion of the belt—the edges, the splice(s), the top cover where it passes under the skirtboard, the tracking—to note any problems. The belt surveyor can also check to see if the belt is centered—when it is both loaded and unloaded.

When to Stop the Conveyor—NOW!

There may come a time when the inspector sees something that poses so much risk to the belt, to the conveyor, or to plant personnel that the belt must be stopped immediately. The problem might be a piece of tramp iron or sharp lump of material lodged into the structure where it could slash the belt. Or it might be an overheating idler, smoldering material buildup, or other condition that could lead to a fire. The belt walker—and the rest of the plant's operating and maintenance crew and management staff—must understand that the essential mission is to ensure safety and preserve the equipment, even at the expense of an unscheduled outage.

What to Look For

It is extremely important that the individual performing the survey has a list of the general components that should be checked during every "walking the belt" conveyor inspection. This guide can be used as a checklist for the worker to complete the inspection. *(See Chapter 28: Maintenance for the list of items to be considered and evaluated.)*

The maintenance recommendations of the manufacturer for various components and subsystems (such as samplers, scales, metal detectors, and magnets) should be added to each plant's specific checklist.

SITE SURVEY

The site survey is a third type of (or reason for) a conveyor walk. The purpose of this inspection is to determine issues that interfere with a conveyor's safe and efficient operation, verify the conveyor's inventory data, and prepare for projects that will enhance performance. The details of this full-blown system analysis may interfere with,

27

or delay, information gathering and routine maintenance activities, so combining this survey with census or maintenance inspections will depend on the time available and is not normally recommended.

In addition to an assessment of the general condition of an existing system and the feasibility of any proposed improvements, a site survey should include a number of other, project-related considerations (**Table 27. 3**). When performing a site survey, attention must be paid to the overall system, the system's structure, and any related equipment that may interfere with a planned project or that may be affected by changes to the conveyor system.

In order to complete a proper site survey, the same tools used when conducting a system census would be used (**Table 27.1**).

CONVEYOR UP, OR CONVEYOR DOWN

The question of walking the belt when it is operating as opposed to walking when the conveyor is NOT running is worthy of some consideration. Walking the system when the belt is moving allows the personnel to see more of the real problems that affect conveyor performance and component life. Problems including material spillage and the influence of the material load on belt tracking are more apparent. If the conveyor is not moving, many key indicators are hidden, from vibrations in the structure and the belt's line of travel to the noises the splices make as they pass over an idler.

Obviously, conveyors are safer when they are not running, and it is safer to do

Site Survey Considerations		_Table 27.3_
Problem	What is the problem to be solved? What are the expectations to be considered a success?	
Hand Offs	Who is responsible for interface issues such as controls, power, or separate supply of key components?	
Unanticipated Results	What effects will the proposed solution have on other processes?	
Training	What general or site-specific training and documentation will be required for workers?	
Code Requirements	Do systems need upgrading to meet current codes or company policies?	
Access	How will new equipment be brought to the point of installation?	
Utilities	Are there utilities available for demolition and construction activities?	
Hazards	Are there hazardous materials or conditions that need special attention?	
Elevated Work	Will special procedures or safety precautions be required because of work at heights?	
Confined Spaces	Will work in confined spaces require specially-trained workers and permits?	
Waste Disposal	Who is responsible for waste disposal?	
Facilities	Are there facilities (e.g., break room, toilets, shower facilities) for project workers?	
Weather	Will weather and time of day (e.g., system available only on second shift) affect the ability to complete the project?	
Deliverables	What deliverable is expected from the survey?	

27

any corrective action on conveyors when they are not running. Some inspections, like inspecting the wear liner inside the skirtboard, cannot be performed safely while the belt is running. However, some inspections and adjustments can be made only with the belt running. The person assigned to walk the belt should be aware of the hazards, experienced with conveyor systems, and trained and authorized to make a limited amount of corrections on an operating conveyor.

So the decision as to whether to "walk the belts" when the conveyors are running or not may depend on external factors, including when manpower is available and the level of service-work expected or required. Either way, a significant amount of caution is advisable.

USING OUTSIDE RESOURCES

A Different Set of Eyes: Using Outside Contractors to Walk the Belt

It is becoming more difficult for regular plant employees—either production personnel or maintenance staff—to find the time for regular service activities like conveyor inspections and routine adjustments.

There can be other problems with "walking the belt"—for whatever purpose—with in-house personnel. A person who is familiar with an operation may be conditioned to see only what they expect to see, rather than what they should be looking for. A plant operations or maintenance worker who has lived with a problem might actually believe that that is standard operating

⚠ SAFETY CONCERNS

Conveyor surveys require workers to thoroughly inspect the bulk-materials handling system. This is good for the equipment, but it can be bad for the employees, as it places them in potentially hazardous situations during the inspection process. Personnel walking the belts must be properly trained in the correct procedures for working on and around the conveyor, and they must maintain a healthy respect for its potential to injure or kill an unwary individual.

The observation of safe work practices should be part of any walk-the-belt routine. The personnel performing the site survey must have the appropriate personal protective equipment (PPE). These may include hardhat, safety glasses, earplugs, and respirator. A means of communication—such as a radio or cell phone—should also be included. And, of course, the appropriate lockout / tagout / blockout / testout procedures are required.

It is wise to use a "buddy system" on a conveyor walk. That will allow one worker to go up each side of the conveyor (assuming there are two walkways) for inspection, in addition to making the inspection safer. If an operator must walk alone—a common occurrence in this era of reduced staff—the worker should maintain radio contact with the control room, maintenance office, or other base of operations. A sort of night watchman's security system should be considered, in which the conveyor patrol "calls in" at specified points along the route—at the head of every conveyor, for example. This provides improved safety for the worker, as well as the opportunity to report unusual conditions or problems that require system shutdown.

27

procedure or an acceptable condition. There is also the danger that a problem may be "missed" in the survey, because the person conducting the survey knows that he or she will be called on to fix the problem: The project will be another "dirty job" added to the employee's list.

One solution is the use of contract personnel for this routine maintenance: data census, site survey, and routine maintenance activities. Outsourcing has several advantages. The first is that a conveyor can be surveyed without requiring the attention of plant personnel: They are free to go about their usual tasks. A second advantage is that the outside surveyor can be an expert in proper conveyor practices and current governmental regulations. This objective resource will be able to tell plant personnel when the "standard" plant practices are not the best way to do things.

Although the belt walks can be useful tools for plant maintenance personnel, they can be even more useful when using an outsourced maintenance service. A plant could give the survey results to a maintenance contractor as a "to do" list. This would give the maintenance contractor focus and direction, while providing much of the useful information needed to correct the problems.

Safety for Outside Personnel

Contract personnel must be safety trained just like regular employees. In fact, it may be more important that they are trained, as they will not have the experience and the reinforcement of everyday exposure to the capabilities and limitations of the equipment. It is imperative to make sure they know.

WHAT TO LOOK FOR IN AN OUTSIDE CONTRACTOR

It is important that outside contractors be familiar with the kind of systems they will be looking at: A plumber should not be asked to fix the plant electrical system.

Unskilled or inexperienced contract laborers may not provide any benefit; they might not know what to look for or what they are seeing. More experience will cost more but provide a real benefit. A basic understanding of conveyors operating in similar industries is helpful. It is important that the expertise be found in the individual(s) who will actually perform the work, and not just in the person who sold the service. It is better if the personnel already have the skill set—and the tool set—to perform the required service when noted, rather than having to get permission and then go back to perform it later.

These outside resources are, by definition, not employees of the operation. But they will need to take ownership of the conveyor system and of the plant's efficient and productive use of that system. They must demonstrate a dedication to the operation and to their responsibilities and duties. Wisely chosen and effectively used, these outside personnel will improve the conveyors and add value to the operation.

WHAT TO DO WHEN THE WALK IS DONE

More important than walking the belt is doing something with the information acquired on the trip. Recording the observations and then submitting the observations to the management is the reason the trip is made.

Once all of the data is collected from a conveyor walk, it should be stored; then it should be acted upon. Each of the observed problems can be evaluated and its root cause(s) identified by plant personnel or outside consultants. When the causes of problems have been identified, solutions can be arranged. Things that can be fixed immediately should be. Remaining concerns should be documented, so proper resources can be allocated to provide the required solutions. A "walk the belt" inspection does not show how to solve the problems, but it is an invaluable tool used in the identification of problems.

27

Figure 27.2

A regular walk of the conveyor belt(s) is an effective way to assess the system, catalog its components, and identify areas that need service or that offer opportunities for improved efficiency.

With good records of the concerns noted on a conveyor walk—or better yet, a regular series of conveyor walks—the operation has the opportunity to prevent problems, boost operating efficiency, and improve profitability. Those are the real points of any walk of the conveyor system.

WALKING THE BELT KEEPS A CONVEYOR HEALTHY

In Closing...

A bulk-material handling operation is a system of linked components. If one component or subsystem stops functioning, processes both upstream and downstream will be affected. In simpler terms, one stopped belt could bring an entire facility to a halt. As a result, a conveyor is a critical piece of equipment in a bulk-materials handling operation (**Figure 27.2**). A regular walk of the conveyor belt(s) is an effective way to assess the system, catalog its components, and identify areas that need service or that offer opportunities for improved efficiency.

Looking Ahead...

This chapter about Conveyor System Survey follows Conveyor Accessibility in the section Conveyor Maintenance. The next two chapters continue this section about maintenance measures to reduce spillage and dust, focusing on Maintenance and The Human Factor.

REFERENCES

27.1 Martin Engineering website: http://www.martin-eng.com

27.2 Conveyor Equipment Manufacturers Association (CEMA). (2005). *BELT CONVEYORS for BULK MATERIALS, Sixth Edition*. Naples, Florida.

27.3 The website http://www.conveyor-beltguide.com is a valuable and non-commercial resource covering many aspects of belting.

27

27

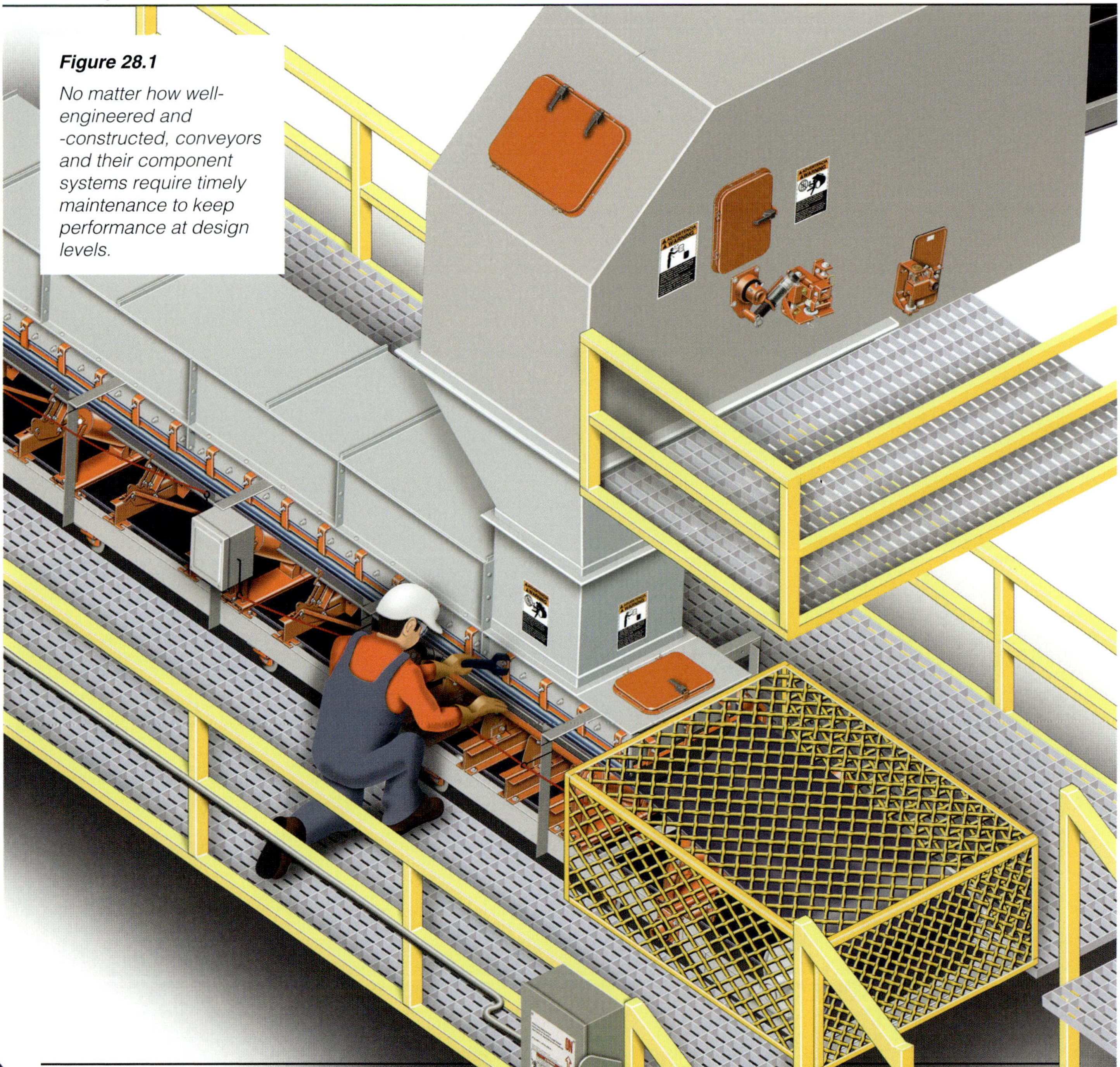

Figure 28.1

No matter how well-engineered and -constructed, conveyors and their component systems require timely maintenance to keep performance at design levels.

Chapter 28

28

MAINTENANCE

In this Chapter...

This chapter will cover recommended maintenance practices to help control fugitive materials, improve safety, and reduce unscheduled conveyor outages. A method of calculating the cost of unscheduled conveyor downtime is suggested.

No matter how well-engineered and -constructed, conveyors and their component systems require timely maintenance to keep performance at design levels (**Figure 28.1**). It is vital that maintenance personnel be properly trained and equipped to detect potential problems as well as perform routine maintenance and repairs.

MAINTENANCE PLANNING

Planning for Maintenance

Experience has shown that, over time, a plant's production requirement increases. Many times management and employee compensation packages are based on achievement of production goals. As a result, the down time allotted for maintenance invariably decreases. In this scenario, the infrastructure of the conveyor is often left un-serviced, and needed repairs are left undone, as the plant struggles to achieve its production goals.

This leads to the conveying system being neglected and run down to the point of catastrophic failure. It is essential the conveyor production schedule allows for adequate down time to perform the necessary maintenance. To allow these maintenance procedures, the conveyor must be shut down—following lockout / tagout / blockout / testout procedures—so scheduled downtime to perform these functions must be kept available. Proper outage time planned into the production schedule to allow maintenance is essential to prevent "crisis management" of conveying systems, in which systems run full time all the time, and the only maintenance provided is when something fails.

The axiom "failing to plan is planning to fail" is true for conveyors. The con- veyor system in which maintenance is not planned is planned for failure.

Designing for Maintenance

Maintenance management needs to be considered when the conveyor system is in the design or planning phase. Many times, the needs of a Maintenance Department are not included during the design phase; in those cases, the new system is not engineered in a way that will allow easy access and maintenance. In the face of industrial conditions—with the continuous stress of day-to-day operations mingled with the demands of changing material conditions and amounts—system service becomes a real requirement. The problem is magnified if the system is not designed from the outset with adequate provisions for maintenance. For many plants, a common assumption is, "If it is difficult, time-consuming, or potentially dangerous, then it is inevitable that shortcuts will be taken." In the case of maintenance activities, this means if a system is difficult to maintain, the service activities will probably not be performed. Or, if the work is performed, it will be done in a superficial, or path-of-least-resistance, fashion. Either way, the increased risk of component failure offers a loss of productivity as its result.

In many cases, the engineering process allows, or even promotes, maintenance problems. Examples include awkward spacing, inaccessible components, "permanent" fasteners, or other non-repairable systems. Designers generally think of, and make provisions for, larger service projects; however, they often neglect provisions necessary to perform routine maintenance easily or effectively. As an example, many conveyors include a frame to lift out and replace the head pulley, an event that may be necessary every five years; yet, no accommodations are made for regular service, such as idler lubrication or belt-cleaner maintenance.

There are ways available to solve these problems by having them included in the design of the system. Examples include adequate walkways, platforms, access,

28

and utilities such as water, electricity, and compressed air readily available to be able to complete tasks in a more timely and efficient manner. Additional examples would include the use of components that are "hammer" adjustable or that are track-mounted for "slide-in/slide-out" adjustment. Having these features "designed-in" will greatly increase the odds of routine maintenance being done properly.

The key is to consider the requirements for maintenance early in the design stage of any project. The Conveyor Equipment Manufacturers Association (CEMA) provides guidelines for conveyor access for maintenance in *BELT CONVEYORS for BULK MATERIALS*, *Sixth Edition*.

Ergonomics and Maintenance

Wherever personnel are involved in a system—whether as designers, operators, maintenance staff, or management— human performance will greatly influence the overall effectiveness and efficiency of that system. Ergonomics (or human-factor engineering, as it is sometimes known) is an applied science concerned with designing and arranging components people use so that the people and components interact most efficiently and safely. For a mechanical system to deliver its full potential, sufficient thought and commitment must be given to optimizing the human role in that system.

Equipment must be designed to improve the reliability and consistency of operators and maintenance personnel in performing regular service.

PERSONNEL AND PROCEDURES

The Maintenance Department

It is important that only competent, well-trained personnel—equipped with proper test equipment and tools—perform conveyor maintenance. For reasons of both safety and efficiency, the maintenance personnel should be skilled, veteran employees empowered with the authority to shut down a conveyor to make a minor repair that will prevent a major outage or equipment expense.

As maintenance crews are reduced, and the demands of conveyor systems increase, it becomes vital to maximize the efficiency of the personnel who are available. One way to maximize efficiency of the personnel is to document procedures for safely performing each task. This will ensure that all workers know both how to perform tasks in the safest, most efficient way and what tools and equipment will be needed to complete the task. It will also allow a plant to better train new employees, as more-experienced personnel retire or transfer.

A computerized maintenance-management system (CMMS) is a good tool for housing the maintenance/repair procedures. The system will administer work orders and manage information, so the maintenance staff can perform chores in a priority-based fashion. Most systems will also track maintenance expenditures for specific equipment: This is essential for justifying upgrading or purchasing new equipment.

A maintenance program that utilizes these types of procedures will prove much more efficient and reliable in the long run.

Contract Services

As plants reduce their head count of employees, many companies are entrusting some or all of their conveyor equipment installation and maintenance to outside contractors. Utilizing contractors allows personnel to be deployed on core activities unique to the operation (**Figure 28.2**).

Figure 28.2

Using contractors for conveyance installation and maintenance allows plant personnel to be deployed on core activities unique to the operation.

28

Contractors prevalent in the bulk-materials handling industry can be categorized as either general or specialty/niche. Both have merit and will bring value to the plant if they are used properly, and the limitations of each one are understood.

A general contractor is willing to do almost any task and has a general understanding of many things; however, the general contractor is not highly skilled in all aspects or components of the conveying system.

Specialty/niche contractors, on the other hand, are experts in certain areas or components; they are often either directly employed by a manufacturer or have been trained and certified by the manufacturer. This training affords them the skills and knowledge necessary to properly install or maintain equipment more efficiently than a general contractor. Specialty/niche contractors should be willing to offer performance guarantees on both labor and the products or components they supply.

Spillage cleanup is another area in which an outside contractor can help free up plant personnel. Contracting spillage cleanup in a plant may actually help to identify the root cause of the spillage, leading to a change in operating procedures or justifying the upgrade, or replacement, of an existing system.

Inspections

Routine maintenance inspections can extend belt- and component-life and improve performance by keeping minor, easily corrected problems from turning into major and costly headaches.

In some of the best operations, maintenance workers routinely "walk" the conveyor system, looking for indications of potential trouble (**Figure 28.3**). *(See Chapter 27: Conveyor System Survey).* It is important that the "conveyor walker" be unencumbered—so the surveyor can be safe and efficient during the inspection tour.

Basic equipment includes:

A. Flashlight

If the conveyor inspector will be looking into closed chutes or underground areas such as chutes and reclaim tunnels

B. Means to record information

Ranging from a pad of paper on a clipboard to a hand-held computer

C. Personal protective equipment (PPE)

Hard hat, safety glasses, hearing protection, and respirator as appropriate to plant conditions

D. Radio or cell phone

To allow communication to the control room or maintenance office

Having an outside opinion often helps to bring attention to areas that plant personnel have come to view as normal. Some suppliers or manufacturers will offer to walk a plant system and provide a "state-of-the-system" report.

The most effective way to detect or troubleshoot problems within a conveyor system is to walk the system while it is operating. Although maintenance or repairs should not be made while the system is operating, watching the system operate and listening to the noises it makes will allow maintenance personnel to identify the components in need of repair/replacement.

Figure 28.3

In some of the best operations, maintenance workers routinely "walk" the conveyor system, looking for indications of potential trouble.

28

It is difficult to conduct a useful belt inspection when the belt is operating at speeds greater than 1,0 meters per second (200 ft/min). Major damage may be spotted, but smaller defects will escape detection above these speeds. One solution to this problem is walking the full length of a stopped belt, checking it section by section. An alternative would be the use of a slow-speed inspection drive. Some conveyors have a supplementary, or "creep," drive operating at 0,1 to 0,25 meters per second (20 to 50 ft/min) to allow a slow-speed inspection.

MAINTENANCE SCHEDULES

Conveyor service intervals and the specific service requirements for a conveyor will depend both on its construction and components and on the volume and nature of materials moving over it. The manufacturers' guidelines for inspection, maintenance, and repairs should be followed. However, if there are no published guidelines, some basic rules to help guide or establish a service schedule are given (**Table 28.1**). Many conveyor components can and should be inspected while the belt is in operation. Care should be taken to only inspect the conveyor components while the belt is in operation, rather than to actually attempt to service them. If items are discovered that are in need of maintenance, these items should be documented. Many maintenance activities can be performed only when the conveyor is not in operation. *(For additional maintenance information, see Martin Engineering's website [Reference 28.1].)*

MAINTENANCE MATTERS

Manuals

It is important to check the owner/operator manual issued by the supplier of any piece of equipment for specific instructions on service requirements, procedures, and timetables. A comprehensive file of equipment manuals should be maintained and accessible to workers on all shifts.

In addition, maintenance personnel should keep careful records of inspections and service performed. This will ensure proper maintenance of the equipment.

Building a File of Information

A file or database of the various components and subassemblies in place on conveyors should be created. This file should include the various manuals, parts lists, and other documents associated with the components included in the material handling system. Items such as manufacturers' part numbers and dates of installation are valuable in ascertaining when service is required and what procedures should be done when that interval comes around.

There should be one central repository for this information. This can be a file cabinet, or, these days, some form of electronic database that holds or links to the various documents, specifications, and supplier websites. The information should be made available from one source, so there are not separate and potentially contradictory files. The decision as to whether that file or database is housed within the bulk-material handling operation or in the maintenance office is a plant decision, but with the availability of electronic networks, this information could and should be shared. The information is then also available to other departments, even those off-site, such as a corporate engineering group.

There are suppliers who can perform this information-gathering and -organizing service for a plant, perhaps as part of their maintenance offering or as a separate service. They will compile and keep these references in a database. This service may be particularly useful if the records of the existing conveyor system have become out of date. With a broader base of experience on equipment and information sources, these services should be able to track down supplier information that might be otherwise unavailable.

28

Spare Parts

Certain repair parts should be available in inventory. This will allow for both routine replacement of worn parts and speedy completion of unexpected repairs, for a faster return to operations. This inventory should include both the parts "likely to be damaged" and replacement "wear parts," such as belt-cleaner blades, impact bars, and idlers. Also included in the maintenance stores should be a supply of "rip repair" belt fasteners for emergency repairs.

By standardizing across the plant on various components, the size of this stockpile of spare parts and, therefore, the expense for these idle parts, can be minimized.

It may be a good idea to keep a "bone yard," where components removed from service can be stored and available to be cannibalized for replacement parts as necessary. Of course, parts taken from used equipment need to be thoroughly cleaned and inspected before re-use.

Lubrication

Due to the large number of bearings present in conveyor systems and their influence on belt tension and power requirements, lubrication is very important. Following the manufacturer's recommendations as to the type, amount, and frequency of lubrication will enhance life-expectancy of the system's rolling components.

Care must be taken not to over-lubricate. Over-lubricating can damage bearing seals, which then allows fugitive materials to enter the bearing, increasing friction and decreasing bearing-life. Excess oil or grease can spill onto the belt, where it can attack the cover, decreasing service-life. Excess grease can also fall onto handrails, walkways, or floors, making them slippery and hazardous.

SAFETY CONCERNS

Conveyor inspection and maintenance can pose significant risks to employees, because these activities bring the workers into close proximity to the conveyor system under potentially dangerous conditions (**Figure 28.4**). There are many systems designed to provide safe conditions for both personnel and equipment. The best approach to safety is a safety program that develops and maintains a healthy respect by engineers, operations personnel, and maintenance staff for the power of the conveyor and the potential risks of its operation.

Some adjustments can be made only with the belt running, and it is advantageous to perform some routine maintenance while the belt is in operation. Most safety regulations recognize this need and provide that "only trained personnel who are aware of the hazards" may perform these routines. There are systems designed to provide safe conditions for both personnel and equipment when inspection and maintenance follow established procedures. Training for these procedures is usually available from suppliers of the systems. It is important that only competent, well-trained personnel, equipped with proper test equipment and tools, perform conveyor maintenance. *(A thorough review of Chapter 2: Safety and Chapter 28: Conveyor System Survey is essential.)*

Figure 28.4

Conveyor inspection and maintenance can pose significant risks to employees, because these activities bring the worker into close proximity to the conveyor system under potentially dangerous conditions.

Belt Conveyor Preventative Maintenance | Table 28.1

SAFETY CONCERNS

Always inspect work area for hazards prior to commencement of inspections or performing work.

LOCATION	PROCEDURE: Weekly	LOCATION	PROCEDURE: Weekly
Bend Pulleys	Ensure belt is centered on pulley	Head Pulley	Inspect belt cleaners for worn or missing blades
Carrying Idlers	Ensure all rolls are turning*		Inspect belt cleaners for cleanliness of frames and blades
	Ensure all idler rolls are free of material buildup*		Check belt-cleaner tension according to manufacturers' recommendation
	Ensure belt touches all three rolls both in loaded and unloaded states*		Ensure belt is centered on pulley
Conveyor Belting	Check for belt damage or abuse:		Check dust-suppression nozzles for pluggage
	Check belt for belt cupping	Loading Zone	Inspect impact idlers for wear
	Check belt for belt camber		Inspect impact bars for top cover wear
	Check for impact damage		Inspect seal-support cradles for wear
	Check for impingement damage		Inspect and adjust dust seals
	Check for chemical damage		Inspect dust-suppression nozzles*
	Check belt for rips or tears	Return Rolls	Ensure rolls are turning freely
	Check belt for junction-joint failure		Inspect rolls for material buildup
	Check belt for top cover cracking		Inspect mounting brackets for wear from belt-tracking problems
Conveyor Drive	Check reducer oil level	Safety Switches	Inspect cables for correct tension
	Check reducer for oil leaks		Ensure flags are free from material buildup
	Inspect drive coupling	Snub Pulley	Ensure belt is centered on pulley
	Check oil level in backstop and inspect for leaks		Inspect pulley for material buildup
	Ensure all safety guards for drive are in place and in good condition	Splices	Mechanical: Check splice and pins for wear
Conveyor Structure	Check for rusted, bent, broken, or missing structural parts		Vulcanized: Check splice for separation
	Check hand rails and toe plates to ensure good condition	Tail Pulley	Ensure belt is centered on pulley
	Check walkways for material spillage or buildup		Check V-plow blade for wear
	Check safety gates to ensure good working order		Check V-plow mounting
Gravity Take-Up	Check take-up carriage for free and straight operation*		Check V-plow tension
	Ensure belt is centered on pulley*	Tracking Idlers	Check for free pivoting of frame*
	Ensure all safety guards are in place and in good condition		Ensure all rolls are turning*
Guards	Check for damage and proper installation.		Check rolls for material buildup

* NOTE: Starred inspections may require the belt to be running.

Always inspect work area and review job procedure for hazards prior to commencement of inspection or performing work. Review local and company safety requirements before conducting any maintenance activities. Some facilities may allow certain specified inspection and/or maintenance activities to be performed on moving conveyors by appropriately-certified and -trained technicians. For those procedures that cannot be safely performed while the belt is in operation and/or those facilities that do not allow maintenance while the belt is in operation, Lockout / Tagout / Blockout / Testout procedures must be followed prior to performing any work.

Belt Conveyor Preventative Maintenance | Table 28.1

SAFETY CONCERNS

Always inspect work area for hazards prior to commencement of inspections or performing work.

LOCATION	PROCEDURE: Monthly	LOCATION	PROCEDURE: Monthly
Bend Pulleys	Check bushings for evidence of movement on shaft	Carrying Idlers	Check lubrication of bearings in rolls
	Check bearing condition and locking collars for tightness	Conveyor Drive	Check lubrication in backstop bearings
	Check for cracks and wear at face and hub ends		Check lubrication in shaft bearings
	Check lubrication in shaft bearings		Inspect drive belts for wear and correct tension
Gravity Take-Up	Check bushing for evidence of movement on shaft	Loading Zone	Inspect chutes and chutewalls for leaks*
	Check bearing condition and locking collars for tightness		Inspect entry seals
	Check for cracks and wear at face and hub ends		Inspect exit seals
	Check lubrication in shaft bearings		Inspect dust-collection pickups for leaks*
Head Pulley	Check bushing for evidence of movement on shaft	Return Rolls	Check lubrication in bearings in rolls
	Check bearing condition and locking collars for tightness	Safety Horns	Test to ensure working properly prior to conveyor start
	Inspect pulley lagging for wear and secure to head pulley	Safety Switches	Emergency-stop switches should be tested in cooperation with management
	Check for cracks and wear at face and hub ends	Tracking Idlers	Check lubrication in rolls and pivot points
	Check lubrication in shaft bearings	**LOCATIONS**	**PROCEDURE: Every 6 Months**
Snub Pulley	Check bushing for evidence of movement on shaft	Brakes/Backstops	Test for proper operation under full load*
	Check bearing condition and locking collars for tightness	Conveyor Structure	Check foundations for settling
	Check for cracks and wear at face and hub ends	Loading Zone	Inspect wear liners for wear
	Check lubrication in shaft bearings	Safety Switches	Test operation for conveyor shutdown*
Tail Pulley	Check bushing for evidence of movement on shaft	Warning Devices/Signs	Test for operation and audible/visual or readable functionality*
	Adjust mechanical take-up for correct belt tension	**LOCATION**	**PROCEDURE: Yearly**
	Check for cracks and wear at face and hub ends	Electrical System	Check for open wiring, damaged conduit, overloading and system grounds
	Check lubrication in shaft bearings	Interlocks	Test to ensure proper interlocking of conveyors*
	Check lubrication in mechanical take-up adjusters	Safety Switches	Test conveyor start circuit with flags pulled
	Check bearing condition and locking collars for tightness	**LOCATION**	**PROCEDURE: Special Circumstances**
		Splices	After new splice: Check for crooked splice*

NOTE: Starred inspections may require the belt to be running.

Note: In all cases, the manufacturers' recommendations for inspection and maintenance should be followed, including recommendations for lubrication. The above list covers most of the common components and systems found on most conveyors. Components and systems specific or unique to the conveyor being inspected should be added at an appropriate frequency. Examples are: scale, inspection doors, rip detection, belt cleaner(s), plow(s), flow aid(s), sampler, level detector(s), lighting, fire protection system, dust-control components and systems, lightning protection, general housekeeping, prohibited items, etc.

28

Today, some plants use idlers and other rolling components equipped with sealed bearings. Sealed bearings require no lubrication and, therefore, reduce maintenance.

Conveyor Startup

A belt is like a new pair of shoes: It needs to be "broken in" gradually and carefully to avoid painful moments. Insufficient attention at conveyor startup—either in the initial running of a new system or following a maintenance outage—can lead to significant and costly damage.

Typically, conveyor operation does not require the attention of many personnel. This is generally one of the selling points in the selection of belt conveyors over the other forms of haulage. However, it would be a mistake to start up a belt, especially on a new conveyor or on a line that has received extensive modifications, without the attention of extra personnel along its route. Spotters should be in place along the belt run where trouble might be anticipated or would be particularly costly. These observers should be equipped with "walkie-talkies" or cell phones and be positioned near emergency shut-off switches. A careful inspection prior to startup should establish that there are no construction materials, tools, or structural components left where they can gouge or cut the belt as it begins to move.

The belt should be run empty, slowly at first, and then at normal operating speed. The belt should then be gradually loaded to full capacity while checking for possible problems.

ADVANCED TOPICS

The Cost of Unscheduled Shutdowns

Conveyors are often the lifeline of an operation, and the availability (or uptime) of these systems has a direct impact on profitability. In many operations, lost production attributed to conveyor downtime results in a lost opportunity that cannot be recovered.

While the relationship between conveyor availability and profits is obvious, the relationships among effective maintenance, component quality, and the fundamental conveyor design assumptions are not so obvious. Because the effects of these conditions and practices are hard to identify, the root causes of conveyor downtime often are not addressed, leading to conveyors with chronic problems. Research published by the Australian Coal Association indicates that when the root cause of conveyor downtime is related to component quality, the cost of downtime is on the order of five times the cost of replacing the component. When the root cause of downtime is related to the basic conveyor design, the downtime cost is approximately two times the cost of the redesign *(Reference 28.2)*. At first these ratios may seem to be incorrect, or even backward, but consider that a component failure often involves maintenance with a relatively short downtime window for replacement, whereas a basic design mistake often involves a significant capital expense and a prolonged outage for correction.

Each situation is different, but it is clear that treating the symptom rather than the root cause results in repetitive downtime costs. With cost of downtime established at two to five times the cost of the corrective action, it is clear that keeping accurate records and performing a thorough analysis of the problem can justify almost any needed corrective action.

When calculating the cost of downtime, common expenses to include are:

A. Lost opportunity cost

B. Purchase cost for replacement components

C. Maintenance labor cost

D. Subcontractor cost

E. Consulting and engineering fees

F. Testing and analysis costs

The opportunity cost can be calculated by multiplying the production rate per hour and number of hours of downtime by the selling price of the product. Since conveyors can transport large quantities per hour, this cost adds up quickly. For example, a coal mine that produces 4000 tons per hour, that it sells for $50 USD per ton, loses the opportunity to sell $200,000 USD of coal for every hour of unscheduled downtime. One must add to that the expense for outside services, replacement components, and the total cost of downtime for the overall cost for that incident.

Using this lost opportunity cost is a good way to capture an operation's direct and indirect costs, because included in the selling price are all the costs of production, management, and administration, as well as the profits. While it can be argued that the coal that was not mined can be sold at a later date, this may or may not be the case, depending on the market and the mine's contract with the customer. In any event, the cash flow for that downtime period has been lost.

THE DOLLARS AND SENSE OF MAINTENANCE

In Closing...

Efficient and effective maintenance lowers costs, not just for the maintenance department, but for the total operation. The goal is to provide quality work with minimal disruption to the production routine. This will produce benefits in operating efficiency, system availability, and, ultimately, the bottom line.

Ironically, plants that cut corners, or "skimp," on maintenance and cleaning activities often end up paying more over the long term in environmental and cash terms to cope with fugitive material in their plant. The key to minimizing these unscheduled downtimes is efficient maintenance.

Unscheduled shutdowns are very expensive. One author listed the cost for service to be three to seven times more in an emergency shutdown than during scheduled downtime. In 2002, another source calculates downtime costs in longwall coal mines at $30,000 USD per hour. Depending on the plant's size, a one-percent difference in system availability for a coal-fired power plant could be worth one to two million US dollars in annual revenue. An unscheduled conveyor outage or failure that decreases generating availability even one-tenth of one percent is a significant cost. The cost of even the shortest unscheduled outage is prohibitive. Obviously, maintenance to prevent unscheduled downtime is a critical factor in the operation's overall profitability.

Looking Ahead...

This chapter about Maintenance continues the section Conveyor Maintenance. The following chapter, The Human Factor, concludes this section about the relationship between timely, regular maintenance and reduced spillage and fugitive dust.

REFERENCES

28.1   Martin Engineering website: http://www.martin-eng.com

28.2   Roberts, Alan. (November 1996). *Conveyor System Maintenance & Reliability*, ACARP Project C3018. Author is from Centre for Bulk Solids and Particulates, University of Newcastle, Australia. Published by Australian Coal Association Research Program; can be purchased at http://www.acarp.com.au/abstracts.aspx?repId=C3018

28

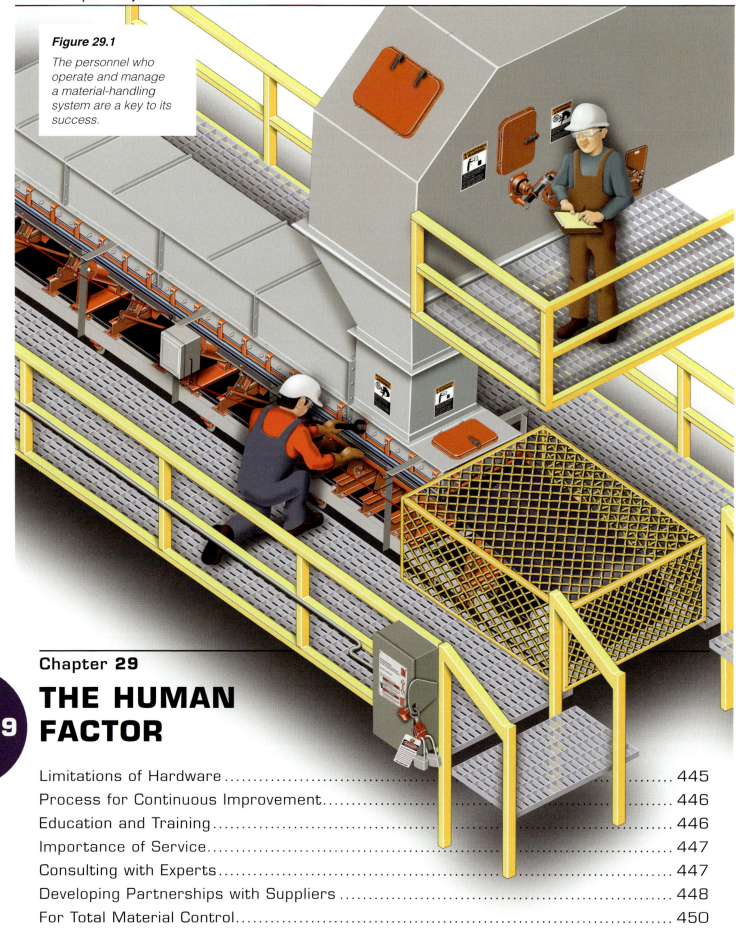

Figure 29.1

The personnel who operate and manage a material-handling system are a key to its success.

Chapter 29

THE HUMAN FACTOR

In this Chapter...

In this chapter, we focus on neither the technology nor the hardware necessary for successful bulk-materials handling; instead, we focus on the human factor—the people who make the difference between a successful and unsuccessful operation. This chapter discusses the need for a commitment to, and a process for, continuous improvement in material control and the education and training required to accomplish it. In addition, this chapter looks at the importance and benefits of consulting with experts and developing partnerships with suppliers.

Industries in established economies are experiencing a trend in which the established workforce is aging, and their base of knowledge and experience is rapidly being depleted.

In many cases, this knowledge base is being replaced with modern technology in the form of more sophisticated monitoring and control systems. The plant personnel who remain have been asked to expand their efforts and capabilities to cover for the lack of manpower.

Much of this book deals with the hardware systems required to provide total material control, including the prevention and control of spillage and dust. New technology and upgraded components for existing systems will enhance the probability for success, but real and sustained progress requires educating the people responsible for the overall performance of the operation. The human factor cannot be ignored in operating and maintaining a material-handling system.

In many ways, a plant is the reflection of the thinking of its personnel: upper management, operations, and maintenance personnel alike (**Figure 29.1**). If these groups see and accept the plant as dirty, inefficient, unpleasant, and unsafe, it will be allowed to become and stay that way. In many ways, allowing dust and spillage to accumulate in the work environment can create fugitive-material "fatigue." Over time, those who work in this environment every day accept the conditions, and they simply no longer see the conditions as a problem. Even when some improvements are implemented, the results quickly become accepted as the norm. Without a process for continuous improvement, the earlier upgrades will soon become accepted as the best that can be achieved.

To make truly beneficial and long-lasting improvements in the performance of hardware systems, it is critical that the attitude and actions of plant personnel, at all levels, be transformed into a desire to improve continuously. Plant personnel who expect higher performance standards, take the steps required to achieve these levels, and commit to maintaining even higher standards will see improvements in efficiency, working conditions, and safety.

The improved hardware systems for material containment and dust management described in this book are important steps in achieving total material control. However, it is the human factor that will provide the ultimate key to success.

LIMITATIONS OF HARDWARE

The goal of achieving total material control requires more than new technology and hardware.

Many times, new components are installed to upgrade a conveyor system in order to generate improvements in material control and enhance plant efficiency. In most cases, these systems perform as expected and provide discernible benefits, at least when the systems are new. However, the solution to controlling fugitive material does not stop with new equipment installations, regardless of how well designed or expensive. New hardware represents only the first step on the road to process improvement.

29

PROCESS FOR CONTINUOUS IMPROVEMENT

Developing a Process for Continuous Improvement

Improving conveyor performance by reducing material spillage and controlling airborne dust is not just a question of buying the latest technology or piece of hardware from the next vendor to walk through the plant's front door. Instead, the solution lies in developing a process that can continuously upgrade the performance of equipment, materials, suppliers, and plant personnel. The process begins with a plant-wide commitment to improve the management of fugitive material. It features education and training of managers and other personnel to understand the value of, and see opportunities for, total material control. It includes optimization and empowerment of the service departments to maintain component performance and improve overall plant operating efficiency. It features the development of relationships with consultants and suppliers who can assist the plant in focusing on material-handling challenges. These steps help develop a process of continuous refinement in pursuit of the elusive goal of total material control.

This process of continuing to improve is not developed overnight; nor can it be accomplished by decree from upper management. While it does require the endorsement of senior officials, it equally requires the commitment and empowerment of operations and maintenance personnel. This process must be carefully nurtured through education and experience.

There are many proven techniques for developing a culture of continuous improvement. The names may be familiar: Lean Manufacturing, Toyota Production System, Business Process Management, Lean, Six Sigma, and others, now and in the future. Regardless of the specific plan selected, the factor most important for success is the commitment of management to that technique.

Commitment to Improvement

Effective material control requires a process for the continuous improvement in material-handling operations. Results are affected by varying material conditions, improper operation of the system, and minimal or non-existent maintenance.

Corporate requirements force the plant to continuously search for opportunities to reduce costs and increase profitability, while maintaining a safe, efficient operation. Pressuring preferred suppliers into price concessions may lead them to find other ways to recover their costs and meet their required margins, so it ends up costing the plant more over the long-term.

Improvement requires a full circle of commitment, beginning at the senior management level and extending through the plant-level operations and maintenance management to the operators, maintenance crew, and every other person in the organization.

Management must acknowledge and demonstrate a commitment to solve problems, while employees must have the access to resources, time, tools, and equipment required to develop, implement, and maintain solutions. It is critical that the plant makes wise investments in selecting systems that will help reduce maintenance; that are maintenance friendly; and that will increase equipment reliability, performance, and safety.

EDUCATION AND TRAINING

A key aspect of the process for continuous improvement is the training of personnel. Employees need to be trained to fully understand the purpose and capabilities of the equipment they will operate, while developing a level of comfort and confidence in their own abilities. This education should include how to identify challenges, how to troubleshoot, and how to take corrective action by adjusting equipment to correct and minimize problems.

29

Education must identify hazards associated with equipment operation and show operators how to use the equipment efficiently and without risk to personnel or the plant.

Education and training can be provided by consultants offering information on the characteristics of bulk materials and overall equipment performance. Qualified suppliers should provide training on the installation, maintenance, and troubleshooting of their systems.

IMPORTANCE OF SERVICE

The axiom "Pay me now or pay me more later" is perhaps most true when it comes to considering maintenance of material-handling equipment. This old saying means that if regular maintenance is neglected, the chance of replacing the component when it fails, probably prematurely, increases.

With the "down-sizing" of companies to cut budgets, along with the early retirement of experienced employees, one area that appears to be hit the hardest is the maintenance department. Hidden expenses from downsizing can come from a variety of causes, including lack of knowledge of the appropriate preventive maintenance, not enough resources to complete early repairs, and longer maintenance cycles as new employees learn the equipment.

The evolving trend to meet the need for this equipment service is to hire outside contractors to perform even routine maintenance tasks. Outsourcing the specialty maintenance procedures is one acceptable way to provide for, and stabilize, the costs of routine service. Typically, these maintenance specialists are experts for a particular system or set of components: They offer improvement in the quality, efficiency, and speed of service. These service specialists can also assist in the specialized training required to improve the skills and knowledge of plant maintenance personnel on complicated procedures and new technologies.

An example would be the copier in the office. If a copier breaks down, the required repair is not usually the responsibility of the building maintenance person: The proper course is to contact an experienced copier repairperson. A copier is a complex piece of equipment that represents a sizeable capital investment. Office maintenance personnel are not familiar with the nuances of a copier, and repairs would go beyond their scope of expertise. Users of the copy machine are discouraged from trying to fix it, as their "repairs" usually make things worse.

Material-handling systems are usually collections of complex pieces of equipment representing a sizeable capital investment. Operation and maintenance personnel may not be familiar with the nuances of specific hardware. The specialty supplier should have the best access to the technical knowledge and nuances of the products being maintained, eliminating the learning curve and allowing the maintenance provider to keep a complete equipment history that can be used to predict intervals and methods of failure. When such material-handling hardware needs service or breaks, experienced repairpersons should be called.

Unfortunately, this is not the usual method used in most operations. The equipment is temporarily patched so the plant can "get by" until the next maintenance cycle. By the time that next cycle rolls around, the patchwork repair has been forgotten, and other, more immediate, problems get the attention.

If equipment suppliers offer maintenance contracts, it may be in an operation's best interest to take full advantage of this resource.

CONSULTING WITH EXPERTS

People sometimes are so busy reacting to daily needs and problems that they are sometimes unable to see the opportunities in front of them. The daily task of operating and maintaining a material-handling

29

system consumes their time and energy, thus limiting their ability to identify challenges or to see the benefits of possible improvements.

Complacency, or the "way we've always done it" attitude, can overcome employees who have become accustomed to a certain condition, certain procedure, certain style, or certain level of performance.

Surveys show that employees say their jobs allow them to do what they do "best" only about one-third of the time *(Reference 29.1)*. Managing in such a way that employees can do what they "do best" most of the time allows and motivates employees to increase their productivity.

Outside resources can make significant contributions to a plant's operation (**Figure 29.2**). They can be experts in a specific industry, specific process, or particular piece of technology as used across the industry. These professionals might be material-handling specialists who can analyze plant systems, identify challenges, and offer recommendations. They provide a broader view of the industry and of material handling and are valuable resources for solutions to the challenges associated with transporting and storing bulk materials. They can offer a "road map" with directions to achieve a cleaner, safer, and more productive plant.

A common excuse is that people fear making improvements because it will reduce their job security or give management a reason to re-assign a key piece of equipment. In fact, just the opposite is true. Companies always have more projects than resources. If workers are open to change and focused on what they do best, those individuals are more likely to be one that the company chooses for a new assignment or the next project. This improves job security. Additionally, if the company becomes more profitable as a result of an individual's contributions, the firm is more likely to remain in business, again increasing job security.

With the philosophy and improved technologies discussed in this book, it is clear that there are "new" ways to do things, ways that will help plants improve their processes and work more efficiently. Often the plant can still use the same basic equipment to increase capacity and output and have a cleaner, safer plant without investing in an entirely new material-handling system.

DEVELOPING PARTNERSHIPS WITH SUPPLIERS

In order to accomplish its goals for process improvements in any area, especially in material handling, the plant must take advantage of the knowledge base available from qualified key suppliers (**Figure 29.3**).

Figure 29.2

Consultants can help plant personnel see the strengths and weaknesses of the operation's material-handling system.

29

Figure 29.3

By making an important contribution to a plant's success, suppliers move from vendors to partners.

An open relationship with key suppliers can be a major competitive advantage while business environments change continuously. Too often, project work has been contracted to installation and maintenance companies that do not specialize, but only generalize, leaving cost overruns and equipment that does not perform to expectations. When suppliers must funnel their products and services through a few master distributors to reach the plant, the plant personnel are isolated from a significant resource for problem solving—those manufacturers and service companies with expertise in specific technical areas.

There is often a fear that outsourcing will mean less work for the employees and thus reduce job security. Service on the systems that control fugitive material is often the lowest priority in the daily maintenance schedule and is almost never a priority during shutdowns. Even though managers will argue that they have the in-house people with the time to do the work, the reality is that employees are usually neither trained nor motivated to take the steps necessary to control fugitive materials and are pressed by the crisis of the moment. As a consequence, the work simply does not get done. Skilled maintenance personnel should focus on what they do best: There is never a shortage of core maintenance work that requires specialized knowledge of a plant's process and its maintenance procedures.

A qualified supplier organization will have experienced and knowledgeable sales personnel and corporate resources, including applications experts, industry managers, product engineers, project engineers, and installation specialists, who can make significant contributions to a plant's program for containing spillage and controlling fugitive dust. The key is for suppliers to become partners with the plant.

Often this partnership concept has failed, because of a failure to recognize the expertise of the supplier and the importance the supplier's knowledge and experience bring to plant personnel.

Relationships will flourish if the supplier is skilled in overcoming the common resistances to change in an organization. One common, but often unstated, problem is sometimes referred to as the "Not Invented Here" (NIH) syndrome. This "condition" refers to an individual's or organization's inability to accept solutions—equipment or ideas—that were not developed internally. An experienced supplier will help find ways to allow changes in the "standard operating procedures" so improvements are made without causing hard feelings.

Suppliers who strive to become part of the team and are capable of adapting to the unique environment of each plant are much more likely to succeed at building a long term, mutually beneficial relationship. Suppliers and sales representatives have to be sensitive to these issues and be prepared to make changes at the first signs of incompatibility.

Total material control requires a great deal of specialization and attention to detail. When choosing a consultant or service company to assist in achieving total material control, careful consideration must be given to the supplier's core competencies and ability to fulfill the commitments made. The capability of the supplier to engineer situation-specific solutions, to manufacture and install them, and then to maintain them to a single performance standard is a strategic advantage for the plant and for the supplier.

A partner will share in the risk and the reward of system improvements. More importantly, a partner earns the right to be there, with energy, effort, and results. The key to success in this partnership is for both the operation and the supplier to establish a long-term relationship centered on improving efficiency and profitability. The most essential ingredient of this process is open and honest two-way communication between vendor and plant leading to mutual trust.

29

FOR TOTAL MATERIAL CONTROL

In Closing...

There are many new and proven technologies and upgraded hardware solutions discussed in this edition. They range from engineered flow chutes to improved belt-cleaning systems, all aimed at achieving total material control. However, the single key ingredient to success in containing spillage and controlling dust is the people—those who operate and maintain the conveyors and the rest of the plant. It is, and always will be, "the human factor" that will ultimately determine a material-handling system's success or failure.

Looking Ahead...

This chapter, The Human Factor, concludes the section Conveyor Maintenance, in which we discussed various aspects of maintenance required to reduce fugitive materials and extend the life of the conveyor system. The following chapter begins the section The Big Picture of Bulk-Materials Handling, starting with the chapter that will focus on Total Project Management.

REFERENCES

29.1 Sullivan, Dr. John. *Increasing retention and productivity: let employees do what they do best!* Article #163. Available online: http://ourworld.compuserve.com/homepages/GATELY/pp15s163.htm

29

29

SECTION 7
THE BIG PICTURE OF BULK-MATERIALS HANDLING

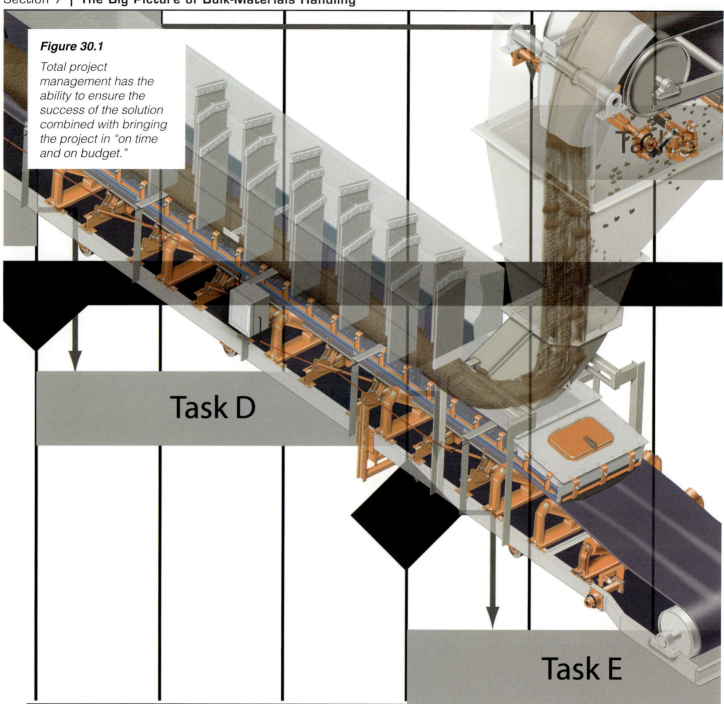

Figure 30.1

Total project management has the ability to ensure the success of the solution combined with bringing the project in "on time and on budget."

Task D

Task E

Chapter **30**

TOTAL PROJECT MANAGEMENT

In this Chapter...

In this chapter, we focus on project management, the value of an experienced project manager, and the work that is done by that person to track the project to its successful completion. The project sequence and the primary tasks that need to be carried out are explained.

Among other benefits, the total-project-management approach has the ability to ensure the success of the solution combined with the most efficient use of capital to provide a mutually beneficial outcome to the customer and the provider (**Figure 30.1**). This chapter provides a brief synopsis of each area of project management. A wealth of books is available on this subject; additional study is recommended.

WHAT TO EXPECT FROM PROJECT MANAGEMENT

Firms that offer total-project-management services provide their clients with a one-stop shop to guide projects from inception to commissioning. This single-source responsibility allows the project manager to package research, concept, quotes, design, procurement, and construction in a manner that effectively controls project risks and streamlines delivery. With just one contract to administer, plant ownership/management benefits from the simplified lines of communication. The integration of the engineering, procurement, and construction functions provides opportunities for fast-track completion.

Project management functions smoothly when accountability, responsibility, and authority are all vested in the project manager. In situations where this is not possible, it is important to define the roles and responsibilities at the beginning of the project, so interested parties know who has responsibility and what procedures must be followed to gain action.

Certification from the International Organization for Standardization (ISO), which ensures rigorous quality control from start to finish of a project, is a desirable feature in a company offering project-management services. ISO 9001 registration certifies that consistent business processes are being applied to the design and manufacturing functions. Additional expertise can be expected if a project manager is certified by the Project Management Institute (PMI). PMI is the world's leading not-for-profit association for the project management profession.

A company that provides totally managed projects typically utilizes a proactive quality management system that provides support and guidance to all levels of engineering, fabrication, and construction. This function operates at all stages of a project and supplements, but does not replace, the existing design, design check, and project review procedures.

PROJECT MANAGER

The project manager is the one person who is responsible for the overall success of a project. By establishing reasonable and clear expectations for co-workers, vendors, and subcontractors, the project manager manages the overall project schedule and budget to ensure that work is assigned and completed to provide a quality solution for the project's (and ownership's) needs, on time, and on budget.

In general terms, the project manager is responsible for:

A. Defining the project goals and objectives

B. Breaking the objectives into tasks

C. Specifying the resources required

D. Allocating the budget among the project objectives

E. Developing the project schedule

F. Implementing the project plan

G. Managing and controlling the work

H. Resolving conflicts

I. Communicating with ownership

30

A qualified and experienced project manager can make the difference between the success and failure of a project. Regardless of whether a job is a retrofit or new construction, a project manager with experience and expertise in the field of bulk-materials handling will be able to successfully implement a transfer-point or conveyor project from analysis and design through installation and commissioning. An experienced project manager is likely to have knowledge of relevant technologies, such as state-of-the-art conveyor components, engineered chutes, and dust minimization.

In addition to knowledge of the field, an effective project manager has good organizational and leadership skills and is able to coordinate communication with ownership/management, vendors, and subcontractors. The project manager maintains liaison with the ownership/management in day-to-day matters and monitors the progress of the job against the project scope, the ownership's expectations, and scheduled outages or other deadlines. Important knowledge areas for a project manager include time management, quality management, human-resource management, and risk management. Effective management of a project requires an analytical person who is well organized, has great follow-up skills, is able to multi-task, and is a good estimator, budget manager, and communicator.

PROJECT TRACKING

As the project moves through its process to completion, all steps must be captured and recorded. All known and estimated time and costs need to be included on the project tracking. Those costs and times that are estimated can be highlighted to be verified or adjusted as additional information is gained. Estimated times can be in the form of lead-time (e.g., "The steel will take two weeks to be delivered after it is ordered,") or it can be in the form of labor (e.g., "It will require 40 hours to assemble the frame using two people.") Predecessor steps can then be identified and placed to generate the project timeline or Gantt chart.

A Gantt chart is a popular type of bar chart that illustrates a project schedule. Gantt charts illustrate the start and finish dates of the terminal elements and summary elements of a project. The various sub-tasks feed into the chart and show the relationships in precedence between activities. Gantt charts can be used to show current schedule status using percent-complete shadings and a vertical "Today" line. Gantt chart development is now available in a number of software programs.

As the project moves toward completion, everything done—not only on the jobsite, but also during the entire project—needs to be tracked and updated on the project plan and/or in a daily project activity log. Material and labor receipts, certifications, and inspection logs need to be organized and filed for control. In many localities, it is a legal requirement to keep daily records of these activities.

PROJECT SEQUENCE

Problem Definition

To define a project, first define the problem. This should be done so that the issue that needs to be addressed is a question asked, not a solution suggested. An example of this could be, "We need a belt conveyor." While this indeed could be the solution, it does not define the issue or opportunity. A better definition of this issue might be, "We need to move our material from point A to point B. How can we do this?" With this open description, there are no expectations for the solution. The best solution is still an option to be identified. This problem, or issue, description needs to be mutually defined and agreed to by both the customer and the project manager. By defining the issue, it opens the opportunity to brainstorm different solutions, rather than committing to a solution that may or may not be optimal.

Initial Project Plan

At this point, the initial project plan should be generated. Even if all details and facts are not defined or understood, a skeleton needs to be put into place. This can be done on paper; there are, also, a variety of project management software packages available that will not only help develop the plan, but also track and update it as well.

Most of this software will allow planning and tracking for the four primary aspects of the project:

A. Timeline

B. Labor

C. Costs

D. Materials

Primary Tasks

To get the project plan started, the primary tasks required in the project need to be listed. This could be done in the software or on an individual piece of paper or file folder (depending on the total project size) for each task. These tasks include:

A. Defining the scope and specifications

B. Creating a conceptual design

C. Defining a preliminary design

D. Securing a budgetary quote

E. Finalizing specifications/design

F. Request for quote (RFQ) and final bid

G. Ownership review

H. Purchase order (PO)

I. Detailing design

J. Manufacturing

K. Installation

L. Operation

M. System performance verification

N. Maintenance

Each of these headings—the primary tasks of the project—can then be supplemented with sub-tasks as more detail becomes available. The project plan must be a living document and continuously updated.

Having the primary tasks defined will help those involved to visualize the bulk of the project for both labor and timing estimates. This improves the quality of the estimates and communication between ownership, project manager, and the various vendors and contractors.

Task 1: Defining the Scope and Specifications

The next step is to define the need and system requirements as general specifications (volume, time, material, distance, constraints) based on customer input and known constraints. An example could be:

> We need to move dry sand (less than 2 percent moisture) 1,2 kilometers (4000 ft) with a continuous 5 percent slope. We do not have a decent road surface. The material path needs to clear a 1,8-meter (6-ft) fence and change angle by 45 degrees.

Other issues addressed at this phase would be to identify both hard and soft criteria. Hard criteria would include such items as how much, how fast, how far. Soft criteria would be the optional, or "We would like to have," items. Examples might include variable speeds, adjustable discharge height, or a specific belt-cleaner selection.

Task 2: Creating a Conceptual Design

The first step in any conceptual design should be to brainstorm solutions. The incorporation of the operation's ownership/management on the brainstorming group will allow them to gain ownership for the solution, have a better understanding of the problems, and have a more positive outlook on the design's success. Brainstorming means identifying all possible solutions, whether practical or not. This exercise ensures that all options are considered. All ideas should be freely accepted and noted for analysis in the next step.

Inherent in this step is to define the design parameters. This step can benefit from material testing to define an acceptable range of design parameters. Another

30

important aspect of the conceptual phase is to identify any field limitations with site verification of workpoints, elevations, and obstructions. In the event that actual field conditions beyond the range of design parameters may come into play, measures to compensate for those conditions need to be considered. This may require the addition of "failsafe features" that allow easy maintenance, as well as the inclusion of methods for mitigating these unforeseen circumstances.

The next step is to list the pros and cons of each idea. Here is where the critical analysis of each idea is performed. A combination of solutions might offer the best value: The issues with one idea might be resolved by the inclusion of a second idea.

After the critical analysis of each idea, the scope and specifications are finalized in a predetermined format. The scope identifies what specifically is to be covered or equipment supplied; the specifications are details of needs. The scope should list the items, concepts, and specifications included in the project; it can also detail items not included in the project. Options can be listed separately, and the use of a predetermined format makes it easier to stay away from "gray" or imprecise specifications.

Conveyor system including:

A. Structure

B. Transfer points

C. Belt

D. Belt cleaners

E. Installation

F. System startup

G. Performance testing

H. Maintenance review with ownership/ management team

I. System handoff

J. Site prep NOT included

K. Electric service to site NOT included

Specifications:

A. Conveyor 1800 millimeters (72 in.) wide

B. 81 meters (265 ft) long

C. 1 transfer point (loading)

D. Adjustable center loading spoon

E. Drive motor
   a. 480 volts
   b. 50 Hz
   c. 25 kW (30 hp)

F. Service guarantee

G. 2400 hours to first major service interval

H. 90 percent "up" time for first year (with service contract)

Task 3: Defining a Preliminary Design

With the scope and specifications in place, a preliminary design can identify system needs for both costing and lead time considerations.

Working with a designer or design engineer, the system identified during the concept stage is sketched. In conveyor work, this is often called a Process Flow Sheet (**Figure 30.2**). It shows the key data and the general conveyor layout. Using this sketch and the other project documents, the parts, or subassemblies, are identified as "make" or "buy." This identification needs to be double checked to ensure all details are included. Items overlooked at this stage will create a delay in delivery and increased costs when the need is discovered during installation or startup.

Task 4: Securing a Budgetary Quote

At this point, the information required for the development of a budgetary quote should be available. A budgetary quote provides a reasonable estimate of the price and delivery of a solution. It is not a firm price and is not intended to be used to secure a purchase order.

30

Task 5: Finalizing Specifications/Design

After determining that the project will fit within budgetary parameters and provide the performance required, the system design and specifications can be finalized.

Task 6: Request for Quote (RFQ) and Final Bid

For each of the items highlighted as an estimate in the plan, the project should obtain hard quotes. There are four areas that require specific action for quoting:

A. Materials

When seeking quotes for materials, lead time needs to be considered, not only for the delivery of the materials, but also for secondary operations that will be needed to convert those raw materials to usable "parts" (e.g., cutting, drilling, and welding brackets).

B. Purchased components

It is important to ensure all necessary specifications are provided to the supplier when the RFQ is issued. If a supplier identifies a need for additional information, the supplier can be told when a reply will be given. Even if it is not known for sure when the information requested will be available, the supplier should be given an expectation of when a reply will be given. There should be follow-up with the supplier. In this follow-up, the supplier can be given an update on the process of getting the answer to the inquiry and told when a response will be given. It must be remembered that delays in response may generate delays in supplier ability to complete the quote.

C. Contract labor

Of the four areas requiring hard quotes, contracted labor is the most difficult for which to obtain accurate estimates or quotes. Any potential suppliers should review the scope, specifications, drawings, and photographs to note any exceptions or revisions required.

D. Direct labor

The project manager will need to discuss direct labor with various resources within the company to outline both requirements and duration. The most accurate method to accomplish the estimate is to break the work into tasks, estimate for each task, and then summarize to generate the total.

Quotes and estimates can be compared to any existing history for similar work, to gain assurance that the costing is reasonable and nothing obvious has escaped consideration.

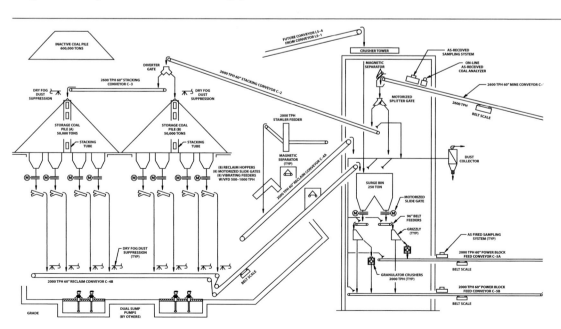

Figure 30.2

A Process Flow Sheet shows the key data and the general conveyor layout.

The project plan and timeline (Gantt chart) need to be updated, with all labor, timing, costs, and cash flows having been given consideration, prior to the quote being finalized for the customer.

The minimum information that needs to be included in the quote is:

A. Scope

B. Specifications

C. Major stages and review points

D. Cost

E. Payment schedule

If additional information is included in the quote, confidentiality statements need to be in place. If sketches or drawings are included, they need to be marked preliminary as appropriate.

The payment schedule in the quote needs to be tied to the completion of specific project stages or milestones.

Task 7: Ownership Review

The quote should be reviewed in detail by the project manager with the project's ownership/management. This task should not be handled via email, fax, or phone, but rather in person, if possible. The understanding of the detailed expectations at this point will significantly reduce misunderstandings along the path to project completion.

Any project preparations to be provided by the plant also need to be reviewed and the hand-over conditions defined. The project manager must verify that the plant ownership/management understands the requirements as well as the fact that when the purchase order (PO) is issued, the plant will be committing to both the work and the schedule.

Task 8: Purchase Order

Once the PO or requisition has been received by the project manager and documented, with a copy of the quote and PO together in the project file, commit-

ments can be made for materials and labor. Commitments cannot be given on a verbal project approval. Until the actual purchase order is received, the timeline and commitments to scope and costs should be flexible and subject to change. If the PO is awarded on time without revisions or changes, no schedule or cost revisions should be necessary.

If the PO is delivered early, certain aspects may be accomplished ahead of schedule (e.g., material orders placed or commitments to sub-contractors to hold position in their schedule), but early project delivery must be dependent upon a careful review of the entire supply chain, including labor commitments and site conditions.

If the PO is received after the date specified on the RFQ, all aspects of the entire schedule—including the work of suppliers, designers, and subcontractors—will need to be reverified by the project manager.

Task 9: Detailing Design

When a project is submitted to the design team, the project manager should ensure that the scope and specifications are reviewed in detail with the design team. Each of these scope and specification documents holds its own importance. The scope is to ensure that all aspects of the quoted work are planned and covered. The specifications ensure that the equipment designed will perform as anticipated and identified in the contract.

A review of the schedule is just as important as the scope and specifications review. Prior to quoting this job, the project manager will have received an estimated timeline from the design team. At this point, the project manager will review this commitment and obtain a renewed commitment from the design team manager. This check is necessary at each step of the project to attain on-time project delivery.

Throughout this process, design reviews are required to achieve four primary design goals: cost, timing, scope, and manufacturability. Each of these should be tracked throughout the design process. A representative of the manufacturing team should participate in the design reviews.

Task 10: Manufacturing

As designs are finalized for components and subassemblies, the handoff between design and manufacturing must take place. As with the design team, the schedule, scope, and specifications should be reviewed by the project manager with all of those involved in the manufacture of the project components and subassemblies.

After a project is underway is not the time to experiment with new vendors for manufacturing. The project manager should utilize manufacturers that have a history of delivering quality products on time and avoid unproven or financially questionable vendors, if possible. Sometimes this may be unavoidable, but in those cases, due diligence would be required. References should be checked closely, asking for statistics on both quality and delivery to schedule, along with a check for credentials, such as ISO quality-system certifications.

As materials, components, and subassemblies are delivered, they need to be inventoried and inspected. Vendors need to be informed immediately of any shortages or nonconforming items and the intended resolution determined, to avoid project delays.

Task 11: Installation

The two most important control issues at installation for a quality, on-time delivery are cash flow and schedule. Actually, if the project manager is dealing with suppliers or subcontractors on the jobsite, cash flow is the primary tool for controlling the schedule as well. As with the project manager's commitment to the customer for delivery, so is the supplier or subcontractor's commitment to the project manager. In the development of the cash flow plan—part of the project plan—a substantial percentage of the total supplier payment should be reserved for the successful delivery of the contracted materials or labor. Delayed, or unreasonably withheld, payment may affect the vendor's capability to maintain future delivery schedules or quality.

Once materials, components, or subassemblies are received and approved at the jobsite, they should be staged by installation order, stored out of the way in a controlled area, and protected. Material shrinkage—through vandalism or pilferage—is a concern in many sites the world over and can affect both project cost and schedule with the time and expense of replacing missing or damaged materials.

The project manager is responsible for controlling all work produced at the installation under the project scope. The manager must watch for, and carefully control, any scope changes, commonly called "scope creep." Changes in the scope for any reason must be documented (change order) and will normally require a contract revision or addendum to ensure validity. They can quickly erode both the project budget and schedule and are one of the primary reasons for unsuccessful project delivery.

Task 12: Operation

Initial startup of any equipment should follow a thorough inspection and adjustment of all components and electrical systems by the project manager, who should review the startup checklist if available.

Before startup, moving equipment should be visually checked to ensure all components are as anticipated, with the project manager identifying and making any adjustments necessary. After startup and initial testing, the equipment should be loaded to check performance, again making any adjustments necessary.

The equipment operation should be demonstrated to the customer's representatives for ownership or management, with a walk-through of each major component, its

30

function, and its maintenance needs. Step-by-step instructions should be provided for both startup and adjustment procedures.

Task 13: System Performance Verification

In conjunction with the startup of the machine and over a period of hours / days / or weeks, as defined in the project scope, the performance of the system should be evaluated to ensure that it meets or exceeds the expectations of the customer as measured against the initial requirements in the scope and specifications phase of the project. In the early phases of a project, when the scope is defined, there are issues and expectations. This phase of the project will utilize those issues and expectations (clearly defined) as metrics or key performance indicators (KPIs) to measure the success of the solution and installation.

Task 14: Maintenance

At the very least, a guide for, and a review of, the maintenance requirements of the system must be provided in detail. The review of the maintenance requirements should be done with a minimum of two customer representatives present to ensure that once the job is handed off, there will be some level of understanding and collaboration of the work necessary for the solution to continue to perform at peak effectiveness.

An improved maintenance position—and thereby improved performance—can be gained if the solution provider also takes ownership of the maintenance of the equipment through a maintenance contract. Through this concept, maintenance costs can be fixed (to the user), the scheduled maintenance will have the priority needed, and equipment downtime can be minimized, maximizing productivity.

HOW SUCCESS IS JUDGED

In Closing...

Finally, when selecting a firm for total project management, it is advised that one be selected that will back its performance—and the performance of the materials-handling solution proposed and installed—with a performance guarantee. The best guarantees are those written by a firm that has a record of success; that covers engineering, procurement, installation, commissioning, and follow-up service; and that has the financial strength to follow through with guarantees and warranties.

A project without a project manager or with an ineffective manager is more likely to encounter excessive costs, equipment failure, and missed deadlines. These projects are also more likely to suffer quality issues due to the lack of project ownership by a trained and experienced project manager. Effective management leads to better projects and better results.

Looking Ahead...

This chapter about Total Project Management is the first chapter in the section The Big Picture of Bulk-Materials Handling. The next three chapters continue this section, beginning with Performance Measurements in Chapter 31, followed by Considerations for Specific Industries in Chapter 32 and Considerations for Specialty Conveyors in Chapter 33.

REFERENCES

30.1 Project Management Institute (PMI). Additional information about project management and the accreditation program for project managers is available from PMI on the organization's website: http://www.pmi.org

30

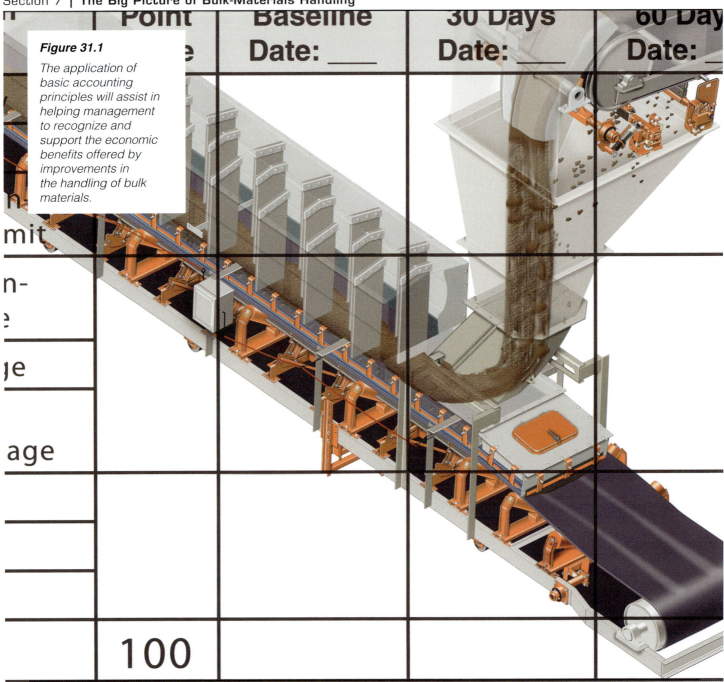

Figure 31.1

The application of basic accounting principles will assist in helping management to recognize and support the economic benefits offered by improvements in the handling of bulk materials.

Chapter **31**

PERFORMANCE MEASUREMENTS

31

In this Chapter

This chapter provides tools to measure the effects of improvements in safety, cleanliness, and productivity of handling bulk materials in industrial operations. It proposes qualitative and quantitative measures to assess the benefits of these improvements. The intent is to help an engineer or manager of a materials-handling operation apply basic accounting principles to the funding and management of systems that improve materials handling through the control of fugitive materials. These measures will assist in helping management recognize and support the economic benefits these improvements will provide (**Figure 31.1**).

MEASURING PERFORMANCE

The Importance of Performance Measurement

A key indicator of a company's management philosophy is its procedure to evaluate the Return on Investment (ROI) on its expenditures. According to a 2007 report by the Aberdeen Group (as highlighted in the February, 2007 issue of *The Manufacturer*), less than 25 percent of companies consistently estimate ROI prior to acting on a proposed project, and only 20 percent measure the actual costs and gains to calculate ROI after the project has been completed. On the other hand, "Best in Class" firms are 88 percent more likely to estimate ROI before initiating projects and 130 percent more likely than their competitors to measure ROI after project completion. The "Best in Class" companies that calculate ROI produce 93 percent more improvement across a variety of performance measures when compared to those firms that do not do this evaluation *(Reference 31.1)*.

Measuring performance in bulk-materials handling involves both quantitative (a precise value based on calculations or measurements) and qualitative (an approximate or intuitive value based on personal opinion and judgment) measurements.

These performance measurements are tools for making decisions, assigning accountability, and determining improvement. These measurements should be part of a strategic process used to assess accomplishment of goals and objectives by a person, team, or company. Another common name for performance measurements is Key Performance Indicators (KPIs). These performance measurements are often displayed in graphs or spread sheets, so trends and progress against the goal can be easily recognized.

The condition and management of bulk-materials handling systems have a direct effect on all aspects of a company's performance. The most common measurement of performance to health and safety standards is the company's compliance with safety and environmental regulations.

The most universal performance measurement is the financial performance of a company, as expressed in Profit and Loss (P&L) and Balance Sheet statements.

Financial Performance

A company must earn a profit to have money to buy materials, pay wages, pay dividends, and re-invest in the business. Sales (or turnover or income) is often confused with profit, leading to misconceptions about how much a company has earned and what they do with that money. A company whose main business involves bulk-materials handling typically makes a profit of less than five percent after taxes. In other words, investors advanced their money, and at the end of a year, they will have earned less than 5 dollars for every 100 dollars risked. If the company does not make a profit, the investors lose money. One reason people are willing to invest in bulk-materials handling operations, rather than putting the cash in the bank and drawing interest, is the potential for higher profits if the company (hence, its bulk-materials handling systems) perform optimally.

To evaluate the performance of a company, both a P&L Statement and a Balance

31

Sheet are needed. A P&L Statement is a collection of income (sales, rents, royalties) minus expenses over a period of time, usually a month, a quarter of the year, or a full year. A Balance Sheet is an accounting of the assets (what is owned by or owed to the firm) minus liabilities (loans and bills the company must pay) and stockholder equity (the accumulated profits and losses) at a moment in time, usually the last day of the month, the quarter, or the year.

It is common to compare financial statements from prior years to the current year to judge the progress of the company in meeting its financial goals. Financial goals are usually set by the board of directors or stockholders. Public companies sell stock in the company to the public as one way of raising money to grow the business. Private companies get their funding from the owners, who use their own funds or borrow money to finance the growth of the business beyond what profits alone can fund.

The money used to operate a business is usually divided into two categories: Capital Funds and Operating Funds. Operating Funds usually come from the continuous process of selling a product, collecting the money, and paying all the bills. To stay in business, the company must sell its products for more than it costs to produce them, or they will soon be in financial trouble. It is important to quickly collect the money from sales, or there will not be enough Operating Funds to run the business. If there are no profits, the business will have to borrow money, or the shareholders will have to invest more.

Capital Funds are used for purchases that will last for a long time, such as land, buildings, and equipment. Capital Funds come from net profits after tax, sale of stock, or borrowing. In order to have money for capital expenditures, the company needs to accumulate profits or borrow money. To borrow money for capital expenditures, the company must have a record of making money year after year, or the interest rate will be higher, reflecting the added risk the bank is taking in lending money to the company. Because Capital Expenditures usually involve large sums of money, the company has to plan ahead and be confident the investment will earn enough profit to pay back the money. For this reason, Capital Expenditures are often planned for a year or more before they are made.

The total Cost of Capital involves a combination of the value of the money retained in the business after expenses and taxes (Total Equity), the value of existing loans (Total Debt), the amount that the shareholders expect to earn on their investment (Cost of Equity), and the amount that the company must pay to borrow money and invest it in the business (Cost of Debt). Quite often, the Cost of Capital is used as the minimum required Return on Investment (ROI) for a project. These numbers vary greatly, depending on how a company is funded, but generally the Cost of Capital is five to ten percentage points higher than the interest rate a bank would charge.

The cost of the money used to invest in new equipment or upgrades is not just the cost of borrowing money at the bank (**Equation 31.1**). The more risky the investment, the higher will be the return that a company must earn to justify spending the money.

The Cost of Capital is typically an average cost established for the whole company by its financial department. It should not be confused with corporate requirements for individual projects, often referred to as ROI or Payback. The ROI value required at a given company could typically range from 10 percent to 33 percent. A major long-term investment such as a complete new conveyor system will usually have an ROI requirement at the lower end of the range. Improvements in conveyors to make them dust and spillage free often have very high ROI requirements, perhaps in excess of 100 percent (**Equations 31.2-4**). In other words, these investments are expected to pay for themselves in less than a year (**Table 31.1**).

31

$$CC = \left(\frac{TE}{TE + TD} \cdot CE \right) + \left(\frac{TD}{TE + TD} \cdot CD \cdot (1 - T) \right)$$

Equation 31.1

Cost of Capital Calculation

Given: *A company is worth $2,000,000 USD and owes $500,000 USD in debt at 8%; the shareholder expected equity is 15% and the corporate tax rate is 35%.* **Find:** *The total cost of capital.*

CC	Cost of Capital	percent
TE	Total Equity	2000000
TD	Total Debt	500000
CE	Cost of Equity	.15 (15%)
CD	Cost of Debt	.08 (8%)
T	Corporate Tax Rate	.35 (35%)

$$CC = \left(\frac{2000000}{2000000 + 500000} \cdot 0.15 \right) + \left(\frac{500000}{2000000 + 500000} \cdot 0.08 \cdot (1 - 0.35) \right) = 0.13$$

CC	Cost of Capital	0.13 (13%)

$$ROI = \frac{TPS}{TPC}$$

Equation 31.2

Return on Investment Calculation

Given: *A project will save $10,000 USD and cost $25,000 USD.* **Find:** *The return on investment.*

ROI	Return on Investment	percent
TPS	Total Project Savings	10000
TPC	Total Project Costs	25000

$$ROI = \frac{10000}{25000} = 0.4$$

ROI	Return on Investment	0.4 (40%)

$$ROI(years) = \frac{1}{ROI}$$

Equation 31.3

Return on Investment in Years Calculation

Given: *A project has a return on investment of 40% (0.4).* **Find:** *The years needed to return the investment.*

ROI(years)	Return on Investment in Years	years
ROI	Return on Investment in Percent (as Decimal)	0.4

$$ROI(years) = \frac{1}{0.4} = 2.5$$

ROI(years)	Return on Investment in Years	2.5 years

$$ROI(months) = \frac{12}{ROI}$$

Equation 31.4

Return on Investment in Months Calculation

Given: *A project has a return on investment of 40% (0.4).* **Find:** *The months needed to return the investment.*

ROI(months)	Return on Investment in Months	months
ROI	Return on Investment in Percent (as Decimal)	0.4

$$ROI(months) = \frac{12}{0.4} = 30$$

ROI(months)	Return on Investment in Months	30 months

31

Table 31.1

ROI Conversions		
ROI	**Payback Years**	**Payback Months**
10%	10.0	120.0
20%	5.0	60.0
30%	3.3	40.0
40%	2.5	30.0
50%	2.0	24.0
60%	1.7	20.0
70%	1.4	17.1
80%	1.3	15.0
90%	1.1	13.3
100%	1.0	12.0

CALCULATING ROI

Data Needed to Calculate Project ROI

It is difficult to justify a project or prove it has an acceptable ROI if the costs of the system's operation have not been recorded. There are dozens of parameters that could be measured to justify a project; not all costs and production numbers need to be collected. Data from a list of common data can be used to justify projects for the control of dust and spillage on conveyor systems (**Table 31.2**).

Having these figures as "hard numbers"—the actual expenses and opportunities lost—provides the ammunition needed to justify improvement projects.

Table 31.2

Data Used in ROI Calculations	
Data	**Units**
Administrative/Operating	
Cost of compliance: record keeping and reporting	currency
Health and liability insurance premiums increase	currency
Reduced life of equipment	currency
Safety/environmental fines	currency
Legal costs	currency
Energy costs	currency
Waste disposal costs	currency
Production	
Throughput: per hour, day, week, or month	tons (st)
Production time	hours
Cost per ton of bulk material	currency/ton (st)
Cost of down time	currency/hour
Cleanup manual (1 ton per hour is average)	labor cost/hour
Cleanup machine (5 tons per hour is average)	labor and machine cost/hour
Lost product due to dust and spillage	0,5% to 3% of production rate is typical
Safety *(Reference 31.2)*	
Cost of recordable incident	currency
Cost of lost-time incident	currency
Maintenance	
New installation: *Estimated cost for labor and materials*	currency
Adjustment: *Estimated labor cost per adjustment*	currency
Replacement Parts: *Cost of parts and labor*	currency
Equipment Wear: *Cost of belt and wear-resistant materials*	currency

31

Accurate and timely records are essential to verifying performance and justifying expenditures. Without record keeping, all that is left are fact-free arguments that too often lead to missed opportunities and lower than average performance.

Using ROI to Justify Equipment

Each plant has different priorities and expectations regarding the treatment of their bulk-materials handling system. While one plant may be dealing with a dry, free flowing material, a second operation may be dealing with a wet and sticky material, and a third plant may be dealing with a hazardous material that can expose workers to health risks and the company to environmental violations, fines, community compensation, and substantial medical costs. Just as plants have different types of material handled, they also have different accounting practices, management styles, and maintenance procedures. The process of determining how much to invest in which system can vary significantly from company to company.

However, given accurate record keeping, there are rational approaches to the consideration of projects that improve bulk-materials handling through the control of dust, spillage, and carryback. The following is an example of how to determine how much investment in belt-cleaning systems is economically sound.

A Sample Calculation: Evaluating the ROI of Belt Cleaners

An operation's success at eliminating carryback can be categorized into arbitrary "Levels." Achievement of these levels would be determined by a measurement of the amount of carryback remaining on a prescribed area (usually, a square meter) of belting. For the purposes of this discussion, the baseline for carryback material remaining on the belt (or "Level 0" cleaning) would be more than 250 grams of material per square meter.

Level I cleaning would be defined as allowing 101 to 250 grams of carryback per square meter to remain on the belt. A typical belt-cleaning system that could achieve Level I cleaning would be a single primary cleaner or a slab-style secondary cleaner.

Level II cleaning is defined as leaving 11 to 100 grams of material per square meter of belting. A typical cleaning system to achieve this level of carryback would be a double or triple engineered cleaner system, composed of a pre-cleaner with a secondary cleaner and sometimes even a tertiary cleaner.

Level III cleaning is defined as leaving carryback levels between 0 to 10 grams per square meter. A cleaning system that might achieve this level of performance in typical circumstances would be a belt-washing system involving one or several water-spray bars, multiple cleaning assemblies, and a method of removing excess moisture from the belt. These more complicated or sophisticated systems achieve improved cleaning performance; they also cost more to purchase and maintain.

As a company increases its expenditure in belt cleaners, its costs for the cleanup of fugitive material (the carryback released from the belt) will decline. The relationship between the purchase (and maintenance) of cleaning systems and the costs for required cleanup can be plotted on a graph (**Figure 31.2**). At some point, cost of the additional cleaners will exceed the savings achieved by reducing cleanup expense. The break-even point will be located at the point where the curve of Cleanup Cost intersects with the curve of Belt-Cleaning System Investment and Maintenance Cost.

The point of optimum economic benefit in this example is at or near Level II cleaning performance. Every time the desired level of carryback decreases (or cleaning effectiveness increases), two related costs will be affected: Investment cost increases, while the cost to clean up carryback decreases. As a company increases its expenditure in belt cleaners, its costs for cleanup should decrease. Carryback can be suppressed to nearly zero, but the investment in a belt cleaning system to do that could be greater than can be justified because of the total cost of installing, maintaining, and operating the system. At some point, the cost invested in additional belt cleaners is greater than the savings resulting from the reduced expense for the cleanup of carryback. Consequently, it might be a bad decision, based on this ROI calculation alone, to add additional belt cleaners beyond that point.

There could be a similar cost versus benefit analysis performed for maintenance and adjustment on belt-cleaning systems. Cleaning performance will be better each time the cleaning blades are cleaned and retensioned against the belt; however, at some point along the line, the labor cost will be greater than the value of the improved performance. As an example, the benefit of weekly inspection and adjustment of cleaning systems might pay for itself, whereas the labor cost of adjustment on an every shift or every day schedule might not.

Of course, this is a purely economic evaluation of the benefit of improvements in belt-cleaning systems. It does not include any variables for health, safety, or community-relations issues. There is no universal answer to these questions: The balance between health and safety, maintenance costs, and return on investment must be evaluated and optimized for each plant.

ROI on Safety-Related Improvements

Accidents are often thought of as bruises, cuts, and broken bones, but the health ef-fects from long-term exposure to dust and spillage can be even more significant. In a worldwide workforce of 2.8 billion workers, it is estimated that there are 270 million lost-time accidents and 2.2 million fatal work related accidents a year. Occupational diseases add another 160 million long-term disabilities. It has been estimated that 95 percent of these accidents and diseases occur in emerging economies *(Reference 31.3)*.

Excluding property damage, the average cost of all workplace lost-time accidents and occupational diseases in the established market economies is approximately $35,500 USD per incident. For emerging economies, the average cost per lost-time accident is approximately $4,700 USD. The cost in the established market economies for a fatal accident is well documented at approximately $1 million USD. There are no reliable estimates of the cost of a workplace fatality in emerging markets, but if the ratio between lost-time accidents between established and emerging economies is maintained, a fatal accident in an emerging economy could cost $132,500 USD.

These values represent the estimated total direct and indirect costs to the company. These values, along with the costs for property and environmental damages, can be used in a company's ROI calculations to help justify improvements in safety and working conditions related to the control of fugitive material.

But they do not represent the cost to the workers and their family, which are often two to three times greater in lost wages, un-reimbursed costs, and inability to return

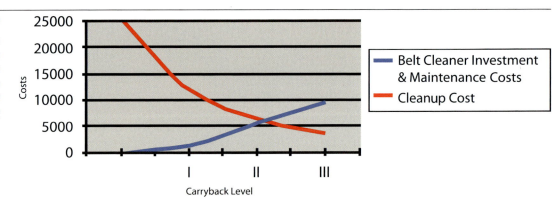

Figure 31.2

Typical costs and production parameters related to projects for the control of dust and spillage versus cost for cleanup.

31

— Belt Cleaner Investment & Maintenance Costs
— Cleanup Cost

Costs

Carryback Level

to the same level of work. It may be easy to rationalize the low cost of an accident to a company in a developing country compared to the cost in developed countries. However, that ignores the cost of human suffering and the fact that any company can benefit by being more productive than their local competitors.

Some have argued that developing countries cannot afford a "health and safety first" attitude—that the emerging nations must do anything to generate income to become economically viable and/or competitive. That puts safety in the back seat, considered only after the nation becomes competitive in the marketplace.

However, the data show something different: In 2003, the International Labour Organization (ILO) of Geneva reported research suggesting the safest-working countries also have the best competitiveness ratings. Noting that the most competitive countries are also the safest, a 2005 report *Decent Work - Safe Work* from the World Congress on Safety and Health at Work says, "There is no evidence that any country would have benefited from low levels of safety and health." The report continued, "Selecting a low safety strategy may not lead to high competitiveness or sustainability" *(Reference 31.3)*.

The ILO report *Safety in Numbers: Pointers for a Global Safety Culture* lists a number of effects of poor safety and health on a company's bottom line *(Reference 31.2)*. These impacts include:

A. Higher absenteeism and more downtime, leading to loss of productivity

B. Underutilization of expensive production plants and a possible decrease in economies of scale

C. Low morale, leading to productivity loss

D. Loss of skilled and experienced employees, as well as loss of the company's investment in their training

E. Difficulty in recruiting high-quality employees

F. Payment of compensation and/or damages to injured or sick workers or to dependents of workers killed

In addition, companies suffer:

A. Associated legal costs

B. Payment of danger bonuses

C. Higher insurance premiums

D. Material damage to equipment and premises due to incidents and accidents

E. Fines

F. Disputes with trade unions, public authorities, and/or local residents

G. Loss of image

H. Reduction in sales

In some cases, the cost will be the complete or partial loss of the company's "license to operate."

When it comes to being competitive, all companies can benefit from being more productive than their local competitors. Being safe is an essential part of being competitive.

MEASURING FUGITIVE MATERIALS

Quantitative and Qualitative

Without both qualitative and quantitative information, discussion about the effectiveness of efforts to control fugitive material becomes a futile debate based on opinion rather than fact. Quantitative measurements of the amount of fugitive material collected provide strong evidence of the nature of an operation's specific problems. These are assessments that capture the actual amount of materials lost, through testing of the system both before and after improvement projects are installed.

The type(s) of fugitive material seen in a bulk-materials handling operation can often be determined by the size of the particles in the bulk material and the shape of the pile of material. Dust is very small particles that tend to become airborne and accumulate by blanketing an area. Spillage

31

is generally granular in nature, representing the average particle size of the bulk material being conveyed, and accumulates in conical piles with a slope equal to the bulk material's angle of repose. In some cases, the hole through which the spillage leaks acts as a screen, and the pile of spilled material will be of a uniform size. In other cases, the process creates a separation of particle sizes via airflow or degradation.

While measuring spillage is relatively straightforward, determining the source of the fugitive material often takes a little detective work. It is important to look up when looking for the source of a spill and trying to deduce where the material could have originated. If there is no obvious source, it could be the spill was caused by an operating problem such as a plugged chute, belt mistracking, an overloaded belt, or a maintenance problem such as a defective indicator.

In addition to quantitative information, there may be occasions when measurements alone do not convey the nature of the problem or signify the amount of improvement required or achieved. In these cases, there is the opportunity for the use of more qualitative standards. However, to be effective and valuable, these standards should be defined in advance and subject to a versatile and well-intentioned scoring system.

The Scale of Fugitive Materials

It is human nature to forget what past conditions were and unconsciously change the definition of what is acceptable. A scale of fugitive materials is an evaluation system, using predetermined point values and comparative photographs of a specific operation or facility as the standard for evaluation. A scale establishes an index that assigns values to a system's performance in control of fugitive materials. It calls for the assignment of scores to the dust, spillage, and carryback control of a given system, using a pre-established, operation-specific

scoring system. Using these scores, an operation can evaluate the performance of its equipment (and its suppliers) in achieving control of fugitive materials.

A scale, when combined with quantitative data collected on production rates, maintenance costs, and worker safety, can be used to prepare a comprehensive evaluation of management's efforts toward continuous process improvement.

Scoring System

The scoring system is based on comparing a visual inspection with ranges of conditions broken down into categories and illustrated with relevant photos. Each category is given a weighted numerical score. Management establishes the weighting in order to focus efforts on a specific problem. The total of all numerical scores gives an overall rating, while a listing of category scores provide a means of breaking the problem into manageable sources of fugitive materials.

Each materials-handling operation should establish its own point system, based on what is acceptable, and use actual photos from the plant for the scoring of system performance.

Definition of the Swinderman Scale

The Swinderman Scale of Fugitive Materials detailed below is one attempt to develop such a scoring system.

The material presented is intended as a sample or demonstration of this scale. Each operation develops its own scoring scale, to fit its specific situation, and uses that scale over time for measuring performance improvement.

The scale can be reviewed periodically, perhaps annually, with increases to the requirements to support continuous improvement as facility cleanliness improves. The scoring system and targets should not be changed too often, or progress will be lost.

The following is a definition of this system

Dust

Definition: particles of fugitive material small enough to become airborne, generally less than 10 microns in diameter. Dust generally distributes itself evenly throughout an area. Dust may be emitted from any process source.

Level D1: Extremely Dusty _____ points (**Figure 31.3**)

- More than 10 milligrams of dust per cubic meter
- Opacity greater than 30 percent
- Able to see less than 15 meters (50 feet) through the dust
- Unable to breathe without a respirator
- Eyes irritated and constantly watering

Level D2: Dusty _____ points (**Figure 31.4**)

- 1,2 to 10 milligrams of dust per cubic meter
- Opacity of 11 to 30 percent
- Able to see less than 50 meters (150 feet) through the dust
- Possible irritation to mouth and/or nose with some minor difficulty in breathing

Level D3: Dust Free _____ points (**Figure 31.5**)

- Less than 1,2 milligrams of dust per cubic meter
- Opacity 0 to 10 percent
- Able to see more than 100 meters (300 feet) through the dust

Figure 31.3
Rated D1—Extremely Dusty

Figure 31.4
Rated D2—Dusty

Figure 31.5
Rated D3—Dust-Free

31

Figure 31.6

Rated S1—Extreme Spillage

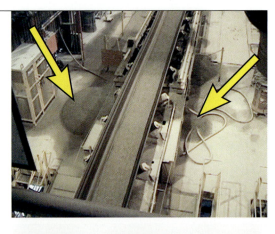

Figure 31.7

Rated S2—Frequent Spillage

Figure 31.8

Rated S3—Spillage-Free

Spillage

Definition: Material that escapes from process equipment or a conveyor belt at unwanted locations. Spillage is usually granular in nature and typical of the bulk material size distribution. Spillage normally accumulates close to and/or below the source of leakage.

Level S1: Extreme Spillage _____ points (**Figure 31.6**)

- Constant rain of material from leaks in chutes and process equipment resulting in buried walkways and equipment

- Weekly accumulations of more than 2 tons of fugitive material

- Difficult to walk along equipment or up conveyor walkways

- Particles become lodged in eyes, ears, and nose

- Constant manual clean up required to maintain production

Level S2: Frequent Spillage _____ points (**Figure 31.7**)

- Layers of spillage with no readily apparent source or repeated operations errors

- Weekly accumulations of up to 2 tons of fugitive material

- Some difficulty walking on walkways and through accumulations

- Manual cleanup required every 1 to 2 weeks

Level S3: Spillage Free _____ points (**Figure 31.8**)

- Spillage that is the result of an occasional process upset or intentionally delayed maintenance

- Characterized by no lumps or granules of material accumulating

- Requires occasional manual cleaning

31

Carryback

Definition: Fugitive material that clings to the belt after the belt has discharged its cargo. Characterized by piles of fine, wet material or dried flakes under the return idlers and gravity take-up; material buildup on bend pulleys and other components. Carryback is a possible source of dust.

Level C1: Dirty ____ points
(**Figure 31.9**)

- 101 to 250 grams per square meter of carryback on the belt surface (Level I Cleaning) (For reference: $1,0 \text{ g/m}^2 = 0.003 \text{ oz/ft}^2$.)

- Characterized by a layer of material 0,5 to 1 millimeter (0.02 to 0.04 in.) thick on the belt

- Material accumulates under the rollers

- Clean up required at least once a week

- Suitable for open cast mining operations where mechanized cleaning up is used

- Can be achieved by a single or dual belt-cleaner system

Level C2: Clean ____ points
(**Figure 31.10**)

- 11 to 100 grams per square meter of carryback on the belt surface (Level II Cleaning)

- Seen as a film or streaks of carryback on the belt slightly discoloring the surface

- Small amount of accumulation under the return rollers—may be in the form of flakes

- Manual clean up required 2 to 4 times a month

- Suitable for most bulk-materials conveying applications

- Can be achieved with a dual or triple belt-cleaning system

Level C3: Very Clean ____ points
(**Figure 31.11**)

- 0 to 10 grams per square meter of carryback on the belt surface (Level III Cleaning)

Figure 31.9
Rated C1—Dirty

Figure 31.10
Rated C2—Clean

Figure 31.11
Rated C3—Very Clean

- Characterized by a mostly slightly damp belt with few to no streaks of carryback

- Manual clean up required less than once a month

- Usually requires use of water sprays and a wash box to achieve this level consistently

31

The plant should determine the maximum possible points for any given category after reviewing their goals for control of fugitive material and enter the point values in the form (**Table 31.3**).

Sample of the Swinderman Scale Procedure: Happy Company

Procedure

The goal is to reduce the amount of cleanup required and to meet the statutory requirement for dust. Management selects an area for performance monitoring. Management and a service provider meet to agree upon the weighting of the three elements, Dust, Spillage, and Carryback (**Table 31.4**).

In this example, supervisors from production and maintenance at the Happy Company get together to develop the definitions and performance levels for their operation based on the problems and the desired outcomes of their operation. Once a month they walk the area as a team and grade the environmental conditions to assign scores that reflect the materials-handling system's performance. In the sample scoring system, Carryback is more heavily weighted, with the value of up to 60 points for a clean operation. Consequently, improvements in control of Carryback will be shown as the greatest improvement in overall score.

Baseline Survey

Management and a service provider tour the area and take representative photos that provide a visual comparison scale from Acceptable to Not Acceptable. A baseline rating is established and a plan for improvement is implemented. In the survey conducted prior to any system improvements, Happy Company's material-handling operations scored a 15 (**Table 31.5**). This score reflects a dusty plant with significant spillage and carryback problems.

30-Day Survey

After 30 days, Happy Company has improved performance to a score of 45 (**Table 31.6**). The installation of belt-cleaning systems improved cleaning performance, increasing the rating for carryback from Dirty to Clean. The plant is still evaluated as Dusty with Extreme Spillage.

60-Day Survey

After 60 days, Happy Company has improved performance to a score of 60 (**Table 31.7**). Improvements in the materials-handling systems have improved the level of spillage from Extreme Spillage to Frequent Spillage. Dust and carryback remain where they were at the previous survey.

90-Day Survey

After 90 days, Happy Company has improved performance to a score of 70 (**Table 31.8**). Dust and spillage have been virtually eliminated, and carryback has been reduced to acceptable levels with the installation of multiple cleaning systems.

Results from Happy Company

By comparing the baseline score to interim results, the analyst can deduce that first the carryback problem was addressed, then the spillage, and finally dust. From the final score, it appears that the original goal was met, but further improvement remains possible. Photos documenting the new level of performance are taken and submitted with the original standards as proof of performance.

31

Table 31.3

Scoring Form for Swinderman Scale

Weighting System for Scoring Area _____ **on Fugitive Material Emissions**

Level	Description	Point Scale	Baseline Date: ___	30 Days Date: ___	60 Days Date: ___	90 Days Date: ___
D1	Extremely Dusty					
D2	Dusty					
D3	Dust Free and Under Statutory Limit					
S1	Extreme and Continuous Spillage					
S2	Frequent Spillage					
S3	Spillage Free or Occasional Spillage					
C1	Dirty					
C2	Clean					
C3	Very Clean					
	TOTAL SCORE	100				

The higher the score the better the performance. Max possible score = 100; min = 0

Table 31.4

Scoring Form for Swinderman Scale (with Management Weighting)

Weighting System for Scoring Area _____ **Tripper 1** _____ **on Fugitive Material Emissions**

Level	Description	Point Scale	Baseline Date: 23 May	30 Days Date: 23 Jun	60 Days Date: 23 Jul	90 Days Date: 23 Aug
D1	Extremely Dusty					
D2	Dusty	20				
D3	Dust Free and Under Statutory Limit					
S1	Extreme and Continuous Spillage					
S2	Frequent Spillage	20				
S3	Spillage Free or Occasional Spillage					
C1	Dirty					
C2	Clean	60				
C3	Very Clean					
	TOTAL SCORE	100				

The higher the score the better the performance. Max possible score = 100; min = 0

31

Table 31.5

Baseline Survey

Weighting System for Scoring Area _____ **Tripper 1** _____ **on Fugitive Material Emissions**

Level	Description	Point Scale	Baseline Date: 23 May	30 Days Date: 23 Jun	60 Days Date: 23 Jul	90 Days Date: 23 Aug
			Rating			
D1	Extremely Dusty					
D2	Dusty	20	15			
D3	Dust Free and Under Statutory Limit					
S1	Extreme and Continuous Spillage					
S2	Frequent Spillage	20	0			
S3	Spillage Free or Occasional Spillage					
C1	Dirty					
C2	Clean	60	0			
C3	Very Clean					
	TOTAL SCORE	100	15			

The higher the score the better the performance. Max possible score = 100; min = 0

Baseline Survey Results			
Area	Tripper 1	Ratings	D2, S1, C1
Date	5/23	Score	15

Table 31.6

30-Day Survey

Weighting System for Scoring Area _____ **Tripper 1** _____ **on Fugitive Material Emissions**

Level	Description	Point Scale	Baseline Date: 23 May	30 Days Date: 23 Jun	60 Days Date: 23 Jul	90 Days Date: 23 Aug
			Rating			
D1	Extremely Dusty					
D2	Dusty	20	15	15		
D3	Dust Free and Under Statutory Limit					
S1	Extreme and Continuous Spillage					
S2	Frequent Spillage	20	0	0		
S3	Spillage Free or Occasional Spillage					
C1	Dirty					
C2	Clean	60	0	30		
C3	Very Clean					
	TOTAL SCORE	100	15	45		

The higher the score the better the performance. Max possible score = 100; min = 0

30-Day Survey Results			
Area	Tripper 1	Ratings	D2, S1, C2
Date	6/23	Score	45

31

60-Day Survey

Table 31.7

Weighting System for Scoring Area ___Tripper 1___ **on Fugitive Material Emissions**

Level	Description	Point Scale	Rating Baseline Date: 23 May	30 Days Date: 23 Jun	60 Days Date: 23 Jul	90 Days Date: 23 Aug
D1	Extremely Dusty					
D2	Dusty	20	15	15	15	
D3	Dust Free and Under Statutory Limit					
S1	Extreme and Continuous Spillage					
S2	Frequent Spillage	20	0	0	15	
S3	Spillage Free or Occasional Spillage					
C1	Dirty					
C2	Clean	60	0	30	30	
C3	Very Clean					
	TOTAL SCORE	100	15	45	60	

The higher the score the better the performance. Max possible score = 100; min = 0

60-Day Survey Results

Area	Tripper 1	Ratings	D2, S2, C2
Date	7/23	Score	60

90-Day Survey

Table 31.8

Weighting System for Scoring Area ___Tripper 1___ **on Fugitive Material Emissions**

Level	Description	Point Scale	Rating Baseline Date: 23 May	30 Days Date: 23 Jun	60 Days Date: 23 Jul	90 Days Date: 23 Aug
D1	Extremely Dusty					
D2	Dusty	20	15	15	15	20
D3	Dust Free and Under Statutory Limit					
S1	Extreme and Continuous Spillage					
S2	Frequent Spillage	20	0	0	15	20
S3	Spillage Free or Occasional Spillage					
C1	Dirty					
C2	Clean	60	0	30	30	30
C3	Very Clean					
	TOTAL SCORE	100	15	45	60	70

The higher the score the better the performance. Max possible score = 100; min = 0

90-Day Survey Results

Area	Tripper 1	Ratings	D3, S3, C2
Date	8/23	Score	70

31

MEASURING EFFICIENCY

Bulk-materials handling is a process; therefore, many of the measurements taken to measure performance are not absolute values but rather individual data points in a statistical process control chart. Most of the results from the process of controlling fugitive material, like dust collection or belt cleaning, follow a classical bell curve. It follows that some of the time the results are above average and some of the time they are below average. It also follows that it is very unlikely it is possible to remove 100 percent of the dust or carryback without extraordinary cost or without unwanted consequences, such as scraping the belt so hard some of the top cover is removed.

As the control of fugitive materials is a continuing process, the more data points obtained, the more representative the results. Therefore, it is best to consider the results from all of the conveyors, rather than just one, to judge the continuing performance of the plant in reducing dust and spillage.

In a similar fashion, it is very inaccurate to refer to efficiency as a value that applies no matter what the condition of the conveyor or the properties of the bulk material. The answer is that it is all relative to the particular bulk material and equipment condition and not necessarily a function of the design of the belt cleaner. If a belt cleaner is rated as 90 percent efficient in removing a layer of carryback, is the remaining layer 100 millimeters (4 in.) thick, or is it 1 millimeter (0.04 in.) thick? Both might be 90 percent effective, but the resulting cleanup cost and operating problems are going to be dramatically different. It would be better to have a cleaner that will remove carryback down to a thickness of 0,1 millimeter (0.004 in.) rather than to the abstract 90 percent efficiency mark.

Because of the large number of variables in handling bulk materials—both in the bulk materials themselves and in the condition of the conveyors—it is physically, financially, and statistically impossible to reduce fugitive emissions to zero over a long period of time. In many operations, an acceptable belt-cleaning performance is one in which the fugitive material can be kept to the level that requires cleaning once a week without causing a safety problem or production loss.

Table 31.9

Profit and Loss Statement (with Level II Cleaning)

Happy Company Profit and Loss Statement For the Period of January 1 to December 31		Currency USD	% of Sales
Income	Sales	$1,000,000	100%
	Total Income	$1,000,000	100%
Cost of Goods Sold	Production Labor	$250,000	25%
	Production Materials	$150,000	15%
	Total Cost of Goods Sold	**$400,000**	**40%**
Expenses	Office Wages & Supplies	$100,000	10%
	Maintenance Wages & Supplies	$250,000	25%
	Water, Gas, & Electricity	$100,000	10%
	Interest, Permits, & Fines	$50,000	5%
	Total Expenses Before Taxes	**$500,000**	**50%**
Profit	Income Minus Expenses	$100,000	10%
	Taxes (50% Tax Rate)	$50,000	5%
	Net Profit After Taxes	**$50,000**	**5%**

31

ADVANCED TOPICS

Sample Calculation: Belt Cleaning ROI and Impact on a P&L Statement

To illustrate the effects of various situations on financial statements, we will use a P&L Statement from the fictional company, Happy Company (**Table 31.9**).

Happy Company: Belt Cleaning ROI

Happy Company is considering the purchase of belt cleaners to reduce cleanup costs, so it estimates the costs for equipment and cleanup labor (**Table 31.10**). A three-year life for the belt-cleaning equipment is used to spread the cost of equipment over time.

The cost of cleanup and the cost of belt cleaning intersect between Level II and Level III Cleaning. Happy Company decides to purchase equipment that will achieve Level II Cleaning (11 to 100 grams per square meter of carryback on the belt after belt cleaning). This cleaning equipment will produce a suitable ROI: 336 percent with payback in 3.6 months (**Equation 31.5**).

Even though the payback is excellent, if Happy Company had kept detailed records, they would be able to look at additional savings resulting from installing and maintaining the belt cleaners. For example, additional savings might come from increased belt and idler life or reduced lost-time accidents in the cleanup crew. Reduced operating expenses, other than cleanup, show a payback of $27,000 USD (**Table 31.11**).

This additional information changes the financial picture, and the graph now shows it is well worth reducing carryback to Level III (0 – 10 grams per square meter) by installing a sophisticated belt-cleaner system (**Figure 31.12** and **Equation 31.6**).

Estimated Costs to Achieve Specified Levels of Cleaning

Acceptable Carryback		Belt Cleaner Purchase & Installation Cost USD (from Supplier)	Belt Cleaner Cost USD per Year (3 Year Depreciation on Equipment	Estimated Belt Cleaning Cost USD per Year Maintenance & Equipment	Total Annual Cost USD for Belt Cleaners Installation & Maintenance	Estimated Clean Up Costs USD
Level	g/m²					
0	>250	N/A	N/A	N/A	N/A	$25,000
I	101–250	$1,500	$500	$1,000	$1,500	$12,000
II	11–100	$6,000	$2,000	$3,500	$5,500	$6,500
III	0–10	$15,000	$5,000	$4,500	$9,500	$3,500

Table 31.10

$$ROI = \frac{SCU}{ACBC}$$

Given: A belt cleaner will save $18,500 USD in annual cleaning. The cost of that belt cleaner is $5,500 USD per year. **Find:** The rate of return.

ROI	Return on Investment in Percent (as Decimal)	
SCU	Annual Savings in Cleanup	18500
ACBC	Annual Cost of Belt Cleaning	5500

$$ROI = \frac{18500}{5500} = 3.36$$

ROI	Return on Investment in Percent (as Decimal)	ROI = 3.36 (336%)(12/3.36 = 3.57 or 3.6 months payback)

Equation 31.5

ROI Calculation for Happy Company with Level II Cleaning

31

Happy Company P&L Statement

Taking the savings from the Belt Cleaner ROI example in which Happy Company installed belt-cleaning equipment to reach Level III cleaning, and putting them in the previous financial statement on the Maintenance Wages & Supplies line, the financial picture of Happy Company changes as shown in the Amended P&L Statement (**Table 31.12**).

By reducing the overall cost of operations with the installation and maintenance of a sophisticated belt-cleaning system, Happy Company's net profit after tax increased by about 30 percent, from 5 percent to 6.5 percent.

Table 31.11

Estimated Costs/Savings to Achieve Specified Levels of Cleaning					
Acceptable Carryback		Total Cost USD for Belt Cleaners Installation & Maintenance	Estimated Clean Up Costs USD	Additional Operating Savings USD	Total Net Cost USD
Level	g/m²				
0	>250	N/A	$25,000	0	$25,000
I	101–250	$1,500	$12,000	($5,000)	$8,500
II	11–100	$5,500	$6,500	($15,000)	($3,000)
III	0–10	$9,500	$3,500	($40,000)	($27,000)

Values in ( ) represent negative numbers.

Figure 31.12

Updated graph shows total net cost is actually a $27,000 USD per year savings with Level III cleaning.

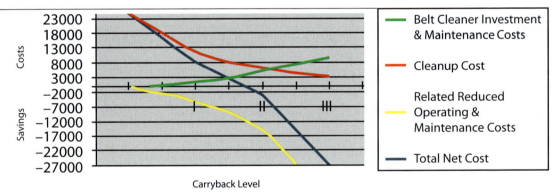

Equation 31.6

ROI Calculation for Happy Company with Level III Cleaning

$$ROI = \frac{SCU + ROC}{ACBC}$$

Given: A belt cleaner will save $21,500 USD in clean up costs and reduce operating costs by $40,000 USD. The cost of that belt cleaner is $9,500 USD per year. **Find:** The return on investment.

ROI	Return on Investment in Percent (as Decimal)	ROI
SCU	Annual Savings in Cleanup	21500
ROC	Reduced Operating Costs	40000
ACBC	Annual Cost of Belt Cleaning	9500

$$ROI = \frac{21500 + 40000}{9500} = 6.47$$

ROI	Return on Investment in Percent (as Decimal)	ROI = 6.47 (647%)(12/6.47 = 1.85 months payback)

31

Amended Profit and Loss Statement (with Level III Cleaning)			Table 31.12	
Happy Company AMENDED Profit and Loss Statement AMENDED For the Period of January 1 to December 31		**Currency USD**	**% of Sales**	
Income	Sales	$1,000,000	100%	
	Total Income	$1,000,000	100%	
Cost of Goods Sold	Production Labor	$250,000	25%	
	Production Materials	$150,000	15%	
	Total Cost of Goods Sold	**$400,000**	**40%**	
Expenses	Office Wages & Supplies	$100,000	10%	
	Maintenance Wages & Supplies	$223,000	22%	
	Water, Gas, & Electricity	$100,000	10%	
	Interest, Permits, & Fines	$50,000	5%	
	Total Expenses Before Taxes	**$473,000**	**47%**	
Profit	Income Minus Expenses	$127,000	13%	
	Taxes (50% Tax Rate)	$63,500	6.5%	
	Net Profit After Taxes	**$63,500**	**6.5%**	

THE PAYBACK OF ROI CALCULATIONS

In Closing...

The discussion and equations in this chapter are not valuable in and of themselves. The key to their value is in their application; they must be used in the evaluation of a system to assess the economic impact of improvement. By applying these economic considerations, one can assess the value to an operation of improvements in bulk-materials handling systems, particularly in the areas of control of fugitive material. With sound record-keeping and analysis of performance, the financial implications of proposed changes can be understood.

The goal of these procedures is to make it easy for management to say "yes" to improvement projects, and then, following completion, to follow up and make them feel good about that decision. This will make it easier to secure approvals for additional projects and make it harder for management to say "no" in the future.

Looking Ahead...

This chapter about Performance Measurements provided tools to assess the need for and benefits of improvements in the control of fugitive materials. Following are the final two chapters, both of which look at conveyors used in specific situations.

REFERENCES

31.1 "Measuring ROI pushes it higher, say Harte Hanks Aberdeen of Enterprise Solutions." (February 12, 2007). *The Manufacturer* (US Edition).

31.2 International Labour Organization. (2003). *Safety in Numbers, Pointers for a Global Safety Culture at Work*. Geneva, Switzerland.

31.3 Takala, J. (18–22 September 2005). *Introductory Report: Decent Work – Safe Work*. XVIIth World Congress on Safety and Health at Work, Orlando, Florida. Available at: http://www.ilo.org/public/english/protection/safework/wdcongrs17/intrep.pdf

31.4 Dorman, Peter. (April 2000). *The Cost of Accidents and Diseases*. Geneva, Switzerland. Available online: http://www.ilo.org/public/english/protection/safework/papers/ecoanal/wr_chp1.htm

31.5 Occupational Safety & Health Administration, U.S. Department of Labor, Mineral Processing Dust Control website: http://www.osha.gov/SLTC/silicacrystalline/dust/

31

Figure 32.1

In the application of belt conveyor systems and components in specific industries, a number of factors related to the material, conditions, equipment, and standards must be considered.

Chapter **32**

CONSIDERATIONS FOR SPECIFIC INDUSTRIES

32

In this Chapter...

In this chapter, we provide an overview of some of the specific conditions that are likely to impact efforts to reduce spillage and dust in a variety of industries conveying bulk materials. General observations and specific information about conveyors and transfers, belt cleaning, and dust management for ten different industries are included.

While there are many things that hold true for all belt conveyors handling bulk materials regardless of industry, numerous factors should be considered for each specific industry. These relate to the materials, conditions, equipment, and standards found in the industry (**Figure 32.1**).

In general, the continuing trends in bulk-materials handling include faster conveyors with higher capacities, greater limits on downtime, and reductions in both maintenance outages and plant workforce.

Due to the capital investment required in bulk-materials handling, most of the industries are risk adverse and reluctant to use unproven technologies. Safety is always an issue, as is the avoidance of regulatory problems. What is basic to all operations is that the materials must be moved in order for the plant to be successful. Added efficiency in materials handling is generally rewarded by added profitability.

The following are capsule looks at some of these industry-related issues. While these observations will hold true in most situations, they might or might not apply to any specific operation.

32

AGGREGATE, CRUSHED STONE, SAND AND GRAVEL

(See also Cement and/or Surface Mining.)

In General

- In many geographic areas, operating permits are difficult to obtain. This means existing operations must be fully exploited while commercial and residential development fills in around them. These operations will be conscious of concerns like dust, spillage, and noise to reduce problems with neighbors.

- In general and around the world, this industry is consolidating. Larger, multinational corporations are acquiring businesses, to expand their geographic territories, as well as integrating on a vertical basis—adding ready-mix concrete and crushed-stone plants to cement production, for example.

- This industry is very dependent on construction and public works (road paving) markets.

- Portable plants are now used for many smaller operations, reducing transportation cost (good for an industry where efficiency is measured by cost per ton of material) and allowing use of smaller deposits.

- Small operations might consist of as few as three or four personnel.

- One advantage in this industry is that the materials handled tend to be very consistent, due to the minimal processing required for the end-product. This allows more plant-wide standardization of accessories than in other industries.

- The recycling of materials—concrete and asphalt pavements, for example—present additional challenges for materials-handling equipment.

- Expense control is a common theme in this industry. Conveyor improvements need to provide obvious cost-effectiveness and prompt payback.

- Maintenance tends to be performed on a breakdown basis, particularly in the smaller plants, and non-essential equipment is often neglected, leading to difficulties controlling fugitive materials.

Conveyors and Transfers

- Plants vary greatly in size and sophistication. The belts tend to be in poorer condition than in most industries, and conveyors may use used belting and belting with numerous mechanical splices.

- Impact can be severe under truck dumps and crushers; loading forces will be lighter on finished-product belts. The use of impact cradles is common, but careful attention to the cradles' duty ratings is essential.

- Chute plugging is a common problem in the screening areas of the plant. Air cannons or vibrators are an effective means to reduce these blockages.

- Spillage of large rocks presents a hazard to pulleys, so winged pulleys are commonly used as tail, bend, and, in some cases, head pulleys. These winged pulleys create belt flap, making cleaning and sealing difficult. The installation of return belt plows and the wrapping of wing pulleys can solve these problems.

- Gravity take-ups are often covered with spilled material. This causes excess or uneven tension and mistracking. Take-up frames are often loose to prevent binding, but this leads to belt mistracking. Take-up frames can be tightened by rebuilding, and "dog houses" can be installed over the take-up pulleys to prevent spilled material from impairing take-up operation.

32

- Skirtboard seals are often rudimentary; the use of used belting as sealing strips is an unfortunate, but common, practice. Chutework is often worn, rusted, or flimsy; and wear liners are located high off the belt, making the installation of engineered skirtboard seals a challenge. However, because of the generally open construction of these conveyors, rebuilding the skirts is generally easily accomplished. Rebuilding the skirtboard-and-seal system to the standards discussed in this volume can control most sealing problems. Self-adjusting seals work well.

- Belts frequently mistrack due to poor loading practices. The installation of multi-pivot training devices before the tail pulley and after the load zone can control most problems.

Belt Cleaning

- The level of belt cleaning required is often minimal, and homemade belt cleaners are often used. When engineered cleaners are used, they typically are used on only difficult belts; even then, they are usually undersized and not well maintained. A typical belt-cleaning system is a single primary cleaner. Carryback is a major source of airborne dust, so belt cleaning will see increased attention in this industry; engineered systems maintained by contractors will become more common, in order to meet regulations consistently.

- Crushed materials on wet belts can be very difficult to clean, because the materials will stick to the belt. Often, particles are hard or sharp-edged, leading to rapid blade wear.

- Using softer urethanes for blade construction, combined with water sprays to help remove the materials, can reduce the wear on the pre-cleaners.

- In some cases, hard-metal secondary blades wear in an uneven "castellated" pattern. This is caused by small particles that wedge between the belt and blade and allow other particles to slip past the cleaning blade. In some cases, the wear

is accelerated by a slightly acidic quality of the water. This wear can be reduced through the use of tungsten carbide blades and the spraying of water onto the belt. Frequent flushing of the cleaner with a water hose is also beneficial.

Dust Management

- Spray-applied water is the typical dust-suppression method. The higher rate of application for "plain" (untreated/no surfactant) water increases problems with materials handling, such as buildup on screens or increased carryback. In addition, some state and local governments have restricted the use (and disposal) of water. As a result, some aggregate operations now use surfactant or foam suppression as an alternative.

- The truck dump is one site in the plant that requires effective dust control. This may require the combination of several dust-control methods to obtain a satisfactory result.

BULK TRANSPORTATION

(See also the listings for specific materials to be handled, e.g., cement, aggregate, coal, forest products.)

This is a review of loading and unloading of bulk-transportation systems—including ships, barges, trucks, and rail cars—and the handling of materials into and from bulk terminals.

In General

- A wide variety of materials—from coal to chemicals, limestone to raw or processed ores, and grain to cement—can be shipped in bulk. It is important that the systems be designed to be flexible and work with a variety of materials. In many cases, different materials are loaded on the same conveyor, and/or the return side of the belt is used for conveying, and contamination between cargos is an issue.

- There are a number of systems used for transporting materials in bulk and for loading and unloading those systems. Materials can be carried by ship, barge, train, or truck; materials can be loaded and unloaded by belt, screw conveyor, pneumatic conveyor, or clamshell. The trend in the industry is higher "tons per hour" rates for loading and unloading, to reduce demurrage charges.

- There is little tolerance for downtime in these facilities, because unexpected outages will result in demurrage charges for delaying the scheduled departure.

- In many cases, these facilities rank high in speed and size of belts.

- Ship unloading is particularly challenging. On self-unloading ships, high-angle conveyors that sandwich the material between two belts are often used. These typically operate at speeds greater than 4,5 meters per second (900 ft/min). Equipment used in ship-unloading systems must be able to withstand rugged conditions, including temperature extremes and exposure to salt water.

- Waterways and ports share use between industrial operations and recreational users. Dust and spillage in these areas will quickly draw complaints to regulatory agencies from non-commercial users.

- Railcar unloading is a particularly difficult situation for effective dust control, as well as material flow maintenance.

Conveyors and Transfers

- Belt tracking is critical to keeping the cargo on the belt and to preventing spillage along the conveyor. The use of multi-pivot belt-tracking devices is effective in controlling tracking problems.

- Access is often sacrificed on shipboard conveyors for space and weight considerations. This will increase the difficulty of maintaining conveyor accessories; therefore, the quality and frequency of service is reduced, leading to excessive dust and spillage. Access is often difficult or impossible on boom discharges, and consideration must be given to portable work platforms or the ability to move the discharge into a location where service can be performed. Incorporating service access will greatly improve the effectiveness of accessories, because they will receive more frequent maintenance.

- The elimination of spillage is especially important (in loading applications), because the materials have a high value per ton and are considered tainted and unusable if they fall on the ground or into the water. In more and more jurisdictions, these materials are considered hazardous waste.

- Spillage on trippers is a common problem, because they must reach minimum loading heights and carry various cargoes. Skirt walls the entire length of the incline and hanging deflectors help control rollback of material. Spillage trays are often placed under the conveyor in critical areas, with provisions for constant flushing or easy cleaning.

- Belts on docks have to be designed with greater than normal edge distance for sealing to control spillage. Since these conveyors are often mounted on structures that must flex, they are prone to mistracking.

Belt Cleaning

- On high-speed conveyors and discharge pulleys that are difficult to reach, long-lasting belt cleaners that automatically maintain cleaning pressure and angle are the best alternative.

- Because some terminal belts are used for multiple materials and/or conveying in both directions, belt washing may be required to reduce cross contamination.

- Wash boxes have proven useful in making sure that belts that travel over bodies of water are very clean. Wash boxes are also effective in reducing contamination when belts are used to handle several different materials. Fresh water must be used in wash boxes and for washdown, or equipment will become corroded.

- Air knives and vacuum systems are effective on very fine materials like alumina. These materials often exhibit a "static cling" tendency, in which the scrapings will flow right back onto the belt; in these cases, a vacuum pickup is needed.

Dust Management

- Dust collection is typically used, ranging from huge central systems to individual point source collectors.

- Conventional methods of dust and spillage control are often not sufficient. The use of wash boxes and maintenance schedules as frequent as once every loading/unloading cycle are necessary.

- Special curved loading chutes are often used to centralize the materials and reduce dust generation.

CEMENT

(See also Aggregate.)

In General

- Consolidation of ownership has made this industry truly global. There is a great deal of information sharing and focus on solving problems in materials handling. This allows new technologies to be quickly adopted at a corporate level. The general trends in this industry directly reflect the global economy.

- Cement (and/or clinker) is readily shipped worldwide in bulk via rail, barge, truck, and ship.

- The cleaning and sealing of very fine, dry, abrasive materials is a continuing challenge.

- Most cement plants also have fuel-handling systems for the coal or pet coke used to fire the kiln. The need for systems to handle waste-derived fuel—including shredded tires, plastics, paint, agricultural wastes, and diapers—will probably continue to grow.

- Materials encountered in this industry vary from large lumps of limestone in the quarry, high-temperature materials

at the clinker cooler, to fine, dry powders that risk fluidization in the packaging/shipping operation.

- Finished cement must be kept dry.

- The elevated temperatures encountered in handling clinker present a common problem.

Conveyors and Transfers

- In the quarry, applications require medium- to heavy-duty ratings for conveyor equipment.

- Primary crusher discharge belts are often subject to huge impact forces that should be evaluated carefully to reduce belt damage, spillage, and leakage.

- On the cement processing side, belts are smaller, and impact is not generally an issue. After crushing, equipment rated for light-duty is generally sufficient.

- Raw and finished cement belts are excellent applications for air-supported conveyors, to reduce spillage and contamination.

- Finished cement tends to aerate, creating problems in transport on belt conveyors from dusting and slideback of material. Belt speeds and angle of inclination need to be carefully selected.

- Clinker is abrasive and handled at high temperatures. This requires special belting and accessories designed for elevated temperatures.

Belt Cleaning

- Cleaning clinker belts may require specialized high-temperature belt-cleaning systems.

- When cleaning belts used for waste-derived fuel, single-blade pre-cleaners are preferred, to reduce the chance of material collecting in the gaps between the blades.

Dust Management

- For dust control in the raw material stockpiles, water has been the suppression system of choice. The use of foam

suppression is effective at the crusher and provides some residual effects.

- On the finish side, the addition of moisture is not allowed, so containment and collection are the only options.

- Air-supported conveyors can be effectively used in this industry for the control of dust.

- Because of the abrasive nature of the clinker and the very fine particle size of the finished cement, leakage from chutes and skirt seals is a common problem. Extra attention to repairing and sealing holes in chutes will provide improved dust control. Belt support and self-adjusting seals are useful in controlling dust at transfer points.

32

COAL-FIRED POWER GENERATION

In General

- Regulation is driving down profitability. To improve reliability and reduce costs, management is seeking to do more with less.

- Control of dust is a major concern for the industry, particularly for plants that have switched to low sulfur sub-bituminous coal, such as Powder River Basin (PRB) coal in the US, or that burn lignite.

- All coal-handling operations must be concerned with fire and explosion issues, including methane gas accumulations and material "hot spots."

- The handling of fly and bottom ash and flue gas desulphurization (FGD) sludge can require additional materials-handling systems and expertise.

- As regulated utilities, many power-generation facilities are required to provide stable productivity by minimizing their risk of unscheduled outages.

- Equipment designed for a particular type of coal can prove problematic when the coal is changed. For example, the lower BTU output of lignite or PRB coal requires additional coal be burned to achieve the same thermal output. This may require changes in the design or operation of the materials-handling system, such as having conveyors increase speed or operate for longer periods. The existing material-chute system might not be able to match the new material flow characteristics and throughput (tonnage) requirements.

- Seasonal changes in climate can result in changes in the performance of the coal as it moves through the materials-handling system.

- Day-to-day changes in material will affect conveyor performance. Due to weather conditions, coal can range from very wet/muddy to very dry/powdery.

- Generally, coal is a relatively low-abrasion material. The exceptions are raw coal or mine refuse as seen at mine-mouth generating plants.

- It is becoming more common to burn auxiliary fuels in combination with coal. These fuels include shredded tires and agricultural waste. The proper introduction of these materials into the coal is critical; if not well metered, this will create spills, plugs, and other operating problems.

- Accumulations of fugitive materials present serious fire/explosion potential from spontaneous combustion, and small events create high dust concentrations and possibly secondary explosions.

- Coal—lignite or PRB coal in particular—is vulnerable to spontaneous combustion of stagnant material: the accumulation of material on chutewalls and underneath conveyors as dust and spillage. Good housekeeping practices and proper cleaning and sealing systems are essential to minimize this risk.

Conveyors and Transfers

- Conveyor belts are generally vulcanized and have a long service-life.

- Conveyor widths are moderate, with widths from 900 to 1800 millimeters (36 to 72 in.) common and speeds of 2,0 to 3,0 meters per second (400 to 600 ft/min) typical. It is common practice to down-rate belt speeds and limit capacity to control fugitive-material issues.

- Crushed-coal conveying is particularly well suited for air-supported conveyors.

- Belt-tracking problems in coal handling can be solved with multi-pivot tracking devices. Standard pivot devices often over-steer on coal-handling belts; that is why these belt trackers are often tied off to one side. Disabling these tracking devices can create even bigger problems that often result in belt damage and material-spillage issues.

- Skirt sealing is important in coal handling. Coal conveyors are particularly well suited for belt-support cradles and self-adjusting sealing.

- The concern for consistent throughput and reduced dust generation leads to the consideration of engineered flow chutes in many coal-handling applications.

Belt Cleaning

- The cleaning of coal-handling conveyors is usually rather straightforward and can be considered the typical application. A standard power plant belt-cleaning system is a dual or triple system, with a urethane primary cleaner and one or two tungsten-carbide-tipped secondary cleaners.

- Some coals contain clays that make cleaning difficult. This material tends to smear on the belt and accumulate as "corn flakes" under return idlers. The normal solution is to operate the belt cleaners at a higher cleaning pressure or to use a more aggressive cleaning angle.

- The use of water is beneficial in maintaining belt-cleaner efficiency, but power plants often have ill-conceived edicts of "no water," due to the BTU penalty. The amount of water required to maintain belt-cleaning efficiency is so small that it cannot be distinguished from other sources of water, such as dust suppression, rain, and even the water absorbed from high humidity.

Dust Management

- Dust-emission regulations affect coal handling, from railcar unloading through the materials-handling system to the bunkers above the boilers.

- Low-sulfur coal burns cleaner, but is typically more friable. As plants move to the cleaner-burning coal, they need to find methods to reduce dust. These might include engineered flow chutes, dust-suppression systems, and the upgrading of existing dust-collection systems (baghouses).

- Water-only dust suppression is not cost-effective, because it reduces the thermal output of the coal. Chemical suppression is the choice of many plants, because its reduced moisture levels minimize the penalty from the added moisture.

- Rotary railcar dumpers for coal unit trains create large problems with dust. Foam or surfactant suppression offers benefits, including a residual effect that stays with the coal as it goes into the stockpile.

- Insertable (modular) dust collectors are well suited for site-specific management of dust if containment is not practical or sufficient.

32

COAL MINING (UNDERGROUND)

(See also Surface Mining and/or Hardrock Mining.)

In General

- Height restrictions are a major factor in this application. This will affect the style of conveyor structure, which in turn affects the accessory systems that can be used. Due to difficulties in the movement and installation of equipment underground, chutes are minimal, and impact cradles are rarely used.

- Modular designs are needed for many components, to counter the limitations on space and access.

- The trend is to use wider conveyors at higher speeds. Main conveyor lines will generally be vulcanized, but other belts may contain a large number of mechanical splices. There is significant use of used belting, as well as use of belts that are well past their true service-life. This means the belts are in rough condition and, therefore, more difficult to clean, seal, and track.

- Belts that feed the main lines are designed to be extended. As the working face moves, additional sections (panels) of belting are installed onto the conveyor. This means these belts incorporate multiple mechanical splices.

- Regulatory approvals (based on safety issues) are a factor in the selection of materials for components. In the United States, Mine Safety and Health Administration (MSHA) sets standards for conveyor belts and the materials in contact with the belt, such as cleaning and sealing systems. Outside the US, British and Deutsches Instit für Normung (DIN/EN) Standards are widely accepted for materials used underground. In most markets outside of North America, aluminum cannot be used underground, because of its low sparking threshold. Regional regulations and agencies (such as ATmospheres EXplosibles [ATEX] or Association for Mining of North Rhine-Westphalia "Landesoberbergamt" [LOBA]) may also need to be considered.

- As of the time of this writing, a final rule published by MSHA in the United States effective December 2008 requires conveyor belts placed in service in underground coal mines to be more flame-resistant than those previously required, beginning December 31, 2009. The rule also requires existing belting to be replaced within ten years.

- It requires damaged rollers or other conveyor components to be repaired or replaced, belts to be properly aligned, materials to not be allowed in the belt entry, and splicing to maintain flame-resistant properties, beginning March 2, 2009. In addition, it lowers the average concentration of respirable dust in the belt air course, also beginning March 2, 2009. MSHA or a reputable belting supplier can be contacted for additional, updated information.

- The use of specialized contractors for accessory maintenance is common in this industry. This is due to the critical role belt cleaning and sealing play in preventing fires, explosions, and production outages.

- Equipment capacities continue to grow, raising the tons of coal produced per miner hour; the downside of this is reduced time for maintenance.

- Underground operations typically use water to suppress dust. This solves one problem but creates others. The moisture will affect the properties of the materials conveyed and the design of equipment, down to type of metal used to reduce corrosion.

32

- With modern mining methods, the lump size is fairly consistent. However, run-of-mine coal, containing rock and clay, presents handling problems.

- Mine safety regulations regarding maintenance procedures make it obvious that quick-change, or service-friendly, systems should be used.

Conveyors and Transfers

- Conveyors are designed to be moved, except for main lines and "slope belts" that carry material out of the tunnels. Belts are larger, thicker, and typically require heavy-duty cleaners and skirting systems. Self-adjusting sealing systems are useful.

- Transfers are often at 90 degrees or have large drop distances, creating heavy impact situations and adding difficulties in sealing. Plugging at transfers is common; air cannons are useful in solving these problems.

- Spillage is difficult to control, due to constantly varying loading conditions creating belt-tracking problems. The use of heavy-duty tracking devices is necessary, because of the constant impact of mechanical splices and generally poor belt condition.

- The detection and removal of foreign objects from conveyors is an issue for underground belts, because the presence of mine tools, roof belts, and other "tramp iron" can lead to belt damage and chute blockages.

- Close quarters and limited headroom present difficulties in the use and replacement of liners inside chutes in underground conveyor applications.

Belt Cleaning

- The materials handled in underground operations contain a great deal of water with the coal. This increases carryback problems. As a result, many operations use three, four, or more cleaners on a belt.

- Many operations use multiple cleaner arrays, with scavenger conveyors to carry the removed materials back from the tertiary cleaners to the main material body.

- Because of the numerous splices, belt cleaners and mounts must be designed to handle repeated impact. For durability, urethane pre-cleaners with heavy blades are preferred. A wide variety of secondary cleaners is used.

- The refuse belt is usually the most difficult to clean and produces the highest rate of blade wear. Aggressive cleaning with frequent maintenance is required, in most cases.

- Cleaners incorporating high-performance urethane blades may be best suited for challenging applications.

- The carrying of lumps on return belts is always a possibilty, due to mistracking, poor belt and splice condition, uneven ground, and/or flooded conditions. Therefore, pulley protection, such as V plows and diagonal plows, is important. This is particularly true on inclined belts, where high moisture content can cause material runback, or what is called a "mud rush" in South Africa.

- The use of maintenance services is common and effective in areas where contract labor costs are low.

Dust Management

- As coal dust is a fire and explosion hazard, all forms of dust management—containment, collection, and various types of suppression—have been used.

- Poor water quality underground may preclude the use of dust-suppression systems incorporating fine-orifice nozzles.

HARDROCK MINING (METALS AND NON-FUEL MINERALS)

(See also Surface Mining and/or Coal Mining.)

In General

- Because metal prices fluctuate, budgets are frequently subject to change, and projects can be accelerated or put "on hold."

- Plants are typically in operation 24 hours per day, 7 days per week, when metal prices are favorable. Outages are scheduled far in advance, and weekly windows for maintenance are short, making contract specialty maintenance an attractive option.

- Ores are often mined by blasting, which results in large lump sizes. Discharge from primary crushers is typically "200 millimeters (8 in.) minus."

- Ores are generally highly abrasive materials, which can shorten the life of the belt and other components.

- The process to make taconite pellets creates situations where high temperatures are common. Taconite dust can become embedded into the edge of the belt, becoming a "grinding wheel" that will saw off a guide roll from a belt-training idler in a matter of weeks.

- Other ores, such as nickel or bauxite, are often found in clay-like formations, resulting in materials with sticky, slick, and/or agglomerating characteristics.

Conveyors and Transfers

- Applications for conveyor equipment are typically on the heavy-duty end of the spectrum, featuring heavy loads and multiple splices on relatively short conveyors. Belt-life is often so short that the belt is considered sacrificial, so more-aggressive cleaning and sealing equipment can be used.

- Overland conveyors are often used for transporting materials and waste. These conveyors are often difficult to access, and they may cross over sensitive areas, such as highways or nature preserves.

- Belt widths of 1800 millimeters (72 in.) and speeds of over 5 meters per second (1000 ft/min) are common. The use of steel cables in the belts introduces new challenges to accessory equipment. It is common for damaged steel cords to protrude from the cover and "whip" belt-cleaner blades and frames.

- Because belts are often loaded to capacity, spillage along the carrying run is common, and the potential is high for large rocks to bounce onto the return side of the belt and end up between the pulleys and the belt. Heavy-duty belt plows are used to prevent damage to the belt and pulleys.

- Abrasion is a significant problem, creating maintenance problems that reduce the effectiveness of accessories and the overall operation of the system. Chute liners and pulley lagging are typically high-wear items. Bolt-on wear liners are commonly used, and the need for access to the bolts limits the skirt-seal options.

- Belt tracking is a common problem on overland conveyors. Heavy-duty multi-pivot tracking devices can be used in place of V-return rollers.

- Due to the stickiness of the materials and the large lump sizes, plugging of chutes is common. Air cannons can be used to reduce this problem.

- Because of the high-capacity transfers and large lump sizes, impact force is extremely high. Normal impact cradles may not be able to withstand the impact forces, and catenary impact systems may be required.

- Chutes with engineered flow can improve the consistency of material movement and help eliminate liner changes resulting from material abrasion against the chutewall. By centering the flow, these chutes can improve the material-crushing phase, by directing materials to the center of the crusher cone.

Belt Cleaning

- In taconite handling, after the bentonite clay is added, the material becomes very sticky. A metal-bladed pre-cleaner is useful for this application.

- In nickel and bauxite applications, the material is thixotropic (gel-like) and difficult to clean from the belt. The use of water to keep belt cleaners free from buildup will enhance performance.

- Secondary cleaners usually have tungsten carbide blades to improve performance and extend life in the face of abrasive material. Heavy-duty cleaners with extra thick tungsten-carbide blades are often used.

- Blade-life is considerably shorter than in other applications, and cleaners require frequent service to remain effective. Cleaners with constant tension for the life of the blade are necessary to keep service intervals to a reasonable period.

- Large lumps can ricochet in the chute and damage belt cleaners, either by direct impact or by becoming lodged in unusual places, disabling the tensioner, or bending the cleaner frame. The use of extra-heavy-duty primary cleaners is required. A system of two primary cleaners often provides acceptable cleaning and is less susceptible to damage.

- High material adhesion may require the use of a crust breaker, installed before the pre-cleaner, to improve overall cleaning performance and extend the life of primary cleaner and secondary cleaning systems.

- Pulley cleaning is important on steel-cord belts, to prevent over-tension and puncture of the belt from material buildup. Arm-and-blade secondary cleaners are often used "upside down" in this application, in combination with a rock guard or deflector plate, to remove the cleaned material from the belt.

Dust Management

- Dust suppression can be used not only in mining, but also in pellet transportation. Dust collection is also common throughout the plant.

- Taconite and other process plants may use a combination of coal and gas to fire the furnace, creating dust, resulting in a need for dust collection above the coal-handling system.

- Chemical suppression is also used to suppress silica dust.

- Dust can also be managed using leading-edge conveyor technologies, including engineered flow chutes and air-supported conveyors.

32

METALCASTING

In General

- As might be expected, high material temperatures and challenging service conditions are common.

- The type of metal cast—ferrous or non-ferrous—is not as significant as the handling of the materials used to form the mold for the casting.

- Sand from the muller that is ready to be formed into a mold is known as prepared sand, or green sand; sand from the shakeout operation (after the casting has been removed from the mold) is called return, reclaimed, or recycled sand.

- Dusty/warm/moist materials can break down urethane products, like cleaner blades.

- While foundry sand is not highly abrasive, its moisture content can lead to corrosion, even in an abrasion-resistant plate.

- Sharp pieces of cast metal will occasionally slip through with the return sand to damage belts or other components.

Conveyors and Transfers

- Belt speeds are not high; they are typically in the range of 0,25 to 1 meter per second (50 to 200 ft/min).

- Applications in foundries are typically light duty, with the exception of reject/return systems, where metal pieces the size of engine blocks will occasionally appear.

- Belt tracking is often affected by spillage of the sand, which tends to build up quickly on return rollers. Multi-pivot tracking devices can be used effectively.

- Belts carry returned, or reclaimed, sand that has already been used in the casting process from the shakeout operation back to sand storage, or the muller, for reuse. The sand here is still hot from the previous molding.

Belt Cleaning

- Softer urethane blades may last longer in cleaning the round particles found in foundry sand from the relatively slow-moving belts.

- Brush cleaners may be effective in removing sand from worn belts.

- Belts carrying return sand usually have a magnetic head pulley, to remove any metal that has remained with the sand. Metal-tipped cleaners should not be used within 300 millimeters (1 ft) of a magnetic pulley.

Dust Management

- Containment and collection are the choices for dust control. It is important to avoid adding moisture to molding sand.

- Foundry dust is often considered hazardous, due to its high silica content.

PROCESS INDUSTRIES

This section looks at "lighter duty" applications in industries including food, chemical, pharmaceutical, fertilizer, grain, and tobacco products.

In General

- While these industries are different in materials handled, there are a number of common standards of design and construction.

- In general, these industries qualify as light-duty applications, with narrower—450 to 900 millimeters (18 to 36 in.)—belts, slower conveyor speeds, and smaller rates of material flow. In many ways, this equipment is a smaller version of the systems used in mine-grade applications. However, because of the limited head pulley size, belt speeds, belt tensions, and special cleaning requirements, they require special components.

- In many applications, food-grade materials are required to be used in the construction of materials-handling equipment and accessories. Food-grade polymers are used in many cases; aluminum blades can be used in applications such as tobacco. The blade type is determined by the belt speed, materials handled, and material temperature.

Conveyors and Transfers

- Flat conveyor belts are more common than troughed belts.

Belt Cleaning

- The smaller size of pulleys, conveyors, and belts can create difficulties in removing built-up materials from belts.

- Specialized washdown systems are used in many industries. Materials-handling equipment and components must be compatible with cleaning processes and chemicals. Belt-cleaning systems must be designed to be easily removed for sanitary requirements.

Dust Management

- Due to the overall limited size of equipment and smaller volume of materials handled, the quantity of dust is not as large as in other industries. However, the value of the materials lost as dust is often greater, adding benefits for its capture and reclaim, or better yet, the prevention of dust escape.

32

PULP & PAPER / FOREST PRODUCTS

In General

- A paper mill is a factory devoted to making paper from wood pulp and other ingredients with a Fourdrinier Machine. A pulp mill is a manufacturing facility that converts wood chips, or other plant fiber source, into pulp that is used at the mill or shipped to a paper mill for further processing.

- Conveyors are used in moving logs to the chipper, carrying chips to the digester, and moving coal and bark to the plant's power-generation system. As the chips are a precious raw material, and the bark is a waste by-product, the chip-handling system is more carefully maintained. The bark-handling system, due to its cargo of oddly sized and stringy materials, is generally more of a mess.

- Plastic can contaminate the pulp made from chips. Consequently, there is a concern with the use of plastics on process and materials-handling equipment. Some plants will forbid the use of urethanes as belt cleaners, for example. This is typically a plant-level decision; there seems to be no set rule in the industry, or even at the corporate level.

- Issues in the industry include the use of additional, or alternative, fuel supplies, including tire chips.

- Other operations in the plant can benefit from improved conveying, including power generation, ash handling, and chemical handling.

- The addition of recycled paper to the papermaking process may affect the plant's wood-processing requirements.

- Most paper mills operate from several sources of chips, including in-house chips and purchased chips.

- The materials conveyed in pulp and paper mills pose some problems. The fibers interlock easily, leading to a buildup of stringy material on equipment and to the plugging of chutes in the wood yard. Wood chips contain sticky resin that builds up on the belt and rollers and is very difficult to remove.

Conveyors and Transfers

- Some applications in this industry are light; other applications are heavy-duty. At the heavy-duty end of the scale are applications near the debarking drum, where tree-length or cut logs are discharged onto conveyors to be moved to the chipper.

- Conveyor belts are typically 500 to 1200 millimeters (24 to 42 in.) wide, and they operate with speeds of 1,3 to 1,8 meters per second (250 to 350 ft/min). In many cases, belts are flat or have chevrons.

- Sticks and chips tend to catch under skirting and can block the loading zone or cause spillage. Careful attention to loading-chute design, with tapered loading chutes and gradually increasing liner height, will help prevent this. Low-pressure, self-adjusting skirt seals tend to be more "self cleaning" than fixed-seal systems.

- The use of air cannons needs to be reviewed carefully, because wood chips and bark often require much larger air cannons than a typical application, due to the porosity of the bulk materials.

- In oriented strand board (OSB) mills, the chips are shaped differently, but they are also coated with a resin prior to the sheets being put into the oven. Depending on the location of the coating process, the rollers can have a severe buildup of resin.

32

Belt Cleaning

- The presence of wood resin, or pitch, on the belt creates cleaning problems. This is difficult to remove by itself, and it causes other materials—bark strands, chips, or fines—to stick as well, complicating the cleaning process. The sticky resin on the conveyor belt can make the urethane pre-cleaner blades chatter and vibrate, causing the cleaner blade to heat up and even, in some cases, to melt. Thin tungsten-carbide-tipped secondaries, operated at higher than normal pressures, are sometimes needed to solve this problem. Running the belt with no load, allowing the cleaner to remove the buildup after each production cycle, can also help prevent the material from hardening.

- The discharge of the digester-feed conveyor contains chemicals that can soften urethane belt-cleaner blades and, therefore, shorten blade-life.

- Cleated belts are often used to move the chips or bark up inclines. Brush cleaners and chevron belt cleaners are required to clean these belts, but these are difficult applications.

- On bark-handling conveyors, "arm and blade" secondary cleaners tend to accumulate the buildup of stringy material that interfere with performance. Using unitary blade-cleaners or in-line cleaners without arms will be more successful.

- Special belt-cleaner blades may be required for white paper production, to prevent the paper's contamination with colored particles from the wear of the blade.

Dust Management

- Fog suppression is common, because plants try to avoid interfering with process chemistry by adding chemicals (like dust-suppression surfactants) to their pulp.

- Dust-collection systems are common; in most cases, these are large, central baghouses.

32

SURFACE MINING (COAL OR OTHER MINERALS)

(See also Coal Mining and Hardrock Mining.)

In General

- The large quantity of materials handled for these operations leads to the use of oversized equipment—from bucket-wheel excavators to haul trucks—and wide, high-speed, high-tonnage belt conveying systems.

- Materials extracted in surface mines range from lignite and low-rank coal to ores for base and precious metals.

- Large quantities of soil and subsoil typically must be removed before the ore is reached. This overburden can change greatly in material characteristics as different layers of strata are removed on the way down to the ore level. After the overburden is removed, the desired material is removed.

- Overburden and ore might be removed by some combination of dragline or bucketwheel excavator, which will feed haul trucks or high-speed conveyors.

- The reclamation of mined areas often involves another high-capacity materials-handling system.

Conveyors and Transfers

- Wide, high-speed, high-capacity conveyors are the rule rather than the exception. For example, German lignite operations use conveyors with belt widths up to 3200 millimeters (124 in.), operating at speeds up to 10,5 meters per second (2100 ft/min). These operations push equipment suppliers for bigger, faster, higher-tonnage systems.

- There are often extreme levels of impact in conveyor loading zones from un-crushed, run-of-mine materials. Loading zones should be designed for these forces with impact idlers and impact cradles, or a combination. To handle this impact, many operations incorporate catenary idlers, creating difficulties in the sealing of load zones.

- Changing material characteristics—from different layers of overburden, for example—might allow accumulations that can choke or plug chutes. The installation of air cannons and/or vibrators on transfer chutes might be useful. Severe belt-cleaning and chute-plugging problems are experienced in the winter or wet seasons.

Belt Cleaning

- High belt speed and material velocity lead to high frictional temperature and high vibration levels. Belt cleaners must be engineered to withstand these conditions. Pre-cleaners with a high volume of urethane are often used to dissipate heat and extend service-life.

- High conveyor operating speeds in some applications may prevent the use of higher-pressure secondary cleaners; however, the large head pulleys on surface mine conveyors may have enough room for two lower-pressure pre-cleaners below the material trajectory.

- High material adhesion may require the use of a crust-breaker, installed before the pre-cleaner, to improve overall cleaning performance and extend the life of pre-cleaner and secondary cleaning systems.

- Return-belt cleaning is important because of the potentially large size or sticky nature of materials. These materials become trapped between the belt and bend pulleys and can damage the belt, by either puncturing it or increasing the tension. The buildup of materials can also quickly cause mistracking.

- Cleaning devices for the inside of the belt must be designed for high impact and to prevent entrapment of materials in the suspension systems. Pulley cleaners are often applied in addition to tail-protection plows.

- Due to the length and carrying capacity of some belts, dual drives (with the second drive a tripper-style booster) are sometimes installed, which can be another source of spillage. It is particularly difficult to install cleaners for these, due to limited space and often small pulley diameters.

- High-performance urethanes can provide superior performance and life in tough cleaner applications.

Dust Management

- Spray-applied water is the typical dust-suppression method; however, high rates of water application will increase problems with carryback or screen blinding. Additionally, the availability of water may be an issue. Surfactant or foam suppression should be considered as an alternative.

- The truck dump leading to the primary crusher will generally require dust-containment systems. Dust from truck dumps is often controlled with the use of a settling shed.

KNOW YOUR "ENEMY"

In Closing...

Each industry that uses belt conveyors for bulk-materials handling has unique conditions that need to be taken into account when determining the configuration of conveyor and accessories. Although there are some things that apply to the use of conveyors in general, differences in materials conveyed, conditions in which the conveyors are located, equipment, and industry standards all impact efforts to limit spillage and dust.

Looking Ahead...

This chapter, Considerations for Specific Industries, continues the section The Big Picture of Bulk-Materials Handling by explaining the importance of knowing the unique conditions of the industries involved in efforts to reduce fugitive materials. The next chapter, Considerations for Specialty Conveyors, concludes this section.

32

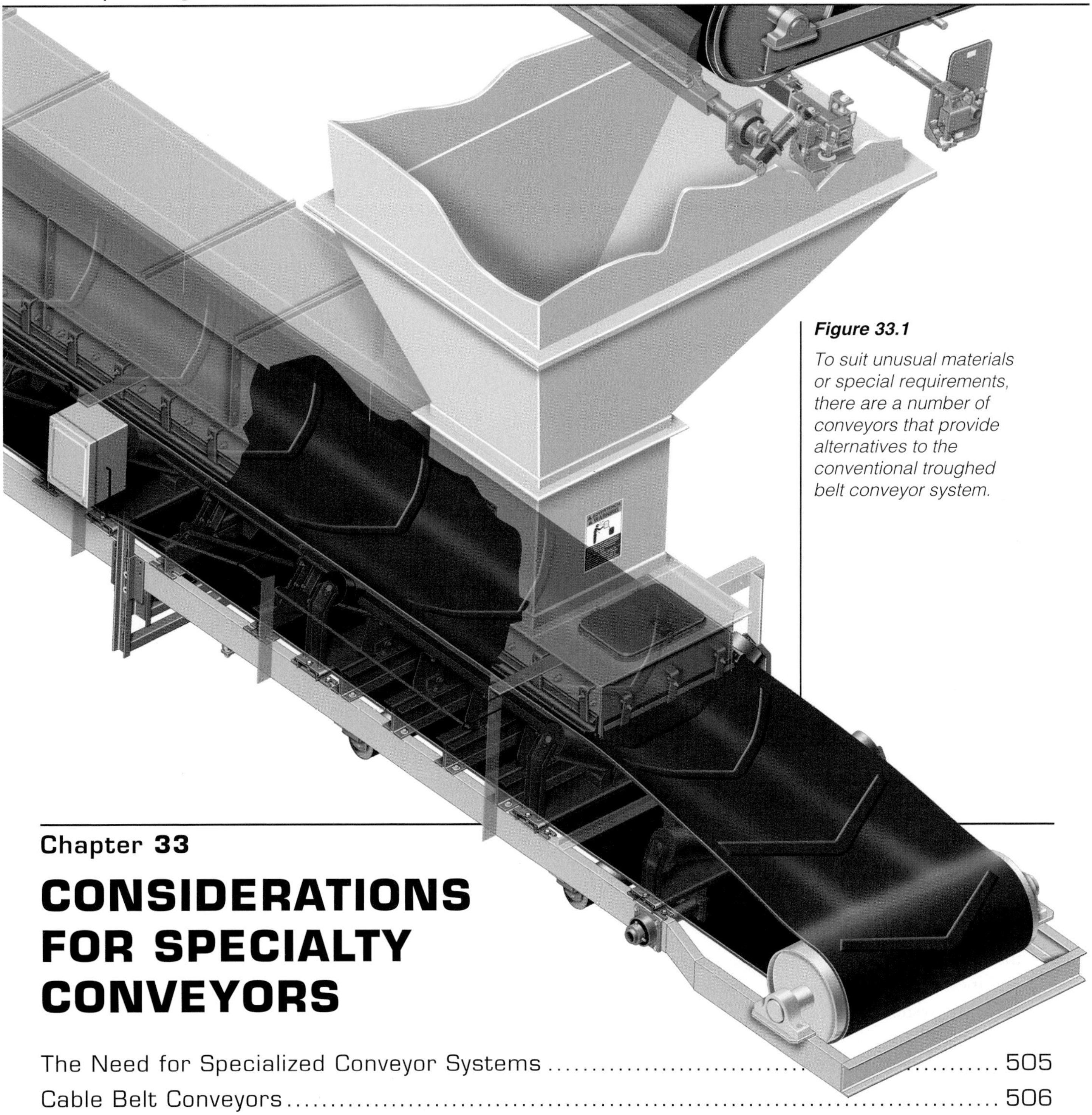

Figure 33.1

To suit unusual materials or special requirements, there are a number of conveyors that provide alternatives to the conventional troughed belt conveyor system.

Chapter 33

CONSIDERATIONS FOR SPECIALTY CONVEYORS

33

In this Chapter...

In this chapter, we describe a variety of alternatives to the conventional troughed belt conveyor used when special circumstances arise. For each alternative technology presented, we provide benefits, drawbacks, and typical applications. We also look at the future of conveyor technology.

While for most applications the conventional troughed-idler belt conveyor dominates the market, there is a growing appreciation of specialized or alternative belt conveyor constructions (**Figure 33.1**). These systems still use a belt to carry the load, but other components have changed. The changes allow these conveyors to provide different capabilities or serve special applications.

THE NEED FOR SPECIALIZED CONVEYOR SYSTEMS

There are several common characteristics among the various alternative technologies described here. The most common are:

A. Improved environmental control

Most state-of-the-art conveying requires clean conveying, i.e., without spillage and dust. Minimizing transfer points, enclosing the belts, and controlling the spillage and dust associated with load-in or load-out operations are essential to obtaining and maintaining permits, maintaining safe working conditions, and increasing the performance of the conveyor system. The unconventional conveyors discussed here may provide general or specific benefits through improved control of material.

B. Reduction in labor and capital costs

These advanced conveyor technologies are often selected to reduce either the size of the conveyor's "footprint" (that is, the floor area required to install the system) or the number of transfer points required to carry the cargo from the loading point to the discharge. Either can reduce capital and maintenance costs. Reducing the size of the foot-

print may be advantageous in situations where there are significant space constraints. It is also cost effective in terms of capital investment in structural components and the operational cost of maintenance. A shorter construction schedule may be another plus, because many of these designs are modular and able to be fabricated off site and "dropped in" when moved to the site.

C. Reduction in number of transfer points

In addition to requiring a smaller investment in transfer-point components, such as idlers, sealing systems, and transfer chutes and liners, these specialized conveyors do not require the man-hours or equipment typically needed to cope with dust and spillage. There is also less material degradation because the material travels through fewer transfer points.

There are a number of factors that are influencing the advancement in conveyor technologies, including the always-present (and conflicting) needs to traverse challenging terrain and minimize capital investment. However, one factor that is emerging as a common denominator in all of the applications of specialty-conveyor technologies is environmental protection: keeping the material contaminant-free and preventing the material from contaminating its surroundings.

These systems offer effective alternatives to the conventional troughed belt conveyor for applications where there are unusual materials, special requirements, or space limitations.

Without reference to trade names or proprietary information, the following summarizes a number of these specialty-conveyor technologies, listed in alphabetical order. Additional information about alternative conveyor technologies is available in Conveyor Equipment Manufacturers Association's (CEMA) *BELT CONVEYORS for BULK MATERIALS, Sixth Edition*, or from the manufacturer of any specific system. *(See Chapter 24: Air-Supported Conveyors for information covering this topic.)*

CABLE BELT CONVEYORS

Instead of idlers, cable belt conveyor systems support the belt with two continuous wire ropes, one near each edge of the belt (**Figure 33.2**). These cables support both the carrying and return sides of the belt and provide the mechanism for moving the belt as the drive power is applied to them.

Cable belt conveyors utilize a special belt with high levels of cross-belt stiffness. The belt is more rigid across its width than typical conveyor belting. This is because as the system uses no idlers, the belt must support both its own weight and the cargo. Tracks (or shoeforms) that hold the cables are molded into both the top and bottom covers of the belt near the outer edges.

Loading and discharge zones resemble a conventional conveyor with the pulleys designed to accommodate the cable tracks in the belt. The ropes and belt are separated at the conveyor's loading (tail) and discharge (head) ends, where each wraps around a separate pulley.

Cable (or wire rope) conveyors are most appropriate for long-distance applications requiring vertical and horizontal curves; they provide reliable transportation over long distances and challenging topography. Typical applications are at least one kilometer (3300 ft) in length. Because they can handle horizontal and vertical curves, cable belts may reduce the need for trans-fer points; the material degradation and maintenance costs associated with transfer points are correspondingly decreased.

Benefits

- Are suitable for horizontal and/or vertical curves

  Cable belt conveyors are well suited to applications over challenging terrain.

- Are ideal for long-distance conveyors

  Typical installation is at least 1000 meters (3300 ft) in length.

- Provide a smooth ride

  Material does not segregate due to the agitation created by idlers.

Drawbacks

- Design trade-off

  Friction available between the belt and cable system is inadequate for using proper belt support in the loading area.

- Difficulty sealing or cleaning the belt

  Hardware required may cause the belt to bunch up or stretch. The placement of the cable drive mechanism at the head pulley makes it difficult to fit cleaners into this location.

- Vulnerability to climate conditions

  Weather can cause a loss of friction between the cable and belt.

Typical Applications

- Long overland systems

  These systems have proved effective used for ore-handling systems over difficult terrain.

Figure 33.2

Cable belt conveyors use a wire rope to support and move the belt.

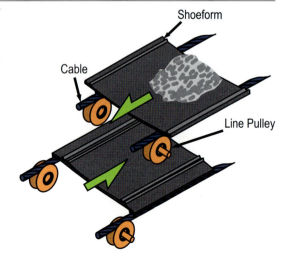

33

BELTS WITH CLEATS OR CHEVRONS

Cleated belts are belts that have large ribs, fins, or chevrons attached to the belt surface (**Figure 33.3**). These cleats can be attached to the belt surface by vulcanization or with mechanical fasteners. The construction of the conveyor is conventional on the carrying side, with the trough angle limited by the stiffness of the belt and the configuration of the cleats. The return idlers must account for the cleats through the use of split rolls, rubber disk rollers, or wing rollers. In some cases, belts have low-profile reverse chevrons—12 millimeters (0.5 in.) or less—attached to or molded into the surface (**Figure 33.4**). Generally, these belts do not require special return idlers.

Benefits

* Allows an increased angle of incline

    These systems can carry material up inclines as steep as 45 degrees. This allows the overall footprint of the conveyor system to be reduced (**Figure 33.5**).

Drawbacks

* Limitations in capacity

    As the angle of incline increases, capacity decreases.

* Susceptibility of cleats to damage

    Objects elevated above the belt surface draw damage from material lumps, skirt seals, and cleaners.

* Difficulty cleaning the belt

    Special devices such as water sprays, air knives, beaters, or specially-designed cleaners with individual fingers must be installed to control carryback. Reverse chevrons make belt cleaning easier but will still require special cleaning devices.

* Difficulty with skirt sealing

    Cleats must be trimmed from the belt's edge in order for sealing systems to control dust and spillage.

* Limited tolerance for belt wander

    Cleated areas will migrate back and forth underneath the sealing system to prematurely wear the seal if the belt does not track true through the loading area.

Typical Applications

* Aggregates

    These systems are especially useful where space is limited and inclines are steep.

* Wood products

    These systems are especially helpful where space is limited/restricted.

* "Round" bulk materials

    Cleats/chevrons are effective where the cargo may tend to roll back down the conveyor.

Figure 33.3

Belt cleats are available in a variety of configurations, to match the conveyor and the cargo.

Figure 33.4

In reverse chevron belts, the chevrons are molded into the surface of the belt.

Figure 33.5

Belts with chevrons are effective in carrying material up inclined conveyors.

33

HORIZONTALLY- AND VERTICALLY-CURVED CONVEYORS

Vertically- and horizontally-curved belt conveyors are used when overland conveyors are unable to connect the loading point and the discharge point in a straight line due to terrain obstructions, right-of-way limitations, industrial or property restrictions, or many other reasons. The curve capabilities of these conveyors may eliminate the requirement for one or more intermediate transfer points (**Figure 33.6**).

Conventional belt conveyors can be made to turn in a horizontal curve by elevating the idlers on the outside of the radius. Other than accommodating the elevated structure, no other major changes are necessary, and the conveyors use commonly-available components.

An engineering analysis is required to specify the correct belt, elevation, and tension for a given situation. The turn radius varies but is commonly on the order of 95 meters (300 ft).

Intermediate drives are used to control belt tensions when they are strategically placed to allow the belt to turn curves with a tight radius. By tilting the idlers, the designers of these curved conveyors have been able to achieve the desired geometry and belt tension. Engineering via computer-modeled simulations has enabled designers to ensure that system components interact together properly.

The best application for this technology allows long overland conveyors to adapt to the terrain, thus eliminating transfer points. Design and construction have been developed to the point where it is common to see overland conveyors up to several kilometers (miles) in length incorporating multiple horizontal curves.

Vertical curves in conveyors can be concave (curved up) or convex (curved down). The down-curved belts usually do not present many problems, but the curve must be gradual so as not to damage the belt by bending it too quickly through the curved area. However, up-curved belts almost always present problems. If the up-curved system is not designed perfectly or there are changes in loading or belt tension, the belt will lift off the idlers in the curved area. This can raise the belt in the loading zone, causing severe damage to the belt and making the loading zone impossible to seal.

Benefits

- Can eliminate the need for a transfer

  Curves can accommodate changes in direction where transfer points would otherwise be required for conventional conveyors.

Drawbacks

- Additional engineering required

  Design of the system calls for more-experienced engineers and, thus, more-expensive engineering.

- Increased intricacy of construction

  Installing the system to achieve a precise path over landscape requires higher expertise.

- Tension at belt edge

  Increased edge tension can cause problems for belt and idlers.

Typical Applications

- Long overland conveyors

  Curves allow the conveyor to go around obstacles without requiring transfer points.

- Elevating conveyors

  The system allows raising material into storage vessels or to process operations at higher elevations in the plant.

Figure 33.6

Some specialized engineering will allow conveyors to incorporate vertical or horizontal curves.

33

POCKET AND SIDEWALL CONVEYORS

Pocket conveyors are similar to cleated conveyors in that they have large center cleats (**Figure 33.7**). Flexible sidewalls are added to the belt, forming a continuous series of pockets similar to a bucket elevator. The belt is of special construction with a high cross-belt stiffness to accommodate the necessary bend. Return pulleys are limited to contact only the outer edges of the belt. The pocket belt conveyor is often configured in an "S" shape and used in situations where there is limited space available. Pocket conveyors provide high conveying capacity over great heights with minimal space requirements (**Figure 33.8**).

Benefits

- Can elevate cargo in relatively small space

  The system can lift cargo vertically with minimal belt support.

- Allows placement of cargo where it needs to be

  The belt can be twisted about the vertical axis to allow offset discharge.

- Forms its own seal

  Side seals are not required.

- Follows a restricted path

  Relatively tight convex and concave bends can be incorporated.

Drawbacks

- Required special belting

  Belt is expensive and must be custom fabricated. There may be long lead times for replacement belting.

- Susceptibility of pockets to damage

  Walls extend above the belt surface; thus, they may catch on obstructions.

Figure 33.7

Pocket sidewall conveyors incorporate central ribs and flexible sidewalls to contain material.

Figure 33.8

Pocket conveyors provide high conveying capacity over great heights with minimal space requirements.

- Difficulty cleaning the belt

  Conventional belt cleaners are ineffective in removing material from inside pockets.

Typical Applications

- Small, dry lumps or dry, fine bulk materials requiring steep inclines

  Material carried in "pockets" will not roll back on inclined conveyors.

- Applications with limited space

  Contained material can be raised up a steep angle, so the conveyor needs minimal floor space.

33

ENCLOSED-ROLLER CONVEYORS

Enclosed-roller conveyors are totally enclosed where the belt is fitted with cleats or ribs on the carrying side. In addition to moving the cargo, these ribs function to drag spillage and dust along the decking under the return run to the loading zone (**Figure 33.9**). There are various methods of self-loading the spillage and dust back onto the belt, usually with paddles attached to the tail pulley. Some designs have the idlers totally enclosed; others have idlers cantilevered or accessible from the outside.

Benefits

• Provides enclosed carrying and return sides of the conveyor

Dust and spillage are contained within the enclosure.

• Uses modular construction

Design and fabrication reduce time required for installation.

• No skirtboards are required

There is no adjustment or replacement of sealing rubber required.

Drawbacks

• Problematic drag system

The drag system may not be effective with all materials. Sticky materials will accumulate.

• Difficult maintenance of idlers

Location of rolling components inside enclosure makes access a challenge.

• Limited speed due to drag system

Returning spillage to the load can slow the entire system.

Typical Applications

• Grain handling

These systems are often seen in grain elevators and processing facilities.

POUCH, FOLD, TUBE, AND PIPE CONVEYORS

This class of specialty conveyors uses a specially-designed belt and carrying system (**Figure 33.10**). In all cases, the belt is formed into a sealed, dust-tight, tube-like shape. In pouch or fold conveyors, the belt has special edges that are captured by an overhead carrying system similar to a trolley conveyor. Tube or pipe conveyors feature a specially constructed belt that is rolled into a tube by a series of radially-placed idlers. The tube is opened at the loading and discharge points with special guides (**Figure 33.11**).

Because these conveyors are completely enclosed, environmental pollution and spillage during transport are eliminated, contamination or theft of cargo is prevented, and product degradation is reduced. The horizontal curves that can be achieved with pipe conveyors decrease the number of transfer points and the number of drives required to power the system. Pipe conveyors can accommodate steep inclines and declines, which may reduce the length of a conveyor while conforming to space or property constraints. The return belt of a pipe conveyor may also be formed into a tube shape to enclose the dirty side of the belt to prevent spillage and dust. In some cases, the return belt may be used to convey material.

Benefits

• Reduces loading zones and transfer points

These systems are able to take tight vertical and horizontal curves, so they can eliminate transfers.

• Provides a contamination-free operation

The load is enclosed, so no material escapes and no contamination enters.

• Allows reasonable flow rates for most bulk materials

The system provides a capacity suitable for most requirements.

33

Figure 33.9

Enclosed-roller conveyors incorporate ribs to pull material to the discharge.

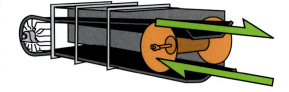

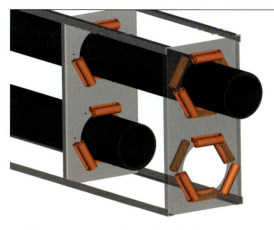

Figure 33.10

Left: Pipe or tube conveyors roll the belt into a tube to contain the material. Right: Pouch or fold conveyors catch the edge of the belt to form a pocket.

- Can carry material in both directions

  The system can carry material to and from a port, for example; therefore, it can eliminate the need for a second conveyor system.

Drawbacks

- Greater cost

  The structure, guides, and belting are more complicated and, thus, more expensive.

- Required specially-designed belting

  Special belting can increase both cost and time when replacement is required.

- Difficulty sealing at loading points

  Belt's transition to final shape makes sealing difficult.

- Required belt cleaning

  Belt cleaning is critical, because carryback can interfere with the hardware that carries the belt.

- Additional required power

  The system can use as much as 30 percent more power than conventional troughed-idler conveyors.

Figure 33.11

The tube is opened at the loading and discharge points with special guides.

Typical Applications

- Ports

  Enclosure of cargo prevents spillage into bodies of water.

- Industrial applications where space is limited

  The system's ability to handle high-incline angle and curved path allows installation in tight spaces.

- Materials requiring contamination-free operation

  Enclosure of the load shields the cargo from contaminants.

33

SANDWICH CONVEYORS

Sandwich conveyors, also known as high-angle conveyors, are generally used to carry materials up steep angles using two smooth-surfaced belts. The cargo is placed between the two belts like the meat in a sandwich (**Figure 33.12**). These systems use conventional troughing idlers to carry the belt sandwich. Radial pressure due to belt tension urges the outer belt onto the material load imparting the hugging pressure that develops the material's internal friction. Sandwich conveyors may be of varying profiles, from C-shaped to S-shaped to multiple-curved and are used to lift or lower materials at angles up to 90 degrees within a small footprint (**Figure 33.13**).

Benefits

* Can transport high capacities up steep inclines

  The system hugs the material, so the belt can climb high angles without material slide-back.

* Is ideal for limited ground space

  The high incline angle holds down the conveyor's footprint requirements.

* Utilizes conventional components

  The system is designed with standard conveyor equipment, including smooth-surfaced rubber belting.

Drawbacks

* Service on rolling components is difficult

  These systems enclose a number of components in narrow spaces.

* Proper loading is critical

  If the material is poorly positioned during loading, the belt sandwich might not seal tightly, negating the system's benefit.

* Belt cleaning is difficult

  Belt cleaning presents special challenges, because the cover belt may need to be cleaned while it is in an upside-down position. In addition, space is typically limited for installation of belt cleaning and tracking equipment.

Typical Applications

* In-pit crusher operations

  These systems are useful where a steep angle is required to raise material from the pit.

* Silo filling

  The system is helpful because of the high-incline angle that can reduce the footprint of the conveyor.

* Self-unloading ships

  These systems make it possible to raise material from the hold of the vessel.

Figure 33.12

Sandwich conveyors place the cargo between two belts to carry it up steep inclines.

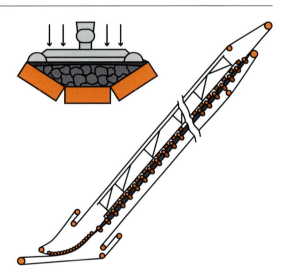

Figure 33.13

Sandwich conveyors are commonly seen in applications like silo filling and self-unloading ships.

33

THE FUTURE OF CONVEYOR TECHNOLOGY

As these advanced technologies continue to evolve, the benefits outlined in the preceding section will become ordinary features of conveyor systems. Thus, belt conveyors will have fewer transfer points and require less maintenance. Other changes will take place, including the adoption of new materials and better maintenance planning.

New Belting Materials

Conveyor belts will be manufactured of newly-developed materials that are lighter and able to withstand high temperatures. Synthetic compounds for splicing may replace mechanical and vulcanized splices. Anti-static, non-stick belt coating will repel material to reduce carryback, in addition to demonstrating improved age and weather resistance.

Predictive Maintenance and Remote Diagnostics

More conveyor systems will incorporate sensors that provide maintenance alerts when equipment requires attention. Predictive maintenance programs that monitor conveyor systems online and generate monitoring reports that predict expected faults have the potential to significantly lower maintenance costs. With this accurate, concise, and practical information on the operating condition of critical plant equipment, maintenance personnel will be able to diagnose problems such as unbalance, misalignment, and bearing defects before they threaten production.

Remote diagnostic devices are available to pinpoint hot spots, detect moisture levels, and predict electrical and mechanical failures in motors and any systems that may be connected to the motors. Belt-cleaning equipment that adjusts to suit changes in the belt surface coefficient of friction is available. Other sensors utilize vibration analysis, laser alignment, and oil and lubricant analysis to alert maintenance personnel of components requiring attention. The prevalence of such devices is likely to increase in the near future. The result will be reduced maintenance costs from timely attention to deteriorating equipment rather than emergency outages to handle catastrophic equipment failures.

THE NEED FOR ALTERNATIVE SYSTEMS

In Closing...

There are numerous examples of bulk materials-handling facilities utilizing advanced conveyor technologies to comply with environmental regulations, to transport material over long distances with challenging topography, and to maximize their return on investment.

These alternative conveyor systems may be suitable for particular installations or solve one particular problem. However, each poses its own set of limitations and drawbacks. These systems exist for reasons: They have achieved some degree of commercial acceptance, because they fit at least one need. For general purposes, the conventional troughed-idler belt conveyor is the performance standard and the value leader against which these other systems must be evaluated. Troughed-idler belt conveyors have a long history of satisfactory performance in challenging conditions.

Looking Ahead...

This chapter about Considerations for Specialty Conveyors is the final chapter in The Big Picture of Bulk-Materials Handling and the final chapter in this book. The following section describes Research, Personnel Development, Services, and Products at Martin Engineering, including the Center for Bulk Materials Handling Innovation and Foundations™ Educational Programs.

33

MARTIN ENGINEERING

MARTIN ENGINEERING CENTER FOR BULK MATERIALS HANDLING INNOVATION (CFI)

To improve the understanding of both the behavior of bulk materials and the performance of the equipment systems that handle those materials, Martin Engineering opened a new corporate research center, the Center for Bulk Materials Handling Innovation (CFI).

Housed in a new 2.100-square-meter (22,600-square-foot) building at Martin's headquarters campus in Neponset, Illinois, CFI is a $5 million USD facility with the goals of innovation, collaboration, and education to improve bulk-materials handling.

The Center for Innovation holds the promise of improved productivity and profitability in industrial operations where clean, efficient handling of bulk material is essential. Martin's new corporate research center is focused on improving the handling of bulk materials in industries such as coal-fired power generation, coal and hard-rock mining, sand and gravel production, pulp and paper, and cement.

Basic Science Meets Product Development

Part pure-science research laboratory and part industrial product-development center, CFI collaborates with partners including corporations, industry associations, and universities for practical research to solve the common problems associated with the handling of bulk materials. These problems lead to added maintenance expenses and reduced productivity.

The full-time CFI staff, including scientists, engineers, and technicians, is dedicated to advancing the understanding of the behavior of bulk materials and the performance of material handling systems.

CFI has both the scientific instruments and the full-scale materials handling equipment, including a three-part recirculating conveyor system, to test bulk materials and prototype components under simulated operating conditions.

A three-part recirculating conveyor system at Martin's Center for Bulk Materials Handling Innovation allows for full-scale materials handling testing.

Laboratories for Specialized Research

Focused laboratories allow for the analysis and testing of characteristics and performance of metals, polymers, and bulk materials, as well as the accelerated testing of components under a variety of harsh environments.

> *"The Center for Bulk Materials Handling Innovation represents a major commitment to the industries we serve."*
>
> *– Edwin H. Peterson*
> *Chairman*

Martin Engineering is the first in the industry to take this basic science-first approach to improving the handling of bulk materials. According to Martin's Chairman Edwin H. Peterson, "The Center for Bulk Materials Handling Innovation represents a major commitment to the industries we serve. Our research partners will benefit from a better understanding of the characteristics of the bulk materials they use and how they should be handled. This will lead to new technologies that make materials handling cleaner, safer, and more productive."

A Resource for Education

CFI offers dedicated training and education resources, including a state-of-the art, 44-seat training room and a video conference center.

CFI also includes a re-circulating three-conveyor process simulation loop. An observation deck allows visitors to view the process simulation loop, through window, or via closed circuit video cameras.

Leadership for Industry

The Center for Innovation will help Martin maintain its position as the leading innovator of systems to make bulk-materials handling cleaner, safer, and more productive.

MARTIN ENGINEERING FOUNDATIONS™ EDUCATIONAL PROGRAMS FOR CLEAN, SAFE, AND PRODUCTIVE BELT CONVEYORS

Martin Engineering's Foundations™ Educational Programs teach "old hands" and "new hires" alike about the operation of belt conveyors used to handle bulk materials.

These non-commercial educational programs present information on preventing damage, controlling fugitive material, reducing maintenance costs, increasing safety, and improving operating efficiency. They are suited for anyone interested in improving the safety, working conditions, performance, and profitability of operations where belt conveyors are a key to success.

Meeting Your Needs, On Your Schedule

Program sessions can be scheduled at your convenience at your facility, a neutral site, or a conference or trade show; as part of a company meeting; or in a focused training session. They can be delivered to—and focused on—personnel from one plant, or they can be arranged to benefit multiple-locations from one company or personnel from several conveniently-located operations. Small class sizes keep the presentations informal and encourage discussion of specific problems.

Foundations™ Educational Programs include opportunities to discuss your facility and specific conveyor problems. By prior arrangement, a workshop at your facility can be preceded by a site survey to document conditions and evaluate equipment. Digital photography from the survey will then be included in the workshop, with discussions centered on specific problems and possible solutions.

Certification

These classes can qualify for Professional Development Hours (PDHs). Certificates of completion are provided to everyone who attends a program. All seminars have a companion "open-book" test to verify understanding of the information presented.

Program Instructors

The Foundations™ Educational Programs are taught by certified Martin personnel who have spent years working on and around belt conveyors.

These experienced professionals have both a theoretical understanding of conveyor principles and practical, "hands-on" experience in operating and troubleshooting belt conveyors. Instructors have seen conveyors handling a variety of materials all over the world, and they have provided innovative solutions to resolve problems and upgrade efficiency.

Workshop instructors draw on an array of training materials and use an interactive style that keeps the sessions lively and interesting.

For More Information

For information on scheduling, registration fees, and the program's money-back guarantee, contact Martin or email workshops@martin-eng.com.

Coming Soon: *3 Levels of the Foundations™ Conveyor Programs*

Level 1: FOUNDATIONS™ BASIC WORKSHOP	Level 2: FOUNDATIONS™ OPERATIONS & MAINTENANCE SEMINAR	Level 3: FOUNDATIONS™ ADVANCED CONVEYOR SEMINAR
Audience New hires with little or no knowledge of belt conveyors	**Audience** Operations and Maintenance Personnel, Supervisors, and Production and Maintenance Managers	**Audience** Conveyor Designers, Plant Engineers, and Plant Managers
Length 1 day or less	**Length** Multi-day	**Length** Multi-day
Emphasis Basic conveyor concepts, safe work practices, and fundamental Foundations™ principals for clean, safe, and productive conveyor operations	**Emphasis** Practical, problem-and-solution approach to the difficult problems faced by those involved in the hands-on conveyor operations	**Emphasis** Technical and commercial aspects of conveyor operations: Topics covered include power requirements, transfer-point design, and return on investment (ROI) calculations
Presentation This is a packaged program using a computer presentation with notes, allowing the course to be taught by in-house personnel or a Martin Engineering subject-matter expert. Concepts are taught in a manner that makes it easy to understand across languages. Depending upon the needs of the audience, the basic workshop can be presented in about 3-5 hours and is designed to augment site-specific training programs.	**Presentation** This seminar typically involves a minimum of a one-day, on-site survey of the local condition and a one-day classroom discussion lead by a Martin subject-matter expert. This discussion focuses on problems and solutions in areas such as controlling fugitive materials, belt damage, mistracking, impact, wear, belt cleaning, dust control, and transfer-point construction and maintenance.	**Presentation** The Operations & Maintenance Seminar is generally considered a prerequisite to this seminar. The advanced Foundations™ seminar is taught by a Martin subject-matter expert who is a qualified engineer and generally lasts 1-5 days, depending on the areas of interest and certification requirements.

MARTIN ENGINEERING: FOCUSED ON IMPROVING THE HANDLING OF BULK MATERIALS

For more than 65 years, Martin Engineering has focused on solving problems in the handling of bulk materials, in all industries around the world. Martin has developed innovative technologies to improve the handling of bulk materials—that boost flow, that reduce dust and spillage, that extend component life and reduce downtime, that improve the operating environment and the bottom line—and backs them all with an Absolutely, Positively, No Excuses Guarantee.

Areas of Expertise

Field Services *Resources to Improve Operations in Plants Handling Bulk Materials*

To solve problems in operating plants, Martin offers a comprehensive array of services focused on improving material handling. Service specialists—certified, skilled, and experienced—will help make a critical difference in the performance of your plant's material-handling systems.

- Equipment Installation
- Specialized Maintenance
- Process Improvement

Site Survey	Flow Modeling (DEM)
Equipment Census	Laser Surveying
Bulk-Material Testing	Silo, Bin, and Bunker Cleaning
Carryback Analysis	Air-Cannon Maintenance
Dust Monitoring	On-Line Asset Library

Personnel Training *Instruction in Why and How to Improve Bulk-Materials Handling*

Through its industry-leading Foundations™ books and its educational programs, Martin helps industry personnel understand the critical importance of bulk-materials handling. The client-driven programs will help new hires; plant operations and maintenance personnel and their managers; and conveyor designers, engineers, and plant managers control the variables that affect the critical factors of conveyor performance. Tailored to meet your needs, on your schedule, programs are provided on-site or at a mutually agreed upon location.

- Conveyor Safety Training
- Foundations™ Educational Programs (3 Levels)
- Certified Conveyor Technician (CCT) Certification Program

MARTIN ENGINEERING: PRODUCT OFFERINGS

Conveyor Products

Systems to Improve Belt Conveyors and Control Fugitive Material

Belt conveyors provide reliable, safe, and effective systems that are capable of moving thousands of tons of material per day. However, problems arise that lead to issues with system performance, component life, operating schedules, and regulatory compliance. Martin Engineering offers solutions designed to help operations where the conveying of bulk materials is a key to overall productivity and profitability.

- Belt-Cleaning Systems
- Tail Pulley-Protection Plows
- Belt-Training Devices
- Dust-Management Systems
    Passive Containment Devices
    Insertable Dust Collectors
    Dust-Suppression Systems
    (Fog, Foam, and Water-Spray)

- Transfer-Point Technologies
    Belt-Support Cradles
    Chutewall and Wear-Liner Systems
    Skirtboard-Sealing Systems
- Leading-Edge Conveyor Technologies
    Engineered Transfer Systems
    Air-Supported Belt Conveyors
    Modern Conveyor Architecture

Flow Aid Products

Systems to Improve the Flow of Bulk Materials from Storage and through Industrial Processes

Martin was born when the company's founder invented the Vibrolator® Ball Vibrator to improve the recovery of molding sand from foundry hoppers. Today, Martin continues its emphasis on the development of systems that enhance the movement of bulk materials from storage bins, hoppers, and silos, and through chutes, screens, feeders, and conveyors.

- Railcar Unloading Systems
- Railcar Unloading Vibrators
- Railcar Connector Boots
- Engineered Vibration Systems
- Pneumatic Rotary Vibrators
- Pneumatic Linear Vibrators
- Rotary Electric Vibrators

- Hopper Gate Openers
- Air-Cannon Systems
    Stand-Alone Air Cannons
    Multiple-Port Air-Cannon Systems
    Air Cannons for High-Temperature Applications

Maintenance Management

Resources to Improve the Maintenance Program in Bulk-Materials Handling Operations

MartinPLUS® Data Manager program is an on-line library of component information. Custom-built for a belt conveyor system, this digital resource can improve maintenance productivity and reduce costs.

APPENDICES

APPENDIX A
REFERENCES

American Conference of Governmental Industrial Hygienists. (1995). *Industrial Ventilation: A Manual of Recommended Practice, 22nd Edition.* Cincinnati, OH.

Arnold, P. C. (September 1993). *Transfer Chutes Engineered for Reliable Performance.* Paper presented at The Institution of Engineers, Australia, 1993 Bulk Materials Handling National Conference, Queensland, Australia. In National Conference Publication No. 93/8, pp. 165–173.

ASTM International. (2006). *Standard Test Method for Shear Testing of Bulk Solids Using the Jenike Shear Cell,* ASTM D6128-06. West Conshohocken, PA. Available online: http://www.astm.org

ASTM International. (2006). *Standard Test Method for Bulk Solids Characterization by Carr Indices,* ASTM D6393-99(2006). West Conshohocken, PA. Available online: http://www.astm.org

ASTM International. (2002). *Standard Shear Test Method for Bulk Solids Using Schulze Ring Shear Tester.* ASTM D6773-02; Work Item: ASTM WK19871 – Revision of D6773-02 Standard Shear Test Method for Bulk Solids Using the Schulze Ring Shear Tester. West Conshohocken, PA. Available online: http://www.astm.org

ASTM International. (2001). *Standard Test Method for Measuring Bulk Density Values of Powders and Other Bulk Solids,* ASTM D6683-01; Work Item: ASTM WK14951 – Revision of D6683-01 Standard Test Method for Measuring Bulk Density Values of Powders and Other Bulk Solids. West Conshohocken, PA. Available online: http://www.astm.org

ASTM International. (2001). *Standard Practice for Coagulation-Flocculation Jar Test of Water,* ASTM D2035-08. West Conshohocken, PA. Available online: http://www.astm.org

Axelrod, Steve. (September 1994). "Maintaining Conveyor Systems," *Plant Engineering,* pp. 56–58. Des Plaines, Illinois: Cahners Publishing Company.

Barfoot, Greg J. (January/March 1995). "Quantifying the Effect of Idler Misalignment on Belt Conveyor Tracking," *Bulk Solids Handling,* Volume 15, #1, pp. 33–35. Clausthal Zellerfeld, Germany: Trans Tech Publications.

Benjamin, C.W. (Jan/March 1999). "Transfer Chute Design: A New Approach Using 3D Parametric Modelling," *Bulk Solids Handling,* pp. 29–33. Clausthal-Zellerfeld, Germany: Trans Tech Publications.

B.F. Goodrich Company. (1980). *Care and Maintenance of Conveyor and Elevator Belting.* Akron, Ohio.

Carter, Russell A. (May 1995). "Knocking Down Dust," *Rock Products,* (pp. 19–23, 40–44). Chicago: Intertec Publishing.

Conveyor Equipment Manufacturers Association (CEMA). (1988). *Conveyor Terms and Definitions, Fifth Edition.* Rockville, Maryland.

Conveyor Equipment Manufacturers Association (CEMA). (1997). *Belt Conveyors for Bulk Materials, Fifth Edition.*

Conveyor Equipment Manufacturers Association (CEMA). (2000). *CEMA Standard No. 575-2000: Bulk Material Belt Conveyor Impact Bed/Cradle Selection and Dimensions.* Naples, FL.

Conveyor Equipment Manufacturers Association (CEMA). (2004). *CEMA SPB-001 (2004) Safety Best Practices Recommendation: Design and Safe Application of Conveyor Crossovers for Unit Handling Conveyors.* Naples, Florida.

Conveyor Equipment Manufacturers Association (CEMA). (2005). *BELT CONVEYORS for BULK MATERIALS, Sixth Edition.* Naples, FL.

Conveyor Equipment Manufacturers Association (CEMA). (2005). "Conveyor Installation Standards for Belt Conveyors Handling Bulk Materials." In *BELT CONVEYORS for BULK MATERIALS, Sixth Edition,* Appendix D, pp. 575–587. Naples, Florida.

Conveyor Equipment Manufacturers Association (CEMA). (2005). "'Universal Method' for Belt Tension Calculation." In *BELT CONVEYORS for BULK MATERIALS, Sixth Edition,* pp. 104–129. Naples, Florida.

Colijn, Hendrik. (1985). *Mechanical Conveyors for Bulk Solids.* Amsterdam, The Netherlands: Elesevier Science Publishers B.V.

Cooper, Paul, and Smithers, Tony. (July 1995). *Air Entrainment and Dust Generation from Falling Streams of Bulk Materials.* Paper presented at 5th International Conference on Bulk Material Storage, Handling and Transportation, Wollongong, Australia.

Cukor, Clar. (Undated). *Tracking: A Monograph.* Scottdale, Georgia: Georgia Duck and Cordage Mill (now Fenner Dunlop).

Density Standards: Aggregates–ASTM C29 / C29M-07, Crushed Bituminous Coal–ASTM D29-07, and Grains–U.S. Department of Agricultural Circular #921.

Dieter, George E. (1999). *Engineering Design: A Materials and Processing Approach, Third Edition.* McGraw-Hill.

Dorman, Peter. (April 2000). *The Cost of Accidents and Diseases.* Geneva. Available online: http://www.ilo.org/public/english/protection/safework/papers/ecoanal/wr_chp1.htm

Drake, Bob. (May 2001). "Cures for the Common Pulley." *Rock Products,* pp. 22–28. Chicago: Intertec Publishing.

Dreyer, E., and Nel, P.J. (July 2001). *Best Practice: Conveyor Belt Systems.* Project Number GEN-701. Braamfontein, South Africa: Safety in Mines Research Advisory Committee (sic) (SIMRAC), Mine Health and Safety Council.

Engineering Services & Supplies PTY Limited. Australian Registration #908273, Total Material Control and Registration #716561, TMC.

Environment Australia. (1998). *Best Practice Environmental Management in Mining: Dust Control,* (ISBN 0 642 54570 7).

Finnegan, K. (May/June 2001). "Selecting Plate-Type Belt Fastener Systems for Heavy-Duty Conveyor Belt Operations," *Bulk Solids Handling,* pp. 315–319. Clausthal-Zellerfeld, Germany: Trans Tech Publications.

Fish, K.A.; Mclean, A.G.; and Basu, A. (July 1992). *Design and Optimisation of Materials Handling Dust Control Systems.* Paper presented at the 4th International Conference on Bulk Materials Storage, Handling and Transportation, Wollongong, Australia.

Friedrich, A.J. (2000). "Repairing Conveyor Belting Without Vulcanizing." In *Bulk Material Handling by Conveyor Belt III,* pp. 79–85. Littleton, Colorado: Society for Mining, Metallurgy, and Exploration (SME).

Gibor, M. (July/September 1997). "Dust Collection as Applied to Mining and Allied Industry," *Bulk Solids Handling,* pp. 397–403. Clausthal-Zellerfeld, Germany: Trans Tech Publications.

Giraud, Laurent; Schreiber, Luc; Massé, Serge; Turcot, André; and Dubé, Julie. (2007). *A User's Guide to Conveyor Belt Safety: Protection from Danger Zones.* Guide RG-490, 75 pages. Montréal, Quebec, Canada: IRSST (Institut de recherche Robert-Sauvé en santé et en sécurité du travail), CSST. Available in English and French as a free downloadable PDF: http://www.irsst.qc.ca/files/documents/PubIRSST/RG-490.pdf or as an html: http://www.irsst.

qc.ca/en/_publicationirsst_100257.html

Godbey, Thomas. (May 1990). "Dust control systems: Make a wise decision," *Chemical Processing*, pp. 23–32. Chicago: Putnam Publishing.

Godbey, Thomas. (November 1989). "Selecting a dust control system (Part II)," *Powder and Bulk Engineering*, pp. 20–30. Minneapolis: CSC Publishing.

Godbey, Thomas. (October 1989). "Selecting a dust control system (Part I)," *Powder and Bulk Engineering*, pp. 37–42. Minneapolis: CSC Publishing.

Goldbeck, Larry J., Martin Engineering (July 2001). "Matching Belt Compatibility to Structures," *Aggregates Manager*, pp. 21–23. Chicago: Mercor Media.

Goldbeck, Larry J., Martin Engineering (July 1988). "Controlling fugitive material at your belt conveyor's loading zone," *Powder and Bulk Engineering*, pp. 40–42. Minneapolis: CSC Publishing.

Goodyear Tire & Rubber Company. (2000). *Handbook of Conveyor & Elevator Belting on CD*, Version 1.0. Akron, Ohio.

Greer, Charles N. (April 1994). "Operating Conveyors in the Real World," *Rock Products*, pp. 45–48. Chicago: Maclean-Hunter Publications.

Grisley, Paul. (February 2002). "Air Supported Conveying in Mines and Process Plants." Paper presented at the 2002 Society for Mining, Metallurgy, and Exploration (SME) Annual Meeting & Exhibit, Phoenix, AZ.

"Hints & Helps: Tips for Tracking Conveyor Belts." (February 1995). Rock Products, p. 25. Chicago: Intertec Publishing.

International Labour Organization. (2003). *Safety in Numbers, Pointers for a Global Safety Culture at Work*. Geneva

Kasturi, T.S. (May 1995). *Conveyor Belting Wear: A Critical Study*. Unpublished study commissioned by Martin Engineering. Madras, India: Jay Kay Engineers & Consultants.

Kasturi, T.S. (1994). *Conveyor Components, Operation, Maintenance*. Failure Analysis. Madras, India: Jay Kay Engineers & Consultants.

Kasturi, T.S. (1992). *Conveyor Belt Cleaning Mechanism*. Madras, India: Jay Kay Engineers & Consultants.

Kestner, Dr. Mark. (February 1989). "Using suppressants to control dust emissions (Part I)," *Powder and Bulk Engineering*, pp. 17–20. Minneapolis: CSC Publishing.

Kestner, Dr. Mark. (March 1989). "Using suppressants to control dust emissions (Part II)," *Powder and Bulk Engineering*, pp. 17–19. Minneapolis: CSC Publishing.

Koski, John A. (March 1994). "Belt conveyor maintenance basics," *Concrete Journal*, p. 5. Addison, Illinois: The Aberdeen Group.

Law, Bob. (August 2000). *Conveyor Belt Cleaner Analysis*. Paper presented at the IIR Conference "Improving Conveyor Performance," Perth, Australia.

Low, Allison and Verran, Michael. (August 2000). *Physical Modelling of Transfer Chutes–A Practical Tool for Optimising Conveyor Performance*. Paper presented at the IIR Conference "Improving Conveyor Performance," Perth, Australia.

Maki, D. Michele, PhD. 2009. *Conveyor-Related Mining Fatalities 2001-2008: Preliminary Data*. Unpublished Report for Martin Engineering.

Martin Engineering website: http://www.martin-eng.com

Martin Marietta Corporation. *Dust Control Handbook for Minerals Processing*, Contract No. J0235005.

Martin Supra Engineering. (2008) *CarrybackTtest/Sum/SBM-001-SBW-05-2008*. Unpublished report for P.T. Martin Supra Engineering: Newmont, Indonesia.

"Measuring ROI pushes it higher, say Harte Hanks Aberdeen of Enterprise Solutions." (February 12, 2007). The Manufacturer (US Edition).

Miller, D. (January/March 2000). "Profit from Preventive Maintenance," *Bulk Solids Handling*, pp. 57–61. Clausthal-Zellerfeld, Germany: Trans Tech Publications.

Mody, Vinit and Jakhete, Raj. (1988). *Dust Control Handbook (Pollution Technology Review No. 161)*, ISBN-10: 0815511825/ISBN-13: 978-0815511823. Park Ridge, New Jersey: Noyes Data Corporation.

Möller, J.J. (September 1985). *Protect Your Conveyor Belt Investment*. Presentation to BELTCON 3 International Material Handling Conference, Johannesburg, South Africa.

Morgan, Lee, and Walters, Mike. (October 1998). "Understanding your dust: Six steps to better dust collection," *Powder and Bulk Engineering*, pp. 53–65. Minneapolis: CSC Publishing.

Morrison, J.N., Jr. (1971). "Environmental Control Applied to Belt Conveyor Transfer Points." In *Bulk Materials Handling*: Volume 1. University of Pittsburgh.

Mine Safety and Health Administration (MSHA). (2004). *MSHA's Guide to Equipment Guarding*. Other Training Material OT 3, 40 pages. U.S. Department of Labor. Available as a free download: http://www.msha.gov/s&hinfo/equipguarding2004.pdf

Muellemann, Alf. (January 2000). "Controlling dust at material transfer points with ultra-fine water drops," *Powder and Bulk Engineering International*, pp. 44–47. Minneapolis: CSC Publishing.

National Industrial Belting Association (NIBA). (1985). *NIBA Engineering Handbook*, Brookfield, WI.

Öberg, Ola. (1986). *Materialspill vid bandtransportörer* (Material Spillage at Belt Conveyors). Stockholm, Sweden: Royal Institute of Technology.

Ontario Natural Resources Safety Association. *Safety Reminder*, newsletter. P.O. Box 2040, 690 McKeown Avenue, North Bay Ontario, B1B 9PI Telephone: (705) 474-SAFE.

Occupational Safety & Health Administration, U.S. Department of Labor, Mineral Processing Dust Control website: http://www.osha.gov/SLTC/silicacrystalline/dust/

Ottosson, Goran. (October 1991). "The cost and measurement of spills and leaks at conveyor transfer points," *World Cement Materials Handling Review*, Berkshire, England.

Padgett, Harvey L. (2001). *Powered Haulage Conveyor Belt Injuries in Surface Areas of Metal/Nonmetal Mines*, 1996–2000. Denver, CO: MSHA Office of Injury and Employment Information.

Planner, J.H. (1990). "Water as a means of spillage control in coal handling facilities." In *Proceedings of the Coal Handling and Utilization Conference: Sydney, Australia*, pp. 264–270. Barton, Australia: Institution of Engineers Australia.

Project Management Institute (PMI). Additional information about project management and the accreditation program for project managers is available from PMI on the organization's website: http://www.pmi.org

Reed, Alan R. (1995). "Contrasting National and Legislative Proposals on Dust Control and Quantifying the Costs of Dust and Spillage in Bulk Handling Terminals," *Port Technology International*, pp. 85–88. London: ICG Publishing Ltd.

Rhoades, C.A.; Hebble, T.L.; and Grannes, S.G. (1989). *Basic Parameters of Conveyor Belt Cleaning*, Report of Investigations 9221. Washington, D.C: Bureau of Mines, US Department of the Interior.

Roberts, Alan. (November 1996). *Conveyor System Maintenance & Reliability*, ACARP Project C3018. Author is from Centre for Bulk Solids and Particulates, University of Newcastle, Australia. Published by Australian Coal Association Research Program; can be purchased at http://www.acarp.com.au/abstracts.aspx?repId=C3018

Roberts, A.W. (August 1999). "Design guide for chutes in bulk solids handling operations," *Centre for Bulk Solids & Particulate Technologies*, Version 1, 2nd Draft.

Roberts, A.W.; Ooms, M.; and Bennett, D. *Conveyor Belt Cleaning – A Bulk Solid/Belt Surface Interaction Problem*. University of Newcastle, Australia: Department of Mechanical Engineering.

Roberts, A.W. and Scott, O.J. (1981). "Flow of bulk solids through transfer chutes of variable geometry and profile," *Bulk Solids Handling*, Vol. 1 No. 4., pp. 715–727.

Sabina, William E.; Stahura, Richard P.; and Swinderman, R. Todd. (1984). *Conveyor Transfer Stations Problems and Solutions*. Neponset, Illinois: Martin Engineering Company.

Scott, Owen. (1993). "Design Of Belt Conveyor Transfer Stations For The Mining Industry." In *Proceedings of the 1993 Powder & Bulk Solids Conference, Reed Exhibition Companies, Des Plaines, Illinois*, pp. 241–255.

Simpson, G.C. (1989). "Ergonomics as an aid to loss prevention," *MinTech '89: The Annual Review of International Mining Technology and Development*, pp. 270–272. London: Sterling Publications Ltd.

Spraying Systems Company (http://www.spray.com) contains a variety of useful material on the basics and options available in spray nozzles.

Stahura, Dick, Martin Engineering. (July 1990). "Ten commandments for controlling spillage at belt conveyor loading zones," *Powder and Bulk Engineering*, pp. 24–30. Minneapolis: CSC Publishing.

Stahura, Richard.P., Martin Engineering. (1987). "Conveyor belt washing: Is this the ultimate solution?" *TIZ-Fachberichte*, Volume 111, No. 11, pp. 768–771. ISSN 0170-0146.

Stahura, Richard P., Martin Engineering. (February 1985). "Conveyor skirting can cut costs," *Coal Mining*, pp. 44–48. Chicago: McLean-Hunter Publications.

Stuart, Dick D. and Royal, T. A. (Sept. 1992). "Design Principles for Chutes to Handle Bulk Solids," *Bulk Solids Handling*, Vol. 12, No. 3., pp. 447–450. Available as PDF: www.jenike.com/pages/education/papers/design-principles-chutes.pdf

Sullivan, Dr. John. *Increasing retention and productivity: let employees do what they do best!* Article #163. Available online: http://ourworld.compuserve.com/homepages/GATELY/pp15s163.htm

Sundstrom, P., and Benjamin, C.W. (1993). "Innovations in Transfer Chute Design." Paper presented at the 1993 Bulk Materials Handling National Conference, The Institution of Engineers, Australia, *Conference Publication No. 93/8*, pp. 191–195.

Swinderman, R. Todd, Martin Engineering. (2004). "Standard for the Specification of Belt Cleaning Systems Based on Performance." *Bulk Material Handling by Conveyor Belt 5*, pp. 3–8. Edited by Reicks, A. and Myers, M., Littleton, CO: Society for Mining, Metallurgy, and Exploration (SME).

Swinderman, R. Todd, Martin Engineering. (February 2002). *Conveyor Belt Impact Cradles: Standards and Practices*. Paper presented at the 2002 Society for Mining, Metallurgy, and Exploration (SME) Annual Meeting & Exhibit, Phoenix, AZ, February 2002.

Swinderman, R. Todd, Martin Engineering. (October–December 1995). "Belt Cleaners, Skirting and Belt Top Cover Wear," *Bulk Solids Handling*. Clausthal-Zellerfeld, Germany: Trans Tech Publications.

Swinderman, R. Todd, Martin Engineering. (July 1994). "Engineering your belt conveyor transfer point," *Powder and Bulk Engineering*, pp. 43–49. Minneapolis: CSC Publishing.

Swinderman, R. Todd, Martin Engineering. (May 1991). "The Conveyor Drive Power Consumption of Belt Cleaners," *Bulk Solids Handling*, pp. 487–490. Clausthal-Zellerfeld, Germany: Trans Tech Publications.

Swinderman, R. Todd; Becker, Steven L.; Goldbeck, Larry J.; Stahura, Richard P.; and Marti, Andrew D. (1991.) *Foundations: Principles of Belt Conveyor Transfer Point Design and Construction*. Neponset, Illinois: Martin Engineering.

Swinderman, R. Todd; Goldbeck, Larry J.; and Marti, Andrew D. (2002). *FOUNDATIONS3: The Practical Resource for Total Dust & Material Control*. Neponset, Illinois: Martin Engineering.

Swinderman, R. Todd; Goldbeck, Larry J.; Stahura, Richard P.; and Marti. Andrew D. (1997). *Foundations2≈: The Pyramid Approach to Control Dust and Spillage From Belt Conveyors*. Neponset, Illinois: Martin Engineering.

Swinderman, R. Todd and Lindstrom, Douglas, Martin Engineering. (1993). "Belt Cleaners and Belt Top Cover Wear," *National Conference Publication No. 93/8*, pp. 609–611. Paper presented at The Institution of Engineers, Australia, 1993 Bulk Materials Handling National Conference.

Takala, J. (18–22 September 2005). *Introductory Report: Decent Work – Safe Work*. XVIIth World Congress on Safety and Health at Work, Orlando, Florida. Available online: http://www.ilo.org/public/english/protection/safework/wdcongrs17/intrep.pdf

Taylor, H.J. (1989). *Guide to the Design of Transfer Chutes and Chute Linings for Bulk Materials*. The Mechanical Handling Engineers' Association.

Thomas, Larry R., Martin Engineering. (1993). "Transfer Point Sealing Systems to Control Fugitive Material," *Conference Publication No. 93/8*, pp. 185–189. 1993 Bulk Materials Handling National Conference of The Institution of Engineers, Australia.

Tostengard, Gilmore (February 1994). "Good maintenance management," *Mining Magazine*, pp. 69–74. London: The Mining Journal, Ltd.

University of Illinois. (2005). *Design of Conveyer Belt Drying Station*. Unpublished study done for Martin Engineering.

University of Illinois. (1997). *High Pressure Conveyor Belt Cleaning System*. Unpublished study done for Martin Engineering.

University of Newcastle Research Associates (TUNRA). Untitled, unpublished study done for Engineering Services and Supplies P/L (ESS).

Weakly, L. Alan. (2000). "Passive Enclosure Dust Control System." In *Bulk Material Handling by Conveyor Belt III*, pp. 107–112. Littleton, CO: Society for Mining, Metallurgy, and Exploration (SME).

Wilkinson, H.N.; Reed, Dr. A.R.; and Wright, Dr. H. (February 1989). "The Cost to UK Industry of Dust, Mess and Spillage in Bulk Materials Handling Plants," *Bulk Solids Handling*, Volume 9, Number 1, pp. 93–97. Clausthal-Zellerfeld, Germany: Trans Tech Publications.

Wilson, Richard J. (August, 1982). *Conveyor Safety Research*. Bureau of Mines Twin Cities Research Center.

Wood, J. P. (2000). *Containment in the Pharmaceutical Industry*. Informa Health Care.

APPENDIX B
GLOSSARY

GLOSSARY

This is a list of belt conveyor-related terms, as they are used in this edition of *FOUNDATIONS*™. It does not pretend to be a complete compendium of all terms used in describing belting, conveyors, and/or systems for handling bulk materials. If a phrase is not shown, first break it down into its component words. Also consider consulting with other references, such as *CEMA Publication #102, Conveyor Terms and Definitions*, as well as the publications and terminologies used by suppliers of specific components.

A

abrasion[1] | Wearing away by friction, as by rubbing or scraping.

access door | Point of entry into an enclosed area, typically with a method of closure.

active dust collection | *See dust-collection system.*

adhesion[1] | The bonding strength between two materials.

aeration device | Device mounted inside a vessel that adds low pressure/ high-volume air to materials that have become compacted and hard to allow them to flow efficiently again, sometimes called aeration diffusers, pads, or nozzles.

agglomeration | Process or act of gathering into a mass; creating larger, heavier groupings of particles.

aging[1] | The exposure to an environment for a period of time.

air cannon[2] | A device that uses periodic blasts of compressed air to clear away material buildup inside pipes or transfer chutes.

air knife | Belt-cleaning system that directs a stream of air to shear off carryback.

air-supported conveyor[2] | A conveyor that uses a conventional belt, pulleys, and drive but is supported on its carrying side by a thin film of air instead of idlers.

air-to-media ratio | Used to describe dust-collection filters, the air-to-media ratio is the flow of air in cubic meters per second (ft³/min) divided by the area of filtration media in square meters (ft²).

amplitude | Half the extent of a vibration, oscillation, or wave; the measurement above or below the base or centerline.

anemometer | Device used to measure air velocity.

angle of attack | The angle at which a cleaning blade is placed against the belt.

angle of repose[2] | The angle or slope that a conveyed material will assume when discharged onto an open pile.

ANSI[2] | Acronym for American National Standards Institute.

apron feeder[2] | A series of overlapping metal plates mounted on a rotating chain that are used to transport heavy, lumpy or abrasive materials.

AR plate[2] | Abrasion-resistant steel plate commonly used for wear liner at transfer points.

aramid fibers | A class of strong, heat-resistant synthetic fibers used in aerospace and military applications, as well as in the carcass of conveyor belting.

arc of contact[1] | The circumferential portion of a pulley engaged by a belt.

ASME[2] | Acronym for American Society of Mechanical Engineers.

aspect ratio | A ratio comparing the thickness of the top and bottom covers of a belt.

ASTM[2] | Acronym for American Society for Testing and Materials.

B

backstop | A mechanical or electric braking device used to prevent a loaded, inclined conveyor belt from rolling backwards if the motor stops. Also referred to as a "holdback clutch" or "clutch brake."

back welding | A method of welding in which at each weld, the bead is drawn back toward the welded end.

backstep welding | A weld applied to the back side of the joint; commonly called back welding.

baghouse[2] | A closed structure that contains a set of filter bags to capture airborne dust.

beater bar | A device (usually a roller device with an external bar) which strikes another object with the object of removing material accumulation.

bed | Some variety of low-friction bars or other flat surface to support the belt profile instead of using an idler's rolling "cans."

belt clamp[1] | Beams or metal plates secured transversely across both belt ends to hold them in a desired position.

belt cleaner[2] | A device that uses one or more tensioned blades mounted on a supporting structure to remove material that clings to the carrying surface of a conveyor belt beyond the normal discharge point.

belt-cleaner effect | Where the pressure of a sealing system against the belt removes residual material from the belt surface, as when the tail seal removes material from the belt where it enters the loading zone.

belt-cleaning system | A belt cleaner, or a group of belt cleaners and associated equipment (such as mounts and tensioners), as located on one conveyor.

belt conveyor[2] | A flexible rubber endless belt, looped over a framework of rollers and pulleys, that is used to transport material from a load zone to a discharge point.

belt fastener[1] | A mechanical device for holding two ends of a conveyor belt together.

belt feeder[2] | A short, flat, variable-speed conveyor belt used to transfer, or "feed," material from one component to another in a material transport system. The material feed rate can be adjusted by speeding up or slowing down the belt.

belt flap[2] | An up and down oscillation of a belt between idlers.

belt grade | A classification of belt cover based on its properties, designed to provide a reference for end users as to what belts to use in different applications.

belt modulus[1] | The force per unit width of belt required to produce a stated percentage of elongation.

belt profile | The shape of the belt, particularly its upper (carrying) surface.

belt runout[2] | A condition where a conveyor belt moves too far to either side of its properly-centered path; also referred to as belt "mistracking" or "wander."

belt sag[1] | The vertical deflection of a conveyor belt from a straight line between idlers, usually expressed as a percentage of the center spacing of the idlers.

belt slip[1] | The speed differential between the belt and the pulley surface.

belt slip switch[2] | A switch that shuts down a conveyor drive motor when it senses the belt moving at a slower speed than the drive pulley.

belt stretch | The increase in belt length that takes place when tension is imposed. Elastic stretch is a temporary change in length that varies directly with the pull. Permanent stretch is the residual change in length after tension has been removed; it generally accumulates over a period of time.

belt-support cradles | A method of belt support without rolling components, using slider or impact beds.

belt-support system | The components below the carrying side of the belt that support the weight of belting and cargo.

belt tracking | The actions a person takes to get the belt to track consistently.

belt training | The actions a person takes to get the belt to track consistently.

belt-cleaner blade | The element of a belt cleaner that comes into contact with the belt.

bend pulley[2] | A pulley used to change the direction of (or "bend") a conveyor belt.

bias cut[1] | A cut of the belt ends made diagonally, that is at an angle less than 90 degrees (usually 22°) to the longitudinal axis.

blockout | A safety procedure involving the prevention of a system from moving by physically holding it in position.

boilover | A problem where material overflows the chute, caused by chute blockages.

booster drive[1] | Used in some long conveyors to reduce the power/tension at the drive pulley.

bottom cover[1] | The non-carrying belt side towards the pulleys.

boundary friction | *See interface friction.*

bow | A concave curve of the belt.

breaker, breaker fabric | An extra ply incorporated in the belt carcass for shock-absorption.

brush cleaner[2] | A belt-cleaning device that uses a rotating brush to clean carryback material from the return run of a conveyor belt.

C

CAD | Acronym for Computer-Aided Design.

camber | A convex curve of the belt *(see bow).*

cantilever | A projecting beam or structure supported at one end.

capacity[1] | The maximum material load on the belt, cargo, or throughput.

capture velocity | The amount of air speed required to gather an airborne dust particle into a dust-collection system.

carcass[1] | The fabric, cord and/or metal reinforcing section of a belt, as distinguished from the rubber cover.

CARP[2] | Acronym for "Constant Angle Radial Pressure," a belt-cleaning blade design concept to maintain cleaning angle as the blade wears.

carryback[2] | Conveyed material that clings to the surface of a belt past the nominal discharge point. If not removed by a belt-cleaning system, these particles become dislodged along the return run and pile up beneath the belt.

carrying idler[2] | Any type of idler that supports the load-carrying run of a conveyor belt.

carrying run[2] | The upper run of a conveyor belt used to transport material from a load zone to a discharge point.

carrying side | The side of the conveyor or belting that would contact the material cargo.

catenary idler[1] | A flexible idler set where the rollers are suspended on a flexible link, rope, or chain structure and the ends are supported in pivoted stands. The tube or rollers sag to form the trough. Also called a Garland idler.

CEMA | Acronym for Conveyor Equipment Manufacturers Association.

center-to-center[1] | The distance between the center of two pulleys or idlers. Sometimes also called centers or center distance.

ceramic-faced wear liner | A lining using ceramic blocks or tiles for improved resistance to abrasion.

CFM or cfm | Abbreviation for "cubic feet-per-minute" in airflow calculations.

chamfer | To cut at an angle, as a bevel.

chatter, blade chatter[2] | The rapid vibration of a belt cleaner that is not aligned properly with a conveyor belt.

chevron, chevron belt | A V-shaped ridge on the carrying side of a belt to keep material from rolling down an incline.

chute[2]**, chutework** | An enclosure that is used to contain material as it is transferred from one piece of equipment to another.

chute wall | The walls of the loading chute and sometimes the transfer-point skirtboard.

chutewall | *See skirtboard.*

classifier[2] | A piece of equipment used to sort and separate material by size.

cleaner[1] | A device for removing adherent material from the belt.

cleat[1]**, cleated belt** | Objects on or raised sections of a conveyor belt, used to stabilize material carried up an incline.

CMMS | Acronym for computerized maintenance-management system, a system that tracks maintenance work and its costs.

coefficient of friction | The ratio of the force required to slide two surfaces to the force pressing them together; equal to the tangent of the interface friction angle.

cohesion | A material's internal strength.

cold splice[2] | A type of belt splice in which the layers of a conveyor belt are overlapped and bonded together with an adhesive compound.

concave | Curved inward; bow is a concave curve in the belt.

confined space | A potentially hazardous enclosed area; access is usually controlled by safety regulations.

consolidated bulk density (ρ_2) | The density of a body of a bulk material after it has been subjected to a compressive force (F) or vibratory energy, sometimes called vibrated bulk density.

convex | Curved outward; camber is a convex curve of the belt.

conveyor[2] | A piece of equipment designed to carry material from one point to another along a predetermined path.

conveyor belt[2] | A length of flexible rubber belt that is stretched over a framework of rollers and pulleys and then made into a single piece by splicing its two ends together.

counterweight[2] | The weight applied to a conveyor belt gravity take-up assembly to maintain proper belt tension.

cover[1] | The outer layer of belting. Also, the lid or roofing structure to protect conveyor and materials from exposure to elements and limit release of material.

creep[1] | The action of a belt alternately losing speed on the driving pulley and gaining speed on the driven pulley.

creeper drive[2] | An auxiliary motor and gearbox that is designed to operate a piece of equipment at a very slow speed. Also referred to as a "pony drive."

crown[2] | The difference between the diameter of a pulley at its center and at its rims.

crowned pulley[1] | A pulley with a greater diameter at the center, or other points, than at the edges.

crusher[2] | A piece of equipment used to crush or shatter larger pieces of material into smaller ones.

crust breaker | A cleaning edge installed on the head pulley just below the material trajectory so it is close to, but does not touch, the belt; serves as a doctor blade to limit the amount of material that gets through to the conventional pre-cleaner installed just below.

cupping[2] | The action of the edges of a belt curving upward on the carrying run and downward on the return run. Also referred to as belt "curl."

cut edge[1] | The uncovered edge of a belt, created by slitting the desired width from wider belting.

cyclone[2] | A high-velocity "whirlwind"-type device that uses centrifugal force to separate dust particles from the air.

D

dBA | Acronym for decibel A scale, a measurement of sound intensity.

deck, decking[2]**, deck plate** | A barrier plate located between the conveyor's stringers to prevent material from spilling off the carrying run onto the return run. Also referred to as "belt pans."

deflector wear liner | A liner installed inside the skirtboard that incorporates a bend toward the center of the belt, which channels material away from the belt edge and sealing system.

deflector[2] | A metal plate installed in a transfer point to change the trajectory of material flow.

delamination[1] | The separation of layers of material.

DEM | Acronym for Discrete Element Modeling, a computer-based technique to analyze and demonstrate the movement of individual particles in or through a structure.

density[1] | The ratio of the mass of a body to its volume or the mass per unit volume of the substance. For practical purposes, density and specific gravity may be regarded as equivalent.

diagonal plow[2] | A device placed at an angle across the surface of a conveyor belt to deflect material off to one side.

DIN | Acronym for Deutsches Institut für Normung, the German Institute for Standardization, which develops norms and standards for industry. DIN standards are used internationally, but still most commonly in Europe.

discharge[2] | The point where material exits from a conveyor or other component in a material handling system.

disk idler[2] | An idler that uses a series of cushioned disks to support a conveyor belt.

displaced air | The air that is pushed out of the chute when the chute is loaded, equal to the volume of materials placed into the chute.

diversion plow[2] | A retractable plow that can be lowered to the carrying surface of a belt to divert material off of a conveyor ahead of the normal discharge point.

downstream | In the direction of the places that the belt has not yet reached, or toward the discharge of the conveyor or system.

drag conveyor | Material-handling system using bars or plates on a chain to pull the cargo to the discharge point.

dribble chute[2] | An angled chute positioned under the head end of a conveyor belt to catch any material that may fall off the return side and drop it into the discharge stream.

drive[2] | An arrangement of electrical and mechanical components that provide motive power to a conveyor or other piece of equipment.

drive pulley[2] | The pulley connected to the drive mechanism of a conveyor belt.

drum pulley[2] | A pulley that is of uniform diameter from side to side.

durometer | A device that measures the hardness of a flexible material (such as an elastomer), accomplished by measuring the resistance to the penetration of an indenter point.

dust bags[2] | Specially designed air-permeable filter bags that trap and collect airborne dust from a material handling system.

dust-collection system(s) | A mechanical system used to remove dust from the air in a material transport system.

dust curtains[2] | Segmented rubber or plastic curtains (baffles) suspended inside an enclosed duct that are used to slow down airflow and allow airborne dust to settle back into the material stream on a conveyor belt before it exits its load zone.

dust-suppression system(s) | A dust-control system using water or enhanced water to reduce the escape of airborne particulates.

dynamometer[1] | An apparatus capable of inducing various loads for evaluation of dynamic belting properties.

E

edge damage[2] | Tears and rips along the edge of a conveyor belt.

edge distance | Dimension between the outside of the skirtboard and the edge of the belt.

edge sealing | *See seal.*

edge-sealing strip(s) | *See sealing strips.*

effective belt width | The measurement of the horizontal width of a troughed conveyor belt that is measured across the dimension parallel to the bottom roller.

effluent | The outflow of water (with material solids) exiting a belt-washing system.

elastomer | A polymer having elastic properties resembling natural rubber; typically rubbers or urethanes.

electrical conductivity[1] | A measure of how well a material accommodates the transport of electric charge, measured in Ohm (Ω).

elongation | An increase in length, usually expressed as a percentage of initial length.

end stop[2] | A clamp equipped with a set screw that is used to secure blades in position on a belt-cleaner mainframe.

entrapment damage[2] | A groove worn into the surface of a belt by material trapped between the moving belt and the skirtboard and/or sealing system.

entrapment point(s) | A point where the two surfaces will allow a material lump to become wedged.

entry, entry point[2] | The point beyond the tail pulley where a conveyor belt passes into the load zone.

EPA | Acronym for Environmental Protection Agency, a branch of the United States government.

exit, exit point[2] | The area of a load zone where the skirtboards come to an end and the main carrying run of the conveyor begins.

F

fatigue[1] | The weaking of a material occuring when repeated application of stress causes permanent strain.

FEA | Acronym for Finite Element Analysis, a computerized numerical analysis technique used for solving differential equations to primarily solve mechanical engineering problems relating to stress analysis, used in bulk-material handling in the design of conveyors and transfers.

feed rate[2] | The amount of material flow that is being transferred on a conveyor at any given time, usually expressed in "tons per hour" (t/h or st/h).

feeder[2] | A device that regulates the flow of material from a bin or storage hopper to a conveyor or other piece of equipment.

feeder belt[1] | A belt that discharges material onto another conveyor belt.

field-trimmed | Cut to the proper size at the point of application (as opposed to being cut at the factory).

fines | Small particles of material.

finger splice | A joint of the belt where the two ends are cut into a number of narrow triangular "fingers" which are interlaced.

flanged pulley[2] | A pulley with a raised rim at the edges for the purpose of keeping the belt contained.

flat belt | A conveyor belt that carries its cargo without being troughed.

flat idler[2] | An idler where the supported belt is flat.

flat roller | *See flat idler.*

flex cracking | A cracking of the surface resulting from repeated flexing or bending.

flight conveyor[2] | A type of conveyor that uses spaced cleats or scrapers (flights) to move material from one point to another through a channeled chute.

flop gate[2] | A pivoted metal plate that can be moved or "flopped" to feed material to either of two different discharge points.

flow aid | Device or method to promote the flow of materials through chutes, including both linear and rotary vibrators, air cannons, aeration systems, chute linings, and soft chute designs.

flush, flush-through[2] | An uncontrolled surge of material through a material handling system component.

footprint | Projected or actual area occupied on the ground.

free-belt edge distance | The non-load carrying portion of the belt's width, toward the belt edges, typically where the skirtboard-sealing system is applied.

friction[1] | The resistance to motion due to the contact of surfaces.

fugitive material[2] | Any stray material that escapes from a material handling system at a place other than its normal discharge point, might originate as carryback, spillage, or airborne dust or from other causes.

full-trough pulley | A tail pulley installed so its top is inline with the top of the center rolls on the first fully-troughed idlers.

G

gauge[1] | The thickness of a belt or of its individual elements.

generated air | Airflow produced by rotating devices that feed the conveyor load zone.

gouging[1] | The effect of sharp heavy material falling onto a conveyor belt cover to damage the surface or tear out pieces of the cover.

grade of belting | A classification of belt cover based on its properties; designed to provide a reference for end users as to what belts to use in different applications.

gravity take-up[2] | A device that adjusts for stretch or shrinkage by using a weighted pulley to maintain tension on the belt.

grizzly[2] | A series of metal bars or grids that are spaced apart to allow small lumps and fines to fall directly through while passing larger lumps on to crushing or breaking equipment.

grooving[2] | *See entrapment damage.*

guards, guarding | Barriers to prevent the entry of personnel into potentially-hazardous areas or equipment.

guide roller[2] | A small outrigger roll on a self-aligning idler. When a conveyor belt mistracks into the guide roll, it causes the pivoted steering rolls to turn inward and force the belt back onto centerline.

gusset | A triangular insert for enlarging or supporting.

H

half-trough pulley[2] | A tail pulley installed so its top surface is inline with the midpoint of the wing rolls on the first fully-troughed idlers, typically used to shorten the conveyor's required transition distance.

hammermill[2] | A type of crusher using multiple rotating hammers mounted on a central shaft to break hard, lumpy materials such as coal or limestone into smaller sizes.

hardness[1] | Degree of resistance to indentation.

head[2] | The discharge end of a conveyor belt.

head load | Pressure from a load on top of an object, such as the weight of the material in a vessel above a belt.

head pulley[2] | The terminal pulley located at the discharge point of a conveyor belt. On many conveyors, the head pulley is coupled to the drive motor to power the conveyor.

heeling[2] | Entrapment point caused by mounting a pre-cleaner mainframe too close to the head pulley.

hold-down roller[2] | An idler used to keep a conveyor belt from raising up, as when traveling unloaded, or used to apply downward pressure on the return run of a conveyor belt to maintain cleaning efficiency by preventing cleaning pressure from changing the belt's line of travel. Also referred to as a "pressure roller."

holdback[2] | *See backstop.*

holdup roller[2] | An idler that is used to increase the effectiveness of a tail protection plow by applying pressure upward to hold the belt flat.

hood | A curved deflector installed at the discharge of a conveyor to direct and confine the moving material stream so it flows smoothly and with minimal induced air.

hydrophobic | Having a high surface tension and averse to combining with water.

hygroscopic | Able to absorb moisture from the air.

I

idler[2] | A non-powered rolling component used to support a conveyor belt on either the carrying run or the return run.

idler-junction failure | See junction-joint damage.

impact[1] | The striking of one body against another; collision. The force or impetus transmitted by a collision.

impact bed, impact cradle[2] | A series of cushioned bars used to absorb loading forces under a conveyor belt load zone.

impact grid[2] | A series of metal bars mounted in a conveyor discharge chute at the point where the material impacts the wall, to reduce wear on the chute liner.

impact idler[2] | A specially constructed idler designed to cushion forces of material impact in the load zone of a conveyor belt.

impact resistance[1] | The relative ability of a conveyor belt assembly to absorb impact loading without damage to the belt.

induced air | air pulled into the voids created as the material stream expands as it leaves the head pulley.

insertable, insertable dust collector, insertable dust filter | A dust-collection system composed of filters designed to be incorporated inside the enclosure of a transfer point or other dust source.

interface friction (Θ) | The friction between the bulk material and the surface(s) that will be in contact with it (e.g. chutewall and belt); can be determined with a shear cell and a sample of the actual interface material; sometimes referred to as wall friction or boundary friction.

intermediate idlers[2] | Idlers placed between impact beds or slider beds to support a conveyor belt when material is not being loaded.

internal friction angle | The angle at which the particles within a bulk material slide over one another within a pile, or failure due to shearing.

ISO | A universal short form of the name of the International Organization for Standardization adopted from the Greek word "isos," meaning equal.

J

jog switch[2] | A manual start switch located near the discharge end of a conveyor used to "jog" or "bump" the belt for short distances for testing purposes or to gradually empty the belt of overloaded material.

joint[1] | The connection of two belt ends.

junction joint[2] | The area between the wing roll and center roll on a set of troughing idlers.

junction-joint damage[2]**; junction-joint failure** | A longitudinal splitting or cracking in a belt caused by insufficient transition distance between the tail pulley and the load zone for the type of belt being used and/or an idler-junction gap of more than 10 mm (0.4 in.) or twice the belt thickness.

K

kicker plate | Deflector to steer the flow of material after it leaves the first point of contact with the transfer chute.

knocking[2] | The process of manually adjusting the cross-structure angle of conveyor belt idlers to train a belt to centerline, accomplished by moving one end of the idler slightly forward or back.

KPIs | Acronym for key performance indicators, performance measurements used as metrics to measure organizational success.

L

lagging[2] | A rubber, fabric, or ceramic covering applied to a pulley shell to improve belt traction against the pulley.

lateral misalignment[2] | The offset of pulleys, idlers, or structure from a designated longitudinal reference line.

leakage | Material that has escaped from the material handling system, spilling from the sides or falling or expelled from openings.

lift | The vertical distance bulk material is moved on a conveyor; the change in height from one end of the conveyor to the other end.

limit switch[2] | An electrical switch used to shut off the drive or actuator of a system component such as a flop gate once it reaches a predetermined set point.

linear tensioner[2] | A type of tensioner that applies direct upward pressure to a belt cleaner.

liner | Material placed on the inside surfaces of an enclosure or vessel, usually to preserve the enclosure by reducing wear.

load out[2] | Area at the discharge of a conveyor where material can be temporarily stored or loaded directly onto a device for transport to another destination.

load zone[2]**, loading zone** | The receiving point where material is dropped or fed onto a conveyor.

loading chute | The enclosure that places the cargo onto the belt.

lockout | A safety precaution of placing a padlock or other control on stored energy sources, the power supply, or control circuit of a machine to prevent its premature resumption of operation or unexpected released energy.

longitudinal[2] | In reference to a conveyor belt, a lengthwise direction that runs parallel with the centerline.

loose bulk density | The weight per unit of volume of a bulk solid, measured when a sample is in a loose or non-compacted condition, (ρ_l).

LRR[1] | Acronym for Low Rolling Resistance, a proprietary rubber formulation.

M

magnetic pulley[2] | A pulley equipped with a permanent or electromagnet, used to remove tramp iron from the material cargo carried on or discharged from the conveyor.

magnetic separator[2] | A device that uses magnetic attraction to pull metal scraps, known as "tramp iron," out of the material stream on a conveyor.

mainframe[2] | The main structural support of a belt cleaner upon which the blades are mounted.

mandrel[2] | A central shaft used for mounting and lateral adjustment of a belt-cleaner mainframe.

manometer | A device used for measuring the pressure of gases or liquids; on conveyors, used for measuring air flow.

maximum tension[1] | The highest tension occurring in any portion of the belt under operating conditions.

mechanical dust collection | Active dust-collection system, typically using fans pulling air through ductwork to a filtration system.

mechanical fastener[1] | A system used to join the ends of belting, typically involving screws or rivets to attach plates connecting the two ends.

mechanical splice[2] | A type of splice in which mechanical fasteners are used to connect the two ends of a belt.

minimum pulley diameter | The minimum pulley size (usually to prevent damage) for a particular belt as specified by the belting's manufacturer.

misalignment switch[2] | A limit switch mounted along the edge of a conveyor belt that will shut the drive motor down if the belt tracks too far to either side of its normal centered path.

mistracking | The off-center travel of a conveyor belt.

molded edge[1] | A solid rubber belt edge formed in a mold, where the belt has been manufacturer to a specific width, rather than slit from a wider piece.

mooning[2] | Uneven wear on a pre-cleaner blade that results from positioning the cleaner mainframe too far out from the head pulley.

MSHA | Mine Safety and Health Administration, a unit of the US Department of Labor.

N

negative rake | Cleaning blades inclined at an angle in the direction of belt travel; also known as the scraping orientation.

O

offset idlers[2] | A troughing idler set where the wing rollers are in a vertical plane different from, but parallel to, the center roller. This permits the wing rollers to overlap the central roller, improving belt support; may also reduce the height of the idler set.

oil resistant | Able to withstand any deterioration of physical properties arising from interaction with petroleum.

operating tension | The tension of a belt while running with a material load.

OSHA | Occupational Safety & Health Administration, in the United States an agency of the United States Department of Labor; the main federal agency charged with the enforcement of safety and health legislation.

outrigger | A projection extending laterally beyond the main structure of a vessel, aircraft, or machine, usually for added stability.

ozone cracking[1] | Cracks in the belt surface caused by exposure to an atmosphere containing ozone.

P

particulates | Fine solid or liquid (other than water) particles found in the air, including dust, smoke, and pollen.

passive dust collection[2] | A dust-collection system that minimizes dust by utilizing efficient transfer-point design and airflow control rather than mechanical devices.

peeling angle | When a cleaner blade is tilted in opposition to the direction of belt travel; also known as positive-rake angle.

pelletizer | A device to form pellets (small lumps) from fines or dust.

permanent stretch | A change in length of a belt seen after tension has been removed; this additional length generally accumulates over a period of time.

picking idlers[2] | A type of troughing idler set with narrow wing rolls and a wide center roll. Idlers of this type are generally used for material that must be picked or sorted as it is conveyed.

pickup velocity | The speed at which air moving over a bed of a given material can pick up dust off the surface and carry it away, typically in the range of 1,0 to 1,25 meters per second (200 to 250 ft/min).

pillow block[2] | A journal bearing enclosed in a bolt-on housing that is used to mount pulleys to a conveyor stringer.

pinch point | A point where a machine element moving inline meets a rotating element in such a manner it is possible to nip, or entrap, a person or object between the members.

pitot tubes | A pressure measurement instrument used to measure the velocity of fluid flow.

PIW | Abbreviation for Pounds per Inch Width, a measurement of a belt's rated capacity for tension.

PLC | *See programmable logic controller.*

plenum | An enclosure in which pressurized air is distributed.

plow | A device stationed across the path of a conveyor to discharge or deflect material.

plug welding | A type of joint made by welding one part to another through a circular hole in the top part.

pluggage | The blocking of the discharge of a chute or hopper.

ply[1], **plies** | A layer of fabric used in the carcass of a belt.

pocket belt | A belt where pockets, formed by the addition of raised cleats and flexible sidewalls, are used to carry the cargo; commonly seen in high-angle applications.

positive pressure | The outward flow of air from the transfer point or other structure.

positive rake | In belt cleaning, a blade tilted in opposition to the direction of belt travel; also known as peeling angle.

pooling[2] | Material that piles up on a belt at the load zone until it reaches belt speed and can be carried away.

PPEs | Personal protective equipment, equipment and attire such as a hard hat, safety glasses, hearing protection, respirators, and steel-toe shoes.

pre-cleaner[2] | A belt cleaner installed on the face of a head pulley to shear off the bulk of any carryback clinging to the belt; primary cleaner.

press | A machine that applies pressure consistently across its surfaces, used for belt splices.

pressure roller | A roller installed to keep the belt in proper position, as above a belt cleaner.

primary, primary cleaner | A pre-cleaner; that is, a belt cleaner installed on the face of a head pulley below the material trajectory to shear off the bulk of any carryback material clinging to the belt. The primary cleaning position is on the face of the head pulley below the trajectory.

primary position | The area around the discharge pulley where primary belt cleaners are usually installed.

profile rip | A form of belt damage to the belt, with a rip running from the edge toward the center.

programmable logic controller (PLC)[2] | A centralized computer system that controls a system's operation and monitoring by communicating with remote input/output circuit boards for each individual system component.

pug mill | Industrial processing machine in which material is simultaneously ground and mixed with a liquid.

pull-cord stop switch | A cable running along the length of a conveyor, connected to one or more switches. In an emergency, a manual pull of the cable at any point will shut down the conveyor system.

pulley[2] | A rotating cylinder mounted on a central shaft that is used to drive, change direction of, or maintain tension on a conveyor belt.

pulley-protection plow | A plow installed so the belt passes under it immediately before the belt enters a pulley (usually, the tail pulley). The plow removes material from the belt to prevent damage to the pulley and belt by entrapment of material between the two.

pulley wrap[2] | The total area of contact where a belt wraps in an arc around the surface of a pulley.

pulverizer[2] | A mechanical device used to grind material down to a fine powder consistency. A ball mill uses heavy steel balls that roll between counter rotating faces to crush the material.

PVC | Acronym for polyvinyl chloride, a material used in the construction of some conveyor belting.

Q

R

radial tensioner[2] | A tensioner that transmits torque through a pivoted extension or torsion spring to a belt cleaner.

rated tension | The minimum breaking strength of a belt in newtons per millimeter (lb_f/in.) of belt width, as specified by the belting manufacturer. In the USA sometimes used as a term for the working tension.

reclaim system[2] | A material handling system use to recover and transport material from a stockpile area to a point where it will be processed or consumed.

regenerative conveyor[1] | A conveyor that discharges at a substantially lower altitude than the tail (so it conveys material downhill), producing electricity rather than consuming it.

relief | A mechanism that allows an item (a cleaner blade, for example) to move away from an obstruction (a mechanical splice, for example). These could include springs in the cleaning-system tensioner.

relieving angle | An incline or opening of surfaces that will allow material to be pulled free by the action of the belt, rather than become more tightly wedged.

residual surfactant | A dust-suppression additive that will continue its agglomeration effect even after the moisture evaporates; also called a binder suppressant.

return idler[2] | An idler used to support the empty, return side of a conveyor belt.

return run[2]**, return side** | The side of a conveyor belt that does not carry cargo, after the discharge, as the belt returns to the loading zone.

reverse-jet | A method of cleaning filters in a baghouse; bags are cleaned by discharging a burst of compressed air into the bags at the top; the compressed-air burst flexes the bag wall and breaks the dust cake off so it falls into the collection hopper.

reversing conveyor[2] | A type of conveyor that can carry material longitudinally in either direction.

ribs | *See cleats.*

rip detector[2] | A system in which an electrical conductor is built into the plies of a conveyor belt that will shut the drive motor down if the belt becomes torn.

RMA | Acronym for Rubber Manufacturers Association, Inc.

rock box[2] | A ledge or shelf inside a transfer chute where material is to accumulate. This allows subsequent material to impact on the accumulated material rather than against the chute, extending the life of the walls.

rock ladder[2] | A series of rock boxes that slow down the velocity of material by cascading it back and forth between ledges.

Rockwell hardness (or scale) | A scale for evaluating the hardness scale of materials, as determined by measuring the depth of penetration of an indenter. Different scales are denoted by a single letter; "B"

and "C" are the most common.

ROI | Return on investment or payback.

roll crusher[2] | A mechanical device that uses a heavy, rotating metal drum equipped with teeth or cogs inside a screened enclosure to crush hard materials.

rollback[2] | Stray pieces of material that roll and bounce backward down an inclined belt after material flow has been shut off. Or, the downhill motion of an inclined conveyor, running backward when the power is shutoff while the belt is loaded.

rolling component(s) | The idlers and pulleys (and other rotating components) of a conveyor system.

ROM | Run-of-mine, the raw mined material that comes directly from the extraction operation prior to crushing, screening, or other treatment.

run | The distance or route covered by a conveyor belt.

S

sacrificial surface | A wear surface that is installed to protect a more valuable structure by absorbing, cushioning, or isolating the abrasion, impact, or other forces.

saddle[2] | An additional short length of belting added to an existing conveyor belt.

safety cable | A restraint used as a safety measure to prevent the fall of an overhead device in the event of the failure of its mounting system.

safety factor[1] | The fraction of a structure's capability over that which is truly required, or a multiplier applied to the maximum expected load (force, torque, bending moment, or a combination) to which a component or assembly will be subjected.

sampler[2] | A mechanical device used to collect small amounts of material at preset intervals from the main material stream for testing or quality-control purposes.

scab plate[2] | A piece of metal plate used to patch over a hole in the wall of an enclosure such as a transfer chute.

scavenger conveyor[2] | A small conveyor or vibrating chute positioned beneath the head of a larger conveyor to capture carryback or material drop-off from a belt-cleaning system and return the discharge to the main material stream.

scraping angle[2]**, scraping position** | A belt cleaner installed so its blade(s) are tilted in the direction of belt travel; also known as negative-rake angle.

screw conveyor[2] | A type of conveyor that uses a rotating auger inside an enclosed tube to convey material from one point to another.

screw take-up[1] | A mechanical take-up to apply tension to a conveyor belt in which movement of a pulley-bearing block is accomplished by means of a screw.

seal | Method to prevent spillage by containing the fines and dust at the edge of the skirtboard.

sealing strip(s) | The elastomer material installed between the skirtboard and the belt to prevent spillage.

sealing system | Elastomer seal and clamping mechanism at the edge of the skirtboard to contain dust and fines and prevent spillage.

secondary belt cleaner, secondary cleaner | A belt cleaner mounted beneath the return side of a conveyor belt to remove any remaining carryback fines that were not removed by the pre-cleaner blade.

secondary position | Position for a belt cleaner, between the point where the belt leaves the head pulley and where it contacts the first snub or bend pulley or return idler.

segregation | The accidental or undesired separation of a material by size.

self-aligning idlers[2] | Idlers that can swivel to the left or right under the influence of the forces of the moving belt to keep the belt traveling on the centerline.

settling zone | An enlarged portion of the covered skirtboard area past the loading zone's impact area; the extra volume designed to slow the airflow and allow airborne dust to return to the main material cargo and cleaner air to escape; also called a stilling zone.

shear cell test | Test to derive flow properties of a bulk material by measuring the force to shear the bulk material.

side-loading forces | Pressure resulting from the energy and weight of material pushing outward from the center.

side-support cradles | Belt support system using slider bars under the skirtboard, to provide a consistent and sealable surface for the sides of the belt.

skim coat | A thin layer of rubber material laid on a fabric but not forced into the weave.

skirtboard[1] | The vertical or inclined plates extending out from a conveyor's loading point and installed closely above the belt to confine the conveyed material.

skirtboard seal, skirting seal | The mechanism (often a strip of elastomer) installed along the bottom of the transfer point's skirtboard to control spillage and keep material on the belt.

skirted area | The area of the transfer point that is enclosed within the skirtboard; the area of the transfer point from the loading point through the exit.

skive[2] | To remove some (or all) of a belt's top cover to recess a mechanical splice; the process of countersinking the fasteners in a mechanical splice closer to the belt carcass to keep the top of the fasteners parallel with the surface of the belt.

slack-side tension[2] | The area of least tension on a conveyor belt; the low-tension areas will vary on the location of the snub and take-up pulleys; they are completely dependent on the individual conveyor and must be identified for each application.

slider bar | A low-friction bar, typically used in the construction of a slider bed belt-support cradle.

slider bed[2] | A series of longitudinal bars assembled in a cradle and placed beneath a conveyor load zone to provide a continuous surface for a loaded belt to ride on.

slider bed conveyor | A conveyor using some variety of low-friction bars or other flat surface, rather than idlers, to support the belt.

slip, slippage | The speed differential between the belt and the pulley surface.

snub, snub pulley[2] | A small pulley used to increase the wrap area of a conveyor belt around a head or tail pulley for improved traction.

spillage | Lost material that has fallen from the side(s) of the conveyor belt; typically in the load zone, but can occur at any point along the conveyor; a general term for all fugitive material.

spiral-wrapped pulley[2] | A wing pulley that is wrapped with a steel band in a spiral pattern to reduce belt vibration while still maintaining the self-cleaning function of the pulley.

splice | The joint where two ends or two pieces of belting are joined together to provide a continuous loop.

splice allowance[1] | Additional belting required to allow a splice to be installed.

splice angle | The angle across the top of the belt at which two pieces of belt are joined.

spoon | A curved trough at the bottom of a transfer chute that directs the placement of the stream of material onto the receiving belt.

spring take-up[2] | A mechanical device that utilizes a variable force spring or springs attached between the conveyor structure and the tail pulley mounting block to maintain tension on the belt.

squeegee blade | A soft urethane blade that wipes the belt to remove water from the belt.

stacker conveyor[2] | A conveyor used to "stack" or drop material onto a stockpile or lowering well. A stacker conveyor can be "fixed," to drop material into a single location, or "rotating," to spread the material in a sweeping motion over a wider area.

stacker/reclaimer[2] | A boom mounted conveyor equipped with a rotating bucket wheel that can "stack" or drop material onto a stockpile for storage or reverse direction and reclaim the material from the stockpile to another destination.

stackout system[2] | A series of conveyors designed to carry material out onto a storage area.

Stahura Carryback Gauge | A method to measure carryback utilizing a collection pan with scraper blades held against the return side of a moving belt to capture residual material; developed by belt-cleaning pioneer Dick Stahura.

steering rolls[2] | A set of rollers (or a set of troughed idlers) mounted on a pivot that can swivel left or right to steer a mistracking conveyor belt toward centerline.

stepped splice[2] | A type of splice in multi-ply belting where the fabric plies on one end of the belt are removed so that it will butt together and overlap adjacent plies of fabric on the other end.

stilling zone | *See settling zone.*

stitch welding | A metal joining technique using a series of spaced welds, with intervals between the welds.

STP | Acronym for Standard Temperature and Pressure; 0°C/32°F, 1 atmosphere (101.325 kPa) (1 atmosphere of absolute pressure).

straight face pulley[2] | A pulley with a flat surface with no crown.

stringer[2] | The longitudinal supporting members of a conveyor structure, between the terminal pulleys.

surcharge angle | The angle to the horizontal which the surface of a body of material assumes while the material is at rest on a moving conveyor belt. This angle usually is 5° to 15° less than the angle of repose, though in some materials it may be as much as 20° less.

surfactant[2] | A surface-acting agent. In dust suppression, this is an additive that is combined with water in a spray or fog to assist in the capture of airborne dust.

Swinderman Scale of Fugitive Materials | A scoring system that assigns values to a system's performance in control of fugitive materials for dust, spillage, and carryback.

T

tagout | The placing of a name tag or other label or sign on a disabled power or control system, to identify that the system is "down" for maintenance and should not be restarted.

tailgate sealing box[2] | An enclosure located at the tail end of a load zone to prevent material from leaking out onto the belt behind the chutework.

tail pulley[2] | A pulley that turns the return run of a conveyor belt 180 degrees back into the carrying run.

take-up[2] | A device used to remove slack from a conveyor belt and maintain tension. Gravity take-ups use a heavy counterweight to maintain belt tension; mechanical take-ups use a hydraulic device or screw adjustment to maintain tension.

take-up travel | The distance the take-up is able to move while the belt is running.

tension | The force along the belt line required to overcome the resistance of components and transport the load.

tensioner, tensioning device | A device used to maintain a belt cleaner's cleaning pressure against the surface of the belt.

terminal pulley | The pulley at either end of the conveyor; the head and/or tail pulleys.

tertiary belt cleaners[2]**, tertiary cleaner** | Any additional cleaners added to a belt after the primary cleaner (pre-cleaner) and initial secondary cleaner; cleaner(s) installed further along the conveyor return than secondary position.

tertiary position | The area after the snub pulley for the installation of additional belt cleaners.

testout | Attempting to operate a device that has been presumably disabled by lockout / tagout / blockout procedures; used as a final safety precaution.

throughput | The amount of bulk material delivered by a material handling system; usually stated as tons per hour (st/h).

tie gum[1] | A thin sheet of unvulcanized rubber inserted between plies in the assembly of a vulcanized belt splice.

tight side tension[2] | The area of highest tension on a conveyor belt, usually located at the point where the belt approaches the drive pulley.

tilt switch[2] | An electrical switch designed to shut off material flow from a conveyor when material backup at the discharge point forces it into a tilted position.

TLV | Threshold limit value, a level of dust to which it is believed a worker can be exposed day after day for a working lifetime without adverse health effects; as expressed in parts per million parts of air (ppm) for gases and in milligrams per cubic meter (mg/m^3) for particulates such as dust, smoke, and mist.

top cover | The carrying surface of the belt.

total material control | Success in containing spillage and carryback and controlling dust, where materials are kept on the belt and within the system.

TPH, tph | Abbreviation for "tons per hour;" a measure of capacity.

tracker[2], **tracking device** | A device used to steer a mistracking conveyor belt back to centerline.

tracking | *See belt tracking.*

training | *See belt training.*

training idler[1], **trainer** | An idler mounted on a pivot or otherwise adjustable base that, when actuated by the mistracking belt moving against it, will automatically adjust its position to steer the belt to the correct path.

trajectory[2] | The arcing path made by conveyed material as it is discharged from the head end of a conveyor.

tramp iron[2] | Pieces of scrap metal that may contaminate the material stream on a conveyor belt.

tramp iron detector | A system to detect the presence of tramp iron in a material stream and either remove the tramp iron or shut down the material handling system.

transfer point | The place (and associated equipment) where a belt conveyor is loaded or unloaded.

transition | The forming of the conveyor belt into a trough to receive its cargo; the area where this change takes place.

transition area[2] | The area between the tail pulley of a conveyor and the start of the load zone where the belt transforms from flat to fully troughed or the area where the belt transforms from troughed onto the discharge pulley.

transition distance | The distance from the centerline of the terminal pulley to the first fully-troughed idler.

transition idlers[2] | Idler sets between the tail pulley and the load zone that gradually transform the belt into the trough for loading.

transverse[2] | The direction from side to side across a conveyor belt.

traveling plow[2] | A plowing device that can be moved back and forth longitudinally over the carrying side of a conveyor belt to deflect material to alternate discharge points along its run.

tripper conveyor[2], **tripper** | A rail-mounted mechanism with a traveling take-up that can move the discharge end of a conveyor to multiple points along a straight line to fill individual hoppers or bins.

trough | the shape of a belt with the edges raised allowing it to carry more material.

trough angle[2] | The angle (from horizontal) at which the belt edges are troughed to help center and contain its load.

troughability[1] | The property of a belt that permits it to conform to the contour of troughing idlers; the amount a belt can be troughed.

troughing idlers[2] | A set of carrying idlers consisting of a horizontal center roll with incline wing rolls on both sides that forms the carrying side of the belt into a trough.

tube conveyor[2] | A conveyor where the belt is formed into a closed tube after it is loaded, typically used to prevent spillage and carry material vertically.

turnover[1] | A system installed in a conveyor that inverts the belt, usually to control carryback by keeping the load-carrying ("dirty") side of the belt up.

U

UHMW[2] | Acronym for Ultra-High Molecular Weight polyethylene, a plastic material commonly used as a chute liner or low-friction belt-support surface.

unidirectional conveyor | Conveyor that carries material in one direction.

upstream | In the direction of the places the belt has already passed, or back toward the loading point.

V

valley angle | The angle between two chute walls created by the side wall joining with the back wall.

vee roller | *See V-return idler.*

vibrated bulk density | Also called consolidated bulk density (ρ_2), achieved by applying a compressive force (F) or vibratory energy to a body of material; used for determining the weight of material conveyed on the belt based on surcharge angle.

vibrating feeder[2] | A type of feeder that uses a suspended or isolated trough with an attached vibrator to move material from a bin or hopper into a transfer chute.

viscosity[1] | Resistance of a material to flow under stress.

V-plow[2] | A "V" shaped device equipped with a rubber or urethane blade that rides atop the return run of a conveyor belt to deflect any stray material away from the tail pulley.

V-return idler[2] | A return idler that incorporates two rolls in a "V" configuration to improve belt tracking on the return run.

vulcanized splice[2] | A type of splice in which the layers of a belt are overlapped and bonded together, using heat and pressure ("hot vulcanization") or a chemical bonding agent ("cold" vulcanization).

vulcanizer[1] | A device to apply heat and pressure for curing a splice; also called a press.

W

wall friction angle | *See interface friction.*

wander | Mistracking.

warp[1] | Lengthwise yarns in a woven fabric.

wash box | An enclosure containing a series of belt cleaners and water-spray nozzles for belt cleaning.

water tensioner[2] | A type of belt cleaner tensioner that uses regulated water pressure to maintain tension on the cleaner blades.

wear liner[2] | A layer of ceramic tiles, AR plate, or other abrasion-resistant material used to line the inside of a transfer chute or skirtboard to improve material flow and prevent abrasive wear and damage to the outer shell and structure.

weft[1] | The crosswise yarns in a woven fabric.

weldment[2] | A fabricated metal component held together by welded joint(s).

wing idler[2] | Either of the outer rollers in a troughed idler set, mounted at an angle to the central roll.

wing pulley[2], **wing-type pulley** | A type of self-cleaning pulley that supports the belt on individual vanes instead of a solid surface. The vanes are mounted on a central section that tapers down from inside to outside to direct stray material out of the pulley and off to the sides.

wing rollers | Rollers on the outside of a troughed idler set. *See wing idler.*

X

Y

Z

zero speed switch | Electrical switches used to detect the stoppage of a rotating shaft, such as on a conveyor drive motor.

zero rake | Belt cleaner angle of attack where blades are installed perpendicular (90 degrees) to the belt line.

SOURCES

[1]**Conveyor Belt Guide**
www.ConveyorBeltGuide.com

[2]**Stahura Coveyor Products**
www.scp-pa.com

Metric		Imperial	
Description	**Abbreviation**	**Abbreviation**	**Description**
centimeter	cm	BTU	British Thermal Unit
cubic centimeter	cm^3	BTU/lb$_m$	British Thermal Unit per pound mass
cubic meters	m^3	ft	feet or foot
cubic meters per hour	m^3/h	ft/min	feet per minute
cubic meters per minute	m^3/min	ft/s	feet per second
cubic meters per second	m^3/s	ft^2	square feet
decibels - audio	dBA	ft^3	cubic feet
degrees Celsius	°C	ft^3/h	cubic feet per hour
gram	g	ft^3/min	cubic feet per minute
hectopascal	hPa	ft^3/s	cubic feet per second
hilohertz	kHz	gal	gallon
kilogram	kg	gal/h	gallons per hour
kilograms per cubic meter	kg/m^3	gal/min/in.	gallons per minute/per inch
kilograms per liter	kg/l or kg/L	gal/s	gallons per second
kilograms per second	kg/s	h	hour
kilograms per square meter	kg/m^2	hp	horsepower
kilojoule	kJ	in.	inches
kilojoule per kilogram	kJ/kg	in.^{3}	cubic inch
kilometer	km	lb/ft^2	pounds per square foot
kilometer per hour	km/h	lb$_f$	pounds-force
kilonewton	kN	lb$_f$/ft	pounds-force per foot
kilonewton per meter	kN/m	lb$_f$/in.^{2}	pounds-force per square inch
kilonewton per cubic meter	kN/m^3	lb$_f$/ft^3	pounds-force per cubic foot
kilopascal	kPa	lb$_f$/in.	pounds-force per inch
kilowatt	kW	lb$_m$	pounds mass
liter	l or L	lb$_m$/ft^3	pounds mass per cubic foot
liters per hour	l/h or L/h	lb$_m$/s	pounds mass per second
liters per minute per meter	l/min/m or L/min/m	mile	mile
liters per second	l/s or L/s	min	minute
megahertz	MHz	mph	mile per hour
meters	m	°F	degrees Fahrenheit
meters per minute	m/min	oz	ounce
meters per second	m/s	oz$_m$	ounce mass
microgram	µg	psi	pounds per square inch
micron	µ	psi	pounds-force per square inch
milligrams per cubic meter	mg/m^3	PWI	pounds per square inch width
milliliter	ml	s	second
millimeters	mm		
newton	N		
newton per millimeter	N/mm		
newtons per meter	N/m		
square meters	m^2		

APPENDIX C
SAFETY LABELS

SAFETY LABELS

The following information and safety label recommendations are published with permission by the Conveyor Equipment Manufacturers Association (CEMA) and are available in CEMA Brochure 201. Labels may be purchased from the CEMA website (www.cemanet.org) or by writing the Conveyor Equipment Manufacturers Association, 6724 Loan Oak Boulevard, Naples, Florida 34109, Telephone 239-514-3441. They also can be obtained directly from many conveyor equipment manufacturers.

CEMA Safety Label Meanings

The CEMA Safety Label program uses color and specific signal words (words that designate a degree or level of hazard seriousness) to identify labels for three classifications of risk:

DANGER LABEL

Labels with "Danger" indicate an imminently hazardous situation, which, if not avoided, will result in death or serious injury. They are to be limited to the most extreme situations.

WARNING LABEL

Labels with "Warning" indicate a potentially hazardous situation which, if not avoided, could result in death or serious injury.

CAUTION LABEL

Labels with "Caution" indicate a potentially hazardous situation which, if not avoided, may result in a minor or moderate injury. They may also be used to alert against unsafe practices.

Note: *"Danger" or "Warning" should not be considered for property damage accidents unless personal injury risk appropriate to those levels is also involved. "Caution" is permitted for property-damage-only accidents.*

Symbol/Pictorial

A graphic representation intended to convey a message without the use of words. It may represent a hazard, a hazardous situation, a precaution to avoid a hazard, a result of not avoiding a hazard, or any combination of these messages.

Word Message

Consists of two parts. The first section describes the hazard. The second section instructs what to do or not do to avoid the hazard.

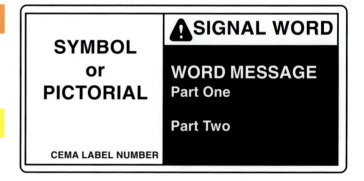

DANGER LABELS

Indicates an imminently hazardous situation which, if not avoided, will result in death or serious injury. This signal word is to be limited to the most exteme situations.

WARNING LABELS

Indicates a potentially hazardous situation which, if not avoided, could result in death or serious injury.

⚠ WARNING
Exposed moving parts can cause severe injury

LOCK OUT POWER before removing guard

CHR930001

⚠ WARNING
Exposed screw and moving parts can cause severe injury

LOCK OUT POWER before removing cover or servicing

CHR930011

⚠ WARNING
Equipment starts automatically - can cause severe injury

KEEP AWAY

CHR930002

⚠ WARNING
Exposed moving parts can cause severe injury

LOCKOUT POWER before servicing

CHS950013

⚠ WARNING
Moving equipment can cause severe injury

KEEP AWAY

CHR931005

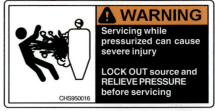

⚠ WARNING
Servicing while pressurized can cause severe injury

LOCK OUT source and RELIEVE PRESSURE before servicing

CHS950016

⚠ WARNING
Servicing moving or energized equipment can cause severe injury

LOCK OUT POWER before servicing

CHR930009

⚠ WARNING
Exposed moving parts can cause severe injury

LOCK OUT POWER before servicing

CHS950021

⚠ WARNING
Exposed conveyors and moving parts can cause severe injury

LOCK OUT POWER before removing cover or servicing

CHR930010

⚠ WARNING
Rotating shaft can cause severe injury

Keep hair and loose clothing away

CHS950023

⚠ WARNING

CVS930010

Exposed conveyors and moving parts can cause severe injury

LOCK OUT POWER before removing

⚠ WARNING

CVS930011

Exposed screw and moving parts can cause severe injury

LOCK OUT POWER before removing cover or servicing

⚠ WARNING
Servicing while pressurized can cause severe injury

LOCK OUT source and RELIEVE PRESSURE before servicing

CHS950022

⚠ WARNING
Guard Removed

Risk of severe injury

DO NOT OPERATE Without guard

CHR000025

⚠ WARNING

CVS930012

Exposed buckets and moving parts can cause severe injury

LOCK OUT POWER before removing cover or servicing

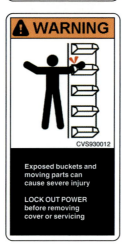

⚠ WARNING

CVS950020

Exposed moving parts can cause severe injury

LOCK OUT POWER before removing guard

⚠ WARNING
Walking or standing on conveyor covers or gratings can cause severe injury

STAY OFF

CHS991026

CAUTION LABELS

Indicates a potentially hazardous situation which, if not avoided, may result in minor or moderate injury. It may also be used to alert against unsafe practices.

⚠ CAUTION
Do not enter
Hazardous Area
Authorized personnel only
CHR930006

⚠ CAUTION
HOT SURFACE
Contact with skin may cause burns
DO NOT TOUCH
CHS950018

⚠ CAUTION
Low Clearance
• Be Alert
CHR930007

⚠ CAUTION
Insure Skirting System is properly adjusted
Failure to properly install, inspect, adjust and maintain this system may result in spillage, dust, downtime, equipment damage, or personal injury.
CHS060028

⚠ CAUTION
Trip Hazard
• Be Alert
• Watch your step
CHR930008

⚠ CAUTION
Insure Belt Cleaning and Scraping System is properly set
Failure to properly install, inspect, adjust and maintain this system may result in spillage, dust, downtime, equipment damage, or personal injury.
CHS060029

⚠ CAUTION
Hazardous environment
Respiratory protection required
Authorized personnel only
CHS950015

⚠ CAUTION
Insure Impact Cradle is properly set
Failure to properly install, inspect, adjust and maintain this system may result in spillage, dust, downtime, equipment damage, or personal injury.
CHS060030

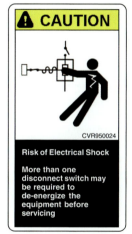

⚠ CAUTION
CVR950024
Risk of Electrical Shock
More than one disconnect switch may be required to de-energize the equipment before servicing

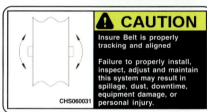

⚠ CAUTION
Insure Belt is properly tracking and aligned
Failure to properly install, inspect, adjust and maintain this system may result in spillage, dust, downtime, equipment damage, or personal injury.
CHS060031

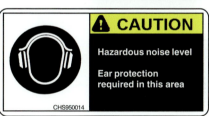

⚠ CAUTION
Hazardous noise level
Ear protection required in this area
CHS950014

Product: **Bulk Handling Equipment** | Equipment: **Bulk Belt Conveyors**

A. *To be placed on removable guards to warn that operation of the machinery with guards removed would expose chains, belts, gears, shafts, pulleys, couplings, etc. which create hazards.*

D. ***Locate at entrance to conveyor walkway.*** *General warning to personnel that a conveyor's moving parts, which operate unguarded by necessity of function, i.e. belts, rollers, terminal pulleys, etc., create hazards to be avoided; in particular, conveyors which stop and start by automatic control near operator work stations would use this label.*

F. *To be placed at entrances to enclosed areas which would expose personnel to operational or environmental hazards which should only be entered by trained and authorized personnel under specific conditions; Examples, lifting conveyors, transfer car aisleways, confined spaces, etc.*

B. ***Locate on inspection door(s).*** *To be located on conveyors where there are exposed moving parts which must be unguarded to facilitate function, i.e. rollers, pulleys, shafts, chains, etc.*

E. ***Space up to a maximum of 50 ft centers (walkway sides).*** *To be placed up to a maximum of 50' centers along the walkway side.*

C. *To be placed on removable guards to warn that operation of the machinery with guards removed would expose chains, belts, gears, shafts, pulleys, couplings, etc. which create hazards.*

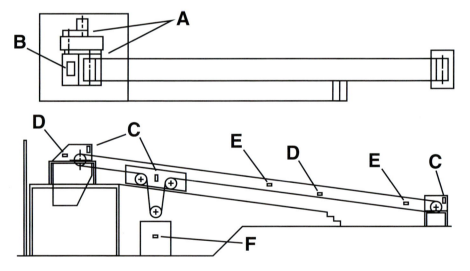

Product: **Bulk Handling Equipment** | Equipment: **Belt Conveyor Accessories**

These labels are to be placed on or near the maintenanced access for the following bulk belt conveyor accessories:

Belt Skirting Systems

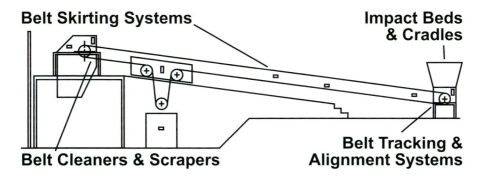

Impact Beds & Cradles

Belt Cleaners & Scrapers

Belt Tracking & Alignment Systems

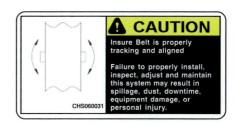

Bulk Handling Conveyors

SAFETY Is In Your Hands

Do Not Climb, Sit, Stand, Walk, Ride, or Touch the Conveyor at Any Time

Do Not Perform Maintenance on Conveyor Until Electrical, Air, Hydraulic and Gravity Energy Sources Have Been Locked Out and Blocked

Operate Equipment Only With All Approved Covers and Guards in Place

Lock Out All Power and Block Gravity Loads Before Servicing

Ensure That All Personnel Are Clear of Equipment Before Starting

Allow Only Authorized Personnel and Trained Personnel To Operate or Maintain Conveyors and Accessories

Keep Clothing, Body Parts, and Hair Away from Conveyors

Clean Up Spillage Around Tail Pulleys, Idlers, and Load Points Only When the Power Is Locked Out and Guards Are In Place

Do Not Modify or Misuse Conveyor Controls

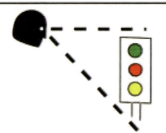

Ensure That ALL Controls and Pull Cords are Visible and Accessible

Do Not Modify or Remove Controls, Guards, Interlocks, Warnings or other Safety Items without Manufacturer's Approval

Report All Unsafe Conditions

POST IN PROMINENT AREA

APPENDIX D
INDEX

AUTHORS & ACKNOWLEDGMENTS

R. Todd Swinderman

Martin Engineering Corporate | Consultant

Todd Swinderman joined Martin Engineering in 1979 as Conveyor Products Engineer before becoming General Manager, President, and CEO. His vision and leadership have focused on developing innovative solutions for bulk-materials handling and expanding the company's capabilities around the world. Swinderman has been active in CEMA as an officer, committee chair, and as the chief editor and driving force behind the updated sixth edition of *BELT CONVEYORS for BULK MATERIALS*.

Andrew D. Marti

Martin Engineering Corporate | Global Technical Writer

Andy Marti has more than 20 years of experience in writing about the problems and solutions in bulk-materials handling. He has served as coordinating author and editor for all four editions of Martin's *Foundations*™ books on improving the performance of belt conveyors and transfer points. Marti holds a B.A. in Journalism from Central Michigan University and an M.A. in Communications Media from the University of Northern Iowa.

Larry J. Goldbeck

Martin Engineering USA | Conveyor Technology Manager

Since joining Martin in 1981, Larry Goldbeck has traveled the world—from Indonesia to Iceland, from Duluth to Delhi—applying solutions to problems in the handling of bulk materials. He combines theoretical knowledge with 40 years of hands-on experience in the operation, maintenance, and troubleshooting of belt conveyor systems. Goldbeck is the developer and lead instructor of Martin's Foundations™ Workshops on Operating and Maintaining Clean and Safe Belt Conveyors.

Daniel Marshall

Martin Engineering USA | Product Specialist

A self-described "numbers guy," Daniel Marshall holds a B.S. in Mechanical Engineering from Northern Arizona University. He joined Martin in 2000 as a Research and Development Engineer. In his career at Martin, Marshall has worked with every conveyor product that Martin has to offer. He is currently instrumental in the design and application of dust-suppression and dust-collection systems.

Mark G. Strebel

Martin Engineering USA | Product Support Manager

Mark Strebel came to Martin after nine years experience as a test and results engineer and operations supervisor with a coal-fired public utility power plant. At Martin, he has focused on the development and application of technologies to improve bulk-materials handling in the positions of R & D Manager and Conveyor Products Manager, before assuming his current position. Strebel holds a B.S. in Mechanical Engineering and an M.B.A. from Bradley University.

John Barickman

Martin Engineering
Corporate

Senior Product Development
Engineer

Greg Bierie

Martin Engineering
USA

Global Project and Technical
Sales Manager

Steve Brody

Martin Engineering
USA

Tool & Die Maker

Jörg Gauss

Martin Engineering
Germany

Germany Operations
Manager

Fred McRae

Martin Engineering
USA

Regional Services Man-
ager/Southeast Territory

Dave Mueller

Martin Engineering
USA

Senior Product Specialist

Tim O'Harran

Martin Engineering
USA

National Business
Development Manager

Frank Polowy

Martin Engineering
USA

Martin Services Project
Estimator

Brad Pronschinske

Martin Engineering
Corporate

Global Product Manager/
Flow Aids

Javier Schmal

Martin Engineering
Brazil

Brazil Managing Director

Andy Stahura

Martin Engineering
USA

Territory Manager

Gary Swearingen

Martin Engineering
USA

Projects Group Estimator

Terry Thew

Equipment Services &
Supplies

Australia Engineering
Manager

Barbara Wheatall

Martin Engineering
USA

Projects Coordinator

Marty Yepsen

Martin Engineering
Corporate

Safety & Risk Manager

Additional Writers

**David Craig, PhD | Larry Engle | David Keil
Roger Kilgore | Stephen Laccinole | Justin Malohn
Arie Qurniawan | Andrew Waters**

D. Michele Maki, PhD

Writing/Proofreading
Consultant

Chelsea Blake

Martin Engineering
USA

Marketing Communications
Administrator

Seth Mercer

Martin Engineering
Corporate

Global Marketing
Communications Specialist

Jared Piacenti

Martin Engineering
USA

Engineering System
Administrator

Bob Tellier

Martin Engineering
Corporate

Global Marketing
Intelligence Specialist

There are always those individuals who, behind the scenes, quietly go about their job to get the task done. Oftentimes, those individuals, the unsung heroes, go unrecognized for the contributions they make to the process. Five individuals have consistently gone out of their way, some for almost two years, to bring the fourth edition *FOUNDATIONS*™ *The Practical Resource for Cleaner, Safer, More Productive Dust & Material Control* to fruition. Without the dedication, hard work, and insight of Martin employees Chelsea Blake, Seth Mercer, Jared Piacenti, and Bob Tellier and writing and proofreading consultant D. Michele Maki, PhD, the fourth edition of *FOUNDATIONS*™ would not have been possible.

This book could also not have been completed without the understanding and assistance of many outside resources and many Martin Engineering employees. These individuals have provided background information, technical expertise, "big-picture" and "detailed-picture" thinking, and "nuts-and-bolts" intricacies. We owe a debt to the following:

Martin Engineering Corporate

Susan Coné, James Daly, Gina Darling, Harry Heath, Michele Ince, Chris Landers, Paul Mengnjoh, Travis Miller, Andrea Olson, Chris Schmelzer, Mark Stern, Kathy Swearingen, Terry Swearingen, Bonnie Thompson, Kathy Thumma, Jim Turner, Tina Usrey, and Ron Vick

Martin Engineering Germany

Reiner Fertig, Dave Harasym, Michael Hengl, and Michael Tenzer

Outside Resources

Mike Braucher, Dave Gallagher, and Frank Hyclak, Goodyear Engineered Products

Joseph A. Dos Santos, Dos Santos International, LLC

Bernd Küsel, http://www.conveyorbeltguide.com

Graham Leason, Tech-A Ltd.

Walt Lynch, FL Smidth

Bob Reinfried, Executive Vice-President, CEMA

Ed Walinski, Pneutech Engineering

Darcy Winn, Winn Conveyor

Ryan Buck and David Pratt, Wethersfield High

Martin Engineering USA

Bob Burke, Jim Burkhart, Jen DeClercq, Julie Derick, Robert Downs, Travis Grawe, Mark Huhn, Sonia Magalhaes, Kevin McKinley, Greg Milroy, Cheryl Osborne, Rachael Porter, Tracey Ramos, Wayne Roesner, Jim Roark, Becky Scott, Richard Shields, and Terry Vandemore

Martin Engineering China

Eric Zheng

Martin Engineering South Africa

Hannes Kotze

Consultants

Charles E. Fleming, *FOUNDATIONS*™, *Fourth Edition*, Project Manager

Paul Grisley, Grisley Conveyors

Bob Law, Engineering Services & Supplies (ESS)

Laurie Mueller

To all who lent a hand, thank you.
RTS, ADM, LJG, DM, MGS

Neponset, Illinois, USA, March 2009